*Alfred Tarski*

*This book is dedicated to Helen Marie Smith, in gratitude for her advice and support, and to Maria Anna McFarland, as she enters a world of new experiences.*

Andrew McFarland • Joanna McFarland
James T. Smith
Editors

# Alfred Tarski

Early Work in Poland—Geometry and Teaching

with a Bibliographic Supplement

Foreword by Ivor Grattan-Guinness

*Editors*
Andrew McFarland
Płock, Poland

Joanna McFarland
Płock, Poland

James T. Smith
Department of Mathematics
San Francisco State University
San Francisco, CA, USA

ISBN 978-1-4939-1473-9     ISBN 978-1-4939-1474-6 (eBook)
DOI 10.1007/978-1-4939-1474-6
Springer New York Heidelberg Dordrecht London

Library of Congress Control Number: 2014945118

Mathematics Subject Classification (2010): 01A60, 01A70, 01A75, 03A10, 03B05, 03E75, 06A99, 28-03, 28A75, 43A07, 51M04, 51M25, 97B50, 97D40, 97G99, 97M30

© Springer Science+Business Media New York 2014
This work is subject to copyright. All rights are reserved by the Publisher, whether the whole or part of the material is concerned, specifically the rights of translation, reprinting, reuse of illustrations, recitation, broadcasting, reproduction on microfilms or in any other physical way, and transmission or information storage and retrieval, electronic adaptation, computer software, or by similar or dissimilar methodology now known or hereafter developed. Exempted from this legal reservation are brief excerpts in connection with reviews or scholarly analysis or material supplied specifically for the purpose of being entered and executed on a computer system, for exclusive use by the purchaser of the work. Duplication of this publication or parts thereof is permitted only under the provisions of the Copyright Law of the Publisher's location, in its current version, and permission for use must always be obtained from Springer. Permissions for use may be obtained through RightsLink at the Copyright Clearance Center. Violations are liable to prosecution under the respective Copyright Law.
The use of general descriptive names, registered names, trademarks, service marks, etc. in this publication does not imply, even in the absence of a specific statement, that such names are exempt from the relevant protective laws and regulations and therefore free for general use.
While the advice and information in this book are believed to be true and accurate at the date of publication, neither the authors nor the editors nor the publisher can accept any legal responsibility for any errors or omissions that may be made. The publisher makes no warranty, express or implied, with respect to the material contained herein.

Printed on acid-free paper

Springer is part of Springer Science+Business Media (www.birkhauser-science.com)

# Foreword

## by Ivor Grattan-Guinness

### 1. Context

It is common in the history of ideas to find that the reception and development of some theories of a historical figure proceeded much more rapidly than others, both during his lifetime and afterward. As a result a historian can enjoy expounding the "neglected" contributions of the figure, and search for causes of this oversight. However, it is rather rare for a body of his work to disappear entirely from sight, especially when it was published at the time of its creation and the figure in question has been subject to considerable attention both by contemporary co-workers and by historians.

Yet this situation pertains to some of the early work of the great mathematician and logician Alfred Tarski (1901–1983). It is well known that he worked as a schoolteacher in Poland from the early 1920s and that he published papers in logic and mathematics regularly, mostly in Polish or Austrian journals; there is even an edition of English translations of versions of many of these papers (Tarski [1956] 1983). However, during this period he also published a substantial number of papers, comments, and problems in Polish journals oriented around mathematics education; but they are not handled in the fine biography, Feferman and Feferman 2004. Further, the journals in question escaped the attention of the German mathematics abstracting journal *Jahrbuch über die Fortschritte der Mathematik*. In addition, Tarski coauthored an elementary textbook on geometry in 1935 that has generally been overlooked. The purpose of the present volume is to bring this material to light in English translation, and also to amplify the historical context in which the teacher Tarski produced it in the first place.

There are two trends in the development of logic and the foundations of mathematics that became of major significance during this period, so that their bearing upon Tarski's work are worth seeking. One was the development of nonclassical logics to complement the classical two-valued logic and, in the excessively polemical case of the intuitionism of L. E. J. Brouwer, to replace it (Mancosu 1998). Tarski does not seem to have been concerned with other logics in the papers here. The other was the growing recognition of the need to work in both logic and mathematics with *hierarchies* of theories, distinguishing mathematics from metamathematics (David Hilbert's program) and logic from metalogic (a central feature of Kurt Gödel's famous paper of 1931 on the incompletability of first-order arithmetic, which stimulated his friend Rudolf Carnap to coin "metalogic" that year as a technical term. Tarski would soon contribute notably to this trend by coining the word "metalanguage" in connection with his semantic theory of truth of the 1930s.

## 2. Cleft

One would have thought that since logic was among other things a study of the theory of deduction and since mathematics deployed deduction in proofs of theorems and definitions of concepts, the two disciplines would happily live side by side. However, this has never been the case (as is already evident in, for example, Euclid's *Elements*); in particular, the rise in interest in foundational subjects from the late nineteenth century onward did not eliminate it at all.

A very important source of maintaining the cleft during the nineteenth century is the founding from the late 1810s onward of the "mathematical analysis" of real variables, grounded upon an articulated theory of limits, by the French mathematician Augustin Louis Cauchy. He and his followers extolled rigor, in particular careful nominal definitions of major concepts and detailed proofs of theorems. From the 1850s onward this aim was enriched by the German mathematician Karl Weierstrass and his many followers—they brought in, for example, multiple-limit theory, definitions of irrational numbers, and an increasing use of symbols—and from the early 1870s, Georg Cantor and his set theory. However, none of these developments explicitly drew upon any kind of logic.

This silence continued among the many set theorists who helped to develop measure theory, functional analysis, and integral equations (Jahnke 2003). Even the extensive dispute over the axiom of choice (a Tarski favorite) from 1904 onward focused mostly on its legitimacy as an assumption in set theory and mathematics and on the use of higher-order quantification (Moore 1982): its need to state an infinitude of independent choices within finitary logic was a trouble for logicians.

The creators of symbolic logics were exceptional among mathematicians in attending to logic, but they made little impact on their colleagues. The algebraic tradition with George Boole, Charles Sanders Peirce, Ernst Schröder, and others from the mid-nineteenth century was just a curiosity to most of their contemporaries. Similarly, when mathematical logic developed from the late 1870s, especially with Giuseppe Peano's "logistic" program at Turin from around 1890, it gained many followers there (Luciano and Roero 2010) but few elsewhere. However, followers in the 1900s included the Britons Bertrand Russell and Alfred North Whitehead, who adopted logistic (including Cantor's set theory) and converted it into their "logicistic" thesis that all the "objects" of mathematics could be obtained from it; G. H. Hardy but not many other mathematicians responded (Grattan-Guinness 2000, chapters 8 and 9). From 1903 onward Russell had also publicized the mathematical logic and arithmetic logicism put forward from the late 1870s onward by Gottlob Frege, which had gained little attention hitherto even from students of foundations and did not gain much more in the following decades. Hilbert's program of metamathematics attracted several followers at the University of Göttingen and a few elsewhere; however, its impact among mathematicians was limited even in Germany.

The situation in Poland is quite poignant; for soon after the re-creation of the country after the Great War, major schools of mathematicians and logicians emerged. Moreover, many of the mathematicians worked in set theory and/or its applications, so that links with logic could be close. In 1920 the Poles launched in Warsaw a journal, *Fundamenta Mathematicae,* to cover both disciplines (Lebesgue 1922). The two coeditors for logic were Jan Łukasiewicz and Stanisław Leśniewski, who had obtained chairs at the University of Warsaw in 1915 and 1919 respectively after taking doctorates at Lwów. They and their associates formed the largest community in the world working on logic and related topics (McCall 1967). I am told that Polish has many properties relevant to logic and set theory, which may have helped stimulate the interest in the first place. However, not many logic papers appeared in the journal (Tarski being the most frequent author), and the two logicians resigned from its board in 1928, with little regret from the mathematicians (Kuratowski 1980, 33–34).

By contrast, a most praiseworthy feature of Tarski's work is that he researched in both logic and branches of mathematics (especially geometry) at the same time, thereby consciously ignoring the cleft. In particular, the material newly come to light treats not only logic and set theory but also elementary geometry and common algebra. But he wrote it mostly in Polish, which will have reduced its market, although some of its contents appeared elsewhere in papers in German or French. Let us now encounter it in English guise.

# Preface

Alfred Tarski is regarded as one of the four greatest logicians of all time—the others being Aristotle, Gottlob Frege, and Kurt Gödel.[1] Most notably, Tarski was largely responsible for designing the infrastructure on which most logical research has been based since 1950. Using that structure, he achieved major technical results in logic, foundations of geometry, and abstract algebra. Outside that framework, Tarski discovered major theorems in set theory and set-theoretic aspects of geometry, and completed some works about teaching geometry.

Tarski was born in Warsaw in 1901, and grew up there in a time of turmoil. He completed doctoral studies in 1924, just after a major expansion of the Polish system of universities. The economic climate was adverse, as was the growing antisemitism. Tarski obtained full-time employment as a secondary-school teacher, and worked part-time as a university assistant and researcher. During the next fifteen years, he gained world renown in the fields of set theory and logic. On a 1939 lecture tour, he was stranded, fortunately, in the United States, when the Germans and Soviets invaded Poland. After some trying years, he secured a professorship at the University of California, Berkeley. There, according to the eminent Polish philosopher Jan Woleński,

> Tarski ... created the great Californian School of logic. He ... had a dominant influence upon the development of logic after World War II.[2]

After four decades of service at Berkeley, Tarski died there in 1983.

Much of Tarski's scientific work has been accessible and has become rather well known. Three years after his death, his colleagues Steven R. Givant and Ralph McKenzie published the remarkably complete *Collected Papers* volumes and a detailed bibliography. An excellent biography by Anita B. and Solomon Feferman is available in English and Polish.[3] Tarski took meticulous care to document his Berkeley research program, and that material is readily available for historical study. But a few of Tarski's early works have been difficult to access, and for some of them, hardly any background was even detectable. This was particularly true for some of his early work about geometry. He had completed that in Warsaw, which was largely destroyed during World War II. There is almost nothing of his left there, and Tarski's archive in Berkeley starts in 1939.

---

[1] Corcoran 1991.

[2] Woleński 1989, 20.

[3] Tarski 1986a; Givant 1986; Feferman and Feferman 2004 and 2009. For descriptions of these works see section 16.2 and chapter 17.

The present book has three main goals:

(1) To publish translations as necessary so that
- Alfred Tarski's works will all be accessible in English, French, or German; and
- his geometric works will all be accessible in English.

(2) To provide scientific and cultural background information about the works translated here: their origin, context, structure, and impact.

(3) To update Givant's 1986 bibliography of Tarski's publications, and include an annotated list of major studies of Tarski's life and work.

By including ample background material in this book, the editors have heeded an opinion expressed succinctly by a leading mathematics historian, David E. Rowe:

> The type of knowledge mathematicians have produced has depended heavily on cultural, political, and institutional factors that shaped the various environments in which they have worked.[4]

The book stresses the connection between Tarski's work as a teacher and the subjects of some of his research. It does not itself pursue other connections between environment and research, but does aim to provide scholars interested in that kind of inquiry a glimpse of the background and routes to deeper and broader study. It does not supplant the Fefermans' 2004 biography, but fills some gaps in their coverage. In some cases, the present editors have included significant background that was apparently not familiar to others who have written about Tarski. This is especially true for the year, 1920, of the Polish–Soviet War, and for Tarski's work as a secondary-school teacher and teacher-trainer.

The present editors expect that further background material specifically about Tarski will come to light only through historians' studies of other students, teachers, and scientists associated with him. This is particularly the case for Tarski's disrupted first university year, 1918–1919. Since articles about mathematics or philosophy are usually edited to exclude material not directly related to their theoretical content, the social context of Tarski's work will probably be better revealed by investigating his activity in teaching and teacher-training. Researchers should adopt a maxim: if Tarski is one of the top four logicians of all time, he must be one of the top $n$ thinkers, and what he thought about almost anything should be interesting!

This book is organized into four parts:

Part One .... *Debut*      Part Three ... *Teaching*
Part Two .... *Geometry*   Part Four .... *Supplement*

The supplement contains some translations not directly related to geometry research or teaching, an update of the 1986 bibliography, and annotated lists of major studies of Tarski and his work.

---

[4] Rowe 2003, 114.

Preface     xi

Readers will find general background material in chapters 1, 3; 4, 8; and 9, 14: the first and last chapters of its three main parts. Background specific to particular translations is located in the chapters devoted to them.

Part One of the book, *Debut*, contains only one translation: Tarski's first published paper, *A Contribution to the Axiomatics of Well-Ordered Sets*, written in 1921 while he was still a student. Two background chapters describe his life through the completion of his doctoral study in 1924. They complement the Fefermans' biography, adding material about the turbulent years 1918–1920 in Warsaw. There remain gaps in the story: for example, it is still not clear why Tarski changed his major interest from biology to logic and mathematics.

Part Two, *Geometry*, is devoted to Tarski's work on equidecomposability. The background chapter 4 summarizes the elementary theory of area and volume, covered to some extent in secondary schools. It then considers the measure-theoretic approach that led Stefan Banach and Tarski to show in 1924 that set-theoretic decomposition alone will not yield a theory in accordance with intuition: a marble and the earth can each be dissected into the same finite number of disjoint sets of points, which are congruent in pairs. Their famous paper is translated in this part, along with Tarski's 1924 paper that showed that this counterintuitive result has no analogue in the plane. The latter paper was published in a journal aimed at secondary-school teachers. Attention is then turned to Tarski's work on the more elementary concept of decomposition of plane polygons into subpolygons whose interiors are disjoint but whose boundaries may overlap. Three elegant little papers are translated, which appeared during 1931–1932 in journals aimed at secondary-school students and their teachers. Two of those are by Tarski; the third, by another schoolteacher, Henryk Moese. Inadequate translations of the three were published obscurely half a century ago, but the translations in the present book are new. The subsequent impact of all five papers is traced in chapter 8, which concludes Part Two.

Part Three of the book, *Teaching*, presents a variety of material. Its first chapter describes Tarski's family situation and his teaching in a secondary school and in university lectures and research seminars. It complements the Fefermans' treatment of these aspects of his life. Tarski's 1929 report to teachers about an important research conference and his 1932 suggestions on teaching about circles are translated. During 1930–1932 Tarski published fourteen exercises to challenge teachers and talented students; they are translated and analyzed here. So are representative sections of his [1935] 1946 coauthored secondary-school text on geometry. The main portion of the present book concludes with chapter 14, which describes some of Tarski's activity during the 1930s that is background for several translations in the supplement, and leads to his 1939 voyage to the New World.

Part Four, the supplement, begins with chapter 15, which consists of translations of the eleven remaining works of Tarski that until now were accessible only in Polish. They are about various subjects not closely related to the earlier chapters, and are all very short. Each is accompanied by a brief discussion that places it in context and renders it intelligible. The most significant one is the report of Tarski's 1930 presentation in Lwów:

the first appearance in print of his celebrated theory of truth. The book concludes with chapters 16–18, which contain annotated lists that update the 1986 bibliography of Tarski's publications and identify major studies of his life and his work.

The material gathered in this book will increase the accessibility of Tarski's early work and explain some of its relationships to the intellectual, political, and social milieu of Poland between the world wars. The present editors hope that it will spur broader investigation into the connections between mathematics and its cultural setting during that era. This would be particularly welcome for the connection between mathematical research and mathematics education, as displayed by Tarski's research on geometry and his practice both in teaching secondary-school students and training secondary-school teachers. This hope is a major reason for including works of such contrasting mathematical sophistication in a single volume of selected translations.

Rough maps on pages xxii and xxiii depict Poland and the surrounding area, with the international borders of 1914 and 1924—before and after the First World War. Cities of importance to this book have been identified. For place names in this region, English versions are used when available. Otherwise, for places that have been in more than one country, the names used in this book are the official names used in 1924.

In 1936 Poland officially adopted some changes in spelling that affect words quoted in this book, particularly involving the letters $i, j$, and $y$. The editors have tried to adhere to the spellings in the original texts.

The translations are meant to be as faithful as possible to the originals.[5] Bibliographic references and personal names have been adjusted to conform with conventions of the present book. Some uses of alternative type styles for emphasis, enunciations, and personal names have been modified. The only intentional modernizations are punctuation and occasional changes in symbols, where Tarski's conflict with others used throughout this book. Those are discussed in the introductions to the individual translations. As an aspect of adjusting punctuation, the editors modified the use of white space to enhance visual organization. All [square] brackets in the translations enclose editorial comments. These are inserted, usually as footnotes, to document changes in technical terms, to note or suggest corrections for occasional errors in the originals, to clarify possibly troublesome translation details, and to explain a few passages that seem opaque.

Polish surnames often have gender-specific suffixes. Thus, the wife or unmarried daughter of Tarski is surnamed Tarska; the wife and the unmarried daughter of Łukasiewicz are surnamed Łukasiewiczowa and Łukasiewiczówna, respectively. Married women often use these names hyphenated with those of their husbands. The order varies; in this book, the husbands' surnames come second.

---

[5] The introduction by Magda Stroińska and David Hitchcock to Tarski [1935] 2002 was helpful in planning the Polish translation process.

In this book, the term *gimnazjum* refers only to Polish secondary schools that prepared pupils for university studies during 1900–1939.

Spaces and diacritical and punctuation marks have been ignored during all alphabetizations, particularly for the bibliography and index. Warning: this resulted in alphabetization different from the Polish standard! Throughout the book, capitalization of Polish titles reflects the conventions of the Chicago 1993 manual, which disagree with those now taught in Polish schools.

The huge bibliography lists all and only works referred to in the book. Each entry indicates where citations occur. The author–date system is employed for citations: for example, *Tarski 1986a* is a citation for a work published under Tarski's name in 1986. The present book mentions more than one author named Tarski; citations that include this surname alone are references to Alfred Tarski. Sometimes an author is to be inferred from the context, so that a date alone may also serve as a citation of a work.

Biographical information about more than sixty individuals involved with Tarski is presented in boxes or notes located in various chapters and cross-referenced, as appropriate, in others. The emphasis is on the years before 1945.[6] Sources for the data are identified by footnotes inside the boxes. The book's index lists both subjects and persons. The latter entries include personal dates when known.

The project culminating in the present book started with brief conversations, years apart, between James T. Smith and Steven R. Givant at Berkeley logic colloquia. Smith recalled that during 1965–1970, he had heard Tarski speak eloquently to general audiences about the degree of equidecomposability of polygons. Although that material had been published, it was nevertheless virtually inaccessible. Smith mentioned this at the fall 2007 meeting of the State of Jefferson Mathematics Congress. Joanna and Andrew McFarland were also attending. Joanna is a teacher of Polish from Płock, Poland. Andrew, for whom Polish was also a first language, was on the mathematics faculty at Sonoma State University. The three discussed the possibility of working together to republish the equidecomposability papers. The critical resources for collaboration had converged: interest in the project and facility with logic, mathematics, English, and Polish. Later, Smith and Givant mused that a few other works from Tarski's early years had suffered the same neglect, and concluded that publishing a volume of translations from Polish might be feasible. The McFarlands, Smith, and Springer editor Ann Kostant discussed this further. It was decided to include translations of representative sections of the virtually forgotten [1935] 1946 secondary-school text *Geometrja* by Tarski and two coauthors. With these translations all of Tarski's geometric work would be accessible in English except the famous 1924 paper, in French, by Banach and Tarski on set-theoretic equidecomposability of geometric figures. This book's selection of translations was rounded out by including that paper as well. Investigation of the background and impact of these works suggested the utility of including an update of the 1986 Tarski

---

[6] Many of these individuals participated in the Polish underground or clandestine educational system during World War II. The famous 1944 book by Jan Karski is a gripping first-hand description of those activities.

bibliography. Springer Science+Business Media (Birkhäuser) agreed to this plan in October 2009.

For the present book, translations from Polish were drafted by the McFarlands, and edited jointly with Smith. Translations from other languages were done by Smith. The McFarlands carried out library and archival research in Poland, particularly on Tarski's teaching. Smith did that for material accessible in libraries in the United States. Smith assembled the background information into the present organization, and the result was edited jointly. Smith was responsible for the design and composition of the book.

The editors wish to acknowledge professors Edith Mendez and Elena A. Marchisotto for inspiration to undertake historical studies, and Helen M. Smith for her patience, generosity, insight, and ingenuity. Alfred Tarski's son, Prof. Jan Tarski, is especially recognized for his assistance to the present editors, and for his editorial work on other publications listed in chapters 16 and 17. On matters of content, the editors are grateful for the advice and assistance of Sheldon Axler, John Corcoran, Stanisław Domoradzki, Steven R. Givant, Jacek Juliusz Jadacki, Anna Jaroszyńska-Kirchmann, Andrzej Jerzmanowski, Anna Kozłowska, Witold Kozłowski, Renato Lewin, Paulo Mancosu, Antony Polonsky, V. Frederick Rickey, Janusz Rudziński, Andrzej Schinzel, James R. Shilleto, and Jan Zygmunt. For help with translation, we are indebted to Grażyna Ula Furman, Arek Goetz, Sergei Ovchinnikov, and Michael Thaler. We are immensely thankful for the library services provided by the Archiwum Akt Nowych, Archiwum Polskiej Akademii Nauk, Archiwum Państwowe Miasta Stołecznego Warszawy, Archiwum Państwowe w Lublinie, Archiwum Uniwersytetu Warszawskiego, Biblioteka Narodowa, Centralna Biblioteka Matematyczna Polskiej Akademii Nauk, Biblioteka im. Zielińskich Towarzystwa Naukowego Płockiego, Książnica Kopernikańska, Narodowe Archiwum Cyfrowe, the United States Holocaust Memorial Museum, the University of California libraries in Berkeley and Richmond, and for the splendid interlibrary loan services of San Francisco State University. This book was made possible by the retirement system of that university.

# Contents

Foreword, by Ivor Grattan-Guinness ................................. v

Preface ............................................................ ix

Illustrations ...................................................... xix

Maps of Central Europe ........................................... xxii

## Part One: Debut

Introduction ....................................................... 1

**1 School, University, Strife** ..................................... 3
   1.1 Coming of Age in Warsaw ...................................... 3
   1.2 Well-Ordered Sets ........................................... 15

**2 *A Contribution to the Axiomatics of Well-Ordered Sets* (1921)** ... 19

**3 Doctoral Research** ............................................ 31

## Part Two: Geometry

Introduction ...................................................... 43

**4 Area, Volume, Measure** ........................................ 45
   4.1 Area and Volume ............................................. 45
   4.2 The Measure Problem ......................................... 51
   4.3 Equidecomposability in the Plane ............................ 58
   4.4 The Banach–Tarski Theorems .................................. 64

**5 *On the Equivalence of Polygons* (1924)** ....................... 77
   5.1 §1 .......................................................... 80
   5.2 §2 .......................................................... 83
   5.3 §3 .......................................................... 84
   5.4 Summary ..................................................... 90

## 6 On Decomposition of Point Sets into Respectively Congruent Parts (1924) .......... 93
   6.1 General Properties of Equivalence by Finite or Denumerable Decomposition .......... 96
   6.2 The Fundamental Theorems about Equivalence by Finite Decomposition .......... 106
      A. Euclidean Space with One or Two Dimensions .......... 106
      B. Euclidean Space of Three (and More) Dimensions .......... 109
      C. The Surface of a Sphere .......... 112
   6.3 The Fundamental Theorems about Equivalence by Denumerable Decomposition .......... 115

## 7 Degree of Equivalence of Polygons .......... 125
   7.1 Introduction .......... 126
   7.2 *The Degree of Equivalence of Polygons* (1931) .......... 135
   7.3 Moese's Contribution (1932) .......... 144
   7.4 *Remarks on the Degree of Equivalence of Polygons* (1932) .......... 151

## 8 Research Threads .......... 159
   8.1 Two Threads in Logic .......... 160
   8.2 Thread: Well-Ordering and Finiteness .......... 161
   8.3 Thread: Cardinal Arithmetic and the Axiom of Choice .......... 161
   8.4 Thread: Application of Set Theory to Geometry .......... 162
   8.5 Equidecomposability in Elementary Geometry .......... 162
   8.6 Equidecomposability in Set Theory .......... 163
   8.7 The Measure Problem .......... 165
   8.8 Generalizing Cantor–Bernstein .......... 167

# Part Three: Teaching

## Introduction .......... 169

## 9 Career and Family .......... 171
   9.1 Employment and Marriage .......... 171
   9.2 Teaching Geometry .......... 179
   9.3 Teaching Teachers .......... 184
   9.4 Research Seminars .......... 194
   9.5 Warsaw/Poznań Congress, 1929 .......... 200
   9.6 Organizations and Journals for Teachers .......... 203
   9.7 *Parametr* and *Młody Matematyk* .......... 204
   9.8 Tarski's Contributions to *Parametr* and *Młody Matematyk* .......... 211
   9.9 Tarski's Coauthored Textbook *Geometrja* .......... 214

Contents　xvii

**10  Congress of Mathematicians of Slavic Countries (1929)** .............. 225

**11  Circumference of a Circle (1932)** ...................................... 229

**12  Exercises Posed by Tarski** ............................................. 243
 12.1 Diluting Wine .................................................. 249
 12.2 Iteration of the Absolute Value Symbol ......................... 250
 12.3 Decomposition into Factors ..................................... 251
 12.4 Interesting Identity ........................................... 253
 12.5 Exercise 118 ................................................... 255
 12.6 Equidistant .................................................... 256
 12.7 System of Inequalities ......................................... 257
 12.8 Another System of Inequalities ................................. 259
 12.9 Cutting a Rectangle Out of a Square ............................ 261
 12.10 The Degree of Equivalence of Polygons ......................... 263
 12.11 Postulate about Parallels ..................................... 264
 12.12 Analytic Geometry of Space .................................... 268
 12.13 Stereometry ................................................... 268
 12.14 Arranging Two Segments in a Plane ............................. 271

**13  *Geometry for the Third Gimnazjum Class* (1935)** ...................... 273
 13.1 §1 Circle and Disk ........................................... 277
 13.2 §2 Construction Problems ..................................... 279
 13.3 §3 Relative Positions of a Circle and a Line Lying in a Plane ....... 281
 13.4 §4 Construction Problems ..................................... 285
 13.5  Summary of §5–§24 ........................................... 291
 13.6 §25 Incommensurable Segments—Irrational Numbers ............. 292
 13.7 §26 Measuring Incommensurable Segments with a Unit ........... 296
 13.8 §27 Rational Approximations of Irrational Numbers
    —Operations on Irrational Numbers ............................ 297
 13.9  Summary of §28–§32 ......................................... 301
 13.10 §33 Equivalent Polygons ..................................... 302
 13.11 §34 On Measuring Areas of Polygons .......................... 304
 13.12 §35 Areas of Similar Polygons ................................ 316

**14  Teaching and Logic—to the New World** ................................. 319
 14.1 Teaching ....................................................... 319
 14.2 Logic .......................................................... 321
 14.3 To the New World .............................................. 330
 14.4 Epilogue ....................................................... 336

# Part Four: Supplement

**Introduction** .................................................. 337

**15 Assorted Contributions** ...................................... 339
    15.1    Discussion of Łukasiewicz 1925, on Understanding Deduction ....... 340
    15.2    Discussion of Greniewski [1927] 1928, on Action ................. 341
    15.3    Discussion of Czeżowski [1927] 1928, on Causality ............... 343
    15.4    Łukasiewicz 1928–1929, on Definitions .......................... 346
    15.5    Tarski 1929, on Polish Pensions ................................ 349
    15.6    Tarski 1930–1931, on the Concept of Truth ...................... 356
    15.7    Tarski 1932, on Banach's Measure ............................... 363
    15.8    Discussion of Ajdukiewicz 1936, on Idealism ..................... 365
    15.9    Discussion of Wilkosz 1936, on the Significance of Logic ........ 367
    15.10   Discussion of Zawirski 1936b, on Synthesis ..................... 370
    15.11   Discussion of Kokoszyńska 1936a, on Relative Truth ............. 374
    15.12   Two Letters to Sierpiński, 1946–1947 ........................... 376

**16 Posthumous Publications** ..................................... 385
    16.1    Papers .......................................................... 385
    16.2    Monographs ...................................................... 390
    16.3    Letters ......................................................... 392

**17 Biographical Studies** ........................................ 399

**18 Research Surveys** ............................................ 411
    18.1    1986, 1988 JSL Surveys ......................................... 411
    18.2    Other surveys .................................................. 413

**Bibliography** ................................................... 421

**Permissions** .................................................... 473

**Index of Persons** ............................................... 477

**Index of Subjects** .............................................. 485

# Illustrations

**Portraits** *section, page*

Alfred Tarski .................................................... frontispiece

Stefan Mazurkiewicz ............................................. 1.1, 6
Zygmunt Janiszewski ............................................ 1.1, 6
Tadeusz Kotarbiński ............................................. 1.1, 6
Leon Trotsky as the Devil ....................................... 1.1, 11
Alfred Teitelbaum (Tarski) ...................................... 1.1, 13
Stanisław Leśniewski ............................................ 1.1, 13
Kazimierz Twardowski ........................................... 1.2, 18

Kazimierz Pasenkiewicz ......................................... ch. 3, 32
Aleksander Jabłoński ............................................ ch. 3, 34
Mordchaj Wajsberg .............................................. ch. 3, 34
Kazimierz Ajdukiewicz .......................................... ch. 3, 37

Wacław Sierpiński ............................................... 4.1, 48
Stefan Banach ................................................... 4.2, 56
Felix Hausdorff ................................................. 4.4, 68
Kazimierz Kuratowski ........................................... 4.4, 72

Henryk Moese ................................................... 7.1, 131

Maria Tarska and Others in the 1919–1920 Polish–Soviet War .... 9.1, 175
Ugo Amaldi ..................................................... 9.2, 182
Federigo Enriques .............................................. 9.2, 182
David Hilbert .................................................. 9.3, 185
Mario Pieri .................................................... 9.3, 187
Jan Łukasiewicz ................................................ 9.4, 196
Mojżesz Presburger ............................................. 9.4, 196
Andrzej Mostowski .............................................. 9.4, 198
Wanda Szmielew ................................................. 9.4, 198
Antoni Marian Rusiecki ......................................... 9.7, 209
Stefan Straszewicz ............................................. 9.7, 209
Zygmunt Chwiałkowski and Others ................................ 9.9, 218
Wacław Schayer ................................................. 9.9, 220

Maria Kokoszyńska and Alfred Tarski ............................ 14.2, 327
Leon Chwistek .................................................. 14.2, 329
Janina Hosiassonówna ........................................... 14.3, 335
Adolph Lindenbaum .............................................. 14.3, 335

Tadeusz Czeżowski .............................................. 15.3, 345
Witold Wilkosz ................................................. 15.9, 369
Zygmunt Zawirski ............................................... 15.10, 373

Alfred Tarski .................................................. ch.16, 384

## Figures

Many figures are not titled, but numbered as in the original versions of translated material. For example, figure 1 in chapter 2 of this book is adapted from figure 1 in the original version of Tarski [1921] 2014.

*section, page*

| | |
|---|---|
| *Alfred Tarski Signature* | frontispiece |
| *Central Europe in 1914* | xxii |
| *Central Europe in 1924* | xxiii |
| *Alfred's Autumn 1919 –Winter 1920 Enrollment Record* | 1.1, 6 |
| *1920 Propaganda* (3 figures) | 1.1, 11 |
| Przegląd Filozoficzny *Cover* | ch. 2, 20 |
| *Figure 1* | ch. 2, 28 |
| *Figures 2–3* | ch. 2, 29 |
| *To the Bursar, Autumn 1923* | ch. 3, 35 |
| *Three Signatures* | ch. 3, 38 |
| Przegląd Matematyczno-Fizyczny, *Masthead* | 4.3, 58 |
| *Hausdorff's Classification of Rotations in G* | 4.4, 68 |
| Przegląd Matematyczno-Fizyczny, *Cover* | ch. 5, 78 |
| *A Square and an Isosceles Right Triangle with the Same Area* | 5.3, 85 |
| Fundamenta Mathematicae *6 (1924), Cover and Table of Contents* | ch. 6, 94 |
| *Parallelograms with the Same Base and Altitude* | 7.1, 128 |
| *Adaptation of Wojtowicz's Argument* | 7.1, 128 |
| *Gimnazjum in Kępno, Where Henryk Moese Taught* | 7.1, 131 |
| *Dividing a Square into Two Polygonal Regions* | 7.1, 134 |
| *Figures 1 to 4* | 7.2, 136 |
| *Figures 5 to 7* | 7.2, 141 |
| *Figure 1* | 7.3, 145 |
| *Figure 2* | 7.3, 146 |
| *Figure 3* | 7.3, 148 |
| *Figures 4 to 6* | 7.3, 149 |
| *Figure 7* | 7.3, 150 |
| *Warsaw, 1925: 140–150 Marszałkowska Street* | 9.1, 173 |
| *51 Koszykowa Street* | 9.1, 178 |
| *4 Sułkowski Street* | 9.1, 178 |
| Hilbert's Foundations of Geometry: *Title Page* | 9.3, 185 |
| Pieri's Point and Sphere Memoir: *Polish Translation* | 9.3, 187 |
| *Tarski's Teaching Load at the University of Warsaw* (table) | 9.3, 191 |
| *First Congress of Mathematicians of Slavic Countries* (group portrait) | 9.5, 201 |
| *General Polish Exposition, Poznań, May–September 1929* | 9.5, 202 |
| *Cover and Contents,* Młody Matematyk *2 (1931, issue 3)* | 9.7, 206 |
| *Masthead and Contents,* Parameter *2 (1931, issues 8–10)* | 9.7, 206 |

# Illustrations

|  | section, page |
|---|---|
| *Advertisement for* Geometrja | 9.9, 215 |
| *Figure 1: Graph of Function* $g \to \log_g n$ | 12.4, 254 |
| *Figure 2: Equidistant Curve* | 12.6, 256 |
| *Figure 3* | 12.7, 258 |
| *Figures 4 and 5* | 12.9, 262 |
| *Figures 6 and 7* | 12.11, 266 |
| *Figure 8: Tarski's Parallel Postulate* | 12.11, 266 |
| *Figure 9* | 12.13, 270 |
| Geometrja *Table of Contents* | ch.13, 275 |
| Geometrja *Cover* | ch.13, 276 |
| *Figure 1, for §1** | 13.1, 277 |
| *Figure 2, for §2* | 13.2, 279 |
| *Figure 3, for §2* | 13.2, 280 |
| *Figures 4 and 5, for §3* | 13.3, 281 |
| *Figure 6, for §3* | 13.3, 282 |
| *Figures 7 and 8, for §3* | 13.3, 284 |
| *Figures 9 and 10, for §4* | 13.4, 286 |
| *Figure 11, for §4* | 13.4, 287 |
| *Figure 12, for §4* | 13.4, 288 |
| *Figure 13, for §4* | 13.4, 289 |
| *Figure 75, for §25* | 13.6, 292 |
| *Figure 76, for §25* | 13.6, 294 |
| *Figure 77, for §27* | 13.8, 298 |
| *Figure 78, for §27* | 13.8, 299 |
| *Figures 106 to 110, for §33* | 13.10, 303 |
| *Figures 111 and 112, for §34* | 13.11, 305 |
| *Figures 113 to 116, for §34* | 13.11, 306 |
| *Figures 117 and 118, for §34* | 13.11, 308 |
| *Figures 119 and 120, for §34* | 13.11, 309 |
| *Figure 121, for §34* | 13.11, 310 |
| *Figures 122 to 124, for §34* | 13.11, 311 |
| *Figure 125, for §34* | 13.11, 312 |
| *Figure 126, for §34* | 13.11, 315 |
| *Figures 127 to 130, for §35* | 13.12, 317 |
| *Fencing* | 14.2, 329 |
| *Headlines from the* Völkischer Beobachter | 14.3, 330 |
| *Gdynia–America Poster* | 14.3, 332 |
| *M/S Piłsudski Menu* | 14.3, 332 |
| *Arrival in New York* | 14.3, 332 |

---

* This and the following § numbers refer to sections of *Geometrja*.

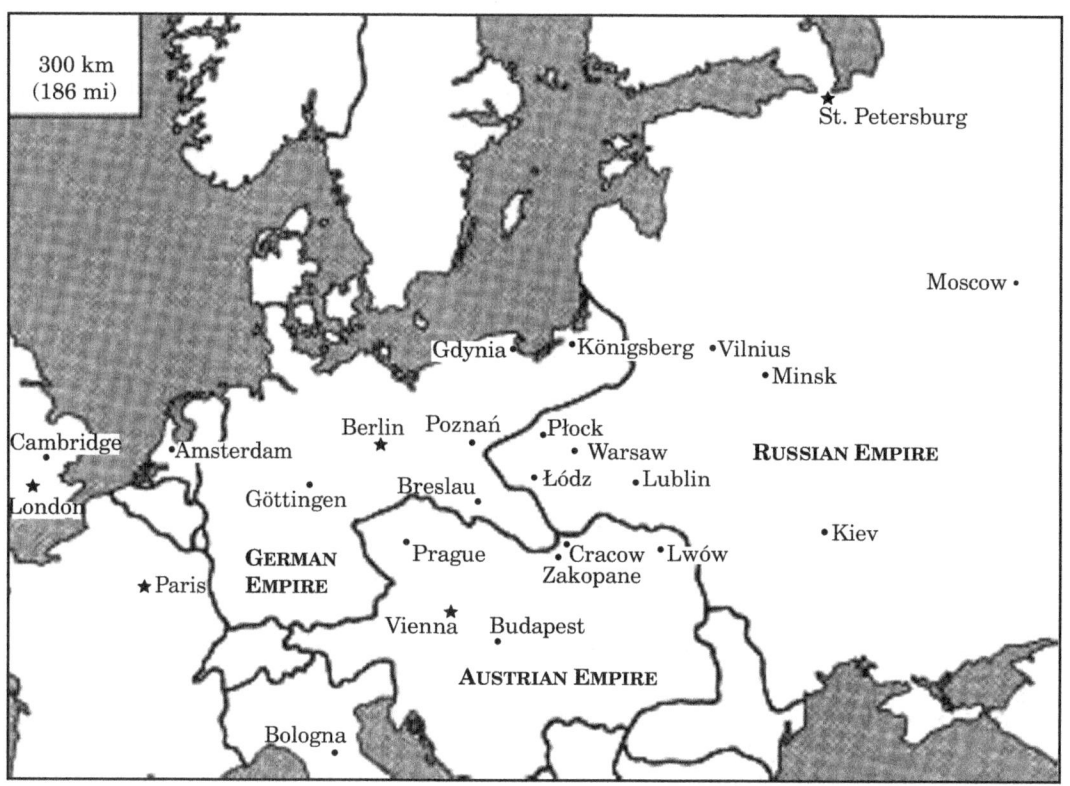

*Central Europe in 1914*

*Central Europe in 1924*

# Part One
# Debut

These first three chapters describe Alfred Tarski's childhood, schooling, and university studies during a time of political chaos and threats of war. Chapter 1 includes background for his first paper, *A Contribution to the Axiomatics of Well-Ordered Sets*, published while he was still a student. The paper is translated in chapter 2. Chapter 3 provides background for Tarski's doctoral research, supervised by Stanisław Leśniewski.

Tarski's life in Poland unfolded amid a chaotic vortex of political, social, economic, and scientific developments. Accounts of events, people, and ideas were merged from several dimensions to form the linear sequence of pages of this book. That is reflected in the background chapters 1 and 3 by the use of boxes interspersed in the main narrative. They contain biographical sketches of some persons associated with Tarski, and informational essays about several other topics. Each box can be read independently: readers are not expected to visit them in sequence. Cross-references refer to them from the main narrative.

# 1
# School, University, Strife

The first section of this chapter describes Alfred Tarski's childhood, schooling, and initial university studies in Warsaw, the city that in 1918 became the capital of the new independent Polish republic. The second section provides background for his first publication, a paper on set theory and logic that he completed while still a student. That paper is translated in the next chapter.

## 1.1 Coming of Age in Warsaw

Alfred Tarski was born Alfred Teitelbaum in January 1901 in Warsaw, Poland, an oppressed part of the Russian Empire.[1] His father, Ignacy, or Izaak, stemmed from Warsaw; his mother, Róża Prussak, from Łódź. Their families were engaged in the lumber and textile businesses. The Teitelbaums had two children: Alfred's brother, Wacław, was two years younger. The family lived at 51 Koszykowa Street, apartment 14, on the second floor of a five-story building that they owned.[2] It was located in the center of the city, about three kilometers from the University of Warsaw. From September 1910 to summer 1915, Alfred attended the State Gimnazjum 4 in Warsaw.[3] The family was Jewish, but secular in outlook. The language at home was Polish; in school, Russian. Alfred was precocious, particularly in languages: he studied French, German, Latin, and Greek at school, and after school went to synagogue to learn Hebrew. Alfred translated a German story into Polish at age twelve as an anniversary gift to his parents. Even earlier, he had shown interest in politics and social justice.

---

[1] For biographical information supporting this section, unless another source is cited, consult the biographies by Anita B. and Solomon Feferman (2004) and by Jacek Juliusz Jadacki (2003a). Also note the descriptions of those works in chapter 17. The Fefermans emphasized personal-interview sources; Jadacki, published records. For historical and sociological information, consult the works by Norman Davies (1982, volume 2, chapters 18–19), Celia S. Heller (1994), and Richard M. Watt (1979).

[2] The spellings of the names are from Tarski [1918] 2014, translated in section 16.4. Jadacki (2003a, 143) verified ownership from 1930 data. Jadacki and the Fefermans (2004, 8) claimed that 15 Koszykowa had been destroyed in the 1940s. But it still stands as 51a Koszykowa: see Golińska, Porębska, and Srebrny 2009a and 2009b. It is pictured on page 176 of the present book.

[3] Tarski [1918] 2014: *rząd gimnazyum*. A Polish secondary school that prepared students for eventual university study was called a *gimnazyum* or *gimnazjum*. The latter term is used in this book to avoid confusion with various types of secondary schools of other times and places.

For more than a century, Poland had been partitioned between Russia on the east, Germany to the west, and Austria to the south. Thoughts of Polish unification and independence had long spawned agitation and intrigue. World War I broke out in August 1914 for other reasons. Its conclusion four years later resulted in a unified Polish republic, but almost by accident. Many Polish military organizations were formed in 1914, and within two years, nearly two million Polish men were involved. Poles had enjoyed greater freedom in the Austrian partition, and the Germans made promises, so most of the Polish armies served, under the leadership of Józef Piłsudski, with Austria and Germany against Russia. Russia's allies Great Britain and France regarded the question of Polish independence as an internal Russian matter. Piłsudski played each side against the other. International posturing and intrigue increased in intensity. Amid this clamor, Alfred became a Polish nationalist.

In September 1915 Alfred transferred to the small, elite Mazowieckie Gimnazjum.[4] Its faculty were highly educated scholars, including two with doctorates in philosophy from the University of Lwów: Stefan Frycz and Bogdan Nawroczyński. Alfred's favorite subject was biology; his teacher, Stanisław Przylecki, had recently earned a medical degree in Zurich. Alfred's brilliance impressed both his teachers and his fellow students. That summer, the Germans entered Warsaw, which they would occupy until the end of the war. They instituted a number of reforms immediately, including permission for Alfred's school to switch its instruction from Russian to Polish.

Since 1870 the University of Warsaw had functioned as a Russian institution, serving the Imperial establishment. Most Polish students had to attend university abroad. After student boycotts and the outbreak of the war, the Russians closed the university altogether in 1914 and moved its faculty back to the homeland. The German occupation supported its autumn 1915 reopening as a Polish university. The new Polish faculty were assembled from various institutes in Warsaw, from universities in the other partitions of Poland, and from exile abroad. The university expanded rapidly during the war years, from one thousand to more than four thousand students. Most belonged to the urban middle or upper class, from central Poland; about 75% had to work to offset expenses. About 65% were male; and about 25%, Jewish. During that time the philosophical faculty grew to about forty, of all ranks.[5] According to the mathematician Kazimierz Kuratowski, who was a student in Warsaw at that time,

> ...the restored institutions of higher education were...a fulfilment of the dreams of many generations, the attainment of the goal of a persistent struggle for Polish education. Therefore, beside young students in classes one could see adult representatives of the Warsaw intelligentsia, for whom a direct contact with the restored Polish universities and colleges was a deep emotional experience.

> ...the atmosphere in which the institutions of higher education in Warsaw began their work...released a great creative potential...which produced a surprising development in many branches of science, including Polish mathematics.

---

[4] Tarski [1918] 2014: *Szkoła Ziemi Mazowieckiej*.
[5] Garlicki 1982, 49, 53, 314–315, 343. See also Manteuffel 1936, 156–175.

## 1.1 Coming of Age in Warsaw

Two mathematicians, the topologists Zygmunt Janiszewski and Stefan Mazurkiewicz, gathered several others, including Samuel Dickstein, Stefan Kwietniewski, and the logician Jan Łukasiewicz, to start building the mathematics faculty.[6]

At the Mazowieckie Gimnazjum, Alfred prepared to enter the university. He was graduated in June 1918 with "excellent, 5" marks in all subjects covered that year:

| Polish | German | History | Logic | Mathematics | Physics |
|--------|--------|---------|-------|-------------|---------|
| Latin  | French | Civics  | Hygiene | Cosmography | |

His mathematics courses had included analytic geometry but not calculus. Two of his nine classmates were too ill to complete the year. Typhus was raging, due to the huge influx of refugees into Warsaw, overcrowding, and poor sanitation. For example, 775 new cases were reported in the city during the week ending 23 February. Nearly three-fourths of these were among the Jewish population; the fatalities, about 9% of the victims, were most common among those over forty, and twice as prevalent among Christians.[7]

Alfred's graduation picture, on page 13, shows his intensity. In October 1918, he enrolled in a broad university curriculum, intending to concentrate in biology.[8] The war ended officially in November. The Germans released Piłsudski, whom they had imprisoned for a year when he stopped cooperating. Immediately, he assumed command of the Polish government in Warsaw. But armed conflict continued in the German part of Poland to the west. The university suspended classes for the academic year 1918–1919 and urged all its students to join the Polish army; about half did so. Alfred was declared unfit for service, but his studies were interrupted by the continuing Polish struggle for independence. The present editors do not know what occupied Alfred during the rest of 1918–1919.[9]

In 1918, Janiszewski published a plan for realizing in the discipline of mathematics the great creative potential that had stemmed from the fulfilment of dreams for restored institutions. According to Kuratowski,

> One of the principal means suggested ... for attaining that end was the concentration ... in a relatively narrow field of mathematics ... one in which Polish mathematicians had common interests and ... achievements which counted on a world scale. This field comprised set theory together with topology, and the foundations of mathematics together with mathematical logic.

---

[6] Kuratowski 1980, 20, 25–26, 28. Łukasiewicz served in the university administration (Łukasiewicz[1953] 1994, 133). For portraits and biographical sketches of Janiszewski and Mazurkiewicz, see pages 6 and 14; for Łukasiewicz, see section 9.4; a biographical sketch of Kwietniewski is in section 9.3.

[7] Szkoła Ziemi Mazowieckiej [1918] 1927. Jadacki 2003a, 141. The typhus report Goodall 1920 presents a vivid account of conditions in Warsaw. Tuberculosis was a comparable threat (Wynot 1983, 340).

[8] Tarski [1918] 2014, translated in section 16.4. Tarski 1924f.

[9] Garlicki 1982, 341; Manteuffel 1936, 26–28. The archive document Warsaw 1918–1919 has top and bottom parts. The top, from the university secretariat, 8 November 1918, certified that Alfred was student number 2909 of the Philosophical Faculty (*Wydział*) and said that he should give it to the military and return it when he was released from service. The bottom, from a military doctor on 5 February 1919, said "By reason of §1D and age, the requirement not applying, [Alfred] was declared unfit for military service."

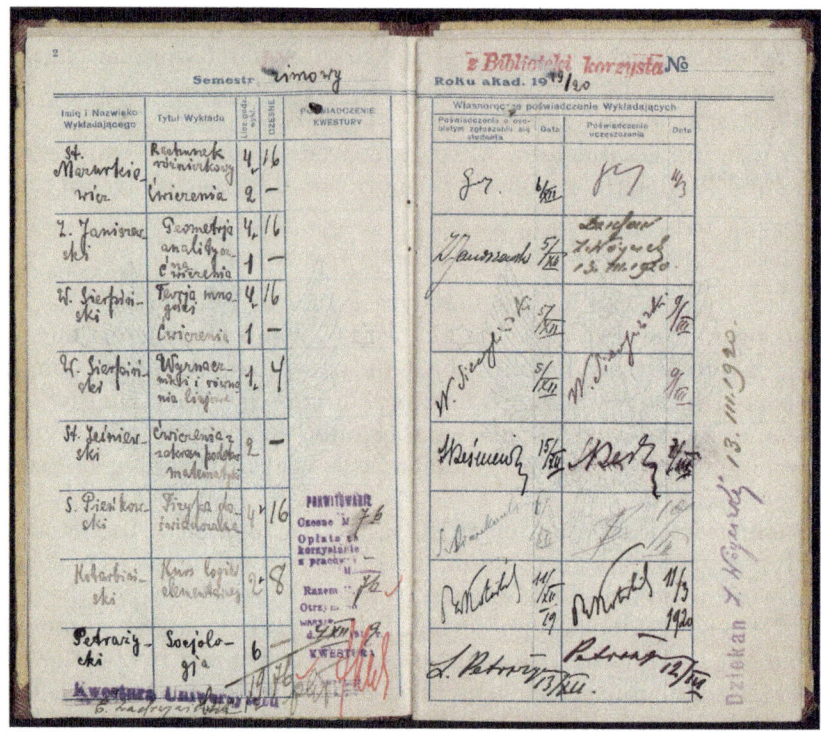

*Alfred's Autumn 1919 – Winter 1920 Enrollment Record*

*Stefan
Mazurkiewicz
around 1930*

*Zygmunt
Janiszewski
around 1915*

*Tadeusz
Kotarbiński
in 1933*

## 1.1 Coming of Age in Warsaw

Janiszewski, Łukasiewicz, and Mazurkiewicz would undertake that development. They were joined by the Polish mathematician of greatest world note, who had come to Warsaw that year: Wacław Sierpiński, a specialist in set theory.[10]

Alfred must have caught a glimpse of this new direction during the disrupted year 1918–1919. Reenrolling for the academic year 1919–1920, he would continue some of his general scientific studies. But his new emphasis was mathematics and logic.[11] The atmosphere was electric:

> ...students could be met more often at political rallies and meetings than in university lecture rooms or laboratories.[12]

Withstanding the distraction, Alfred signed up for thirty-one hours of classes per week. The corresponding entries of his enrollment record are displayed on the facing page.[13] They show that Alfred attended

lecture/exercises courses by
- Mazurkiewicz on differential calculus
- Janiszewski on analytic geometry, and
- Sierpiński on set theory;

lectures by
- Sierpiński on determinants and linear equations;

exercises with
- Stanisław Leśniewski on foundations of mathematics; and

lectures by
- Stefan Pieńkowski on experimental physics,
- Tadeusz Kotarbiński on elementary logic, and
- Leon Petrażycki on sociology.

The stamp near the bottom of the left-hand page is the bursar's receipt for payment of 76 Polish marks. Alfred was charged no laboratory fees, and only for the nineteen hours of lectures checked in the third column. (Footnotes will lead to biographical sketches and portraits of many of Alfred's teachers.)[14]

---

[10] Janiszewski's proposal ([1918] 1968) was published by the Mianowski Fund, which supported many academic activities described in this book. For more information about the fund, consult a box in section 9.3. Kuratowski 1980, 29–31. For a portrait and biographical sketch of Sierpiński, see section 4.1.

[11] Givant (1991, 28) mentioned this change but presented no background for its occurrence in 1918–1919.

[12] Garlicki 1982, 341.

[13] Tarski 1924f. The legend at the top reads "Semestr *zimowy*. Roku akad. 19*19/20*." When Alfred's enrollment booklet was issued in 1918, academic years consisted of winter and summer semesters. In 1919–1920, the university converted to three trimesters: autumn, winter, summer (*jesień, zima, letni*). Its documentation placed data for the first two trimesters in the space for the former winter semester. The headings identify columns for lecturers' names, lecture titles, hours, tuition, bursar's certification, and lecturers' signatures and dates to certify enrollment and attendance. Summer data are on the following pages. Jadacki (2003a, 142) reported some of this information, but inaccurately.

[14] Biographical sketches of Kotarbiński and Leśniewski are on page 9; portraits, on pages 6 and 13.

During the summer trimester in 1920, Alfred continued with twenty-nine hours of classes per week: with Mazurkiewicz on differential and integral calculus and analysis; with Sierpiński on number theory and measure theory; with Leśniewski on foundations of set theory; with Pieńkowski on physics, with laboratory; with Kotarbiński on logic and Francis Bacon's methodology; and in his philosophy seminar. Alfred paid a laboratory fee of 10 Polish marks and 116 Polish marks for tuition.

The German withdrawal, the Russian Revolution, the collapse of Austrian hegemony, various independence movements, and the internal politics of the new Polish nation brought about strife and chaos in Eastern Europe. Polish armies fought six wars in three years. Eventually the Polish and Soviet armies confronted each other, under military and political control of Piłsudski and Leon Trotsky. (See the box on page 10.) Poland was immersed in propaganda to stir up support and recruit troops for the war. A million Poles were marshalled along the front. Battles moved back and forth for months. In summer 1920 a huge Soviet advance threatened the very outskirts of Warsaw. Upper-class citizens left town, the rest were mobilized but remained strangely calm, and

> in July, the students again adopted a resolution to join the army en masse...for the second time lectures were suspended.[15]

The posters depicted on page 11 give an idea of the atmosphere in Warsaw. Alfred served with a military supply and medical unit.[16] His professors Leśniewski, Mazurkiewicz, and Sierpiński had been working with the military on a project that had decoded the Soviet army's communications. With this advantage, the Poles routed the invaders on 18 August, in a battle known as the *Miracle of the Vistula*.[17] Piłsudski's troops quickly pushed the invaders out of Poland. This victory brought to a close the chaotic Polish struggle for unified, independent nationhood. Poles settled down to work, to put into action the many plans set out during the previous years, and to let Polish culture blossom.

Alfred forwarded his 1919 certificate of military leave to the university administration with a note signed "Alfred *Tajtelbaum*." It seems to be his first recorded use of that Polish spelling. His patriotic fervor might have been damped somewhat had he realized the extent to which the war propaganda had heated Poland's antisemitism. The new Polish government actually imprisoned some units of volunteers from Jewish communities.[18]

---

[15] N. Davies 1982, volume 2, 294–295. Garlicki 1982, 341.

[16] Feferman and Feferman (2004, 26) reported that Alfred's service was in 1918, but the earlier account in this section contradicts that. Requesting enrollment in autumn 1919, Alfred submitted the first form in Tarski 1919–1920. He described two attachments: his certificate of military leave with a clause indicating unsuitability for military service, and a certificate recognizing community service. In autumn 1920, Alfred submitted the second form, indicating that a certificate of military service was attached.

[17] Nowik 2004–2010, volume 1, 20–27, 39, 231–233, and volume 2, 77 and section IX.5. Volume 1 also serves as a guide to the confused historiography of this subject, and volume 2 displays numerous signed examples of decoding. See also Czyż 1990, Pepłoński 1995, 42–43, and the overview Bury 2004. (Much information about the 1920 war became accessible to the public only after the Communist regimes fell around 1990.) According to *Gazeta Warszawska* 1920, Kotarbiński and Łukasiewicz also volunteered for military service. For more information about the professors, see boxes on pages 9 and 14 and in sections 4.1 and 9.4.

[18] Tarski 1919–1920. Jadacki 2003a, 143. Polansky 2012, 53–54; Heller 1994, 51 and 314.

## 1.1 Coming of Age in Warsaw

**Tadeusz Kotarbiński** was born in 1886 in Warsaw, then a part of the Russian Empire. His father was a professional artist; both parents were musically gifted. After completing school in Warsaw in 1906, Tadeusz entered the University of Lwów. He studied logic with Jan Łukasiewicz, then turned to philosophy with Kazimierz Twardowski and earned the doctorate in 1912. Kotarbiński then returned to Warsaw to teach classics in a gimnazjum and lecture on cultural subjects to the public. He was a cofounder of the Warsaw Philosophical Institute. In 1919 he was appointed professor of philosophy at the University of Warsaw. He was a major inspiration for Alfred Tarski, and became a leader of world significance among analytic philosophers. In 1929 Kotarbiński became dean of the Faculty of Humanities. His activity in liberal political circles increased as well. During World War II he taught clandestinely. Afterward he married his former student Dina Sztejnbarg-Kamińska, who had been a neighbor of Tarski's parents. She became known as the philosopher Janina Kotarbińska. Kotarbiński served for four years as rector at the University of Łódź, then during 1951–1961 as professor of philosophy at Warsaw. During 1948–1978 he was president of the Polish Philosophical Society, and during 1957–1962, of the Polish Academy of Sciences. Kotarbiński died in 1981.*

**Stanisław Leśniewski** was born in 1888 near Moscow, to Polish parents. His father was a railroad engineer. His mother died soon after his birth, and his father remarried. Stanisław was educated in Siberian schools, German universities, and then the University of Lwów, where he earned the doctorate in 1912 under the supervision of Kazimierz Twardowski. He was also inspired by the work of Jan Łukasiewicz. Leśniewski's dissertation included the germ of much of his later work, in particular, insistence on extreme precision of language, and distinction between language and metalanguage. Around 1913, Lesniewski married the well-to-do Zofia Prewysz-Kwinto; they had no children.

After research, teaching, and socialist political activity in Moscow, St. Petersburg, and Warsaw, Leśniewski was awarded the *venia legendi* by the University of Warsaw in 1919 and appointed to its chair of philosophy of mathematics. During the next year, with colleagues Stefan Mazurkiewicz and Wacław Sierpiński, he helped the Polish army decode Soviet military communications. Their work was instrumental in the decisive 1920 defeat of Soviet invading forces in the suburbs of Warsaw.

Leśniewski initiated several fundamental logical ideas used in the analysis of language—for example, semantic categories. He became known for his obsessive precision and perfectionism in philosophical writing and discussion, and his sharp criticism of the work of most others, who did not attain that standard. His long-time Warsaw colleague Tadeusz Kotarbiński reported these characteristics:

> There was either the ultimate *yes* or the ultimate *no*, there was an aversion to half-measures, loathing of pettiness ... a bent for sudden elations, sharp turns, radical breaking off of friendships, vehement antipathy towards insincere feelings, adherence to principles and intolerance of exceptions, the talent and tendency to carry on endless discussions....

Mazurkiewicz wrote that Leśniewski

> as a researcher had a high sense of and need for monumentality, and made no [small] contributions or fragments, but an idiosyncratic system of foundations of logic and mathematics.

That system, whose three parts were called *mereology*, *prototethic*, and *ontology*, is still studied, but it has never come into common use. Kotarbiński explained that Leśniewski's

> teaching was never aimed at a mass audience: on the contrary, he preferred to deal with the most exquisitely intelligent.

Leśniewski's lectures generally attracted only tiny audiences, and Alfred Tarski was Leśniewski's only doctoral student. Nevertheless, Leśniewski "attached great importance to his university lectures and he lectured almost entirely about his own work." Kotarbiński remembered his colleague's research habit of walking to and fro deep in thought, all in vaporous clouds of nicotine. Leśniewski died of lung cancer at age 51, in May 1939. Assessing his contribution to science will probably be a subject of discourse for years to come.†

---

*Kotarbiński 1977, Woleński 1990b, Golińska-Pilarek et al. 2009a

†For more information, consult the box on page 10 and these works, particularly the pages indicated: Betti 2004, 272; Jadczak 1993b, 312, 316; Kotarbiński 1966, 156, 160; Leśniewski 1992, xii; Mazurkiewicz 1939; Pasenkiewicz 1984, 4.

**Polish–Soviet War.** During 1792–1914 the region called Poland was divided between three empires. Russia controlled the center and parts in the east, including Warsaw; Austria controlled parts in the south, including Cracow and Lwów; and Germany controlled parts in the west and north. (See the 1914 and 1924 maps on pages xxii–xxiii.) The eastern areas included many minorities, particularly Lithuanians, Belarusians, Ukrainians, and Jews.

The eastern campaigns of World War I started with Russian invasions of German and Austrian territory in August 1914. Within a year, however, the Germans had driven the Russians entirely out of Poland. The Russian Revolution began in March 1917. The Bolsheviks seized power in October, and withdrew from hostilities in March 1918. The German and Austrian regimes began to collapse that summer, and the war officially ended in November 1918. Within a week Poland declared independence, and General Józef Piłsudski became its chief of state. Thousands of refugees entered the country from the east, and there was large-scale migration, particularly of Jews, into Warsaw. Polish armed forces fought within the country to secure internal control, and against Ukrainian and Czechoslovak armies to secure its boundaries. Even the 1919 Versailles peace treaty left the question of Polish borders unsettled.

The Polish army recognized the advantages of intercepting and decoding enemy messages, and recruited university mathematicians to help. Stanisław Leśniewski, Stefan Mazurkiewicz, and Wacław Sierpiński played vital roles in that project.

Piłsudski aspired to leadership of a community of Slavic border nations: Lithuania, Belarus, and Ukraine. Strong opponents favored expanding Poland itself into those regions. The Bolsheviks aspired to restore Russia's prewar borders and to spread revolution to Germany.

During 1919 Piłsudski led the Polish army into the Soviet Union. By May 1920 the Poles controlled Vilnius, Minsk, and Kiev. These places suffered greatly during 1918–1920, changing hands several times. Because of the great distances, however, the Poles could not sustain their effort. Moreover, the British and French stopped supporting Poland. Even though the Bolsheviks still did not have complete control of their own country, their army, under Leon Trotsky, drove the Poles back, deep into Poland, in summer 1920. The Soviet army, controlled by political commissars, was highly effective and disciplined. The high visibility of some Jews among those leaders, especially Trotsky, inflamed the antisemitism that was rife in Poland.

Warsaw remained surprisingly calm under threat of conquest. The Soviet army penetrated within twenty-five kilometers to the east and north and reached the Vistula River. The city mobilized totally to support the Polish army.

Like the Polish army a few weeks earlier, the Soviet armies were overcome by the great distances. Moreover, the Poles intercepted and understood *all* Soviet communications. In August 1920, Piłsudski's army used that intelligence to locate and attack a weak point. They routed the invaders and soon drove them back through Poland and German East Prussia all the way to Lithuania. In September, the Soviet army began to disintegrate, and sued for peace. This great victory, the *Miracle of the Vistula*, put an end to Bolshevik ambitions toward the West.

There was no great celebration in Warsaw. Exhausted, Poles just resumed the slow work of constructing their new country. During 1918–1920 its boundaries were established as shown on the 1924 map, except that Vilnius and its surrounding area did not become Polish until 1922. Antisemitism continued to rise: Poland even imprisoned some of its own Jewish volunteer units after the war. And its incorporation of large territories in the East gave it a large population of resentful minorities.

On the facing page are depicted three large colorful 1920 posters. The captions translate their texts. Similar propaganda was spread throughout Poland to stir up support and recruit troops for the war. These display common themes. Trotsky is caricatured stereotypically as a bright-red Jewish devil. The sky above the ruined city on the right burns red, and Bolsheviks are characterized as anti-Christian. Contrast the red devil with the manly grenadier, advancing the Polish shield to protect the future of his country! *

---

*For more information see page 8, footnote 17; N. Davies 1982, chapter 13, and 1972; Szczepański 1995; and Watt 1979, chapter 6. The Trotsky poster's artist was named Skabowski (Fuchs 1921, 280).

## 1920 Propaganda

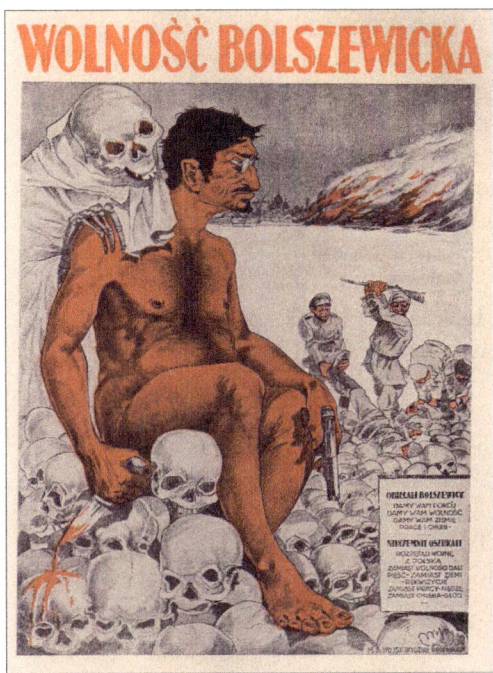

**BOLSHEVIK FREEDOM**

*THE BOLSHEVIKS PROMISED TO*
  *give you peace,*
  *give you freedom,*
  *give you land,*
  *work and bread.*

*THEY BASELY CHEATED, AND*
  *unleashed a war with Poland.*
  *Instead of freedom, the fist.*
  *Instead of land, requisition.*
  *Instead of work, misery.*
  *Instead of bread, hunger.*

*To Arms! This Is How a Polish Village Occupied by the Bolsheviks Looks*

*To Arms! Save the Fatherland! Always Think of Our Future.*

In October 1920 the university reopened, and Alfred returned to his studies, perhaps even with greater excitement and vigor. He continued in the same vein, with courses from Leśniewski on foundations of arithmetic and on algebra of logic, Mazurkiewicz on analytic geometry, and with Sierpiński on higher algebra and on set theory. Łukasiewicz had returned to the faculty after serving during 1919 as the first Polish minister of higher education,[19] and Alfred enrolled in his seminars and courses on philosophical logic. It is possible to discern three intellectual threads emerging from Alfred's studies during his first two years at the university: logic, set theory, and measure theory. They would extend far into his research career. Repeatedly during 1920–1924, Alfred participated in the seminars of Kotarbiński, Leśniewski, and Łukasiewicz. These professors shared a similar background: they were all students of Kazimierz Twardowski, the founder of the famed Lwów–Warsaw school of logic.[20]

In 1921, while still a student, Alfred published his first research paper: *A Contribution to the Axiomatics of Well-Ordered Sets*. A subtitle indicated that it stemmed from his work in Leśniewski's seminar.[21] Its mathematical content is discussed in the next section, and it is translated in its entirety in chapter 2. The paper appeared in *Przegląd filozoficzny*, a philosophical journal founded in 1897 and published at that time by the Warsaw Philosophical Institute, with the support of the Mianowski Fund.[22] The journal's cover is displayed on page 20. Its table of contents listed Alfred with the surname *Tajtelbaum*, with the Polish spelling that he had adopted the previous year.

The journal was devoted to material from the whole field of philosophy. For example, the other papers published in the same issue with Alfred's had to do with philosophy of law, experimental psychology, history of philosophy, and developmental psychology. Their authors were established scholars, in their thirties; but Alfred was only twenty, just a student. The next page after the title, not shown here, lists the editors: Twardowski and Marjan Borowski. Soon after publication, Borowski wrote to Twardowski,

> ...I inform you—discreetly—that *Przegląd filozoficzny* doesn't have many papers of any real worth in the editorial office. Warsaw coryphaei write little, being afraid of Leśniewski!— although the "scourge of God" has also risen upon him in the person of his pupil, Tajtelbaum.[23]

---

[19] Łukasiewicz [1953] 1994, 133. The prime minister was the internationally famous pianist Ignacy Paderewski.

[20] See Tarski 1924f. Givant 1999 describes these and other threads in detail. Łukasiewicz earned the doctorate in 1902; Kotarbiński and Leśniewski, in 1912. Alfred's philosophy teacher at the Mazowieckie Gimnazjum, **Bogdan Nawroczyński**, had also been a student of Twardowski; he later became professor of education at the University of Warsaw. For biographical information on Twardowski and a portrait, see page 18.

[21] Tarski [1921] 2014. The present editors looked for other papers of that era that might claim such a relationship to the seminar, but found none.

[22] *Philosophical Review*, *Warszawski Instytut Filozoficzny*. For information about the Mianowski Fund, consult a box in section 9.3.

[23] Borowski 1922. A *coryphaeus* is a leader of a dramatic chorus. Borowski had also been a student of Twardowski.

*Alfred Teitelbaum
(Tarski) in 1918*

*Stanisław Leśniewski
around 1915*

**Zygmunt Janiszewski** was born in Warsaw in 1888; his father was a financier. After completing school in 1907 in Lwów, then part of the Austrian Empire, Zygmunt studied at Zurich, Munich, Göttingen, and Paris. At Zurich he displayed his social talent by organizing a support group for Polish students. In 1911, he earned the doctorate in Paris with a dissertation on point-set topology supervised by Henri Lebesgue. Janiszewski produced several significant works on that subject, particularly a simplification of the Jordan curve theorem, and authored some influential articles explaining modern abstract mathematics to educated Poles. Janiszewski taught briefly in Warsaw and Lwów, then joined Józef Piłsudski's Polish Legion for service during 1914–1915 against the Russians. He soon left it, refusing to sign an oath of loyalty to Germany and Austria. When the University of Warsaw reopened as a Polish university in 1915, Janiszewski was selected as one of its initial professors. His seminal [1918] 1968 report guided development of a world center of mathematical research in Poland, through specialization in set-theoretic topology, logic, and foundations of mathematics. He was a principal founder of the journal *Fundamenta Mathematicae*, devoted to those subjects. Janiszewski died at age 31 in the influenza epidemic in January 1920. According to his student Bronisław Knaster, "For Janiszewski teaching was a mission and the student a comrade...." Janiszewski donated all his possessions to educational charities.*

**Stefan Mazurkiewicz** was born in 1888 in Warsaw, then part of the Russian Empire. His father was a noted attorney. After graduating from secondary school in 1906 in Cracow, Stefan attended university courses there, in Munich, in Göttingen, and then briefly in Lwów, where he earned the doctorate in 1913 with a dissertation on area-filling curves, supervised by Wacław Sierpiński. Mazurkiewicz began a very extensive research career in probability theory, topology, and analysis.

In 1915 Mazurkiewicz became the youngest of the founding mathematics faculty of the newly reconstituted Polish University of Warsaw. He played a major role in developing the Warsaw school of mathematics, particularly by leading frequent faculty meetings, formal and informal, on current research work and on strategies for expanding that activity in the future. Alfred Tarski enrolled in his courses nearly every semester of his Warsaw studies.

Although best known for theoretical results, Mazurkiewicz's mathematical activity had major applied aspects. In Göttingen he had earned a diploma on insurance mathematics. In 1919, together with his Warsaw colleagues Sierpiński and Leśniewski, he began work with the Cipher Bureau (*Biuro Szyfrów*), the Polish military agency devoted to decoding enemy communications. In 1920, the group broke the code used by the invading Soviet army, enabling General Józef Piłsudski to avert catastrophe through a decisive victory at the very approaches to Warsaw. Mazurkiewicz continued working with the Bureau as a consultant at least until 1930. The Bureau's later success in breaking German codes played a major role in World War II.

Mazurkiewicz married twice. In 1920 he reportedly fought a duel with a military officer over an affair of the heart. After a divorce, Mazurkiewicz's former wife married Jan Kowalewski, the officer who headed the Cipher Bureau in 1920. Mazurkiewicz served several terms, about ten years in all, as dean of the Faculty of Mathematics and Science and as prorector of the university. Nevertheless, he kept up a stream of research in topology, analysis, and the foundations of probability theory. Most of his eight or so doctoral students became leading researchers; two of them—Kazimierz Kuratowski and Bronisław Knaster—play roles in this book.

During the 1939–1944 German occupation, Mazurkiewicz taught clandestinely and helped his colleagues plan the postwar resurgence of Polish mathematics. But he became seriously ill and died in a hospital near Warsaw a month after the armistice. He had been reconstructing a destroyed manuscript for a major work on probability theory; it was published a decade later, in 1956.†

---

*Knaster 1973; Kuratowski 1980: 158–163.

†Pawlikowa-Brożek 1975. Kuratowski 1981, 62, and 1980. See also the references in footnote 17 on page 8, especially Czyż 1990.

## 1.2 Well-Ordered Sets

This section is devoted to the technical background and impact of Alfred Tarski's first research paper, *A Contribution to the Axiomatics of Well-Ordered Sets*, which is translated in its entirety in chapter 2. It was published in 1921 under the surname Tajtelbaum. Alfred did not change that until 1924. Nevertheless, this section, on his mathematics and its presentation, will use the name Alfred Tarski.

Tarski's paper is about the interrelationships between various basic properties of well-ordered sets. Georg Cantor had introduced those in 1883: the elements of a well-ordered set should be generated by a process that

- selects some *first* element, and
- always identifies a *unique next* element, no matter how many have already been generated, provided only that it has not yet generated all elements.

That description is somewhat vague, particularly in its allusion to time via the words "already" and "not yet." But it reflects how mathematicians of that era considered a set and its elements simultaneously, and it shows the centrality of the "unique next" idea in Cantor's conception. Cantor evolved this notion, replacing that description in 1897 with the notion of a *well-ordering* for a set $M$. This should be an antireflexive antisymmetric transitive binary relation $R$ on $M$ such that

- $M$ has an element $m$ such that $m\,R\,s$ for every $s \in M$ different from $m$;
- if $M$ contains an upper bound for a subset $S$ —that is, an element $u \in M$ such that $s\,R\,u$ for every $s \in S$ —then it contains an upper bound $v$ for $S$ such that no upper bound $u$ for $S$ satisfies the condition $u\,R\,v$.

It is easy to see that these latter two conditions are equivalent to the one most commonly used today: $M$ must be nonempty, and each nonempty subset $S$ of $M$ must contain an element $m$ such that $m\,R\,s$ for every $s \in S$ different from $m$.[24]

Tarski's methodology was that of postulate theory. During the 1880s and 1890s, mathematicians had become increasingly aware of the usefulness of applying the axiomatic method precisely in presenting a theory:

- stating clearly the notions that will be left undefined,
- defining solely from these primitive notions all other terms to be used,
- stating clearly the principles that will be left unproved, and
- deriving all theorems of interest solely from these axioms and the primitive notions and definitions.

---

[24] Cantor 1883, §2; [1895–1897] 1952, §12, and note 16. The words "already" and "not yet" in this paragraph approximate Cantor's use of *Sukzession*.

Postulate theory evolved to facilitate investigation of the selection of primitive notions and axioms and the phrasing of definitions and theorems.[25] It stressed their economy, simplicity, gracefulness, and ease of use. These are often conflicting criteria: for example, making individual axioms simpler often requires increasing their number, and simpler primitive notions, like some primitive hand tools, are often awkward to use. Several applications of postulate theory involved similar details: studying the order of points on a line, that of the natural numbers, that of infinite ordinal numbers, and other types of ordering closely related to those. By about 1900, these studies had become the theory of ordered sets, a coherent part of postulate theory. Tarski's paper lies there.

Tarski denoted by $A_1$, $A_2$, and $A_3$ the familiar trichotomy, antisymmetry, and transitivity axioms that characterize a strong[26] linear ordering $R$ of a set $Z$. He then considered the following sentences:

$$(\forall U \subseteq Z)(\phi \neq U \Rightarrow (\exists a \in U)(\forall u \in U) \neg u\, R\, a) \tag{B}$$

$$(\forall U \subseteq Z)(\phi \neq U \Rightarrow (\exists a \in U)(\forall u \in U)(u \neq a \Rightarrow a\, R\, u)) \tag{C}$$

$$(\forall U \subseteq Z)(\phi \neq U \Rightarrow (\exists! a \in U)(\forall u \in U) \neg u\, R\, a) \tag{D}$$

$$(\forall U \subseteq Z)(\phi \neq U \Rightarrow (\exists a \in U)(\forall t,u \in U)(t\, R\, a\ \&\ u\, R\, a \Rightarrow t = u))^{27} \tag{E}$$

$$(\forall U \subseteq Z)(\phi \neq U \neq Z \Rightarrow (\exists a \in U)(\forall u \in U)(u \neq a \Rightarrow \neg u\, R\, a)) \tag{F}$$

Tarski showed that equivalent axiom systems result by adding $B$, $C$, $E$, or $F$ to $\{A_1, A_2, A_3\}$, and those systems characterize well-ordering relations $R$. Moreover, the latter two systems—with $E$ or $F$—are independent, whereas the former two—with $B$ or $C$—are not. Finally, he proved that these systems are all equivalent to that consisting of the *single axiom D*.

Tarski's paper was reviewed in detail by Leon Chwistek, a noted artist and philosopher, in the *Jahrbuch über die Fortschritte der Mathematik*. The previous paragraph closely echoes his report. Then a secondary-school teacher of mathematics, Chwistek would later become professor at the University of Lwów and a leader in Polish logic.[28]

Tarski's axiom $D$ is close to Cantor's original 1883 definition of a well-ordered set. Cantor did not stress the properties stated by $A_1$, $A_2$, and $A_3$, but did emphasize the existence and uniqueness of the next element to be selected from those not yet generated, which can be construed as the elements of the set $U$.

---

[25] Consult Russell 1903 for an overview of early work, and Scanlan 2003 for a retrospective from the point of view of modern logic.

[26] An analogous characterization of a weak linear ordering would employ the transitivity axiom with "dichotomy" and "weak antisymmetry" axioms. Analogous results would ensue.

[27] The symbol $\exists!$ in $D$ stands for "there exists a unique." The terminal clause in $E$ following the universal quantifier $\forall$ says that at most one element $t$ of $U$ satisfies $t\, R\, a$.

[28] Chwistek omitted the condition $U \neq Z$ in Tarski's axiom $F$. For more information on Chwistek, consult the biographical sketch in section 14.2.

## 1.2 Well-Ordered Sets

The most involved arguments in Tarski's paper are the proofs that the axiom systems with $E$ and $F$ are equivalent to the others. Tarski included no references to the literature, and only one less-formal reference, attributing a certain logical method to Jan Łukasiewicz. That is Tarski's sole discussion in this paper of a logical method, independent of the mathematics under consideration.

Tarski revisited this subject in his 1924c abstract *Sur les principes de l'arithmétique des nombres ordinaux (transfinis)*. There he presented, without proof, a system of six independent axioms for the notion of ordinal number in Zermelo set theory without the axiom of choice. Here and there in later discussions of well-ordered sets, Tarski's axioms appear, usually without attribution. For example, Patrick Suppes used the system $\{A_1, B\}$ in his 1960 set-theory text.[29] The most important impact of Tarski's paper is not its mathematical content, but its reflection of personal style. It displays Tarski's practice, in lectures and most research papers, of providing extreme detail in proofs, and of kneading the formulations of definitions and axioms to achieve great concision without sacrificing grace. He probably acquired that habit from his teachers Tadeusz Kotarbiński, Stanisław Leśniewski, and Łukasiewicz: recalling those times, the historian of logic Józef M. Bocheński (1994, 7) attributed this trait to their common teacher, Kazimierz Twardowski.

---

[29] In 1960, 75–76, Suppes derived $A_2, A_3, C$ from $A_1, B$.

*Kazimierz Twardowski
around 1918*

**Kazimierz Twardowski** was born in 1866 in Vienna, to an aristocratic family from the part of Poland in the Austrian Empire. He attended an elite school in Vienna, then studied mathematics, science, and philosophy at the University of Vienna, with a year at Leipzig and Munich. Twardowski earned the doctorate in Vienna in 1892, worked for a while in insurance and as a writer, and soon obtained the *venia legendi*. The guiding light of his studies was the renowned Franz Brentano, who had recently left the priesthood and resigned from the regular faculty. His associate, Robert von Zimmermann, officially supervised Twardowski's dissertation.

Twardowski was appointed professor of philosophy at the University of Lwów in 1895, and soon built an extremely strong program of scientific philosophy there, which became the Lwów–Warsaw School. According to the historian of philosophy Arianna Betti,

> Twardowski laid emphasis on "small philosophy," namely on the detailed, systematic analysis of specific problems—including problems from the history of philosophy—characterised by rigor and clarity, rather than on the edification of whole philosophical systems and comprehensive worldviews.

Most of Alfred Tarski's philosophy teachers at Warsaw were students of Twardowski. Twardowski was wildly popular: his lectures were often held in a concert hall, and one year at seven in the morning at a cinema near the university! He spent great effort to establish an infrastructure for philosophy in Lwów, for example, various societies, a psychology laboratory, and the journal *Ruch filozoficzny*. Betti noted,

> All these activities cost him a lot of time: in fact, Twardowski's choice to be most of all an educator and an organizer left him very little time for academic writing.

Twardowski retired in 1930 and died in Lwów in 1938.[†]

---

[†]Betti 2011.

# 2
# *A Contribution to the Axiomatics of Well-Ordered Sets* (1921)

This chapter is devoted to an English translation of Alfred Tarski's first published paper, *Przyczynek do aksjomatyki zbioru dobrze uporządkowanego*, [1921] 2014, written while he was still a student at the University of Warsaw using the surname Tajtelbaum.[1] It appeared in volume 24 of the journal *Przegląd filozoficzny*. This is its first translation. Background for the paper and a summary are provided in sections 1.1 and 1.2.

The translation is meant to be as faithful as possible to the original. Its only intentional modernizations are punctuation and some changes in symbols where Tarski's conflict with others used throughout this book. As an aspect of adjusting punctuation, the editors greatly increased use of white space to enhance visual organization of the paper. [Square] brackets in the Polish original have been changed to braces or parentheses. All [square] brackets in the translation enclose editorial comments. Those are inserted, usually as footnotes, to indicate or suggest corrections for occasional errors in the original, and to explain a few passages that seem obscure.

---

[1] Alfred's changes of surname from *Teitelbaum* and to *Tarski* are discussed in section 1.1 and chapter 3.

# PRZEGLĄD FILOZOFICZNY

założony przez Władysława Weryhę.

Wychodzi kwartalnie pod redakcją
WARSZAWSKIEGO INSTYTUTU FILOZOFICZNEGO.

## Rocznik 24 (1921).

ZESZYT I i II.

Wydane z zasiłku Ministerstwa W. R. i O. P.

---

CZ. ZNAMIEROWSKI: O przedmiocie i fakcie społecznym.
A. DRYJSKI: Źródła cenestezji.
A. TAJTELBAUM: Przyczynek do aksjomatyki zbioru dobrze uporządkowanego.
B JASINOWSKI: Konflikt rozumu i wiary a rozwój dziejowy filozofji.
S. BALEY: Uwagi psychologiczne o genezie poematu Słowackiego „W Szwajcarji".
XIV Sprawozdanie Polskiego T-wa Psycholog. w Warszawie.

---

WARSZAWA — 1921.
Adres Redakcji i Administracji: ul. Piękna 44.
Sp. Akc. Zakł. Graf. „Drukarnia Polska", Szpitalna 12.

*Journal Containing the First Paper of Alfred Tarski (Tajtelbaum)*

ALFRED TAJTELBAUM

# A Contribution to the Axiomatics of Well-Ordered Sets

## From the Seminar of Professor Stanisław Leśniewski at the University of Warsaw

According to the traditional definition accepted in set theory, a set $Z$ is ordered with respect to a relation $R$ if and only if the following three "order axioms" are satisfied:

$A_1$  For all $x$ and $y$, if $x$ and $y$ are distinct elements of the set $Z$, then $x$ has the relation $R$ to $y$, or else $y$ has the relation $R$ to $x$. (Or, in an equivalent formulation, if $x$ and $y$ are distinct elements of the set $Z$ and $x$ does not have the relation $R$ to $y$, then $y$ does have the relation $R$ to $x$.)

$A_2$  For all $x$ and $y$, if $x$ and $y$ are elements of the set $Z$ and $x$ has the relation $R$ to $y$, then $y$ does not have the relation $R$ to $x$.

$A_3$  For all $x$, $y$, and $t$, if $x$, $y$ and $t$ are elements of the set $Z$, and $x$ has the relation $R$ to $y$, and $y$ has the relation $R$ to $t$, then $x$ has the relation $R$ to $t$.

I shall write

$xRy$   instead of "$x$ has the relation $R$ to $y$"
$xR'y$  instead of "$x$ does not have the relation $R$ to $y$"
$x \neq y$ instead of "$x$ is different from $y$"
$x = y$  instead of "$x$ is not different from $y$" [that is,] "$x$ is identical with $y$."

A relation $R$ with respect to which a given set $Z$ is ordered is often called a precedence relation; instead of "$xRy$," we read "$x$ precedes $y$." Axiom $A_2$ is called the axiom of antisymmetry; axiom $A_3$, the axiom of transitivity.

As is well known, we call an ordered set well-ordered if and only if each of its nonempty subsets has a first element. Thus, in order to obtain a system of axioms for a well-ordered set $[Z]$, we append to the three order axioms a fourth "well-ordering axiom," which in precise formulation takes one of the following shapes:

B. For every $U$, if
1. $U$ is a set,
2. for every $x$, if $x$ is an element of the set $U$, then $x$ is an element of the set $Z$, and
3. for some $k$, $k$ is an element of the set $U$,

then for some $a$,
1. $a$ is an element of the set $U$,
2. for every $y$, if $y$ is an element of the set $U$, then $y$ does not precede $a$.

C. For every $U$, if
1. $U$ is a set,
2. for every $x$, if $x$ is an element of the set $U$, then $x$ is an element of the set $Z$, and
3. for some $k$, $k$ is an element of the set $U$,

then for some $a$,
1. $a$ is an element of the set $U$,
2. for every $y$, if $y$ is an element of the set $U$ different from $a$, [that is] $y \neq a$, then $a$ precedes $y$.

We see that these two axioms differ in their two distinct, inequivalent interpretations of the term "first element."

On the other hand, it is nearly obvious that axiom systems $\{A_1, A_2, A_3, B\}$ and $\{A_1, A_2, A_3, C\}$ are equivalent: from the first of these it is possible to deduce axiom $C$ as a theorem; from the second, axiom $B$.

Indeed, let us consider the first system. Every set $U$ satisfying the hypothesis of axiom $C$ also satisfies the identically phrased hypothesis of axiom $B$; thus it has an element $a$ such that if $y$ is an element of the set $U$, then $y$ does not precede $a$. If in addition $y$ is different from $a$, then $a$ precedes $y$, according to axiom $A_1$. In other words, axiom $C$ is satisfied.

Similarly, it is straightforward to show that axiom $B$ can be deduced from axioms $A_2$ and $C$ (and the so-called theorem of antireflexivity of the relation $R$, which follows from axiom $A_2$).

However, neither the first nor the second axiom system for a well-ordered set is an independent axiom system. In the first system, axioms $A_2$ and $A_3$ can be deduced from the rest; in the second, $A_1$ and $A_3$.

In fact, to prove axiom $A_2$ from the axiom system $\{A_1, B\}$ let us consider a set $U$ consisting of any two elements $x$ and $y$ of the set $Z$ that satisfy the hypothesis $x R y$ of axiom $A_2$. The set $U$ satisfies the hypothesis of axiom $B$; therefore, it has an element that no element of $U$ precedes. Since $xRy$, this element is not $y$. Therefore, it must be $x$, and because $y$ is an element of the set $U$, it follows that $yR'x$, which is exactly what was to be proved.[2]

To prove axiom $A_3$ [from the axiom system $\{A_1, B\}$], let us consider a set $U$ consisting of any three elements $x$, $y$, and $t$ of the set $Z$ that satisfy the conditions $xRy$ and $yRt$. Again, the set $U$, satisfying the hypothesis of axiom $B$, has an element that no element of this set precedes. Since $xRy$ and $yRt$, [this element] is neither $y$ nor $t$. Therefore, it must be $x$, and thus

$$tR'x. \tag{1}$$

On the other hand,

$$t \neq x. \tag{2}$$

Indeed, $yRt$ entails $tR'y$ by virtue of the previously proven axiom $A_2$. Thus $x$ precedes $y$, while $t$ does not precede $y$, [and] therefore $x$ and $t$ are distinct. But if conditions 1 and 2 are satisfied, then $x$ stands in the relation $R$ to $t$ by virtue of axiom $A_1$, Q.E.D.

To prove axiom $A_1$ from the axiom system $\{A_2, C\}$, let us consider a set $U$ consisting of two distinct elements $x$ and $y$ of the set $Z$. (Should the set $Z$ not contain two distinct elements at all, axiom $A_1$ would clearly be satisfied,[3] since its hypothesis would not hold for the given set.) The set $U$ satisfies the hypothesis of axiom $C$, [and] therefore has an element that precedes every element of the set $U$ that differs from [it]. This element can be $x$ or $y$, and therefore $xRy$ or $yRx$, Q.E.D.

To prove axiom $A_3$ [from the axiom system $\{A_2, C\}$], let us first of all observe that the so-called theorem of antireflexivity of the relation $R$ follows from the axiom of antisymmetry:

> T. For all $x$ and $y$, if $x$ and $y$ are elements of the set $Z$, and $xRy$, then $x \neq y$ (or, in an equivalent formulation, for every $x$, if $x$ is an element of the set $Z$, then $xR'x$).

Let us now consider a set $U$ consisting of any three elements $x$, $y$, and $t$ of the set $Z$ that satisfy the hypothesis of axiom $A_3$. The set $U$ has an element that precedes every element of this set that differs from it. This element is not $y$, since $xRy$ implies $y \neq x$ and $yR'x$, by virtue of axiom $A_2$ and theorem $T$. Neither is it $t$, since $yRt$ entails $t \neq y$ and $tR'y$. Therefore, this element is $x$. On the other hand, however, $t \neq x$ because both $xRy$ and $tR'y$ hold. Therefore $x$ precedes $t$, as an element different from it in the set $U$, which is exactly what was to be proved.

Thus, to define a well-ordered set, each of the two axiom systems $\{A_1, B\}$ and $\{A_2, C\}$ is sufficient.

---

[2] [This paragraph actually proves $A_2$ from $B$ alone.]
[3] [In error, Tarski wrote $A_2$ in place of $A_1$ in this sentence.]

Formulation of an axiom that by itself is equivalent to each of the previous systems is not difficult. Such, for example, is the following axiom:

D.   Every nonempty subset $U$ of the set $Z$ has one and only one element that none of its elements precedes.

More precisely: for every $U$, if
1. $U$ is a set,
2. for every $x$, if $x$ is an element of the set $U$, then $x$ is an element of the set $Z$, and
3. for some $k$, $k$ is an element of the set $U$,

then for some $a$,
1. $a$ is an element of the set $U$,
2. for every $y$, if $y$ is an element of the set $U$, then $y$ does not precede $a$,
3. for every $t$, if $t$ is an element of the set $U$ different from $a$, then for some $b$, $b$ is an element of the set $U$ that precedes $t$.

I shall give a proof of the equivalence of axiom $D$ with the axiom system $\{A_1, B\}$.

Axiom $D$ directly implies axiom $B$. (It suffices to disregard the third part of the conclusion.)

But to prove axiom $A_1$ [from axiom $D$], let us consider a set $U$ consisting of any two elements $x$ and $y$ of the set $Z$ that satisfy the conditions $x \neq y$ and $xR'y$. The set $U$ satisfies the hypothesis of axiom $D$, [and] therefore has an element that no element of the set $U$ precedes. From axiom $B$, and thus also from axiom $D$, as I already proved above, axiom $A_2$ can be deduced, and therefore also theorem $T$ on antireflexivity. That implies $yR'y$: that is, the element $y$ of the set $U$ is just the one that no element of this set precedes. On the other hand, $x \neq y$; therefore, according to the third point in the conclusion of axiom $D$, $x$ is not such an element, and some element of the set $U$ precedes it: $xRx$ or $yRx$. But $xR'x$ (according to theorem $B$), and therefore $yRx$, Q.E.D.

And conversely, axiom $D$ follows from the axiom system $\{A_1, B\}$. Indeed, every set $U$ satisfying the hypothesis of axiom $D$ also satisfies the identically phrased hypothesis of axiom $B$; therefore, it has an element $a$ that no element of the set $U$ precedes. This element is the only one having that property: in fact, if $t$ is an element of the set $U$, then $tR'a$; if, in addition, $t \neq a$, then $aRt$ according to axiom $A_1$, and thus element $t$ does not satisfy the conclusion of axiom $B$.[4]

Thus, axiom $D$, just like axiom systems $\{A_1, B\}$ and $\{A_2, C\}$, suffices for defining a well-ordered set.

---

[4] [The last clause of this sentence seems unnecessary to the proof.]

Similar definitions cannot, however, have greater significance, in view of the role that the theory of well-ordered sets plays in modern set theory, being only a part of the theory of ordered sets in general. In my opinion, a more interesting problem is replacing axiom $B$ or $C$ by a "well-ordering axiom," logically weaker than them, which, together with the three ordering axioms, would form an independent axiom system equivalent to each of the previous systems.

I shall give two formulations of such an axiom.

$E$. Every nonempty subset $U$ of the set $Z$ has an element that at most one element of the subset $U$ precedes.

Precisely: for every $U$, if
1. $U$ is a set,
2. for every $x$, if $x$ is an element of the set $U$, then $x$ is an element of the set $Z$, and
3. for some $k$, $k$ is an element of the set $Z$,

then for some $b$,
1. $b$ is an element of the set $U$,
2. for all $y$ and $t$, if $y$ and $t$ are elements of the set $U$ that precede $b$, then $y$ is not different from $t$ ($y$ is identical to $t$, $y = t$).

$F$. Every nonempty proper subset $U$ of the set $Z$ has an element that no element different from it in the subset $U$ precedes. (In this axiom the term "first element" thus occurs with a third meaning, weaker than what is expressed by axiom $B$.)

More precisely: for every set $U$, if
1. $U$ is a set,
2. for every $x$, if $x$ is an element of the set $U$, then $x$ is an element of the set $Z$,[5]
3. for some $k$, $k$ is an element of the set $U$, and
4. for some $l$, $l$ is an element of the set $Z$, and $l$ is not an element of the set $U$,

then for some $a$,
1. $a$ is an element of the set $U$,
2. for every $y$, if $y$ is an element of the set $U$ different from $a$, then $y$ does not precede $a$.

In this way, I have obtained two new axiom systems for a well-ordered set: $\{A_1, A_2, A_3, E\}$ and $\{A_1, A_2, A_3, F\}$.

---

[5] [At this point there is an editorial error in the reprint of Tarski 1921 in the 1986a *Collected Papers*: omission of the clause *x jest elementem zbioru Z*, which corresponds to the last clause of condition 2. See footnote 8 on page 28.]

I turn to the proof of the equivalence of each of these systems with any of the previous systems: for example, with the system $\{A_1, B\}$. We know already that from this last system, axioms $A_2$ and $A_3$ can be deduced. What remains to be proved is that (1) axioms $E$ and $F$ follow from the system $\{A_1, B\}$ as theorems, (2) from each of the new systems axiom $B$ can be deduced.

The first problem is completely straightforward: axioms $E$ and $F$ follow directly from axiom $B$, weakening it. Indeed, if every subset $U$ of the set $Z$ has an element $a$ that no element in the subset precedes, then it also has an element that at most one element in the subset precedes: such an element is in fact exactly that element $a$. Therefore, axiom $E$ is proved. Similarly, from axiom $B$ follows axiom $F$: if every subset of the set $Z$ has a first element, then every proper subset has one as well; if this element is such that no element of the subset precedes it, then certainly, no element different from that one precedes it, which is exactly what was to be proved.

I now give proofs that axiom $B$ can be deduced (I) from the system $\{A_1, A_2, A_3, E\}$, (II) from the system $\{A_1, A_2, A_3, F\}$.

I. Every set $U$ satisfying the hypothesis of axiom $B$ also satisfies the hypothesis of axiom $E$, [and] thus has an element $b$ that at most one element of $U$ precedes. There are two cases to consider:

1. There does not exist an element of the set $U$ that precedes $b$. Then, clearly, element $b$ satisfies the conclusion of axiom $B$: no element of the set $U$ precedes it.
2. There does exist an element of the set $U$ —call it $a$ —that precedes $b$. Then $a$ satisfies the conclusion of axiom $B$.

Indeed, if any $y$ is an element of the set $U$, then either $y = a$ or $y \neq a$. If $y = a$, then $yR'a$ (by virtue of theorem $T$, which follows from axiom $A_3$). If $y \neq a$ on the other hand, then $yR'b$ (by axiom $E$), [and] therefore $y = b$ or $bRy$ (by axiom $A_1$). If $y = b$, then $aRy$ (according to the definition of element $a$), [and] thus $yR'a$ (by axiom $A_2$). Alternatively, if $bRy$, then since $aRb$, the axiom of transitivity ($A_3$) yields $aRy$, and thus also $yR'a$. Thus, element $a$ indeed satisfies the conclusion of axiom $B$: no element of the set $U$ precedes it.

II. [This proof extends all the way to a matching editorial comment in brackets on the next page.] A set $U$ satisfying the hypothesis of axiom $B$ might satisfy the hypothesis of axiom $F$ or not satisfy it—specifically, it might not satisfy the third premise.

If that is satisfied (that is, if the set $U$ is a nonempty proper subset of the set $Z$), then [$U$] has an element $a$ such that if $y$ is an element of the set $U$ and in addition $y \neq a$, then $yR'a$. But if $y = a$, then also $yR'a$, and therefore the element $a$ satisfies the conclusion of axiom $B$.

On the other hand, if the set $U$ does not satisfy the third premise of axiom $F$, then the set $U$ is not different from the set $Z$ (it is the improper subset of the set $Z$). I shall consider here two possibilities.

1. The set $U$, and thus the set $Z$, has only one element: denote it by $k$. Since $kR'k$ (by virtue of theorem $T$), axiom $B$ is satisfied.

2. The set $U$ has more than one element. Let one of these elements be $l$. I take under consideration the set $W$ consisting of all elements of the set $U$ with the exception of $l$ ($x$ is thus an element of $W$ if and only if $x$ is an element of the set $U$ and in addition $x \neq l$).

This set certainly satisfies the hypothesis of axiom $F$, [and] thus has an element $a$ such that if $y$ is an element of the set $W$, then $yR'a$ (by virtue of theorem $T$, the restriction "if $y \neq a$" does not apply). It remains to consider, however, the relationship of element $a$ to that one element of the set $U$ that is not an element of the set $W$: that is, to $l$. Since $a \neq l$ by virtue of axiom $A_1$, two possibilities then occur here: $aRl$ or $lRa$.

If $aRl$, then $lR'a$ (by virtue of axiom $A_2$), [and] therefore element $a$ satisfies the conclusion of axiom $B$.

On the other hand, if $lRa$, then element $l$ satisfies the conclusion of this axiom. Indeed, if $y$ is an element of the set $U$, then either $y = l$ or $y$ is an element of the set $W$.

In the first case, $yR'l$ (by virtue of Theorem $T$). Alternatively, if $y$ is an element of the set $W$, then $yR'a$, [and] therefore, by axiom $A_1$, either $a = y$ or $aRy$. If $a = y$, then $lRy$, [and] so $yR'l$ (by axiom $A_2$).

On the other hand, if $aRy$, then $lRa$ and $aRy$ both occur, [and] therefore $lRy$ by virtue of axiom $A_3$), and thus $yR'l$, Q.E.D.

Having exhausted all the possibilities, we have in general proved axiom $B$. [This ends the proof that started after the matching editorial comment in brackets on the previous page.]

Next I come to the proof of the independence of the axioms in both systems $\{A_1, A_2, A_3, E\}$ and $\{A_1, A_2, A_3, F\}$. In order to prove the independence of an axiom in a given system, it suffices to give an interpretation satisfying all the axioms in the system with the exception of the one whose proof of independence is in question. I will give in turn proofs of the independence of each of the axioms, in parallel for each system.

(1) Axiom $A_1$. As the set $Z$ I choose the set of natural numbers, and define a relation $R$ for the elements of this set in the following way: $xRy$ if and only if $x < y - 1$. Axiom $A_1$ is not satisfied: for example, $1 \neq 2$ but also $1R'2$ and $2R'1$. On the other hand, the remaining axioms are satisfied. [For] $A_2$: if $x < y - 1$, then $y > x + 1$, [and] thus

$y > x - 1$, $yR'x$. [For] $A_3$: if $x < y - 1$ and $y < z - 1$, then $x < z - 1$. Finally, $E$ and $F$: the conclusions of these axioms are in fact satisfied by the smallest number in the given subset.

The simplest interpretation is the set consisting of two consecutive natural numbers, for example 1 and 2, with the previous definition of the relation $R$; it satisfies axioms $A_2$ and $A_3$ because for no values of the variables $x$, $y$, and $t$ are the hypotheses of these axioms satisfied.[6]

(2) Axiom $A_2$. I choose the same set as before and define $xRy$ if and only if $x < y$ or $x = y$ ($x \leq y$). Axiom $A_2$ is not satisfied: $xRy$ and $yRx$ occur together only if $x = y$. On the other hand, [these] axioms are satisfied: $A_1$ (if $x \neq y$, then $x < y$ or $y < x$, therefore $xRy$ or $yRx$); $A_3$ [similarly]; and $E$ and $F$ —these last two are satisfied, as before, by the smallest number belonging to the given subset.

The simplest interpretation is the set consisting of a single number with the previous definition of the relation $R$.[7]

(3) Axiom $A_3$. As the set $Z$, I choose a set consisting of three distinct points on a circle, following one another[8] in a given direction: for example, points $a$, $b$, and $c$ (figure 1). The relation $R$ between the elements of this set I denote with the aid of arrows, as indicated in the figure; thus $aRb$, $bRc$, and $cRa$.

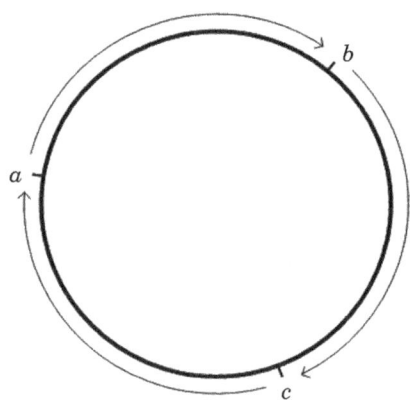

**Figure 1**

Axiom $A_3$ is not satisfied: $aRb$, $bRc$, and simultaneously $aR'c$. On the other hand, the remaining axioms are satisfied: $A_1$ and $A_2$ are direct consequences of the definition of the relation $R$; and for $E$ and $F$, it is easy to check separately each of the seven nonempty subsets of the set $Z$.

(4) Axioms $E$ and $F$. As an interpretation, any ordered set that is not well-ordered will do: for example, the set of rational numbers, ordered according to size.

---

[6] [This paragraph seems unrelated to the rest of the paper. Its relation $R$ is empty; axioms $B$, $E$, and $F$ are thus valid, but not axioms $C$ and $D$. Tarski's use of *simplest* (*najprostszą*) is unclear: an interpretation with a single element and empty $R$ satisfies all the axioms except $A_1$.]

[7] [This paragraph also seems unrelated. Its relation $R$ satisfies all the axioms except $A_2, B, D$.]

[8] [At this point there is an editorial error in the reprint of Tarski 1921 in the 1986a *Collected Papers*: insertion of the clause *x jest elementem zbioru Z*, unrelated to the surrounding text. See footnote 5 on page 25.]

In this way, I have carried out the proof of the independence of the axioms in both systems.

It is worthwhile to note that axioms $E$ and $F$ are not equivalent, and neither of them follows from the other. In order to prove this, it suffices to give two such interpretations that would, alternatively, satisfy one of the axioms while not satisfying the other.

Thus, if to the set, already mentioned, of three points $a$, $b$, and $c$ on the circle we append the center $o$ of the circle and specify new relationships $oRa$, $oRb$, and $oRc$ (figure 2), we obtain a set that does not satisfy axiom $F$: for example, the subset $U$ consisting of points $a$, $b$, and $c$ does not satisfy it. This set, on the other hand, satisfies axiom $E$: in fact, if a subset contains point $o$, this point satisfies the conclusion of axiom $E$; otherwise, any element of the subset satisfies this conclusion, because at most one element of the subset bears the relation $R$ to it.

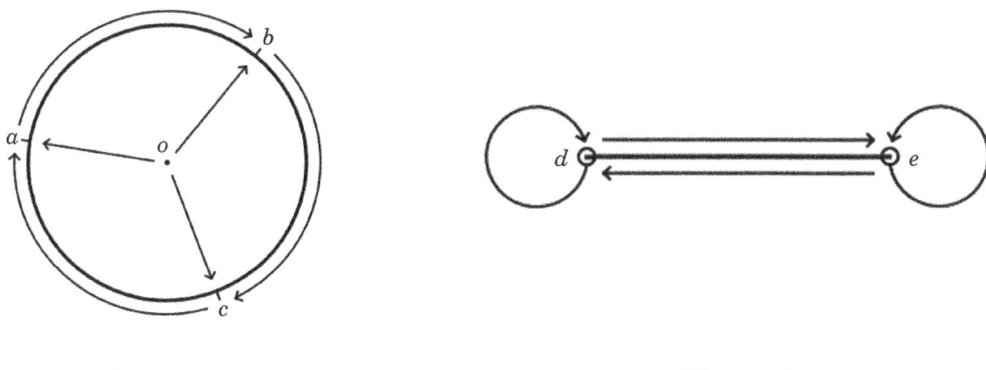

**Figure 2**      **Figure 3**

On the other hand, a set consisting of two points $d$ and $e$, in which we have specified the relationships $dRd$, $dRe$, $eRd$, and $eRe$ (figure 3), does not satisfy axiom $E$: in fact, the improper subset of the set does not satisfy it. But [this set] satisfies axiom $F$: every subset consisting of a single element satisfies this axiom, and every proper subset of the given set is composed of only one element.

Neither the system of order axioms, nor any of the systems obtained that define a well-ordered set, is a relatively weakest system, that is, one in which it is not possible to obtain an axiom system equivalent to the given system by replacing any axiom by an axiom weaker than it without changing the remaining axioms.

Thus, in each of these systems that include axioms $A_2$ and $A_3$ it is possible to replace axiom $A_2$ by the weaker axiom $T$, of antireflexivity. The system obtained is equivalent to the previous one, because $A_2$ follows from axioms $T$ and $A_3$. And yet in this way we will not obtain relatively weakest systems.

A certain theorem of Prof. Jan Łukasiewicz gives us a method for obtaining such systems. By virtue of this theorem, if $\{T_1, T_2, \ldots, T_n\}$ is an independent system of axioms, then a relatively weakest system equivalent to the given one is the following system, written in symbolic form:

$T_1$         (1)   $T_1$
$T_1 \supset T_2$        (2)   if $T_1$, then $T_2$
$T_1 \,\&\, T_2 \supset T_3$      (3)   if $T_1$ and $T_2$, then $T_3$
$\vdots$              $\vdots$
$T_1 \,\&\, T_2 \,\&\, \cdots \,\&\, T_{n-1} \supset T_n$   ($n$)   if $T_1$, $T_2$, …, and $T_{n-1}$, then $T_n$.[9]

In this way, we would obtain the following axiom system for ordering:

$A_1$     $A_1 \supset A_2$     $A_1 \,\&\, A_2 \supset A_3$.

Moreover, the well-ordering axiom would take on one of two forms, easily proved equivalent:

$A_1 \,\&\, A_2 \,\&\, A_3 \supset E$   or    if axioms $A_1$, $A_2$, and $A_3$ are satisfied,
$A_1 \,\&\, A_2 \,\&\, A_3 \supset F$       then axiom $E$ or $F$ is satisfied.

However, I do not give a complete formulation of the axioms constructed in this way, because despite rewording I was not able to put them into an aesthetic form.

---

[9] [This theorem is easy to prove. The present editors have not been able to locate any other source for it.]

# 3
# Doctoral Research

The account of Alfred Tarski's life begun in chapter 1 continues here.[1] This chapter describes the latter years of Alfred's studies at the University of Warsaw, and various other aspects of his life during 1921–1924. These include some personal details, his part-time employment as a teacher during his student years, his doctoral research, and his early participation in professional meetings. It provides a setting for both the detailed mathematics of Part Two of the book, and for Tarski's professional career during 1924–1939 as a researcher and secondary-school teacher and teacher trainer, which are the subject of Part Three.

Chapter 1, section 1.1, described Alfred's exciting first year of courses, 1919–1920; the Soviet invasion in summer 1920, the Polish victory that August—the Miracle of the Vistula—in the very suburbs of Warsaw; and Alfred's subsequent re-enrollment for autumn 1920, with courses from Stanisław Leśniewski, Jan Łukasiewicz, Stefan Mazurkiewicz, and Wacław Sierpiński. Three intellectual threads in Alfred's studies had taken form during those semesters: *logic, set theory*, and *measure theory*. They would extend far into his research career.

From autumn 1920 though autumn 1923, Alfred enrolled in many of the same classes as the future politician and logician Kazimierz Pasenkiewicz. Decades later, Pasenkiewicz described[2] the atmosphere that they shared:

> In the year 1920–1921, around ten thousand persons enrolled at the University of Warsaw. This number exceeded by many times the didactic possibilities of the university. In the lecture halls and laboratories it was very crowded; but not in all. Not many students enrolled in the theoretical directions, especially in mathematics, philosophy, and logic. ...
>
> This was a period of great liberalism and tolerance. The bureaucratic requirements were reduced to the most essential. ... The course of studies was very liberal. The only condition for receiving credit for a semester was to inscribe in a register a definite number of lecture-hours for subjects lectured in the department, and to obtain from the professor a confirmation of attendance. Very often these signatures were obtained through the janitors.
>
> *[continued on the next page]*

---

[1] For biographical information supporting this chapter, unless another source is cited, consult the biographies by Anita B. and Solomon Feferman (2004) and Jacek Juliusz Jadacki (2003a). Also note the descriptions of those works in chapter 17. The Fefermans emphasized personal-interview sources; Jadacki, published records. For historical and sociological information, consult the works by Norman Davies (1982, volume 2, chapters 18–19), Celia S. Heller (1994), and Richard M. Watt (1979).

[2] Pasenkiewicz 1984, 2–3. A portrait and biographical sketch of Pasenkiewicz are on page 32. The following quotation and those on pages 33 and 36 were translated by Jan Tarski, then lightly edited by the present editors to conform with the conventions of this book.

*[quotation continued from the previous page]*

There were no required lectures, nor examinations. One could, however, pass a discussion section for an attended lecture course. For a passed discussion section one received a confirmation with a grade.

After attending six semesters, one could take the final examination. Its successful result gave a diploma of completion of university studies and the right to lecture in secondary schools. The procedure for obtaining doctorates remained unchanged. It was necessary to obtain a positive evaluation of the doctoral dissertation and to pass through a traditional routine.

*Kazimierz Pasenkiewicz  
in 1920*

---

**Kazimierz Pasenkiewicz** was born in 1897 in Kiev, and schooled there. He fought with the Russian army in World War I, then against it in the Polish–Soviet War of 1920, in which he lost a leg. Pasenkiewicz then entered the University of Warsaw to study mathematics and logic, and attended lectures and seminars alongside Alfred Tarski. Pasenkiewicz earned the doctorate there in 1933, with a dissertation supervised by Tadeusz Kotarbiński. He continued that study and research until World War II. During that war he worked with the Polish underground and with socialist organizations. Afterward he became very active in the Communist Party, continuing until its dissolution in 1991. During 1948–1968 Pasenkiewicz served as lecturer, professor, and dean in the philosophy faculty at the University of Cracow. He was an expert on the logical work of Leon Chwistek. Pasenkiewicz died in 1995 in Cracow.*

---

* Kutta 1997.

## 3 Doctoral Research

During the period from winter 1921 through autumn 1923, Alfred enrolled in courses following much the same pattern as that of his previous semesters:

- in nearly every term, two or three courses or seminars with Leśniewski;
- in nearly every term, courses from Mazurkiewicz—on analytic geometry, Jordan continua, topology of the plane, and entire functions;
- in most terms, seminars with Łukasiewicz and his courses on logic and philosophy;
- from winter 1921 through winter 1922, courses from Sierpiński on higher algebra and analysis;
- an autumn 1921 course on mechanics from Czesław Białobrzeski;
- in winter 1922 and autumn 1923, courses from Kazimierz Kuratowski on topology and set theory.[3]

Alfred was thus continuing to explore the *logic* and *set theory* threads that would extend into his research career. Moreover, with the courses on topology and continua he added a new strand that, with his previous study of measure theory, would become a thread of *applications of set theory to geometry*.

Pasenkiewicz enrolled in the lectures of Sierpiński and Mazurkiewicz that involved set theory, those of Leśniewski on logic, some lectures on physics, and lectures by Łukasiewicz on history of philosophy. Pasenkiewicz recalled,

> To the lectures on set theory there came then a fair number of students. These were the times of Émile Borel, Henri Lebesgue, Ernst Zermelo. The theory of sets was rather fashionable.... To the [very popular] lectures on theoretical physics of Białobrzeski from Kiev came students of mathematics and of philosophy; those were the times of Albert Einstein and the atom.
>
> In contrast, few came to the lectures of Leśniewski ... regularly three persons: Jan Drewnowski, Aleksander Jabłoński, and I.... A few came irregularly, among them Alfred Tarski. He sat in the last row and read newspapers. After a lecture, or during intermission, he conferred with the professor; he did not enter into conversations with fellow students. During 1922–1923, Leśniewski acquired a few new students: there appeared Adolf Lindenbaum, Mordchaj Wajsberg, and a few others. Tarski livened up. At the seminars he had the possibility to evaluate those who spoke; he took interest in some of these.
>
> [Tarski] appeared from time to time also at the lectures of Sierpiński, primarily ... when he had some matters to speak with him about. Tarski did not come to [the lectures on physics or] history of philosophy. He was consistent in his interests, did not distract himself.... In his surroundings Tarski did not notice those who did not show particular abilities or interests in logic. On the other hand, he valued people with whom he shared these interests. In discussions with them he did not take on attitudes or a position of authority.[4]

---

[3] Tarski 1924f.

[4] Pasenkiewicz 1984, 2–3. For information about Lindenbaum, see a box in section 14.3; for the other students mentioned, see page 34. The present editors suggest the word *question* in place of *evaluate* in the second paragraph of this translation by Jan Tarski.

*Aleksander Jabłoński
in 1935*

*Mordchaj Wajsberg
around 1925*

---

**Jan Franciszek Drewnowski** was born in Moscow in 1886. He attended lectures and seminars at the University of Warsaw during the early 1920s, alongside Alfred Tarski, and earned the doctorate in philosophy there in 1927, supervised by Tadeusz Kotarbiński. Drewnowski had studied finance and economics as well. He worked in industry and served as a government official both before and after World War II. Drewnowski was a member of the Cracow Circle of Catholic philosophers, and published in both economics and analytical philosophy. He died in Warsaw in 1978.*

**Aleksander Jabłoński** was born in 1898 in Woskresenówka, in eastern Poland, then in the Russian Empire. He served Poland as an engineer officer during the 1920 Polish–Soviet War. Afterward, he worked as a professional musician while studying at the University of Warsaw. He attended lectures alongside Alfred Tarski. Jabłoński earned the doctorate at Warsaw in physics in 1930, and received the *venia legendi* in 1934. As an engineer officer again in 1939, he was interned in the Soviet Union, left with the army of General Władysław Anders via Central Asia and the Middle East, then served as a lecturer in Scotland. After the Second World War, he helped build the Department of Physics at the University of Toruń. He was a world authority on the physics of light. Jabłoński died in 1980 in Skierniewice, in central Poland.†

**Mordchaj Wajsberg** was born in 1902 in Łomża, in central Poland, then in the Russian Empire. His schooling was interrupted by World War I, then by military service in the 1920 Polish–Soviet War. He finally completed secondary school in Łomża in 1923, and entered the University of Warsaw, to study mathematics and logic. He attended lectures and seminars with Alfred Tarski. Wajsberg soon began a stream of research results, concentrating on axiomatics of three-valued logic and modal logic. This area was intimately tied to the work of Tarski and of Jan Łukasiewicz, who supervised the research for Wajsberg's master's and doctoral degrees, awarded in 1928 and 1931. Wajsberg pioneered the use of algebraic and model-theoretic techniques to study those logics. In 1933 he began work as a secondary-school teacher, first in the eastern Polish town of Kowel (now Kovel, in Ukraine), then in Łomża. There he extended his techniques to include general multivalued and intuitionistic logic. Wajsberg perished in the Holocaust, probably during the Nazi massacre of Jews in Łomża in 1942.‡

---

*Majdański and Lekka-Kowalik 2001. Note: there was another scholar of the same era with a similar name.

† Frąckowiak 1998
‡ Surma 1977

The tuition costs entered in Alfred's enrollment record reveal a major ordeal for Polish citizens in the early 1920s: *hyperinflation*. The entries for autumn 1919 were pictured in section 1.1. Figures for subsequent autumn trimesters, shown in the table at right, illustrate exponentially increasing costs for 1919–1922, followed by a year of hyper-exponential growth. The bursar's certificate corresponding to that last line is pictured below. Poland's currency soon collapsed. In 1924, the government instituted reforms, and the new Bank Polski introduced the currency used today, the *złoty*.[5]

| *Autumn Trimester* | *Tuition Costs (Polish Marks)* |
|---|---|
| 1919 | 76 |
| 1920 | 600 |
| 1921 | 2,800 |
| 1922 | 14,500 |
| 1923 | 32,300,000 |

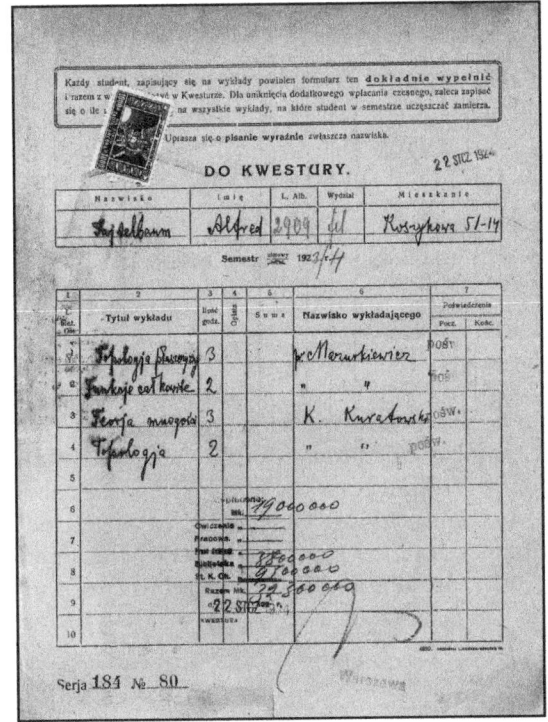

*To the Bursar*

Tajtelbaum, Alfred; Student 2909;
Philosophical Faculty; Koszykowa 51-14
Winter 1923–1924

| Topology of the Plane | 3 hr | Mr. Mazurkiewicz |
| Entire Functions | 2 hr | " " |
| Theory of Sets | 3 hr | K. Kuratowski |
| Topology | 2 hr | " " |

Payment:
    Mk. 19 000 000
    ⋮    ⋮
Library    "   8 800 000
St. K. Ch.   "   4 500 000
Total    Mk. 32 300 000
    22 January 1924

---

[5] Tarski 1924f, 8–11, and 1921–1926. It would be difficult to make precise year-to-year comparisons, because different aspects of tuition were billed in different ways. Moreover, Alfred evidently obtained partial tuition waivers for spring 1921 and spring 1922. But the exponential growth outweighs such details. For information about the Polish economy, see N. Davies 1982, volume 2, 307; Watt 1979, 202–207; and Wynot 1983, 46–47. *Złoty* is a Polish word for *gold*; the English spelling is *zloty*. The abbreviation *St. K. Ch.* in the caption means *Studencka Kasa Chorych* (Student Health Fund).

Alfred continued his development as a researcher in logic, guided by Łukasiewicz, Leśniewski, and Tadeusz Kotarbiński. All three had studied with Kazimierz Twardowski in Lwów. In 1920 yet another student of Twardowski had arrived in Warsaw to enrich its research climate: Kazimierz Ajdukiewicz.[6] At an April 1921 meeting of the Logical Section of the Warsaw Philosophical Institute, Alfred presented his first public scientific address, entitled *On the Notion of Proof (In Response to the Dissertation of K. Ajdukiewicz)*. No copy of its text is known, but Alfred wrote later that it contained the first discussion of the deduction theorem, a logical tool that would play a fundamental role in his organization of metamathematics. In December 1921 Alfred presented another paper to the same group: *On the Problems of Extensionality in Logistic and in Ontology*. No copy of this text is known either, but the title suggests that it pertained to Leśniewski's developing system of foundations of mathematics, the subject of Alfred's doctoral research.[7]

At the age of twenty, Alfred was learning the etiquette of scientific discourse. Pasenkiewicz reported about Leśniewski's autumn 1920 seminar:

> The base of the seminar's activities was the book by Louis Couturat, *Algebra of Logic*. The criticism of this book was shattering. At one point Tarski stood up and asked whether it was at all worthwhile to busy oneself with this. Leśniewski felt a bit slighted, but asked with humor, "Do you think that my seminar is a waste of time?" Tarski sat down.

Pasenkiewicz attended the meetings at the Institute, too. Tarski's patience was improving, but it had limits:

> Tarski was unusually efficient intellectually. At the meetings...a few times papers were read...by Tarski....I admired...[him for answering] the voices in the [subsequent] discussions without notes, but in the order in which they were presented....[Unlike another speaker, who] answered to all the raised issues with equal attention and friendliness, Tarski, on the other hand, [responded] only to those which deserved an answer.[8]

Alfred still lived with his parents, and would do so until he married. According to his biographers Anita B. and Solomon Feferman, he felt a need to establish some financial independence. Around 1920, Alfred found a job teaching geometry at a girls' secondary school. After two years' service, according to his account many years later, he was fired for being Jewish. In 1921 Alfred obtained an appointment to teach logic at the National Pedagogical Institute. This organization had been founded in 1920 as an adjunct of the

*[continued on page 38]*

---

[6] For more information about Ajdukiewicz and Twardowski, see the box on page 37 and one in section 1.2.

[7] The two talks were listed in Warszawski Instytut Filozoficzny 1921–1922 as A. Tajtelbaum, *O pojęciu dowodu (z powodu rozprawy K. Ajdukiewicza)* and *O zagadnieniach ekstensjonalnych w logistyce i ontologii*. The dissertation mentioned in the parenthetical phrase was [1921] 1966, which Ajdukiewicz had submitted for the *venia legendi*. According to Coniglione and Betti 2001, and Betti 2008, 61, that dissertation introduced to Poland the structural definitions of *proof, theorem, consequence, logical theorem,* and *logical consequence* that Alfred would later organize precisely and forcefully. See Tarski [1930] 1983b, 32.

[8] Pasenkiewicz 1984, 5–6. The seminar date is from the enrollment record, Tarski 1924f. Couturat [1908] 1918 had recently been translated from French into Polish by Łukasiewicz and Bronisław Knaster, supported by a grant from the Mianowski Fund.

# 3 Doctoral Research

*Kazimierz Ajdukiewicz
around 1920*

**Kazimierz Ajdukiewicz** was born in 1890 in Tarnopol, then in the Austrian Empire (now Ternopil in Ukraine). His father was a government official. He was schooled in Cracow, then in Lwów, and entered the university there in 1908. He studied with Jan Łukasiewicz and Wacław Sierpiński, and earned the doctorate in 1912 with a thesis on Kant's philosophy of space, under supervision of Kazimierz Twardowski. Tadeusz Kotarbiński and Stanisław Leśniewski also earned doctorates from Twardowski that year. Ajdukiewicz qualified for a teaching credential, but spent the next year in Göttingen, where he could study the work of David Hilbert, Edmund Husserl, and Leonard Nelson. During World War I and its aftermath, he served in the Austrian and Polish armies, taught in a gimnazjum, and married Twardowski's daughter, a classical philologist. During this time Ajdukiewicz wrote his influential [1921] 1966 booklet on the methodology of deductive science, which won him the *venia legendi* and served as a basis of Alfred Tarski's later work in that area. During 1922–1928, Ajdukiewicz served in Lwów as gimnazjum teacher and university dozent, and in Warsaw as philosophy teacher at the same gimnazjum as Tarski and as professor at the university. He was appointed to the permanent faculty at the University of Lwów in 1928 and promoted to full professor in 1934. Ajdukiewicz was the originator of many concepts and techniques now familiar in the philosophy of science; he adapted and extended Tarski's approach to several areas of philosophical logic. During World War II, he worked as a teacher during the Soviet occupation, and as a clerk and clandestine teacher under the German oppression. Afterward, he became professor at the University of Poznań, where he served as rector during 1948–1952. From 1953 on he edited the journal *Studia Logica*. In 1955 Ajdukiewicz returned to Warsaw as head of the Division of Logic of the Polish Academy of Sciences and professor at the university. He retired from the latter position in 1961, and died suddenly in 1963.*

---

* Giedymin 1978, Raabe 1926.

*[continued from page 36]*

University of Warsaw through the efforts of Łukasiewicz, then a university and government administrator. Its purpose was to train schoolteachers; its instructors were selected and supervised very seriously by university faculty.[9] Alfred continued with the institute for about four years. During that same period, he also taught at Zofja Kalecka's Gimnazjum, a girls' school that emphasized mathematics and physical science.[10]

During 1922–1924 Alfred was baptized a Roman Catholic, and he changed his surname legally to the Polish-sounding *Tarski*. At right are his signatures from this period. These acts may reflect a change in religious conviction, a desire to be more Polish, or an assessment of the disadvantages of being Jewish. Jewish scientists from Poland have painted a bleak picture of the situation facing Tarski there in academia. Alfred's decision has been discussed at some length in the literature; writers disagree about his motivation. In the early 1970s, he told the Israeli logician Menachem Magidor,

> I considered myself, at that point, to be Polish, culturally and nationally; and I didn't care about religion at all, so it had nothing to do with religious belief.

Alfred's former Berkeley colleague John Corcoran recently reported:

> Tarski and I talked about religion quite a bit. It was easy and natural.... Tarski never revealed a wavering from the atheistic humanism he came to as a young man... He never mentioned that he had ever converted to Catholicism; I think that "conversion" was never of any inner significance to him.[11]

---

[9] New educational regulations permitted schools to hire teachers after an oral examination, a sample lesson, and some university studies or completion of a program at the institute (Manteuffel 1936, 163). This permitted Alfred to teach at a secondary school, and created a demand for instruction at the institute.

[10] The biographies by Steven R. Givant (1999, 50), Jacek Juliusz Jadacki (2003a, 142–143), and the Fefermans (2004, 54) disagree about the details of this paragraph. It follows Givant's most closely. Only Givant mentioned the first school, with no name. Jadacki corrected the others about the name of the institute: Państwowy Instytut Pedagogiczny. The Fefermans described the institute as a women's school whose headmistress fired Tarski. But its director, Paweł Sosnowski, described it in 1923 as coeducational; he listed "A. Tarski" as instructor for logic during 1921–1922 (Sosnowski 1923, 78, 81). Alfred must have notified him of the impending name change just before that publication. Tarski was still a staff member there in autumn 1923 (Zagórowski 1924–1926, volume 1, 106); Jadacki suggested that Alfred remained so until the institute closed in 1925. Only Jadacki, among the biographers, mentioned the Kalecka Gimnazjum; a Jewish school, it was not likely the one that fired Alfred. (That is confirmed in Zagórowski 1924–1926, volume 1, 151, and Shneiderman 1995.)

[11] The signatures are from Tarski [1918] 2014, 1919–1920, and [1924] 2014a; the first change was discussed in section 1.1. See Feferman and Feferman 2004, 38–40, 269; this includes a reproduction of the application for change of name, Tarski [1924] 2007. See also Jadacki 2003a, 142–145, Woleński 1995a, note 7, and Woleński 2008. The autobiographies of physicist Leopold Infeld and probabilist Mark Kac record their impressions as students in Cracow and Lwów during 1920–1935 (Infeld 1980, chapters 1–2; Kac 1985, 28). Heller (1994, chapter 6) discusses assimilation and conversion in depth. Corcoran 2011a, 2011b.

## 3 Doctoral Research

Alfred was awarded the doctorate in March 1924, as recommended by Leśniewski.[12] His dissertation, *On the Primitive Term of Logistic*, was published in *Przegląd filozoficzny* in 1923, under the surname Tajtelbaum-Tarski. The term *logistic* referred to a system of logic whose features included material equivalence $\leftrightarrow$ and perhaps other Boolean connectives, propositional variables $p, q, \ldots$, variables $f, g, \ldots$ for truth functions, and universal and existential quantifiers over these variables. A *truth* function $f$ is one that maps truth values to truth values, such that if $p \leftrightarrow q$ is valid, then so is $f(p) \leftrightarrow f(q)$.[13] Alfred showed, roughly speaking, that negation and conjunction can be defined in terms of the equivalence connective $\leftrightarrow$ and universal quantifier $\forall$ by regarding $\neg p$ and $(p \mathbin{\&} q)$ as abbreviations for $(p \leftrightarrow (\forall q) q)$ and $\forall f (p \leftrightarrow (f(p) \leftrightarrow f(q)))$. This can be justified intuitively,[14] but the steps that suggest themselves do not all have counterparts in Leśniewski's system. Alfred's achievement was to adapt and fit such definitions into the system; he had to use a more complicated one for conjunction. Alfred immediately published a two-part French translation of his dissertation in the journal *Fundamenta Mathematicae* during 1923 and 1924. He published the first part under the surname Tajtelbaum; the second, as Tajtelbaum-Tarski.[15] From that time on, he published all his work as Alfred Tarski.

Leśniewski worked on his logical system through the 1920s, and finally published its details at the end of the decade. In good humor, he emphasized his debt to Tarski:

> The system of the foundations of mathematics I have constructed owes a number of important improvements to Alfred Tarski... "my" Doctor in the year 1924.... I will endeavor to show them explicitly; but because of the nature of things I cannot show properly all of Tarski's occasional critical remarks, which undermined this or that link of my theoretical conceptions at the different stages in the building of my system, and all the subtle and sympathetic counsel and often impalpable suggestions, from which I had the opportunity to profit in numerous conversations with Tarski.[16]

Tarski regarded Leśniewski's system as equivalent to other better-known systems, but referred to its use in just one more publication, his 1927 report about geometry based on

---

[12] The official certificates are displayed in Jadacki 2003, 168; the official critics are identified there on page 144 as Łukasiewicz and Sierpiński. For a detailed description of the presumably similar doctoral defense of mathematician Jerzy Neyman in Warsaw the same year, see Reid 1983, 53.

[13] Truth values are values of propositional variables. See Church 1956, §§06, 28, for more information on this sort of logical system.

[14] To see this, suppose first that $p, q$ should both be valid; that would imply validity of $p \leftrightarrow q$, $f(p) \leftrightarrow f(q)$, and $p \leftrightarrow (f(p) \leftrightarrow f(q))$, no matter what $f$ might be. Should $p$ be valid but $q$ not, then letting $f$ denote the identity function, $f(p) \leftrightarrow f(q)$ and hence $p \leftrightarrow (f(p) \leftrightarrow f(q))$ would be invalid. If $p$ were invalid, then letting $f$ be any constant truth-function, $f(p) \leftrightarrow f(q)$ would be valid and thus $p \leftrightarrow (f(p) \leftrightarrow f(q))$ invalid, no matter what $q$ might be. The definition of negation works because $(\forall q) q$ is invalid. The remaining connectives and the existential quantifier can then be defined as in elementary logic.

[15] See Tarski 1923a, 1923b, and 1924d.

[16] Leśniewski [1927–1930] 1992, 180. Leśniewski included a translation of the 1927 Polish version of this quotation in the German original version of Leśniewski [1929] 1992, 414. His Polish and German words for *impalpable* were *nieuchwytny* and *ungreifbar*: *elusive* and *ungraspable*.

the notion of solid body. Arianna Betti has analyzed Tarski's major scientific debt to Leśniewski, which would appear most clearly in Tarski's work of the 1930s:

> ...among the things which Tarski was taught by Leśniewski were...how to analyse quotation marks, the language/metalanguage distinction, the idea that truth is language-relative, the notion of a closed language, and that natural language is such a language, namely that natural language is universal.[17]

During 1922–1923, Alfred had enrolled only in lectures and seminars on logic and philosophy, ostensibly concentrating on his doctoral research project in logic. Alongside those studies, however, he pursued intellectual threads that had evidently captivated him even earlier: set theory and its applications to geometry. As early as 1919–1920, he had studied geometry and calculus with Zygmunt Janiszewski and Stefan Mazurkiewicz, and set theory and measure theory with Wacław Sierpiński. Tarski had continued with the latter two, studying analysis and algebra, and during spring 1922 and autumn 1923, he also attended Kazimierz Kuratowski's topology courses. Sierpiński influenced Tarski to study in detail Felix Hausdorff's 1914a book *Grundzüge der Mengenlehre*.[18] Sierpiński had just published his major 1918 survey of the use in set theory and analysis of the axiom of choice, the basic principle according to which,

> for any family $F$ of nonempty sets, there is a set that contains exactly one element from each member of $F$.

Moreover, for several years, Sierpiński and Mazurkiewicz had been studying seemingly paradoxical constructions in set theory. For example, in a 1914 paper, they described two disjoint unbounded subsets $A$ and $B$ of the plane, each of which is congruent to the union $A \cup B$. Their excitement about such results surely infected their lectures. Tarski immersed himself in that research area, too. His work would involve its most advanced aspects.

One filament of this set-theory thread involved delicate consideration of the axiom of choice. Tarski established striking relationships between this idea and that of the finiteness of a set, and discovered that certain other set-theoretic propositions, some quite familiar, were equivalent to the axiom. For example, in 1914a, Hausdorff included the theorem that for any infinite set $A$, there is a one-to-one correspondence between the elements of $A$ and those of the set $A \times A$ of all ordered pairs of elements of $A$. The proof made essential use of the axiom of choice. Tarski showed that some use of that principle was inescapable, because that theorem actually implies the axiom.[19] Tarski published this work simultaneously with his dissertation, and continued to work on this thread in collaboration with Adolf Lindenbaum. Their joint papers would start appearing in 1926.

---

[17] Lindenbaum and Tarski 1926, 299; Tarski [1927] 1983. Betti 2004, 278–279. Concerning the relationship between Leśniewski and Tarski, see also Betti 2008. For a brief account of Tarski's research in logic up to 1939, see section 8.1.

[18] See Feferman and Feferman 2004, 48, and Tarski 1924f. For biographical information about Sierpiński and Hausdorff, consult boxes in sections 4.1 and 4.4. Tarski always recommended Hausdorff 1914a to his own students, even decades later.

[19] Hausdorff 1914a, 127. Tarski 1924b and 1924e, 150. Lindenbaum and Tarski 1926. This theorem about $A \times A$ is due to Gerhard Hessenberg (1906, 108).

That research is not considered further here, because it is relatively accessible and lies beyond the scope of this book.

Soon after his graduation, Tarski applied to the dean of the faculty, the chemist Wiktor Lampe, for permission to pursue further study in mathematical sciences, particularly theoretical physics. His letter is translated in chapter 16, section 4. Tarski's signature included the title *Dr.*; it is reproduced on page 38. Tarski signed up for nine physics and astronomy courses in spring 1925 and five in spring 1926. There was evidently no tuition charge; he must have been going to sample those offerings.[20]

In 1925, probably on the basis of his 1924b research on the definition of finiteness, Tarski was awarded the *venia legendi* by the faculty: this allowed him to use the title *docent* and to give courses, receiving no salary but simply a fraction of his students' fees. For a year starting in February 1925, Tarski gave a course on deductive logic at the Free University of Poland, a private postsecondary institution in Warsaw; the course met for two hours each week.[21]

During 10–13 May 1923, Tarski had performed one of the duties of an advanced student: he had served as secretary of the logic section of the First Polish Philosophical Congress in Lwów. He did not present a paper, but participated in the discussions after talks on the principle of contradiction, type theory, axiomatization of physics, truth functions, and the logic of adjectives. At that meeting he met a new Lwów professor of mathematics, Stefan Banach, who was also fascinated by the intricacies of set theory and geometry.[22] They began a collaboration that led to their famous 1924 paper *On Decomposition of Point Sets into Respectively Congruent Parts*. Part Two of the present book is devoted to Tarski's research on decomposition of point sets, and chapter 6 contains a full translation of that paper.

---

[20] Tarski [1924] 2014a; Tarski 1922–1926, documents 18 RP 2909 and 20 RP 2909. Jadacki 2003a, 144.

[21] Wolna Wszechnica Polska 1925, 8, 25. Many other University of Warsaw faculty taught part-time at the Free University; this was apparently Tarski's only involvement with it. For his service at the University of Warsaw, see section 9.3.

[22] Polski Zjazd Filozoficzny 1927, 266, 284–294, 361. Tarski's comments were not recorded. For biographical information about Banach, consult a box in section 4.2.

# Part Two
# Geometry

Part Two of this book is devoted to a single thread of Alfred Tarski's research: *applications of set theory to geometry*. Five research papers are translated here, all concerned with the problem of decomposing two geometric figures of equal measure into equal finite numbers of components that are congruent in pairs. In the first two papers, both published in 1924, the components may be sets of any sort, but they must all be disjoint. Chapter 5 contains the translation of Tarski's paper *On the Equivalence of Polygons*, in which the original figures are constrained to be polygons in the plane. In chapter 6, the translation of the famous paper by Tarski and Stefan Banach, this restriction is relaxed: the figures need only be bounded and contain interior points, and all dimensions are considered. Both of these results depend on the most advanced results in set theory at that time. The remaining three papers make up chapter 7, and include one by Henryk Moese that is inextricably tied to Tarski's work. These results are cast in the framework of secondary-school geometry: the components must be polygons, and only their interiors must be disjoint. With these translations and others published elsewhere, all of Tarski's geometric research has become accessible in English.

Chapter 8 summarizes some other filaments of this thread of Tarski's research, which can be described as technical results in set theory, and which are accessible elsewhere in French or German. It discusses the impact of all this work on the mathematics of subsequent years. Chapter 4, the first chapter of Part Two, provides the mathematical background for the papers translated here, and discusses some details of their original publication. The subject matter of Part Two is all related to the subjects that Tarski taught in his position as secondary-school teacher and teacher trainer. That activity is the subject of Part Three of this book.

Tarski's research in Poland unfolded amid a chaotic vortex of political, social, economic, and scientific developments. Accounts of events, people, and ideas were merged from several dimensions to form the linear sequence of pages of this book. That is reflected in the background chapter 4 by the use of boxes interspersed in the main narrative. They contain biographical sketches of some persons associated with Tarski, and informational essays about several other topics. Each box can be read independently: readers are not expected to visit them in sequence. Cross-references refer to them from the main narrative.

# 4
# Area, Volume, Measure

The synthetic approach to the theory of area and volume initiated by Euclid is described in the first section of this chapter. The area $a$ of a polygonal region $P$ is computed by decomposing $P$ into a finite number of polygonal components with disjoint interiors, which can be reassembled to form a rectangle $R$ with unit base: $a$ is then the altitude of $R$. The volume of a polyhedral region can be reckoned in a similar way, but for that it is necessary to use some form of Eudoxus's method of exhaustion. Perfected over the centuries, these methods are employed even today in many secondary schools. This area theory provides the mathematical background for the research papers by Alfred Tarski ([1931] 2014a and [1932] 2014d) and Henryk Moese ([1932] 2014) on the degree of equivalence of polygons, which are introduced and translated in chapter 7. The intended audience for those papers was in fact talented and highly motivated secondary-school students and their teachers. The earlier research papers by Tarski ([1924] 2014b) and by Stefan Banach and Tarski ([1924] 2014), translated in chapters 5 and 6, are concerned with decomposition of geometric figures into more general components that are entirely disjoint. They require more complicated set theory and the notion of *measure*, which is a generalization of area and volume. The necessary additional background is described in section 4.2; the papers themselves are summarized in sections 4.3 and 4.4.

## 4.1 Area and Volume

Studying geometry in school, in informal discussions at university, in teaching his own students, and in training teachers, Alfred Tarski confronted the theories of area and volume of geometrical figures. There are several very different ways to develop them. In his paper translated in chapter 5, Tarski referred first to the approach expounded in a popular text by the noted Italian mathematicians Federigo Enriques and Ugo Amaldi.[1]

In that presentation, two polygonal regions are called *equivalent* if they are unions of the same finite number of polygonal subregions with disjoint interiors, such that corresponding subregions are congruent. This relation is reflexive, symmetric, and transitive. Parallelograms with congruent bases and altitudes are equivalent. That is visually apparent if the bases are superimposed and the edges opposite them overlap. Using the Archimedean axiom, Enriques and Amaldi reduced to this case the one in which the opposite

---

[1] Tarski [1924] 2014b (see page 79); Enriques and Amaldi [1903] 1916, chapter 5. For biographical information about Enriques and Amaldi, and portraits, see section 9.2.

edges are collinear but do not overlap. From there they pursued a chain of arguments to show that every polygonal region is equivalent to a rectangle with a prescribed base.[2]

The simplest result in the opposite direction is that equivalent parallelograms $P$ and $Q$ with congruent altitudes have congruent bases. To prove this, Enriques and Amaldi assumed the contrary, that the base of $P$ might be congruent to a proper part of that of $Q$. On that assumption, they showed how to construct within $Q$ a parallelogram $P'$ congruent to $P$, which would then be equivalent to the whole parallelogram $Q$. At the analogous step in the classical development, Euclid had noted that this situation would contradict the (vague) common notion, "the whole is greater than the part," and thus he rejected the contrary hypothesis. The Italian mathematics teacher Antonio De Zolt analyzed this argument, and devised a more precise geometric assumption to complete it:

> *If we divide a given polygon into polygonal parts, then the union of all but one of these parts is not equivalent to the given polygon.*

Enriques and Amaldi stated and used that as an axiom.[3]

From this basis, Enriques and Amaldi derived various familiar results in the theory of area, including the Pythagorean theorem and its converse. Like Euclid, they avoided defining area as a function with numerical values. They would only do so three chapters later, after using the Archimedean and continuity axioms to establish the use of real numbers to measure lengths of segments: the area of a polygonal region $P$ should be the length of the altitude of a rectangle with base of length 1 and equivalent to $P$.

Soon after citing Enriques and Amaldi [1903] 1916, Tarski suggested that readers should compare the related exposition in David Hilbert's book *Grundlagen der Geometrie*.[4] Without using any Archimedean or continuity axiom, Hilbert set up an arithmetic of segments. He constructed an area function by defining the area of a triangular region to be half the product of an edge and the corresponding altitude, and that of a polygonal region $P$ to be the sum of the areas of the subregions in a triangulation of $P$. This required proving first that the familiar triangle-area formula does not depend on which edge is selected as the base, and that any two triangulations of $P$ determine the same area sum. The former results from consideration of similar triangles; the latter requires a complicated but elementary process that Hilbert only sketched. Equivalent polygonal regions clearly have equal areas: that implies De Zolt's axiom.

---

[2] The Archimedean axiom says that given two segments $S$ and $T$, one can always find a segment $U$ consisting of congruent copies of $S$ laid end to end such that $T$ is congruent to a subsegment of $U$. Details of the arguments depend on those of the definition of *polygonal region*. The present editors have this definition in mind: a *triangular* region is the union of a triangle and its interior; a *polygonal* region is a union of a finite family $\mathcal{T}$ of triangular regions whose intersections consist of common edges or vertices. Such a family $\mathcal{T}$ is called a *triangulation* of the region.

[3] Euclid [1908] 1956, volume 1, 155, 336 (proposition 39); De Zolt 1881 and 1883. The displayed wording is from Enriques and Amaldi [1903] 1916, chapter 5; a requirement that the interiors of the parts be disjoint was evidently implicit in their notion of *divide*. For information about De Zolt, see the box on page 47.

[4] Hilbert [1899] 1922, chapter 4.

## 4.1 Area and Volume

> **Antonio De Zolt** was born in 1847 in Conegliano, a town north of Venice, then part of the Austrian Empire. He earned the laureate at the University of Turin in 1872. De Zolt taught for forty-four years at the Reale Liceo Parini in Milan, and was a long-time member of the Italian Mathesis society. He published several mathematical works, including 1881 and 1883 booklets on the theory of area and volume. De Zolt died in Milan in 1926.*
> 
> ---
> *Tricomi 1962, 45.

Following Euclid, Hilbert also introduced a weaker version of the equivalence relation: two polygonal regions $A$ and $B$ are *equicomplementable*[5] if there exist equivalent polygonal regions $A_i$ and $B_i$ for $i = 1, \ldots, n$, such that $A$ and all the $A_i$ have disjoint interiors, $B$ and all the $B_i$ have disjoint interiors, and the union of $A$ and all the $A_i$ is equivalent to that of $B$ and all the $B_i$. It is visually apparent that parallelograms with congruent bases and altitudes are equicomplementable. Clearly, equicomplementable polygonal regions have equal areas. Hilbert showed that the converse is true as well, using an argument based entirely on his planar incidence, order, congruence, and parallel axioms, without appealing to the Archimedean principle.

Hilbert's argument that polygonal regions with equal areas are equicomplementable can be modified easily to show that they are in fact equivalent, provided the Archimedean axiom is used to show that parallelograms with congruent bases and altitudes are equivalent. The first known statement of this result was a question posed in 1814 by William Wallace; it was solved the same year by John Lowry. However, it is often called the Bolyai–Gerwien theorem, referring to the work of the Hungarian and Prussian mathematicians Bolyai Farkas and P. Gerwien in the 1830s.[6] Hilbert showed that proving it actually *requires* appeal to the Archimedean principle.

Much of this area theory has an analogous counterpart for volumes of polyhedral regions. But the argument that polyhedra with equal volumes should be equicomplementable fails. Hilbert did not discuss that in the 1899 first edition of *Grundlagen der Geometrie*, but posed it the next year as the third of his famous list of problems for the twentieth century: to find inequivalent tetrahedra with equal base areas and congruent altitudes. His student Max Dehn did so, using advanced algebraic methods. Hilbert cited Dehn's work in all later editions of the *Grundlagen*.[7]

---

[5] Hilbert's term was *inhaltsgleich*; for equivalence, he used *Flächengleichheit*. In later editions, he employed other words.

[6] Lowry and Wallace 1814; F. Bolyai [1832] 1904; Gerwien 1833a. This theorem is discussed in detail in Bartocci 2012, 29–39, and in Amaldi [1900] 1914, §3. It could have been proved easily in Enriques and Amaldi [1903] 1916 by noting that if polygonal regions $P$ and $Q$ have the same area, then their equivalent rectangles with a specified base must have the same area and hence the same altitude: they must be congruent, and so $P$ and $Q$ must be equivalent. In 1833b, Gerwien derived an analogous theorem for spherical polygons.

[7] Hilbert [1900] 2000. Dehn 1901–1902. For a particularly simple proof, consult Benko 2007, §2.

These considerations show that advanced mathematical techniques can yield deep results about the theories of area and volume. As noted in the previous chapter, Tarski had acquired the background to consider them in that light. His university course work had provided essential background for the mathematics that he would soon develop, in collaboration with Stefan Banach, for investigating advanced aspects of those theories. Tarski had studied advanced geometry and analysis with Zygmunt Janiszewski and Stefan Mazurkiewicz; topology with Kazimierz Kuratowski; and higher algebra and analysis, set theory, and measure theory with Wacław Sierpiński. Sierpiński evidently influenced Tarski to study in detail Felix Hausdorff's 1914a book *Grundzüge der Mengenlehre*. Sierpiński had recently published a survey of the use of the axiom of choice in set theory and analysis. For a decade he and Mazurkiewicz had been considering seemingly paradoxical constructions in set theory. For example, they had discovered two disjoint unbounded subsets $A$ and $B$ of the plane, each of which is congruent to the union $A \cup B$ —this shows that De Zolt's axiom does not necessarily hold when the restriction to polygonal regions is weakened. Their excitement about such results surely influenced those who attended their lectures. Tarski would soon immerse himself in this research area. His work would involve the most advanced results in measure theory.[8]

*Wacław Sierpiński*
*in 1928*

---

[8] See Feferman and Feferman 2004, 48, and Tarski 1924f. Tarski always recommended Hausdorff 1914a to his own students, even decades later. Sierpiński 1918; Mazurkiewicz and Sierpiński 1914. For information about Hausdorff, Janiszewski, Kuratowski, Mazurkiewicz, and Sierpiński, consult the boxes on pages 69, 14, 72, 14, and 49, respectively; their portraits are displayed near those pages.

**Wacław Sierpiński** was born in 1882 in Warsaw, which was then in the Russian Empire. The son of a physician, he attended Russian-language schools and was graduated in 1899 from the highly regarded Fifth Gymnasium. He then entered the University of Warsaw, also a Russian institution, studied with the number theorist Georgy F. Voronoy, and won a prize for his first research project. Sierpiński earned a degree there in 1904 and became a teacher of mathematics and physics at a girls' secondary school. He participated in the school strikes associated with the 1905 Revolution, resigned, moved to Cracow—then in the Austrian Empire—and entered the university there to complete work for the doctorate under Stanisław Zaremba. Sierpiński returned to Warsaw to teach in a Polish-language school permitted as a concession by the government. He continued research in number theory, and became very active in the movement that would soon establish academic and scientific institutions for an independent Poland.

In 1908–1909 Sierpiński became a docent at the University of Lwów, also in the Austrian Empire. There he switched his emphasis to set theory, and gave one of the very first university courses on that subject. He took Austrian citizenship and was appointed professor in 1910. That same year, he married Anna Kazimiera Leśniewska, who hailed from Belarus; they had one child, their son Mieczysław. During the next years, the first three of Sierpiński's many monographs appeared, on number theory, on irrational numbers, and on set theory—one of the first comprehensive works on that subject. He brought Zygmunt Janiszewski to Lwów as his assistant, and supervised the doctoral research of Stefan Mazurkiewicz. At the outbreak of war in 1914, Sierpiński and his wife were visiting her family, and were interned in Moscow. There he met and began a long research collaboration with Nikolai N. Luzin on set theory.

Immediately after the war, Sierpiński returned to Lwów, then soon became professor at the new Polish university in Warsaw. With Janiszewski, Mazurkiewicz, and several others, he had been planning to make Poland a world center of mathematical research, specializing in logic, topology, and foundations of mathematics. During the next few years, they succeeded. Alfred Tarski was one of their first students. During 1919–1920, Sierpiński joined other mathematicians in a successful project to decode enemy military communications. They played a major role in the decisive defeat of the Bolshevik army in summer 1920 on the banks of the Vistula River only a few miles from Warsaw.[*]

During 1920–1939 Sierpiński continued research, publication, and academic leadership at a frantic pace. He was a pioneer in set theory, and one of the most prolific of all mathematicians: he published more than seven hundred research papers and fifty books. For his research on equidecomposability, Tarski cited as principal background source the 1923 second edition of Sierpiński's set theory text.[†] Later French and English versions have served as standard references for several generations of mathematicians. By 1939 Sierpiński was world-renowned.

That year, the German invasion and occupation destroyed the Polish academic community. Sierpiński survived as a minor office worker, but continued research and gave private clandestine mathematics courses. Late in the war, his own home and library were destroyed, and he was exiled to Cracow. After the war, Sierpiński returned to Warsaw as a leader in the reconstruction of Polish academic and mathematical institutions. He continued research, and received many honors, national and worldwide. Sierpiński died in Warsaw in 1969.[‡]

---

[*] See section 1.1 of the present book for more details and references about this period.
[†] Tarski [1924] 2014b, translated in chapter 5: see page 79.
[‡] Fryde 1964; Kuratowski 1974; Kuratowski 1980, 167–173; Schinzel 1974, 2009.

## 4.2 The Measure Problem

Questions about area, volume, and measure were in the air not only of Tarski's circle in Warsaw, but also among the mathematicians in Lwów. In 1919, as part of his application for the *venia legendi* at the Polytechnic School in Lwów, Antoni Łomnicki had published a paper on axioms for magnitude. Using techniques of postulate theory, he investigated the interdependence of some familiar statements involving a system of three relations $\prec, \equiv, \succ$ —for example, the transitive and trichotomy laws. There are many common interpretations of these symbols. For example, the statement $A \equiv B \ \& \ B \prec C$ might mean that geometric figures $A$ and $B$ should be equidecomposable and $B$ equidecomposable with a proper part of figure $C$, or that propositions $A$ and $B$ should be logically equivalent and $B$ a logical consequence of proposition $C$. In recent decades, mathematicians had considered questions about this sort of statement in many contexts.[9]

In January 1922, Łomnicki, by then a professor, presented a teacher-training course, *On Equivalence of Plane Figures*, and was preparing a new edition of his secondary-school geometry text, which would be published a year later. That same month, at a meeting of the Polish Philosophical Society in Lwów, he presented a research paper on this subject.[10] He claimed that with the equidecomposability interpretation, all required theorems about area can be derived from a certain system of axioms of the sort mentioned in the previous paragraph, which included De Zolt's axiom in the form $A \prec B \Rightarrow A \not\equiv B$. Łomnicki noted that the remaining axioms in that system, with the propositional interpretation, are all valid in logic, but De Zolt's is not. Therefore, De Zolt's axiom is independent of the others, and deriving it from geometric principles should require a method different from those used to derive the others.

Among the attendees who discussed Łomnicki's paper was his assistant, Stefan Banach, a self-educated but already noted mathematician who would be awarded the doctorate from the University of Lwów later that year. Soon, Banach considered this subject from the point of view of measure theory, and investigated the family $\mathfrak{M}$ of point sets to which De Zolt's postulate could apply. (By the example of Mazurkiewicz and Sierpiński 1914 mentioned in section 4.1, some restriction is necessary.) The next paragraphs display the connection with measure theory.

Suppose relations $\equiv$ and $\prec$ are defined between members of a family $\mathfrak{M}$ of point sets, so that

$$A, B \in \mathfrak{M} \ \& \ A \prec B \ \Rightarrow \ (\exists C \in \mathfrak{M})(A \equiv C \ \& \ C \subsetneq B \ \& \ B - C \in \mathfrak{M}), \tag{1}$$

---

[9] Among the works cited in Łomnicki 1919 was a major 1916 paper by Jan Łukasiewicz. Apparently, neither of those scholars was aware at that time of the similar work of the Peano school in Italy a generation earlier: for example, Bettazzi 1890. For more information on Łomnicki, see the box on page 53.

[10] Łomnicki 1923; Maligranda 2008, 105. Describing his teacher-training course, Łomnicki surveyed the material covered in section 4.1 and cited for background the Italian source Enriques and Amaldi [1903] 1913 and its Polish translation [1903] 1916.

and a real-valued function $m$, applicable to all members of $\mathfrak{M}$, enjoys the following analogues of familiar properties of the measures of length, area, and volume:

$$A, C \in \mathfrak{M} \ \& \ A \equiv C \Rightarrow mA = mC \tag{2}$$

$$D \in \mathfrak{M} \ \& \ D \neq \phi \Rightarrow mD > 0 \tag{3}$$

$$C, D \in \mathfrak{M} \ \& \ C \cap D = \phi \Rightarrow m(C \cup D) = mC + mD \tag{4}$$

Property (4) is called *finite additivity* because it entails that the sum of the measures of any finite number of disjoint sets is the measure of their union. De Zolt's axiom for the family $\mathfrak{M}$ can then be derived by contraposition: given $A, B \in \mathfrak{M}$ such that $A \prec B$, find $C$ by property (1) and let $B - C = D$, so that $D \in \mathfrak{M}$, $D \neq \phi$, $C \cap D = \phi$, $C \cup D = B$, and by (2) to (4),

$$mA = mC < mC + mD = m(C \cup D) = mB;$$

thus, $mA \neq mB$ and by (1), $A \not\equiv B$.

In his pioneering lectures early in the century, Henri Lebesgue had introduced a central focus of measure theory, the *measure problem*: to assign, to *all bounded* subsets of the line, nonnegative numbers called their *measures*, such that two congruent sets always have the same measure, the unit interval has measure 1, and the measure of the union of *any finite or infinite* bounded sequence of disjoint sets is the sum of their individual measures. (This last requirement, called *full additivity*, is stronger than the finite additivity mentioned in the previous paragraph.) Analogous problems can be stated for higher-dimensional point sets. Such a measure would coincide with the length, area, or volume of any set for which one of those notions had already been defined. Clearly, any solution of the measure problem would yield a verification of De Zolt's axiom for the family $\mathfrak{M}$ of *all* point sets. However, that verification would use only finite, not full, additivity.

Lebesgue had defined for each dimension a fully additive measure function that applied to all bounded *measurable* point sets, which was sufficient for many arguments in analysis. Almost immediately, using the axiom of choice, Giuseppe Vitali showed that no fully additive measure function can apply to *all* bounded point sets in any dimension. Through an even more complicated construction, Felix Hausdorff showed that no *finitely* additive measure can apply to all bounded point sets in any dimension greater than two. The corresponding statements for dimensions one and two remained undecided.[11]

In a 1923 paper, Banach settled those two cases of the measure problem that Hausdorff had left open: finitely additive measures $m$ applying to all bounded point sets are indeed possible in dimensions one and two. This result is discussed in detail in the following paragraphs.

---

[11] Lebesgue 1904, 103; Vitali 1905; Hausdorff 1914a, section 10.1 and its appendix, 399–403, 469–472. See also Chatterji 2002, §4, and the boxes on pages 67 and 75 about Hausdorff's "paradox" and a nonmeasurable set.

## 4.2 The Measure Problem

> **Antoni Łomnicki** was born in 1881 in Lwów, then in the Austrian Empire. His father was a secondary-school teacher. Antoni completed schooling there and studied at the University of Lwów, earning the doctorate in 1903 with a dissertation on hypergeometric functions. Until 1919 he taught in secondary schools. He obtained a government stipend to study at Göttingen during 1913–1914, and began publishing research on axiomatics and in analysis. On that basis he won the *venia legendi* at the Polytechnic School at Lwów, and became a docent, then professor there in 1921. Stefan Banach served briefly as his assistant while completing doctoral studies at the university. Łomnicki became an expert on mathematical cartography and radio-navigation, and wrote popular secondary-school texts. He was a gifted organizer, and served several years in the administration of the Polytechnic. In 1930 Łomnicki won eight months' paid research leave to visit mathematicians in Rome, Paris, Göttingen, and Berlin. With many other leading citizens, he was murdered by the Germans in 1941, immediately after they invaded Lwów.*
> 
> ---
> *Jakimowicz and Miranowicz 2007, 90; Maligranda 2008.

Alfred Tarski and Stefan Banach met in Lwów during May 1923, at the First Polish Philosophical Congress.[12] Each one interested in applications of set theory to geometry, they began a collaboration. In his [1924] 2014b paper, Tarski showed that two polygonal regions $A$ and $B$ are equivalent precisely when they have the same area. *Equivalent* means that $A$ and $B$ can be divided into the same finite number of subfigures such that corresponding subfigures of $A$ and $B$ are congruent. Departing from the conventions of elementary geometry, Tarski allowed the term "figure" to encompass *all* plane point sets, and required the subfigures to be *entirely* disjoint. The collaborative paper Banach and Tarski [1924] 2014 pursued this notion of equivalence in much greater detail, particularly in three dimensions. It contained a very startling theorem, often termed "paradoxical": two balls *with any radii whatever* are equivalent—for example, the earth and a marble. Chapters 5 and 6 contain the first published English translations of Tarski [1924] 2014b and Banach and Tarski [1924] 2014.

An essential step in Tarski's argument in [1924] 2014b involved appeal to the following result in Banach 1923:

> *Theorem III\**. To each bounded set $A$ of points in the plane, one can assign a nonnegative real number $m(A)$, so that if $A$ and $B$ are bounded point sets, then
> 
> (1) $m(A) = m(B)$ if $A = \varphi[B]$ for some isometry $\varphi$,
> (2) $m(A \cup B) = m(A) + m(B)$ if $A$ and $B$ are disjoint,
> (3) $m(A)$ is the Lebesgue measure of $A$ if $A$ is Lebesgue measurable.

Property (1) is called *invariance under the group* of isometries. Property (2) is *finite additivity*. Banach's theorem III* was preceded by his analogous theorem I, for subsets of the circumference of a circle. These are cases of the Lebesgue–Hausdorff *measure problem* mentioned earlier.[13]

---
[12] See the last paragraph of chapter 3. For information about Banach and a portrait, see pages 57 and 56.
[13] Tarski [1924] 2014b, §2: see section 5.2. Banach 1923, 30–31.

Tarski noted hyperbolically that the proof of Banach's theorem used "the entire apparatus of contemporary mathematical knowledge...."[14] It did require virtually all of measure theory, and the rudiments of functional analysis. An essential step in Banach's argument was his theorem 16, an early version of what is now called the Hahn–Banach theorem; for its proof he used well-ordering and recursion on infinite ordinals, where later mathematicians would use a set-theoretic maximal principle. Thus, the proof of Banach's theorem required the axiom of choice.[15] Tarski and others investigated various aspects of Banach's theorem in several later papers (see sections 8.6–8.8). Nevertheless, Wacław Sierpiński remarked in 1954, "the demonstration is still very troublesome."[16] Banach's proof of his theorem I, about a circle, is understandable. It has been presented a bit more clearly by Adriaan C. Zaanen, preceded by an excellent exposition of the required measure theory and analysis. Sierpiński was probably lamenting the fact that Banach had specified a construction for the finitely additive function $m$ in his theorem III*, about the plane, but no argument that it satisfies his requirements (1) to (3). The boxes on this page and the next contain an outline of such an argument, based on material in the books by Zaanen and by Stan Wagon.[17]

---

**Banach's Theorem III***: outline of proof. For the required framework in measure theory and the Hahn–Banach theorem, consult Zaanen 1958. This discussion is based on Zaanen's proof (§28) of Banach's theorem I, his remark that it can be generalized to a partial proof of theorem III*, and on ideas in Wagon 1993, chapter 10, especially theorems 10.4(a), 10.4(e), and 10.8. In turn, Wagon's discussion was based on Banach 1923, Neumann 1929, and Mycielski 1979.

For any dimension $n$, consider the set $T = \{x \in \mathbb{R}^n : \forall i \, (0 \leq x_i < 1)\}$; regard it as an $n$-torus, with the left and right ends of the $x_i$ intervals identified when convenient. Let $B$ be the vector space of all bounded real-valued functions on $T$. The subspace $L$ of its Lebesgue-measurable functions is invariant under all translations $\tau$ of $T$: $f \circ \tau \in L$ whenever $f \in L$. The Lebesgue integral $\int$, a nonnegative linear functional on $L$, is also invariant under all such $\tau$: $\int_T (f \circ \tau) = \int_T f$ for all $f \in L$.

Given $f \in B$ and $c_1, \ldots, c_k \in T$, note that the set $A(f, k, c_1, \ldots, c_k)$ of all averages $(1/k) \Sigma_j f(x + c_j)$ for $x \in T$ is bounded because $f$ is. Define

$$Jf = \inf_{c_1, \ldots, c_k \in T} \sup_{x \in T} A(f, k, c_1, \ldots, c_k).$$

Following Banach 1923, Zaanen showed that the functional $J$ is nonnegative and subadditive: for every $f \in B$, if $f(x) \geq 0$ for all $x \in T$, then $Jf \geq 0$; and for all $f, g \in B$ and any scalar $t \geq 0$,

$$J(f + g) \leq Jf + Jg, \qquad J(tf) = tJf.$$

*[continued]*

---

[14] Tarski [1924] 2014b, introduction (chapter 5, page 79).

[15] Banach 1923, 19. For the Hahn–Banach theorem, see Zaanen 1983, 112.

[16] Sierpiński 1954, 97.

[17] Zaanen 1983, chapters 1–9, especially §28; Wagon 1993, chapter 10. The argument in the last paragraph of the box can be used in any dimension. Applied just before the specialization to the case $m = 2$, with $\nu$ in place of $\nu'$, it would yield an additive extension of $\mu$ to the family of all bounded subsets of $\mathbb{R}^m$, invariant under translations and under reflections across hyperplanes perpendicular to the coordinate axes. In the case $m = 1$, that is theorem I of Banach 1923.

## 4.2 The Measure Problem

*[continued]*

Moreover, $\int_T f \leq Jf$ for all $f \in L$. By the Hahn–Banach theorem, $\int$ can be extended to a nonnegative linear functional $I$ on $B$ such that $If \leq Jf$. Zaanen used this inequality and properties of the functional $J$ to show that $I$ is invariant under all translations of $T$.

Tesselate $\mathbb{R}^n$ with copies of $T$ translated parallel to the coordinate axes. Any bounded $S \subseteq \mathbb{R}^n$ is the union of finitely many disjoint pieces $S_j$, each one the intersection of $S$ with the image of $T$ under a unique translation $\tau_j$ of $\mathbb{R}^n$. Let $\nu(S)$ be the sum of the values $If_j$, where $f_j$ is the characteristic function of the set $\tau_j^{-1}[S_j]$. It is straightforward to show that $\nu$ is finitely additive. If $S$ is Lebesgue measurable, then so are each set $S_j$ and function $f_j$, and hence

$$If_j = \int f_j = \mu(\tau_j^{-1}[S_j]) = \mu(S_j),$$

the Lebesgue measure; and since that is finitely additive, $\nu(S) = \mu(S)$. A similar argument shows that this finitely additive extension $\nu$ of $\mu$ to the family of all bounded subsets of $\mathbb{R}^n$ is invariant under each translation $\varphi$ of $\mathbb{R}^n$. (Subdivide $S$ more finely, so that each set $\varphi[S_j]$ also lies wholly within some tesellation cell.)

Now set the dimension $n = 2$. Let $C$ be the circle with circumference 1 about the origin. The previous discussion, with $n = 1$, implies the existence of a nonnegative linear functional $H$ on the space of all bounded real-valued functions on $C$, invariant under rotations of $C$ and agreeing with $\int$ on Lebesgue-measurable functions. In particular, if $f$ is a constant function on $C$, then $Hf = c$.

For each $t \in C$, let $\rho_t$ be the counterclockwise rotation about the origin that maps the point $\langle 1/(2\pi), 0 \rangle$ to $t$. For each bounded $S \subseteq \mathbb{R}^2$, define $f_S : C \to \mathbb{R}$ by setting $f_S(t) = \nu(\rho_t^{-1}[S])$; then let $\nu'(S) = Hf_S$. If $S$ is Lebesgue measurable, then so is $\rho_t^{-1}[S]$, and hence $\nu(\rho_t^{-1}[S]) = \mu(S)$, the Lebesgue measure. Thus, $f_S$ is constant and $\nu'(S) = Hf_S = \mu(S)$: that is, $\nu'$ extends $\mu$. Its finite additivity follows from that of $\nu$ and the linearity of $H$. Invariance of $\nu'$ under translations follows from the fact that the translations $\tau$ form a normal subgroup of the isometry group of the plane: for each bounded $S \subseteq \mathbb{R}^2$ and each $t \in C$,

$$f_{\tau[S]}(t) = \nu(\rho_t^{-1}\tau[S]) = \nu(\rho_t^{-1}\tau\rho_t\rho_t^{-1}[S]) = \nu(\rho_t^{-1}[S]) = f_S(t)$$

because $\nu$ is invariant under the translation $\rho_t^{-1}\tau\rho_t$. Invariance of $\nu'$ under rotations $\varphi$ about the origin follows from that of $H$ under $\varphi^{-1}$: for each $S$ and $t$,

$$f_{\varphi[S]}(t) = \nu(\rho_t^{-1}\varphi[S]) = \nu\left(\rho_{\varphi^{-1}(t)}^{-1}[S]\right) = f_S(\varphi^{-1}(t)),$$

and therefore

$$\nu'(\varphi[S]) = Hf_{\varphi[S]} = H(f_S \circ \varphi^{-1}) = Hf_S = \nu'(S).$$

Invariance under these rotations and all translations implies that $\nu'$ is invariant under all direct isometries.

One more step yields an extension $\nu''$ of $\mu$ that is invariant under *all* plane isometries. Choose a reflection $\sigma$ across some line. For each bounded $S \subseteq \mathbb{R}^2$, set

$$\nu''(S) = \tfrac{1}{2}\nu'(S) + \tfrac{1}{2}\nu'(\sigma[S]).$$

That $\nu''$ is finitely additive, extends $\mu$, and is invariant under $\sigma$ follows straightforwardly. Invariance under direct isometries $\varphi$ follows from the fact that they also form a normal subgroup:

$$\nu''(\varphi[S]) = \tfrac{1}{2}\nu'(\varphi[S]) + \tfrac{1}{2}\nu'(\sigma\varphi[S]) = \tfrac{1}{2}\nu'(\varphi[S]) + \tfrac{1}{2}\nu'(\sigma\varphi\sigma\sigma[S])$$
$$= \tfrac{1}{2}\nu'(S) + \tfrac{1}{2}\nu'(\sigma[S]) = \nu''(S),$$

because $\nu'$ is invariant under the direct isometries $\varphi$ and $\sigma\varphi\sigma$.

*Stefan Banach in 1919*

**Stefan Banach** was born in 1892 in Cracow, then part of the Austrian Empire.* His parents were Stefan Greczek, a soldier and office worker, and Katarzyna Banach, a maid. Several months after his birth, the parents entrusted the boy's care to a foster mother, Franciszka Płowa, who ran a successful laundry business. Greczek provided financial support until 1910, when Stefan Banach was graduated from Cracow's classical Henryk Sienkiewicz Gimnazjum. A family friend had given him French lessons during those years, and he had taught himself real analysis. Banach then moved to Lwów to study engineering at its Polytechnic School. He had to support himself by tutoring mathematics, and consequently required four years to finish a two-year program. At the onset of war in 1914 he returned to Cracow. Physically disqualified from military service, he worked in construction, continued tutoring, and pursued his private study of mathematics. Banach married Łucja Braus, a secretary, in 1919. They had one child, a son born in 1922.

By chance, in 1916 Banach met the mathematician Hugo Steinhaus, who recognized his mathematical talent. In 1918 Banach began publishing original research results, particularly involving measure theory, and in 1919, he helped found the Cracow Mathematical Society. Steinhaus soon became a professor at the University of Lwów, and secured for Banach a position as academic and personal assistant to the mathematician Antoni Łomnicki at the Lwów Polytechnic. Although Banach lacked a university degree, he was allowed to submit a dissertation to the university, which awarded him the doctorate in 1922. That work introduced the structures now called Banach spaces, and presented their fundamental properties, including the famous Banach fixed-point theorem. A period of intense research led to Banach's 1923 paper on the measure problem, a position at the University of Lwów, and almost immediate promotion to full professor. Banach and Steinhaus founded the Lwów school of mathematical research. Banach met Alfred Tarski at a conference in Lwów. A year later, they published their famous 1924 paper on seemingly paradoxical decompositions of solid bodies, translated in chapter 6.

During the 1920s and 1930s, Banach continued major research activity and teaching in real analysis and mathematical physics. A series of successful school and university textbooks, some written with Wacław Sierpiński and Włodzimierz Stożek, provided extra income.† Banach's [1931] 1932 monograph on functional analysis made him famous as the founder of that discipline. Several of his doctoral students themselves achieved worldwide reputations in that and related fields.

After the Soviet Union invaded Poland in 1939, Banach served as dean at the university. He escaped the assassination of leaders when the Germans invaded in 1941, but lost his livelihood, and for three years survived only as a flea-feeder for Rudolf Weigl's laboratory in Lwów, which manufactured typhus vaccine for the Germans. Infected fleas were enclosed in a matchbox taped to his body with a screened opening next to the skin. They would feed for a week and then were harvested for the antibodies.‡ This and hunger destroyed his health. After the war he tried to return to professional life, but died of lung cancer in 1945.

---

* Consult Kałuża 1996, and Jakimowicz and Miranowicz 2007 for further information, and Duda 2009, which corrects some errors. See also Albiński 1976.

† For example, Banach, Sierpiński, and Stożek 1933.

‡ Baumslag 2005, 133–134; Szybalski 1999; Waszyński 1996. On the one hand, Weigl collaborated closely with the Nazis and used coerced human subjects. On the other, his flea-feeders were exempt from close Nazi scrutiny, and he managed to save from extermination many Lwów intellectuals and Jews, including some other noted mathematicians. After the war he was honored by the State of Israel with the designation "Righteous among the Nations."

## 4.3 Equidecomposibility in the Plane

In 1924, the year of his doctorate, Alfred Tarski published two abstracts and five research papers. One of the abstracts was related to his paper on well-ordered sets, translated in chapter 2. The other began a new research thread, axiomatics of arithmetic. One of the papers was directly related to his dissertation. Two others were concerned with deep questions in pure set theory.[18] The remaining two papers, applications of set theory to geometry, are translated in chapters 5 and 6. The latter, his joint paper with Stefan Banach on surprising counterintuitive decompositions of solid figures, quickly became famous.

This section is concerned with the former: the paper Tarski [1924] 2014b, *On the Equivalence of Polygons*, translated in chapter 5. It appeared in volume 2 of *Przegląd matematyczno-fizyczny*. That journal, published by the commercial firm Książnica-Atlas, had been founded the previous year by editors Władysław Wojtowicz and Stefan Straszewicz. At a time when the relationship of schools and society was undergoing significant transformation, they aimed

> to establish a link between secondary schools and higher education and research. On the one hand, changes in basic concepts and theories of science cannot be without influence on secondary-school methods and curriculum. On the other, nothing raises the teaching level in schools as much as a teacher's scientific investigations, however modest in scope.... Our second goal, no less important, will be to elicit an exchange of thoughts between teachers...[19]

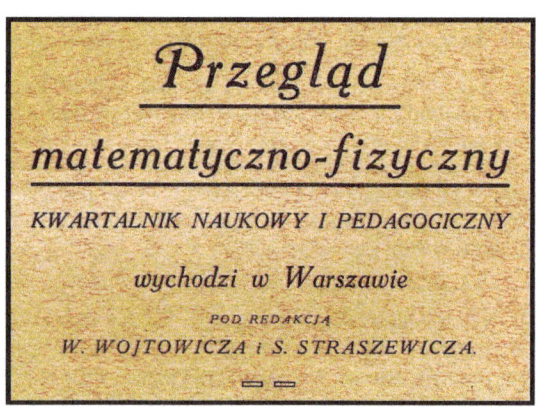

*Mathematical-Physical Review*

*Scientific and Pedagogical Quarterly
Published in Warsaw
under the Editorship of
W. Wojtowicz and S. Straszewicz*

---

[18] The five publications just mentioned are Tarski 1924c, 1924a, 1924d, 1924b, and 1924e, respectively, all in French. The first was discussed briefly in section 1.2, and the last three in chapter 3.

[19] Wojtowicz and Straszewicz 1923. Wojtowicz was a senior member of the Polish mathematics community; Straszewicz, a new professor at the Warsaw Technical University. This and other Polish mathematics-education journals of the day are discussed in sections 9.6 and 9.7. The latter includes a portrait and biographical sketch of Straszewicz. For more information on Wojtowicz see a box in section 9.3.

## 4.3 Equidecomposibility in the Plane

Besides Tarski's paper, this volume of the journal included articles by

- government educator Antoni M. Rusiecki, on methods for teaching arithmetic;
- recent Lwów physics doctorate Leopold Infeld, on presenting the theory of similar triangles in secondary school;
- the noted Italian mathematician Giuseppe Peano, on presenting $n$-place logarithms without using irrational numbers;
- Lwów professor Hugo Steinhaus, on reasoning about plane areas;
- and several other teachers, professors, and laboratory scientists.[20]

There were also reviews of physics and mathematics texts from elementary to research level. Tarski's article was probably the most advanced in this volume. He selected material from his joint work with Banach that, with patient and utterly clear exposition, would be accessible to teachers and students and introduce them to new and advanced aspects of the material they dealt with in classes.

Tarski's paper is about the decomposition of plane figures. According to the theory presented in elementary geometry texts and described in section 4.1, two polygonal regions have the same area if and only if they are unions of the same finite number of *polygonal subregions with disjoint interiors* such that corresponding subregions are congruent. Tarski's goal was the same result for a different notion of decomposition: two polygonal regions should have the same area if and only if they are unions of the same finite number of *disjoint subsets* such that corresponding subsets are congruent. Tarski permitted subsets of any sort, not just polygonal regions; but he imposed the stronger requirement that these subsets themselves be disjoint, not just their interiors. Tarski's illustrated discussion of the difference between these notions of decomposition is at the top of page 85 of the present book.

As mentioned earlier, Tarski assumed familiarity with the elementary geometry described in section 4.1. Because he employed complicated dissections into *arbitrary* disjoint subregions, he made sophisticated use of elementary set theory. Although common in the mathematical research of the previous two decades, these techniques had been presented in only a few monographs. Tarski cited just one for this background material: Sierpiński 1923, the recent, greatly expanded second Polish edition of a book first published in 1912. For his own study, Tarski had been also been using Hausdorff 1914a. Reporting on all the set-theory texts available at that time, Walter Purkert has written,

---

[20] Rusiecki 1924—Rusiecki plays a major role later in this book: see section 9.7. Infeld 1924—Infeld became an influential physicist both in Poland and worldwide; for a description of his situation as schoolteacher, see Infeld 1980, book I. Peano 1924—late in his long career at the University of Turin, Peano was heavily involved in mathematics education: see Marchisotto and Smith 2007, subsection 5.1.5. Steinhaus 1924.

...even through the first decade of the twentieth century no independent textbook had appeared that systematically and comprehensively presented the entire subject of set theory as then understood.... Even the little Polish book published in 1912 by Wacław Sierpiński cannot be considered such a work.... It was above all intended for beginners. Finally, with Hausdorff 1914a the first textbook was available that treated systematically all subjects then regarded as part of set theory.... Felix Hausdorff's work is also *the last such book*, at least in a commonly understood language.[21]

Tarski also used one recent advanced result in set theory, Banach's 1923 solution of the measure problem in dimensions one and two. Tarski noted that its proof had required "the entirety of the apparatus of contemporary mathematical knowledge...."[22] That result was described in detail in the previous section. But Tarski's paper used only the result, not Banach's argument.

After defining two point sets $A$ and $B$ to be *equivalent*—abbreviated $A \equiv B$ —if they are unions of the same finite number of disjoint subsets, such that corresponding subsets are congruent, Tarski proved some elementary properties of this notion. For the one troublesome property, transitivity of equivalence, he presented the usual "double network" argument in complete detail. He showed easily that disjoint unions of equal numbers of pairwise equivalent sets are equivalent. From Banach's theorem he inferred the existence of a finitely additive function $m$ that assigns, *to all subsets* of the plane, nonnegative real numbers called their *measures* so that congruent sets have equal measures and the measure of a polygonal region is its area. Half of Tarski's target result followed directly from this: by the finite additivity of $m$ and its invariance under congruence, two equivalent polygonal regions have equal measures, and thus equal areas.

Tarski then turned to the converse result, that two polygonal regions with equal areas should in fact be equivalent. He used sophisticated set-theoretic manipulation, but not the advanced mathematics, including the axiom of choice, required to justify Banach's theorem. The crucial parts of his argument are two lemmas:[23] if $A$ is a plane set with an interior point, and $B$ is a plane set disjoint from $A$, then $A$ and $A \cup B$ are equivalent if

(1) $B$ consists of a single point, or
(2) $B$ is a segment, with or without one or both endpoints.

---

[21] Feferman and Feferman 2004, 47. Purkert 2002, 47–49. Hausdorff treated both general set theory and point-set theory. Purkert considered the 1913 book by Arthur Schönflies somewhat comparable to Hausdorff 1914a, but noted that for Schönflies, general set theory was just one of many parts of mathematics, whereas Hausdorff regarded it as the foundation for the whole discipline.

[22] Tarski [1924] 2014b, §2: see page 79 of the present text.

[23] Closely related to, but not identical with, lemmas I and II in section 5.2, these results have been rephrased to facilitate this brief summary.

## 4.3 Equidecomposibility in the Plane

These can be extended recursively to apply when $B$ is a finite union of such points and segments. Now, if two polygonal regions $V$ and $W$ have the same area, then they are equivalent in the elementary sense by the theory in section 4.1; each is thus the union of the same finite number of polygonal subregions with disjoint interiors, and corresponding subregions are congruent. The interiors of corresponding subregions must also be congruent, so the unions $A$ and $B$ of the interiors of the subregions of $V$ and of $W$ must be equivalent. Therefore,

$$V = A \cup (V-A) \equiv A \equiv B \equiv B \cup (W-B) = W$$

by the recursive extension of lemma 2, because each of $V-A$ and $W-B$ is the union of finitely many segments.

Tarski's proofs of the two crucial lemmas used the same technique, as follows. First, given a set $A$ with an interior point, choose a disk $K \subseteq A$, a point $p_0$ on its circumference, and a rotation $\rho$ about its center through an irrational number of degrees. For $k = 1, 2, \ldots$ let $p_k = \rho^k(p_0)$. To prove lemma 1, translate the single point in $B$ to $p_0$, map the set $C$ of all points $p_k$ to the set $C - \{p_0\}$ by the rotation $\rho$, and apply the identity mapping to the rest of $A$; these mappings show that $A \equiv A \cup B$. To prove lemma 2, first assume that $B$ is an open segment congruent to $R_0$, the radius of $K$ through $p_0$, without its endpoints. Map $B$ to $R_0$ by an appropriate direct motion, map the union $C$ of all the sets $\rho^k[R_0]$ to the set $C - R_0$ by the rotation $\rho$, and apply the identity mapping to the rest of $A$: again, $A \equiv A \cup B$. The cases in which $B$ contains one or both of its endpoints can then be covered by referring to lemma 1. The general case is covered by applying this one recursively: divide $B$ into a finite number of segments, each sufficiently short.

Presumably to make contact with contemporary presentations of elementary geometry (see section 4.1), Tarski arranged the argument outlined in the previous two paragraphs to include a proof of an analogue of De Zolt's axiom in this context; that was not necessary for the main points just emphasized.

Toward the end of his paper, Tarski noted that the theory of polyhedral volume is quite different: some three-dimensional statements analogous to major results in plane geometry are false. He alluded to the 1901–1902 result of Max Dehn already mentioned in section 4.1—not directly, but by citing a section about it in Ugo Amaldi's major [1900] 1914 expository paper, *On the Theory of Equivalence*.

Next, Tarski considered the area of a disk $C$. He referred again to Amaldi [1900] 1914 for a proof that $C$ is not equivalent in the elementary sense to any polygon $P$. Just what sort of subregions Amaldi allowed is not clear, but his argument was essentially this: if $C$ and $P$ were the unions of equal finite numbers of subregions with disjoint interiors, and all corresponding pairs of subregions were congruent, then for the subregions of $P$ the sum of the lengths of boundary segments that are concave from within should be equal to the analogous sum for the convex segments, because the boundary segments of $P$ itself are straight; but for the subregions of $C$ the concave sum would exceed the convex sum by the circumference of $C$, contradicting their congruence with the corresponding subregions of $P$. Amaldi's argument evidently stemmed from work of the Hungarian mathematician Móricz Réthy about equivalence of figures bounded by curves of a more general nature, and Amaldi described some controversy concerning that.[24] By allowing arbitrary subregions in the paper under discussion, Tarski sidestepped part of this difficulty. He posed the question, *can this theory be extended to arbitrary plane regions bounded by closed curves*?

In conclusion, Tarski posed another question, his *circle-squaring problem*,[25] which became famous and remained open for decades:

*Can a disk and a polygonal region with equal areas be equivalent in Tarski's sense?*

As mentioned in section 4.1, questions about equidecomposability in the elementary sense attracted attention of Euclid in ancient times, of William Wallace, John Lowry, Bolyai Farkas, and P. Gerwien in the early 1800s, and of Réthy, David Hilbert, and Dehn around 1900. It has since spawned a large literature, including the works of Tarski and Henryk Moese described and translated in chapter 7.[26] The set-theoretic version of this notion emerged in the early twentieth century via Hausdorff, Stefan Mazurkiewicz, and Sierpiński, and solidified in the 1924 papers of Banach and Tarski. An even larger and more varied literature stems from those. In major part, it is concerned with extending the theory of "paradoxical" decompositions in solid geometry as presented in Hausdorff 1914a and in the Banach–Tarski paper discussed in more detail in section 4.4 and translated in chapter 6.

---

[24] Amaldi [1900] 1914, §6–7. Réthy 1891; see also Kötter 1891.

[25] Tarski also published this question as Tarski 1925a, in the problem section of *Fundamenta Mathematicae*. The *Collected Papers* volume Tarski 1986a included the wrong text for it. Here is the complete original:
> 38) Un carré et un cercle dont les aires sont égales peuvent-ils être décomposés en un nombre fini de sous-ensembles disjoints respectivement congruents? Problème de M. Tarski.

[26] Tarski [1931] 2014a and [1932] 2014d and Moese [1932] 2014.

## 4.3 Equidecomposibility in the Plane

Later studies of set-theoretic decomposition of plane figures, particularly those concerned with the circle-squaring problem, really stem from Tarski's paper translated in chapter 5.[27] That work was not reviewed in the *Jahrbuch über die Fortschritte der Mathematik*, probably because that publication accorded low priority to journals intended primarily for teachers. In any case, *Przegląd matematyczno-fizyczny* was not a familiar journal, nor its language widely understood. In the subsequent literature, Tarski [1924] 2014b has been mentioned only rarely, and then obliquely. For example, in his elegant little book on this subject, Sierpiński merely suggested that readers compare Tarski [1924] 2014b when following a reference to Banach and Tarski [1924] 2014. Tarski's use of a rotation through an angle with irrational degree measure is echoed precisely in Herbert Meschkowski's well-known book on geometry problems, but without mentioning Tarski. Each of these authors stated the circle-squaring problem, but without specific reference to Tarski.[28]

For a survey of related results in the six decades after Banach's and Tarski's work, consult sections 8.6–8.8 and the literature cited there, particularly the excellent exposition published in 1993 by Stan Wagon. In 1990, to considerable acclaim, Miklós Laczkowicz published an affirmative solution to the circle-squaring problem, making heavy use of results in graph theory and number theory. Remarkably, his proof showed that any two polygonal or circular regions with the same area are the unions of the same finite number $n$ of subsets such that corresponding subsets are related by *translations*. No rotations or reflections are required! Laczkowicz estimated that with his method, $n \approx 10^{50}$. For the background and methods of proof of this result, and related open questions, consult another excellent exposition, Gardner and Wagon 1989, entitled

*At Long Last, the Circle Has Been Squared!*

---

[27] That problem is not mentioned in Banach and Tarski [1924] 2014.

[28] Sierpiński 1954, 260. See section 5.3 of the present book and compare Meschkowski 1966, 139–141. Its title *Ist die Quadratur des Kreises lösbar?* notwithstanding, Menger 1934 does not mention Tarski's problem explicitly.

## 4.4 The Banach–Tarski Theorems

This section describes the 1924 paper by Stefan Banach and Alfred Tarski, translated in chapter 6: *On Decomposition of Point Sets into Respectively Congruent Parts*. It appeared originally in volume 6 of *Fundamenta Mathematicae*.

This journal was founded as part of Zygmunt Janiszewski's plan for establishing Poland as a world center for mathematical research. He had proposed

> to establish a journal, a strictly scholarly one, devoted primarily to one branch of mathematics, in which we have many truly creative and distinguished mathematicians.... This journal... would accept articles in each of the four languages recognized as international in mathematics.... A publication such as this... would find readers everywhere.... The very existence and distribution of such a journal published in Warsaw would bear testimony to our cultural life.[29]

The specified branch was foundations of mathematics, together with set theory, including its applications. These aims were controversial. Clearly they were nationalistic, but the insistence on publishing in international languages, not Polish, could be interpreted as counter to that. And it was not clear to some that enough material would be submitted to ensure high quality of the published papers.[30] The first volume of the journal appeared in 1920, published by Państwowe Wydawnictwo Naukowe (the State Scientific Publisher) with Janiszewski as editor-in-chief. It was the very first specialized mathematics journal. One of its twenty-four articles was concerned with pure set theory; the others, with applications in general topology, measure theory, and real analysis. The eight authors were all Polish: this volume served to introduce them to the world. Five were established professors; the other three were emerging scholars. A problem section in this volume presented ten questions on these same subjects, all of which led to substantial later research. Their proposers included two additional young Polish scholars and a Russian.[31] Unfortunately, Janiszewski died during the 1920 influenza epidemic. Stefan Mazurkiewicz and Wacław Sierpiński succeeded him in the editorship.

Besides the Banach–Tarski paper, the 1924 volume 6 of *Fundamenta Mathematicae* included Tarski's 1924b study of set-theoretic definitions of finiteness, and twenty-two additional articles and three problems whose subject matter was distributed like those in volume 1. Their seventeen authors included four from Russia, two from the United States, and one each from France and Germany. The journal had quickly become international, premier in its field. It is still publishing in 2014.[32]

---

[29] See section 1.1. Janiszewski [1918] 1968, 116–117. The languages were English, French, German, and Italian.

[30] Kuratowski 1980, 32–37.

[31] The third problem, due to the Russian Mikhail Suslin, became famous. For example, see the conclusion of section 15.12. The problem solution cannot be derived from standard set theory. An appendix to the celebratory 1935 "second edition" of volume 1 of *Fundamenta Mathematicae* reviewed later work arising from these articles: see Braun, Szpilrajn, and Kuratowski 1935.

[32] For further information about *Fundamenta Mathematicae*, consult Duda 1996.

## 4.4 The Banach–Tarski Theorems

The Banach–Tarski paper extended to the context of general set theory the underpinnings of the decomposition theory for planar point sets presented in Tarski [1924] 2014b, the paper translated in chapter 5. The planar theory showed that results familiar in elementary geometry about decompositions of *polygonal regions* into equal finite numbers of *subregions with disjoint interiors* such that corresponding pairs of subregions are congruent, continue to hold when the notion of decomposition is modified. The Banach–Tarski paper is about the decomposition of *arbitrary point sets* into equal finite numbers of *disjoint subsets* such that corresponding pairs of subsets are congruent. These sets are assumed only to lie in a space $S$ equipped with a *metric*: a real-valued function[33] on $S \times S$. Sets are called *congruent* if they correspond under a transformation of $S$ that preserves the metric. In parallel, Banach and Tarski developed the corresponding theory for decompositions into denumerable infinities of subsets.

In contrast to the planar case, the results for finite decompositions in Euclidean spaces of three and more dimensions are surprising. Banach and Tarski first presented Felix Hausdorff's startling "paradoxical" decomposition of the surface $S$ of a sphere into four disjoint parts: one that is countable and another that is congruent to the remaining two *and* to their union. This is the construction that Hausdorff had used to show that the measure problem has no solution in three or more dimensions (see section 4.2 and the box on page 67). By elaborating this construction, Banach and Tarski discovered for general figures in three or more dimensions results that are utterly counterintuitive. In particular, they showed that *any two* bounded point sets with nonempty interiors are finitely equidecomposable in the manner just described: a ball the size of an atom is equidecomposable with one enclosing our galaxy.[34] By Banach's positive solution of the measure problem for lower dimensions, the analogous statements for the line and the plane are false. But Banach and Tarski did prove the corresponding statements for *denumerably infinite* decompositions of linear and planar sets.

In their introduction these authors emphasized that use of the axiom of choice was crucial in proving these results, even for the special cases of polygonal and polyhedral regions. Tarski was then heavily involved in other research specifically about that axiom: his 1924e paper *On Some Theorems Equivalent to the Axiom of Choice* had appeared in the preceding volume of *Fundamenta Mathematicae*. The axiom was controversial, and some mathematicians regarded the counterintuitive results of Hausdorff, Banach, and Tarski as evidence against it. For example, the eminent French mathematician Émile Borel railed about Hausdorff's construction:

---

[33] Banach and Tarski developed much of this theory without assuming additional properties of the space $S$ and its metric.

[34] According to a conversation reported in Moore 1982, 264, Banach and Tarski independently discovered instances of this theorem, but the general result was due to Tarski.

Thus if one designates by $a, b, c$ the probabilities that a point of $S$ should belong to $A, B, C$ and admits that the probability that a point should belong to a set $E$ does not change under rotation, ... one has the contradictory equations $a + b + c = 1$, $a = b + c$, $a = b$, $a = c$. This contradiction has its origin in the application ... of [Ernst] Zermelo's *axiom of choice*. ... The paradox arises because $A$ *is not defined*, in the logical and precise sense of the word *defined*. If one scorns precision and logic, one is led to contradictions.[35]

But the axiom was just as heavily involved in showing the *absence* of "paradoxical" results in one and two dimensions. Reflecting on the proofs, Banach and Tarski even suggested that it was more so. Decades later, logicians investigated deeply the extent to which the axiom was required: chapter 13 of Stan Wagon's 1993 monograph is devoted to that investigation.

Banach and Tarski began §1 of their paper by defining the relations $\cong$ of congruence, and $\overline{\overline{=}}_f$ and $\overline{\overline{=}}_d$ of equivalence by finite and by denumerable decomposition, and describing their elementary properties. They presented theorems in pairs: for example, theorems 1 and 1' stated analogous properties of the relations $\overline{\overline{=}}_f$ and $\overline{\overline{=}}_d$. To guide readers in constructing proofs, they presented several in detail, including proofs of the transitivity of $\overline{\overline{=}}_f$ and of the equivalence of disjoint unions of equal finite numbers of pairwise equivalent sets. This section of their paper constitutes a kit of tools for gracefully describing and analyzing complicated mappings in later arguments. For example, directly applying a hypothesis $A \overline{\overline{=}}_f B$ would involve specifying subdivisions of $A$ and of $B$ and the separate isometries under which they correspond. That would lead to an awkward argument if a previously specified subset $C$ of $A$ had to be related to $B$. From this hypothesis, however, their corollary 7 would provide in one step a proper subset $D$ of $B$ equivalent to $C$.

Theorems 8 and 8' are analogues of the familiar Cantor–Bernstein theorem of set theory, in the new contexts of equivalence by finite and denumerable decomposition. Like the proof of the familiar theorem, those of the new results are surprisingly complicated. Banach noticed that analogous theorems are useful in other contexts, too. Therefore, he published a generalization separately in 1924, about relations with two properties that he called $\alpha$ and $\beta$. The Banach–Tarski paper merely referred to that. For a statement and proof of Banach's theorem about $\alpha, \beta$-relations, see the box on page 70. Various familiar corollaries of the Cantor–Bernstein theorem also have analogues in the context of $\alpha, \beta$-relations.

*[The narrative continues after the box on page 70.]*

---

[35] Borel 1914, 255–256. Hausdorff concluded instead that probabilities cannot be measured in this way. Just as Hausdorff's construction was included at the very end of his 1914a treatise, so are these the last words of Borel 1914.

## 4.4 The Banach–Tarski Theorems

**Hausdorff's "Paradox."** In 1914, Felix Hausdorff reported* a startling decomposition of the surface $S$ of the unit sphere into four disjoint parts $Q, A, B, C$: the first is countable, and the remaining three all congruent, and congruent to $B \cup C$ as well. This is often termed "paradoxical"; it implies that no finitely additive extension $m$ of the familiar area function, which is invariant under rotations of $S$, can assign nonnegative values—*measures*—to *all* subsets of $S$. To see this, note first that such an $m$ would have to be monotonic: $X \subseteq Y \subseteq S \Rightarrow m(X) \le m(X) + m(Y-X) = m(Y)$. If $m(Q) > 0$, a contradiction would ensue: there is a rotation $\rho$ of $S$ such that the sets $\rho^n[Q]$ for all natural numbers $n$ are disjoint; but they all have the same measure, and therefore the measure of a union of more than $m(S)/m(Q)$ of them would exceed $m(S)$. To find such a $\rho$, choose an axis $a$ through the origin but not intersecting $Q$, and a rotation $\rho$ about $a$ through an angle incommensurable with the differences in longitude, with respect to $a$, between points of $Q$. By the congruences and additivity, $m(A) = m(B \cup C) = m(B) + m(C) = m(A) + m(A)$: that is, $m(A) = 0$. The contradiction $m(S) = m(Q) + m(A) + m(B) + m(C) = m(Q) + 3m(A) = 0$ then shows that no such $m$ can exist. As a corollary, there can be no solution of the three-dimensional measure problem discussed in section 4.2, for otherwise, the measure $m(X)$ of an arbitrary subset $X \subseteq S$ could be computed as three times the "volume" of the union of the radii through the points in $X$.

To construct the decomposition, Hausdorff worked in three-dimensional coordinate geometry with conventional $x, y, z$ axes. He considered 180° and 120° rotations $\varphi$ about the $z$-axis and $\psi$ about an axis in the $xz$-plane inclined at an angle $\theta$ from the $z$-axis. These generate a group $G$ of rotations represented by formulas such as $\iota, \varphi, \psi, \psi^2$, products of any two of those save the first, $\psi^{m_1} \varphi \psi^{m_2} \varphi \cdots \varphi \psi^{m_n}$ with $m_1, \ldots, m_n = 1$ or 2, and others like the last, preceded and/or succeeded by $\varphi$. By meticulously considering the matrices corresponding to these formulas, Hausdorff determined that unless $\theta$ should belong to a particular countable set of angles, $G$ will be a free group: no two of these formulas will represent the same rotation. He then selected any $\theta$ outside that set.

Hausdorff presented the following algorithm to partition $G$ into three disjoint subsets $\bar{A}, \bar{B}, \bar{C}$. First, put $\iota \in \bar{A}$ and $\varphi, \psi \in \bar{B}$ and $\psi^2 \in \bar{C}$. If the formula for a rotation $\rho \in G$ begins with $\psi$ or $\psi^2$, put $\varphi\rho$ in $\bar{B}$ or $\bar{A}$ according to whether $\rho \in \bar{A}$ or not. If the formula begins with $\varphi$, put $\psi\rho$ in $\bar{B}, \bar{C}$, or $\bar{A}$, and $\psi^2\rho$ in $\bar{C}, \bar{A}$, or $\bar{B}$, according to whether $\rho \in \bar{A}, \bar{B}$, or $\bar{C}$. This classifies products of any two of $\varphi, \psi, \psi^2$ as shown in Hausdorff's table reproduced on the facing page. That classification can be used with a simple recursive argument to show that $\{\varphi\rho : \rho \in \bar{A}\} = \bar{B} \cup \bar{C}$, $\{\psi\rho : \rho \in \bar{A}\} = \bar{B}$, and $\{\psi^2\rho : \rho \in \bar{A}\} = \bar{C}$.

Let $Q$ be the (countable) set of intersections of $S$ with the axes of the nontrivial rotations in $G$. The family of all orbits $P_u = \{\rho(u) : \rho \in G\}$ for $u \in S - Q$ partitions that set. Hausdorff appealed to the axiom of choice to form a set $M$ consisting of a single element from each orbit, then defined the desired sets $A, B, C$ as the unions of the images $\rho[M]$ for $\rho \in \bar{A}, \bar{B}, \bar{C}$, respectively. Each pair of these sets is disjoint. To see this, suppose $u \in A \cap B$, for example; then $u = \rho(v) = \sigma(w)$ for some $v, w \in M$, $\rho \in \bar{A}$, and $\sigma \in \bar{B}$; therefore, $v$ and $w$ would belong to the same orbit, and hence coincide, so that $\rho(v) = \sigma(v)$, $v = \rho^{-1}\sigma(v)$, and thus $v \in Q$, contradicting $M \subseteq S - Q$. From the construction of $\bar{A}, \bar{B}, \bar{C}$, it follows that $\varphi[A] = B \cup C$, $\psi[A] = B$, and $\psi^2[A] = C$: for example,

$$\psi[A] = \{\psi\rho(u) : u \in M \ \& \ \rho \in \bar{A}\} = \{\rho(u) : u \in M \ \& \ \rho \in \bar{B}\} = C.$$

Thus, $A, B, C$ are all congruent, and congruent to $B \cup C$ as well.

---

* Hausdorff 1914a, appendix, 469–472; 1914b. Hausdorff achieved this result at the very time his 1914a book was going into print. The main points of his proof are presented here. For further details, consult Stromberg 1979 and Chatterji 2001. For a biographical sketch of Hausdorff, see the box on page 69.

*Felix Hausdorff around 1918*

| | | | | | |
|---|---|---|---|---|---|
| $\bar{A}$ | $\iota$ | | $\varphi\psi, \varphi\psi^2, \psi^2\varphi$ | $\varphi\psi\varphi$ | ... |
| $\bar{B}$ | | $\varphi, \psi$ | | $\varphi\psi^2\varphi, \psi\varphi\psi, \psi\varphi\psi^2$ | ... |
| $\bar{C}$ | | $\psi^2$ | $\psi\varphi$ | $\psi^2\varphi\psi, \psi^2\varphi\psi^2$ | ... |

*Hausdorff's Classification[36] of Rotations in G*

---

[36] Hausdorff 1914a, 472. See the box on page 67. The original row labels were $A, B, C$. The order of the products has been reversed here and in that box, to agree with today's practice.

**Felix Hausdorff** was born into a wealthy Jewish commercial family in 1868 in Breslau, which was then in Prussia, but now is called Wrocław, in Poland. He grew up in Leipzig. Although the family was seriously religious, Hausdorff became a serious and critical agnostic. His first major interest was music; he started to study composition. However, during university studies in Leipzig, Freiburg, and Berlin, he switched to astronomy. Hausdorff earned the doctorate in 1895, then spent a year in military service, and two as mathematician for the observatory back in Leipzig. He habilitated there in astronomy, to be qualified to teach. His first publications, in that subject, were of little consequence. In 1899 Hausdorff married Charlotte Goldschmidt, the daughter of a physician. She was Lutheran; their only child, a daughter, was born in 1900.*

During these years Hausdorff's scientific interests changed to mathematics. He published research on probability theory, including the first explicit treatment of conditional probability and some fundamental applications to insurance problems. Then he entered the field of set theory and topology, which would be his main concern for many years. In 1901 Hausdorff taught one of the first courses ever in set theory, and started a research program that would provide much of the structure of these new subjects. He was appointed professor in 1901, but outside the regular system of ranks, and against major, explicitly antisemitic, opposition.

Hausdorff was very active in artistic and literary circles. Under the pseudonym Paul Mongré he published philosophical studies, poetry, two novels, and a popular play that satirized the aristocratic honor code. The main German encyclopedia after World War I featured him equally as mathematician and writer.[†] He pursued research and teaching as an amateur—teaching was financially unnecessary. Never comfortable at Leipzig, and realizing that his scientific future required a conventional career path, he won regular professorships at Bonn in 1910, then at Greifswald, on the Baltic, in 1913.

There he completed his 1914a book on set theory, one of the first on that subject. A recent appraisal characterized it as "a pioneering achievement that paved the way for the development of modern mathematics." Its appendix presented the complicated "paradoxical" decomposition of a three-dimensional point set that provided the negative solution to Henri Lebesgue's measure problem in dimensions beyond two and led to even more striking decompositions in Banach and Tarski [1924] 2014.[‡]

Greifswald was isolated, and Hausdorff sometimes its only mathematician. After World War I he turned somewhat to research in analysis. His 1919 work on the dimension and outer measure of point sets led decades later to the development of fractal geometry. He returned to Bonn in 1921. There he applied measure theory in his 1923 course on axiomatic probability theory. Hausdorff's research papers and lectures were famous for clarity and elegance, and his colleagues treasured his critical but genial nature. During the 1920s and 1930s he published new editions of his set theory book, changing with the times but undiminished in influence.

Hausdorff retired in 1935. Reluctant to leave his familiar surroundings, he made no major attempt to emigrate, but remained in Bonn as his colleagues and former students departed, died in battle, or just disappeared. Finally, under threat of deportation to a death camp, Hausdorff, his wife, and her sister committed suicide together in 1942.

---

* Czyż 1994, 1. For further information, consult Dierkesmann et al. 1967, Mehrtens 1980, and Purkert 2008.

† *Mon gré* means *my taste*. Grosse Brockhaus 1931.

‡ Purkert 2008, 45. Hausdorff 1914a is described in section 4.3, and the "paradox" in the box on page 67. Banach and Tarski [1924] 2014 is translated in chapter 6.

> **$\alpha, \beta$-Relations.** In 1924, Stefan Banach introduced* a new general set-theoretic notion and a related modification of the familiar Cantor–Bernstein theorem. This enabled major simplifications to the decomposition theory presented in Banach and Tarski [1924] 2014 and translated in chapter 6. Similar benefits would accrue in pure set theory and topology. Banach considered two properties that might be satisifed by relations $R$ between sets:
> 
> $\alpha$. whenever $A\,R\,B$, there exists a bijection $\varphi : A \to B$ such that $X\,R\,\varphi[X]$ for each $X \subseteq A$;
> 
> $\beta$. if $A_1 \cap A_2 = \phi = B_1 \cap B_2$ and $A_1\,R\,B_1$ and $A_2\,R\,B_2$, then $(A_1 \cup A_2)\,R\,(B_1 \cup B_2)$.
> 
> *Theorem* (Cantor–Bernstein–Banach). Should $R$ have property $\alpha$ and there exist $A' \subseteq A$ and $B' \subseteq B$ such that $A\,R\,B'$ and $A'\,R\,B$, then there will exist $A_1, A_2 \subseteq A$ and $B_1, B_2 \subseteq B$ such that $A_1 \cap A_2 = \phi = B_1 \cap B_2$, $A_1 \cup A_2 = A$, $B_1 \cup B_2 = B$, $A_1\,R\,B_1$, and $A_2\,R\,B_2$. Thus, if $R$ should also have property $\beta$, then $A\,R\,B$.
> 
> *Proof.* By property $\alpha$, there exist bijections $\varphi : A \to B'$ and $\psi : A' \to B$ such that $X\,R\,\varphi[X]$ and $Y\,R\,\psi[Y]$ whenever $X \subseteq A$ and $Y \subseteq A'$. Set $C_0 = A - A'$ and $C_{n+1} = \psi^{-1}\varphi[C_n]$ for each natural number $n$, then let $C$ be the union of all of the sets $C_n$. It is straightforward to show that $A - C \subseteq A'$ and $\psi[A-C] = B - \varphi[C]$. Finally, set $A_1 = C$, $A_2 = A - C$, $B_1 = \varphi[C]$, and $B_2 = B - \varphi[C]$. The restriction of $\varphi$ is a bijection from $A_1$ to $B_1$ and that of $\psi$, from $A_2$ to $B_2$.
> 
> ---
> 
> * Banach 1924. The proof given here is adapted from Wagon 1993, 25.

*[continued from page 66]*

With theorems 10 and 10′, Banach and Tarski departed from intuitively comfortable material. They showed that if a set $A$ is equivalent by decomposition to a union $A \cup B_k$ for each $k = 1, \ldots, n$, then $A$ is equivalent to $A \cup B_1 \cup \cdots \cup B_n$. That result is analogous to one about *cardinal* equivalence of infinite sets. Under the same analogy, the familiar theorem that $2m = 2n$ implies $m = n$ for all cardinals $m$ and $n$ corresponds to theorems 11 and 11′: if disjoint pairs of sets $A_1, A_2$ and $B_1, B_2$ are each equivalent by decomposition, and so are their unions, then $A_1, B_1$ are equivalent. In 1922, Sierpiński had provided for the theorem on cardinals a proof that appeared to be generalizable; in 1924, Kazimierz Kuratowski modified that to prove a corresponding result in the context of the $\alpha, \beta$-relations of Banach 1924. To prove theorem 11, Banach and Tarski referred to Kuratowski 1924. For a statement and proof of the result that they used,[37] see the box on page 71. Theorem 11, in turn, is used just once, much later, in the proof that two point sets on the surface of a sphere are equivalent if they contain interior points.

Section 1 of the Banach–Tarski paper concludes with several results relating equidecomposability and certain families of "small" sets in a Euclidean space: the families $\mathscr{B}$ of bounded sets, $\mathscr{N}$ of nowhere dense sets, $\mathscr{P}$ of sets with zero Peano–Jordan content, $\mathscr{F}$ of sets of first category, and $\mathscr{L}$ of sets with zero Lebesgue measure.[38] Each of these families contains every subset of and any set congruent to any of its members. Thus, if

---

[37] Although Banach and Tarski described Kuratowski's theorem in terms of $\alpha, \beta$-relations, they made no use of that idea in their proof.

[38] For these concepts see Birkhoff 1948, chapter 11; compare the German and French terminology in Hausdorff 1914a, §§7.8, 10.2–10.3, and in Kuratowski 1958, §36.

## 4.4 The Banach–Tarski Theorems

a set $A$ belongs to one of them and is equidecomposable with a set $B$, the subsets of $A$ and $B$ in the decomposition all belong to that family. Moreover, these families are all closed under formation of finite unions, and $\mathscr{F}$ and $\mathscr{L}$, under denumerably infinite unions. Thus the set $B$ will belong to the same family as $A$, as long as, for families $\mathscr{B}, \mathscr{R},$ and $\mathscr{P}$, the sets $A$ and $B$ are *finitely* equidecomposable.

Section 2A of the Banach–Tarski paper contains its main results for dimensions one and two. These had been presented for polygons already in Tarski [1924] 2014b, described in section 4.3 and translated in chapter 5. The first main result is theorem 16: bounded Lebesgue-measurable subsets of the line or plane, if equivalent by finite decomposition, must have the same measure. Its proof depended on theorem III* of Banach 1923, described in the box on page 54, and thus made essential use of the axiom of choice. Banach and Tarski noted that its converse is false. In fact, Hausdorff had defined a

*[The narrative continues on page 73.]*

---

**Kuratowski's Theorem.** Kazimierz Kuratowski provided* a technical result for the theory of decomposition of point sets presented in Banach and Tarski [1924] 2014 and translated in chapter 6: given
$$E = M \cup N = P \cup Q, \quad M \cap N = \phi = P \cap Q, \text{ and bijections } \varphi : M \to N \text{ and } \psi : P \to Q,$$
there exist disjoint $M_1, M_2, M_3, M_4$ and disjoint $Q_1, Q_2, Q_3, Q_4$ such that
$$M = M_1 \cup M_2 \cup M_3 \cup M_4, \quad Q = Q_1 \cup Q_2 \cup Q_3 \cup Q_4,$$
$$Q_1 = M_1, \quad Q_2 = \psi[M_2], \quad Q_3 = \varphi[M_3], \quad Q_4 = \psi\varphi[M_4].$$

To prove this Kuratowski first extended $\varphi$ and $\psi$ to involutory permutations of $E$: each extension is the union of the given function and its inverse. Next he considered for each $x \in E$ the set $C_x$ of all images of $x$ under elements of the permutation group generated by the extended $\varphi$ and $\psi$. Each $C_x$ intersects each of $M, N, P, Q$. The family of all sets $C_x$ is a partition of $E$. By the axiom of choice, there is a function that associates to each $C_x$ a member of $C_x \cap M$ —call it the *chosen* one. Each set $C_x$ consists of these members:

$$\vdots$$
$$x_3 = \varphi\psi\varphi(x_0)$$
$$x_2 = \psi\varphi(x_0)$$
$$x_1 = \varphi(x_0)$$
$$x_0 \text{ —the chosen member}$$
$$x_{-1} = \psi(x_0)$$
$$x_{-2} = \varphi\psi(x_0)$$
$$x_{-3} = \psi\varphi\psi(x_0)$$
$$\vdots$$

By a recursive argument, Kuratowski verified that if $x_k = x_l$ then $k$ and $l$ have the same parity and $x_{k+j} = x_{l+j}$ for all integers $j$. Now consider an arbitrary $m \in M$. It belongs to a unique set $C_x$, with chosen member $x_0$, and thus $m = x_k$ for some $k$. Kuratowski partitioned $M$ as follows:

if $k$ is odd and $m \in Q$, set $m \in M_1$;      if $k$ is even and $x_{k+1} \in Q$, set $m \in M_3$;
if $k$ is odd and $m \notin Q$, set $m \in M_2$;      if $k$ is even and $x_{k+1} \notin Q$, set $m \in M_4$.

Then he defined $Q_1$ to $Q_4$ as in the statement of this result. It is straightforward to show that these sets have the desired properties.

---
*Kuratowski 1924.

*Kazimierz Kuratowski
around 1915*

**Kazimierz Kuratowski** was born in 1896 in Warsaw, then part of the Russian Empire, the son of a prominent lawyer. Kazimierz completed secondary school there in 1913, then started engineering studies at the University of Glasgow. He was home for vacation in August 1914 when World War I began. That disrupted his studies, but a year later, he was able to enter the new Polish University of Warsaw, to study mathematics and logic. His first published work stemmed from the logic seminar of Jan Łukasiewicz, but in 1917 he began to work with Stefan Mazurkiewicz and Zygmunt Janiszewski on topology; the latter mentored his doctoral research. Wacław Sierpiński assumed that role after Janiszewski died in 1920. After earning the doctorate the next year, Kuratowski was soon appointed docent. During the next years he originated many concepts now fundamental in set theory, general topology, and graph theory, including the now familiar definition of ordered pair, formulation of a maximal principle equivalent to the axiom of choice, axiomatization of topology in terms of the closure operator, and a simple characterization of planar graphs. Alfred Tarski attended his lectures and collaborated with him over many years.

In 1927 Kuratowski was appointed full professor at the Lwów Polytechnic University. In 1933 he published a text, *Topologie I*, that served for several decades as a standard source. The next year, he returned to the University of Warsaw and soon became head of its mathematics program. He helped plan further expansion of Polish mathematical research, but that was thwarted by World War II. During the German occupation he taught in the clandestine academic system. Afterward, he was instrumental in the rebirth of Polish mathematics, and helped found the Polish Institute of Mathematics, which he served as director during 1948–1967. It is now part of the Academy of Sciences. Author of the 1980 history of Polish mathematics that is a principal source for the present book, Kazimierz Kuratowski was a mathematical statesman of world stature. He died in 1980.*

*Arboleda 1990, Kuratowski 1981.

## 4.4 The Banach–Tarski Theorems

*[continued from page 71]*

certain subset $G$ of the closed unit interval $I$ that is a countable union of closed intervals, and showed that it has Lebesgue measure less than one.[39] Its complement $F = I - G$ is thus nowhere dense, with positive measure $m$. An interval with length $m$ is not nowhere dense; by the previous paragraph, it cannot be equivalent to $F$ by finite decomposition. Further, the Cartesian product $F \times I$ has planar measure $m$, but is nowhere dense and thus not equivalent by finite decomposition to any rectangle with area $m$.

Lemma 17 in §2A of the Banach–Tarski Paper is the crucial step in proceeding from equivalence by decomposition in the elementary sense to the set-theoretic decomposition employed there: if $A$ is a planar set with an interior point, and $B$ is the union of a finite number of segments, then $A \equiv_f A \cup B$. This entails that a polygonal region and its interior are equivalent by finite decomposition. Immediately afterward, the authors remarked that an analogous one-dimensional result can be obtained by replacing the words *planar* and *segments* by *linear* and *points*. They must have intended that to be theorem 18, because there is no result with that number! In fact, it is lemma I in Tarski [1924] 2014b and section 5.3; a proof is provided there.[40] Section 2A reaches the same conclusion as Tarski [1924] 2014b: two polygonal regions are equivalent by (set-theoretic) finite decomposition if and only if they have the same area.

Section 2B of the Banach–Tarski paper contains its main results for dimensions greater than two. The authors presented them for dimension three, but described an easy way to generalize them to higher dimensions. Their argument started with Hausdorff's "paradoxical" 1914a decomposition of the surface of a sphere into four disjoint parts: a countable set and three congruent parts, one of which is also congruent to the union of the remaining two. (See the box on page 67.) The authors then divided the *solid* sphere $S$ into *five* disjoint parts: the center $p$ and four others, $B, C, D, E$, obtained by adjoining to the surface parts the half-open radii containing their points. In summary,

$S = B \cup E \cup \{p\} \cup D \cup C$      $E$ is a union of countably many half-open radii.

$D \cup C \cong B \cong C \cong D$      $B$, $E$, $\{p\}$, $D$, and $C$ are disjoint.

Of the uncountably many rotations of $S$, only countably many map one of those radii to another. Tarski and Banach let $\rho$ be one that does not, so that

$E \cong \rho[E] \subsetneq B \cup C \cup D.$

Then they applied corollary 7, mentioned earlier, to determine a set $G$ and a point $q$ such that

$\rho[E] \equiv_f G \subsetneq C$      $q \in G - C.$

---

[39] Theorem 16 is a generalization of theorem 7 in Tarski [1924] 2014b (section 5.2); their proofs are the same. Hausdorff 1914a, 418.

[40] Lemma 17 is the pair of lemmas II, III of Tarski [1924] 2014b (section 5.3). The proofs there are more detailed than those in the Banach–Tarski paper.

Finally, they defined

$$A_1 = B \cup E \cup \{p\} \qquad A_2 = D \cup G \cup \{q\}.$$

From the five lines of formulas just displayed and the transitivity of equivalence by finite decomposition, Banach and Tarski derived the remarkable lemma 21, $A_1 \overline{\overline{\underset{f}{}}} S \overline{\overline{\underset{f}{}}} A_2$: every solid sphere $S$ contains two disjoint subsets, each equivalent to $S$ by finite decomposition.

Again applying corollary 7 and a corollary of theorem 8, the analogue of the Cantor–Bernstein theorem mentioned earlier, Banach and Tarski quickly deduced lemma 22: every solid sphere $S$ is equivalent by finite decomposition to its union with another sphere congruent to $S$. Using theorem 10, the tool presented earlier for handling finite unions, they deduced lemma 23: every bounded set containing a solid sphere $S$ is equivalent to $S$ by finite decomposition. Finally, they presented their culminating result, any two sets that contain interior points are equivalent by finite decomposition: they must contain congruent solid spheres, to which they are equivalent by lemma 23.

In §2C of their paper, Banach and Tarski derived a theorem for the surface $S$ of a sphere completely analogous to their main three-dimensional result: any two subsets of $S$ with interior points are equivalent by finite decomposition. The arguments are similar, with two exceptions. First, lemma 27, which says that $S$ can be decomposed into two disjoint sets, each equivalent to $S$, is both stronger and easier to prove than the analogous lemma 21 for three dimensions, because the simpler definition $A_2 = S - A_1$ is available. Second, lemma 30, which says that every subset of $S$ with an interior point is equivalent to $S$ itself, has no analogue in three dimensions. Its proof required Gerwien's theorem that two spherical polygons on $S$ are equivalent if they have the same area, and the theorem of Kuratowski that was presented as a tool in §1 of their paper.[41]

Section 3 of the Banach–Tarski paper had two goals:

- *theorem 35*—in *any* dimension, *any* two sets with interior points are equivalent by *denumerable* decomposition; and
- to relate decomposibility questions to Lebesgue measure.

Their first step toward theorem 35 was to recall the main feature of Hausdorff's construction of a nonmeasurable set: the half-open unit interval $A$ is the union of a denumerable infinity of disjoint subsets equivalent to each other by finite decomposition. (See the box on page 75.) Applying an affine transformation to this $A$ shows that *every* half-open interval $A$ has that property. Next, Banach and Tarski applied that result in a straightforward manner to prove lemma 32: the union of a denumerable infinity of disjoint intervals each congruent to a single interval is equivalent to that interval by denumerable decomposition. They presented that result in one dimension, then indicated how to generalize it to higher dimensions, where an interval is a Cartesian product of one-dimensional intervals.

---

[41] Gerwien 1833b; see Boltyanskiĭ 1978, §§7–8, for a related discussion. For Kuratowski's theorem, see the box on page 71.

## 4.4 The Banach–Tarski Theorems

> **Nonmeasurable Set.** Early in the twentieth century, Henri Lebesgue posed the *measure problem*: to assign, to all bounded subsets of the line, nonnegative numbers called their measures, such that two congruent sets always have the same measure, the unit interval has measure 1, and the measure of the union of any finite or infinite bounded sequence of disjoint sets is the sum of their individual measures. In an obscure 1905 publication, Giuseppe Vitali showed that no measure for *all* bounded subsets is possible. A decade later, Felix Hausdorff published a version of Vitali's argument, as follows (1914a, 401–402).
>
> Let $\alpha$ be any irrational number in the half-open unit interval $A$. The sets
>
> $$P_x = \{\, (x + k\alpha) \bmod 1 : k \in \mathbb{Z}\,\}$$
>
> for $x \in A$ form a partition of $A$. By the axiom of choice, there is a subset $A_0 \subseteq A$ containing exactly one element of each $P_x$. For each $k \in \mathbb{Z}$ define
>
> $$A_k = \{\, (x + k\alpha) \bmod 1 : x \in A_0\,\}.$$
>
> These sets also partition $A$. The elements of $A_k$ are obtained from those of $A_0$ by adding either $k\alpha$ or $k\alpha - 1$; therefore, $A_0$ and $A_k$ are either congruent or equivalent by decomposition into pairs of mutually congruent subsets. That is, $A$ is the union of a denumerable infinity of disjoint subsets $A_k$, equivalent to each other by finite decomposition. Lebesgue's requirements would entail that these $A_k$ should all have the same measure $m$. If $m = 0$, then $A$ would have measure zero, contrary to assumption. Thus $m > 0$, which would contradict the finiteness of the measure of $A$.

Banach and Tarski quickly reached their target, theorem 35, as follows. Considering a tiling of the *entire* space $E$, they showed that the union of the sets in the previous paragraph is equivalent to $E$ by denumerable decomposition. Next, they noted that any set $A$ with an interior point must include a nonempty interval, which is equivalent to the entire space. By Banach's adaptation of the Cantor–Bernstein theorem (see the box on $\alpha,\beta$-relations on page 70), $A$ must be equivalent to $E$. It follows immediately that any two sets with interior points must be equivalent by denumerable decomposition.

In their concluding discussion of measure-theoretic considerations, Banach and Tarski used the extension of Lebesgue-measure theory formulated by Hausdorff to accommodate unbounded sets. They defined two point sets $A$ and $B$ to be *almost equivalent by denumerable decomposition*—abbreviated $A \underset{p}{=} B$ —if denumerably equivalent sets can be obtained by removing subsets with Lebesgue measure zero from $A$ and $B$. They proved that this reflexive and symmetric relation on point sets is also transitive.[42]

Their main result about this notion, theorem 41, states that in any dimension, any two sets with positive Lebesgue inner measure are almost equivalent by denumerable decomposition. Aiming toward this, Banach and Tarski first proved lemma 37: a Lebesgue-measurable set $A$ and an open set $B$ of equal measure can be approximated arbitrarily closely (with regard to their measures) by closed subsets that are equivalent by finite

---

[42] Hausdorff 1914a, 416. Zaanen 1983, chapters 1–9, is an excellent general introduction to this material. The letter $p$ in the abbreviation stands for the French word *presque* for *almost*.

decomposition.⁴³ They iterated that process, approximating each of $A$ and $B$ by unions of sequences of disjoint subsets, each corresponding pair of which are equivalent by finite decomposition. They showed that the relative complements of these unions have measure zero, and hence that $A$ and $B$ are almost equivalent by denumerable decomposition. They applied this result in the case in which $B$ is the entire space $E$, and made some minor adjustments, to prove lemma 39: to any bounded set $A$ with positive Lebesgue measure corresponds a set $C$ of measure zero such that $A \cup C$ and $E$ are equivalent by denumerable decomposition. Some further adjustments led to their main results, lemma 40 and theorem 41: every set $A$ with positive Lebesgue inner measure is almost equivalent to $E$ by denumerable decomposition, and hence any two such sets are almost equivalent to each other.⁴⁴

According to the final theorem 42 of the Banach–Tarski paper, any two sets are almost equivalent by denumerable decomposition into Lebesgue-measurable components if and only if they are Lebesgue measurable with the same measure. The authors presented no proof of that result, but indicated that it could be obtained by analyzing the previous arguments.

---

[43] Their demonstration used the *regularity theorem*: any Lebesgue-measurable set $A$ can be approximated by an open set $G \supset A$ and a closed set $F \subseteq A$ such that the measure of $G - F$ is smaller than any previously selected positive number. This stems from the fact that the measure of $A$ is simultaneously the supremum of the (outer) measures of the closed polygonal or polyhedral subsets of $A$ and the infimum of the (inner) measures of the open sets containing $A$. See Hausdorff 1914a, 408–413.

[44] Wagon has presented a different proof that applies just as well to subsets of the surface of a sphere (1993, theorem 9.17, 140–141).

# 5

# *On the Equivalence of Polygons* (1924)

This chapter contains an English translation of Alfred Tarski's paper *O równoważności wielokątów*, [1924] 2014b, written as he was completing his doctoral studies. It appeared in volume 24 of the journal *Przegląd matematyczno-fizyczny*. This is its first translation. A description of the journal, background for the paper, and a summary are provided in sections 4.1–4.3.

The translation is meant to be as faithful as possible to the original. Its only intentional modernizations are punctuation and some changes in symbols, where Tarski's conflict with others used throughout this book. Bibliographic references and some personal names have been adjusted to conform with the conventions used here. The original paper sometimes employed barely discernible a u g m e n t e d   l e t t e r s p a c i n g to emphasize a phrase.[1] In some cases the translation uses italics instead; in others, this emphasis has been suppressed. As an aspect of adjusting punctuation, the editors greatly increased use of white space to enhance visual organization of the paper. All [square] brackets in the translation enclose editorial comments. Those are inserted, usually as footnotes, to indicate changes in notation and explain passages that seem obscure.

---

[1] In the letter translated in section 15.12, Tarski conveyed his dislike of that style.

ROK II NR. 1 i 2

# PRZEGLĄD MATEMATYCZNO - FIZYCZNY

TREŚĆ:

M. *Grotowski.* Spór o istnienie elementarnego naboju elektrycznego.
H. *Steinhaus.* O mierzeniu pól płaskich.
R. *Witwiński.* Warunek arytmetyczny podzielności wielomianów.
S. L. *Ziemecki.* Proste doświadczenia z dziedziny promieniotwórczości.
A. *Tarski.* O równoważności wielokątów.
S. *Guzel.* O wyznaczaniu wartości przybliżonej $\sqrt{N}$.
A. *Rajchman.* Dowód twierdzenia Hadamard'a o wyznacznikach.
A. *Dominikiewicz.* Elementarny sposób wyznaczenia środka ciężkości łuku kołowego.
Z. *Rutkowski.* Z teorji równań pierwiastkowych.
L. *Infeld.* O dowodzie twierdzenia Talesa w szkole średniej.
Z literatury.
Miscellanea.

SOMMAIRE:

M. *Grotowski.* Sur l'existence de la charge élémentaire.
H. *Steinhaus.* Sur la quadrature des aires.
R. *Witwiński.* Condition arithmétique de la divisibilité des polynomes.
S. L. *Ziemecki.* Expériences simples en radioactivité.
A. *Tarski.* Sur l'équivalence des polygones.
S. *Guzel.* Calcul approximatif de $\sqrt{N}$.
A. *Rajchman.* Une demonstration du théorème de M. Hadamard sur le maximum d'un déterminant.
A. *Dominikiewicz.* Détermination élémentaire du centre de gravité d'un arc de cercle.
Z. *Rutkowski.* Sur les équations irrationnelles.
L. *Infeld.* Le théorème de Thales dans l'enseignement secondaire.
Revue des livres.
Communications diverses.

KSIĄŻNICA-ATLAS
ZJEDNOCZONE ZAKŁADY KARTOGRAF. I WYDAWNICZE
TOW. NAUCZ. SZKÓŁ ŚREDN. I WYŻSZ. – SP. AKC.
LWÓW – WARSZAWA
1924

*Journal Containing Alfred Tarski's Paper*
On the Equivalence of Polygons

ALFRED TARSKI

# On the Equivalence of Polygons

In elementary geometry,[2] we call two polygons *equivalent* if it is possible to divide them into the same finite number of respectively congruent polygons not having common interior points. In the theory of the equivalence of polygons, the following statement, usually accepted without proof in elementary geometry and sometimes called *De Zolt's axiom*, plays a fundamental role:

*If polygon V is a part of polygon W, then these polygons are not equivalent.*

As is well known, David Hilbert showed[3] that the preceding statement can be proved with the help of axioms usually cited in elementary geometry textbooks. Because of the difficulty of that proof, however, one does not make use of it in a secondary-school class.

Relying on De Zolt's axiom, among others, it is possible in the theory of mensuration to prove the following theorem, which provides a necessary and sufficient condition for the equivalence of two polygons:

*In order for polygons V and W to be equivalent, it is necessary and sufficient that they have equal areas.*

The question arises, do the [italicized] formulations of both statements above remain true sentences if equivalence is understood in a broader sense than it usually is in elementary geometry: that is to say, if two geometric figures (thus in particular, two polygons) are called equivalent when it is possible to divide them into the same finite number of respectively congruent *arbitrary* geometric figures not having *any* common points.

In the present article I show that this question ought to be given an *affirmative answer*. It is also interesting that the proofs of both of these very straightforward statements, the first of which may seem almost obvious, rely on results obtained by Prof. Stefan Banach with the aid of the entire apparatus of contemporary mathematical knowledge: in particular, with the help of the so-called *axiom of choice*.[4]

---

[2] The definitions and theorems of elementary geometry to which I refer in the present article can be found, for instance, in the textbook Enriques and Amaldi [1903] 1916.

[3] See Hilbert [1899] 1922.

[4] Banach 1923 [discussed in section 4.2 of the present book]. All those notions and principles of set theory to which I refer in the present article—just a few—are contained in the book Sierpiński 1923.

## Notation

By means of letters $p, q, s, \ldots$ I denote *points*, while the letters $A, B, K, P, V, \ldots$ [denote] *geometric figures*—that is, *point sets*.

The symbol $A \cup B$ denotes the *union of sets* $A$ and $B$: that is, the set consisting of all those points that belong either to set $A$ or to $B$. The notion of the union of sets can be extended with ease to an arbitrary finite number of components; it is even possible to consider the union of all sets that are terms of a certain infinite sequence. We use the symbols

$$\bigcup_{k=1}^{n} A_k \quad \text{and} \quad \bigcup_{k=1}^{\infty} A_k,$$

respectively, as well.[5]

The symbol $A - B$ denotes the *difference of sets* $A$ and $B$: that is, the set consisting of all those points of the set $A$ that do not belong to $B$.

As an expression that sets $A$ and $B$ *are identical*—that is, that they have all points in common—I write $A = B$. To express that the sets $A$ and $B$ *are disjoint*—that is, that they do not have any points in common—I will write $A \,][\, B$. Finally, to express that the set $A$ *is a proper part of* set $B$ —that is, that every point belonging to $A$ also belongs to $B$, but not conversely —I will write $A \subsetneq B$.[6]

In this article I do not distinguish a point $p$ from the set consisting solely of that same point $p$. In this way, for example, the symbol

$$\bigcup_{k=1}^{n} p_k$$

denotes the set consisting of points $p_1, p_2, \ldots, p_n$.

## §1

I begin by recalling the familiar definition of the congruence of two arbitrary geometric figures, based on the notion of *equal distances between two pairs of points*. (This notion should, of course, either be defined earlier or assumed as a primitive notion.)

**Definition 1.** Point sets $A$ and $B$ are *congruent*— $A \cong B$ —if between their points a perfect (one-to-one) correspondence can be established that satisfies the following condition: if $p$ and $q$ are arbitrary points of set $A$, while $p'$ and $q'$ are their corresponding points in set $B$, then the distances between the pairs of points $p$ and $q$, and $p'$ and $q'$, are equal.

In my following discussions, I will assume familiarity with elementary properties of the congruence relation.[7]

---

[5] [For *union* Tarski used the Polish equivalent of the English term *sum*, and for $\cup$ and $\bigcup$ he used + and $\Sigma$, respectively.]

[6] [In the original, Tarski used the symbol $\subset$ for *is a proper part of*. He employed it only once.]

[7] [The correspondence between $p, q, \ldots$ and $p', q', \ldots$ need not be a *direct* isometry; it may reverse orientation. Moreover, Tarski did *not* require that it be a restriction of an isometry of the entire plane.]

**Definition 2.** Point sets $A$ and $B$ *are equivalent*— $A \equiv B$ —if there exist sets $A_1, A_2, \ldots, A_n$ and $B_1, B_2, \ldots, B_n$, where $n$ [is] a natural number, that satisfy the conditions

(a) $A = \bigcup_{k=1}^{n} A_k$ and $B = \bigcup_{k=1}^{n} B_k$,

(b) $A_k \cong B_k$ whenever $1 \leq k \leq n$,

(c) $A_k \, ][ \, A_l$ and $B_k \, ][ \, B_l$ whenever $1 \leq k < l \leq n$.

To express that the figures $A$ and $B$ *are not equivalent* I shall write $A \not\equiv B$. In the following five theorems, I shall present several elementary properties of the relation of equivalence.

**Theorem 1.** If $A \cong B$, then $A \equiv B$. In particular, an arbitrary point set $A$ satisfies the condition $A \equiv A$.

**Theorem 2.** If $A \equiv B$, then $B \equiv A$.

Both of those theorems follow directly from definition 2.

**Theorem 3.** If $A \equiv B$ and $B \equiv C$, then $A \equiv C$.

*Proof.* For the proof I shall apply a method similar to the one that we use in the proof of an analogous theorem in elementary geometry, the so-called "method of double networks."

In view of the equivalence of sets $A$ and $B$ as well as of $B$ and $C$, there exist point sets $A_1, A_2, \ldots, A_n$ and $B_1, B_2, \ldots, B_n$, as well as $B'_1, B'_2, \ldots, B'_m$ and $C_1, C_2, \ldots, C_m$ that satisfy all the conditions of definition 2. Let us denote by $B_{k,l}$ the set of all those points that belong simultaneously[8] to $B_k$ and $B'_l$. Since every point in set $B_k$ belongs to one of the sets $B'_l$, [where] $1 \leq l \leq m$, and conversely, it is thus easy to check that

(1) $B_k = \bigcup_{l=1}^{m} B_{k,l}$ when $1 \leq k \leq n$,

(2) $B'_l = \bigcup_{k=1}^{n} B_{k,l}$ when $1 \leq l \leq m$.

In addition, according to condition (c) of definition 2, we have

(3) $B_{k,l} \, ][ \, B_{k_1,l_1}$ whenever $k \neq k_1$, or $k = k_1$ but $l \neq l_1$.

In accordance with condition (b) of definition 2, figures $A_k$ and $B_k$ are congruent [when] $1 \leq k \leq n$. From (1), (3), and the general properties of congruence, we thus infer

---

[8] Of course, the possibility is not excluded that some of the sets $B_{k,l}$ may be empty: that is, that they should not contain any points. A small modification to the proof would permit removing such sets from our consideration.

with ease the possibility of dividing each of the sets $A_k$ into parts $A_{k,1}, A_{k,2}, \ldots, A_{k,n}$ that satisfy the conditions

(4) $\quad A_k = \bigcup_{l=1}^{m} A_{k,l}$ when $1 \le k \le n$,

(5) $\quad A_{k,l} \cong B_{k,l}$ when $1 \le k \le n$ and $1 \le l \le m$,

(6) $\quad A_{k,l} \,][\, A_{k_1, l_1}$ whenever $k \ne k_1$, or $k = k_1$ but $l \ne l_1$.

Similarly, from the congruence of figures $B'_l$ and $C_l$ [when] $1 \le l \le m$, [and] from (2) and (3), follows the possibility of analogous division of each of the sets $C_l$ into parts $C_{1,l}, C_{2,l}, \ldots, C_{n,l}$:

(7) $\quad C_l = \bigcup_{k=1}^{n} C_{k,l}$ when $1 \le l \le m$,

(8) $\quad C_{k,l} \cong B_{k,l}$ when $1 \le k \le n$ and $1 \le l \le m$,

(9) $\quad C_{k,l} \,][\, C_{k_1, l_1}$ whenever $k \ne k_1$, or $k = k_1$ but $l \ne l_1$.

Since according to condition (a) of definition 2 the set $A$ is the union of sets $A_1, A_2, \ldots, A_n$, and set $C$ [is] the union of sets $C_1, C_2, \ldots, C_m$, we may thus conclude from (4) and (7),

(10) $\quad A = \bigcup_{k=1}^{n} \bigcup_{l=1}^{m} A_{k,l}$,

(11) $\quad C = \bigcup_{l=1}^{m} \bigcup_{k=1}^{n} C_{k,l} = \bigcup_{k=1}^{n} \bigcup_{l=1}^{m} C_{k,l}$.

Moreover, from (5) and (8) we immediately obtain

(12) $\quad A_{k,l} \cong C_{k,l}$ when $1 \le k \le n$ and $1 \le l \le m$.

Equations (10) and (11) show that each of the sets $A$ and $C$ can be divided into $n \cdot m$ parts, which in view of (6) and (9) have no points in common, and in accordance with (12) are respectively congruent to each other. Therefore, according to definition 2, $A \equiv C$, Q.E.D.

Theorems 1–3 express that the relation of equivalence is reflexive, symmetric, and transitive.

**Theorem 4.** If

(1) $\quad A_k \equiv B_k$ (possibly $A_k \cong B_k$ or $A_k = B_k$) whenever $1 \le k \le n$,

(2) $\quad A_k \,][\, A_l$ and $B_k \,][\, B_l$ whenever $1 \le k < l \le n$,

then $\bigcup_{k=1}^{n} A_k \equiv \bigcup_{k=1}^{n} B_k$.

*Proof.* [To facilitate typesetting, those two unions will be denoted by $A$ and $B$, respectively.] Taking theorem 1 into account, we can restrict ourselves in the proof to considering the hypothesis that $A_k \equiv B_k$ for each value of $k$ [such that] $1 \le k \le n$. Now according to definition 2, for each pair of sets $A_k$ and $B_k$ there exist sets $A_{k,1}, A_{k,2}, \ldots, A_{k,m_k}$ and $B_{k,1}, B_{k,2}, \ldots, B_{k,m_k}$ that satisfy the conditions

(1) $\quad A_k = \bigcup_{l=1}^{m_k} A_{k,l}$ and $B_k = \bigcup_{l=1}^{m_k} B_{k,l}$ when $1 \le k \le n$,

(2) $\quad A_{k,l} \cong B_{k,l}$ when $1 \le k \le n$ and $1 \le l \le m_k$,

(3) $\quad A_{k,l} \,][\, A_{k_1,l_1}$ and $B_{k,l} \,][\, B_{k_1,l_1}$ whenever $k \ne k_1$, or $k = k_1$ but $l \ne l_1$.

From (1) we immediately obtain

(4) $\quad A = \bigcup_{k=1}^{n} A_k = \bigcup_{k=1}^{n}\bigcup_{l=1}^{m_k} A_{k,l}$ and $B = \bigcup_{k=1}^{n} B_k = \bigcup_{k=1}^{n}\bigcup_{l=1}^{m_k} B_{k,l}$ .

From statements 2–4 it follows that the point sets $A$ and $B$ can be divided into the same finite number of respectively congruent parts without common points. From this, in accordance with definition 2, $A \equiv B$, Q.E.D.

The preceding theorem can be expressed in words in the following way:

*If two given point sets can be divided into the same finite number of respectively equivalent parts having no common points, then these sets are equivalent.*

**Theorem 5.** If sets $A$ and $B$ consist of the same finite number of points, then $A \equiv B$.

For the proof, it suffices to note that according to definition 1, two arbitrary points are congruent figures, hence definition 2 can be applied directly.

## §2

I turn now to the proof of theorem 6, which can be regarded as a *generalization of De Zolt's axiom*. The proof will rely on the theorem of Banach mentioned already in the introduction, which for our purposes can be adequately formulated in the following way.

**Banach's Theorem.** Each point set $A$ that is part of any polygon can be assigned some nonnegative real number $m(A)$, called the *measure* of that set. Moreover, the following conditions are satisfied:

(1) if $A \cong B$ then $m(A) = m(B)$,
(2) if $A \,][\, B$ then $m(A \cup B) = m(A) + m(B)$,
(3) if $W$ is a polygon then $m(W)$ is its area.[9]

---

[9] To the word *area* one ought to append throughout the words, *in relation to some square, chosen as the unit of area*. [Banach's theorem and its proof are discussed in section 4.2.]

**Theorem 6.** If $V$ and $W$ are polygons and $V \subsetneq W$, then $V \not\equiv W$.

*Proof.* Suppose, contrary to the conclusion of the theorem, that

(1)   $V \equiv W$;

then there would exist point sets $A_1, A_2, \ldots, A_n$ and $B_1, B_2, \ldots, B_n$ that satisfy all the conditions of definition 2 [with point sets $V, W$ in place of $A, B$].

Further, in accordance with Banach's theorem, let us assign to every bounded planar set—that is, part of any polygon—its measure. Condition 2 of the theorem mentioned can be extended with ease, by applying the principle of mathematical induction, to a sum of an arbitrary finite number of sets having no common points. In view of conditions (a) and (c) of definition 2, we infer from this that

(2)   $m(V) = \sum_{k=1}^{n} m(A_k), \quad m(W) = \sum_{k=1}^{n} m(B_k)$.

From condition 1 of Banach's theorem and condition (b) of definition 2 we obtain

(3)   $m(A_k) = m(B_k)$ whenever $1 \le k \le n$.

Equations (2) and (3) entail immediately

(4)   $m(V) = m(W)$.

Therefore, according to condition 3 of the cited theorem, polygons $V$ and $W$ must have equal areas, which contradicts the assumption of our theorem, since $V$ is a proper part of $W$.

Assumption (1) thus leads to a contradiction, and we must accept that $V \not\equiv W$, Q.E.D.

Reasoning in an analogous way, [we can] prove the more general

**Theorem 7.** If $V$ and $W$ are polygons with different areas, then $V \not\equiv W$.

## §3

We now take up the proof of the theorem converse to the one just presented. First of all, we note that despite what might at first glance be supposed, this theorem does not follow directly from an analogous theorem of elementary geometry. I will illustrate this circumstance with a straightforward example.

Let $V$ be an arbitrary square and $W$, an isosceles right triangle with base twice as long as the edge of the square. Having equal areas, $V$ and $W$ are thus equivalent in the sense of elementary geometry. In fact, each of these polygons can be divided into two

triangles without common interior points, respectively congruent. (See the figure.) From this subdivision, however, the subdivision that would satisfy definition 2 cannot be obtained via a direct route. Although the interiors of the [smaller] triangles are in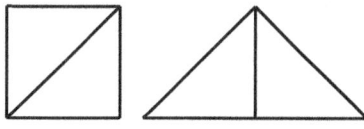
fact congruent, nevertheless the [parts] that stand out in the figure are broken lines, [and] since these unions of the boundaries of the triangles have different length, it is not hard to demonstrate that they are not equivalent in the sense that we established in the present article.[10]

The proof of the theorem that interests us will rely on several lemmas.

**Lemma I.** If $A$ is a plane set having interior points,[11] whereas set $B$ consists of a finite number of points, and $A \,][\, B$, then $A \equiv A \cup B$.

*Proof.* Certainly there exists some disk $K$ that is part of the set $A$; let us denote its center by $s$. Let us choose some positive irrational number $\alpha$ and some point $p_0$ lying on the circumference of the disk $K$. For each natural number $k$ let us denote by $p_k$ the point resulting from the rotation of point $p_0$ about point $s$ through an angle whose degree measure is the number $k \cdot \alpha$ (or a number differing from $k \cdot \alpha$ by a multiple of 360) —and we always carry out the rotation in some specified direction. From this, since the angle of $\alpha$ degrees is incommensurate with a full angle, we infer with ease that no two points $p_k$ and $p_l$ with different indices are identical.

Let $n$ be the number of points in the set $B$. Set

(1) $\quad B' = \bigcup_{k=0}^{n-1} p_k,$

(2) $\quad C = \bigcup_{k=0}^{\infty} p_k,$

(3) $\quad C' = \bigcup_{k=n}^{\infty} p_k,$

(4) $\quad D = A - C.$

From (1) to (4) and the definition of the points $p_k$ we immediately obtain

(5) $\quad A = C \cup D = B' \cup C' \cup D,$

(6) $\quad A \cup B = B \cup C \cup D.$

---

[10] [If the correspondence of the interiors of the triangles were extended by somehow subdividing the segments shown on the left and rearranging the parts to form those on the right, the total length of the left-hand segments would equal that on the right. But they differ: $4 + \sqrt{2} \neq 3 + 2\sqrt{2}$.]

[11] We call point $p$ an *interior* point of a plane set $A$ if there exists a disk with center $p$ that is a part of the set $A$.

According to theorem 5, since each of the sets $B$ and $B'$ consists of $n$ points, this statement follows:

(7)  $B \equiv B'$.

With ease, we also convince ourselves that

(8)  $C \cong C'$.

In fact, if the set $C$ is rotated about an angle of $n \cdot a$ degrees, then it covers the set $C'$. In other words, if we assign to an arbitrary point $p_k$ of the set $C$ the point $p_{k+n}$ of the set $C'$, then we define a perfect correspondence between the points of these sets, that satisfies the conditions of definition 1.

Statements 5 to 8 show that the sets $A$ and $A \cup B$ can be divided into the same finite number of parts, respectively equivalent, or even congruent or identical. It is also easy to check, [by] relying on statements 1 to 4, by the way of specifying the points $p_k$, and [by] the hypothesis of the theorem, that no two of the three parts into which we divide each of these sets have common points. From this, in accordance with theorem 4, we infer that $A \equiv A \cup B$, Q.E.D.

**Lemma II.** If $A$ is a plane set having interior points, while the set $B$ consists of all points of some segment except at most the end points, and $A \,][\, B$, then $A \equiv A \cup B$.

*Proof.* The ideas behind the proofs of lemmas I and II are similar to each other. Let us denote by $\delta$ the length of the segment from which $B$ differs by at most the absence of the end points. Certainly there exists a natural number $n$ large enough that some disk $K$ that has a radius of length equal to $\delta/n$ is part of the set $A$.

Clearly, the set $B$ can be divided into $n$ segments of length $\delta/n$ without common interior points, and two of these might have only one endpoint each. Let us denote the interiors of these segments by $C_0, C_1, \ldots, C_{n-1}$ and set

(1)  $C = \bigcup_{k=0}^{n-1} C_k,$

(2)  $D = B - C.$

It is easy to see that $D$ is a set consisting of a finite number of points.[12]

Let us choose some positive irrational number $\alpha$ [and] denote by $C_0'$ the interior of some radial segment of the disk $K$; when $k$ is an arbitrary natural number, [denote] by $C_k'$ the set formed by rotating the set $C_0'$ through an angle of $k \cdot \alpha$ degrees about the center of the disk $K$ in a certain specified direction. As in the proof of lemma I, we convince ourselves that no two of the sets $C_k'$ and $C_l'$ with different indices have common points. Set

---

[12] [This number is] equal to $n+1$, $n$, or $n-1$, depending on whether the set $B$ has both endpoints, or just one, or, lastly, does not have any.

(3) $\quad C' = \bigcup_{k=0}^{n-1} C'_k,$

(4) $\quad E = \bigcup_{k=0}^{\infty} C'_k,$

(5) $\quad E' = \bigcup_{k=n}^{\infty} C'_k,$

(6) $\quad F = A - E.$

From (1) to (6), we immediately obtain

$$B = C \cup D, \quad E = C' \cup E', \quad A = E \cup F = C' \cup E' \cup F,$$

from which [follow]

(7) $\quad A \cup D = C' \cup D \cup E' \cup F,$

(8) $\quad A \cup B = C \cup D \cup E \cup F.$

As the interiors of segments of the same length $\delta/n$, the sets $C_k$ and $C'_k$ are congruent. Therefore, from (1) and (3) we infer

(9) $\quad C \equiv C'.$

Furthermore, reasoning as in the proof of lemma I, we reach the conclusion that

(10) $\quad E \cong E'.$

Finally, as it is not difficult to be convinced, no two of the sets $C', D, E', F$ nor of $C, D, E, F$ have common points. In view of this, we can apply theorem 4; by virtue of statements 7 to 10 we have

(11) $\quad A \cup D \equiv A \cup B.$

On the other hand, the set $D$, as we already noticed, consists of a finite number of points. Therefore, according to lemma I,

(12) $\quad A \equiv A \cup D.$

From (11) and (12) it follows, in accordance with theorem 3, that $A \equiv A \cup B$, Q.E.D.

**Lemma III.** If $A$ is a plane set having interior points, while $B$ [is] the union of a finite number of segments, and $A \,][\, B$, then $A \equiv A \cup B$.

*Proof.* It is nearly obvious that the set $B$ can be regarded as a union of a finite number of segments $B_1, B_2, \ldots, B_n$ without common interior points. Let us set $B'_1 = B_1$, and when $2 \le k \le n$ denote by $B'_k$ the set differing from the segment $B_k$ in at most the absence of one or two endpoints and that of the points belonging to any of the segments $B_1, B_2, \ldots, B_{k-1}$ [—that is,]

$$B'_k = B_k - \bigcup_{l=1}^{k-1} B_l.^{13}$$

We obtain

(1) $\quad B = \bigcup_{k=1}^{n} B'_k$,

(2) $\quad B'_k \mathbin{][} B'_l$ whenever $1 \le k < l \le n$.

On the other hand, there certainly exist sets $A_1, A_2, \ldots, A_n$ having interior points and also satisfying the conditions

(3) $\quad A = \bigcup_{k=1}^{n} A_k$,

(4) $\quad A_k \mathbin{][} A_l$ whenever $1 \le k < l \le n$.

In fact, if we divide into $n$ circular sectors some disk $K$ that is part of the set $A$, denote by $A_1, A_2, \ldots, A_{n-1}$ the interiors of all but one of these sectors, and set

$$A_n = A - \bigcup_{k=1}^{n-1} A_k,$$

then we will at that time obtain sets with the desired properties.

In view of (1), (3), and the condition $A \mathbin{][} B$ given in the hypothesis of the theorem, we have

(5) $\quad A_k \mathbin{][} B_l$ whenever $1 \le k \le n$ and $1 \le l \le n$.

Thus, we can assert with ease that every pair of sets $A_k$ and $B_k$, where $1 \le k \le n$, satisfies the conditions of lemma II. Therefore,

(6) $\quad A_k \equiv A_k \cup B'_k$ when $1 \le k \le n$.

From statements (2), (4), and (5) we conclude further that

(7) $\quad A_k \cup B'_k \mathbin{][} A_l \cup B'_l$ whenever $1 \le k < l \le n$.

Moreover, from (1) and (3) also follows

(8) $\quad A = \bigcup_{k=1}^{n} A_k, \quad A \cup B = \bigcup_{k=1}^{n} (A_k \cup B'_k)$.

In accordance with (6) to (8) the sets $A_1, A_2, \ldots, A_n$ and $A_1 \cup B'_1, A_2 \cup B'_2, \ldots, A_n \cup B'_n$, [whose unions[14] are] sets $A$ and $A \cup B$, satisfy all the conditions of theorem 4. Thus, we finally obtain $A \equiv A \cup B$, Q.E.D.

Lemma III now enables us [to give] a direct proof of the theorem converse to theorem 7.

---

[13] [This sentence and the next might not fully explain the first sentence of the proof. One can apply mathematical induction as follows: if $B$ is the union of a finite number of segments without common interior points, and $C$ is a segment, then $C - B$ is also such a union, and $B \cup C = B \cup (C - B)$.]

[14] [In the original, the phrase here in brackets was vague: *w stosunku do*.]

**Theorem 8.** If $V$ and $W$ are polygons with the equal areas, then $V \equiv W$.

*Proof.* As is well known, polygons $V$ and $W$ are equivalent in the sense of elementary geometry. Thus, they can be divided into the same number of polygons having no common interior points. Let $V_1, V_2, \ldots, V_n$ and $W_1, W_2, \ldots, W_n$ be the interiors of the polygons obtained as a result of such a division. Certainly we have

(1) $V_k \cong W_k$ when $1 \le k \le n$,

(2) $V_k \;][\; V_l$, and also $W_k \;][\; W_l$, whenever $1 \le k < l \le n$.

According to definition 2 we infer from (1) and (2) that

(3) $$\bigcup_{k=1}^{n} V_k \equiv \bigcup_{k=1}^{n} W_k.$$

Set

(4) $$A = V - \bigcup_{k=1}^{n} V_k, \quad B = W - \bigcup_{k=1}^{n} W_k.$$

From this we immediately obtain

(5) $$A \;][\; \bigcup_{k=1}^{n} V_k, \quad B \;][\; \bigcup_{k=1}^{n} W_k,$$

and

(6) $$V = A \cup \bigcup_{k=1}^{n} V_k, \quad W = B \cup \bigcup_{k=1}^{n} W_k.$$

In view of (4) it is easy to see that $A$ and $B$ are broken lines, the unions of the boundaries of the polygons that we obtained by the subdivision of $V$ and $W$; each is thus the union of a finite number of segments. Moreover, since the sets

$$\bigcup_{k=1}^{n} V_k \quad \text{and} \quad \bigcup_{k=1}^{n} W_k$$

certainly have interior points, after applying lemma III [and] in accordance with (5) and (6), we thus obtain

(7) $$\bigcup_{k=1}^{n} V_k \equiv A \cup \bigcup_{k=1}^{n} V_k = V,$$

(8) $$\bigcup_{k=1}^{n} W_k \equiv B \cup \bigcup_{k=1}^{n} W_k = W.$$

From statements (3), (7), and (8), according to theorem 3, it follows that $V \equiv W$, Q.E.D.

Theorems 7 and 8 immediately entail

**Conclusion 9.** In order for polygons $V$ and $W$ to be equivalent, it is necessary and sufficient that they have equal areas.

Theorem 6 and conclusion 9 settle the question posed at the beginning of the present article.

The question arises here whether the statements are true [that are] analogous to the theorems proved in this article but relate to polyhedra instead of polygons. As it happens, such statements are false. Specifically, the following theorem can be proved:

*Two arbitrary polyhedra are equivalent.*

The proof of this statement is complicated enough not to include it here: it is contained in a joint article by Banach and me, entitled *On Decomposition of Point Sets into Respectively Congruent Parts*.[15]

We are easily made aware of how greatly the above theorem contradicts our intuitions, if we consider so much as the following conclusion that flows from it:

*An arbitrary cube can be divided into a finite number of parts without common points, which then can be rearranged to form a cube with an edge twice as long.*

The theorem becomes even more striking when we recall that, as Max Dehn showed,[16] even two polyhedra with equal volumes may not be equivalent in the sense of elementary geometry.

In conclusion, I pose here the following problem, which as far as is known, is to this day not settled:

*Can theorem 8 be extended to arbitrary plane regions bounded by closed curves? Specifically, can a disk and a polygon with equal areas be equivalent in the sense of definition 2?*[17]

### Summary[18]

### On the Equivalence of Polygons

In elementary geometry two polygons (or polyhedra) are called *equivalent by decomposition* if they can be decomposed into the same finite number of respectively congruent polygons (or polyhedra) that have no common interior points. In the theory of equivalence of polygons the following theorem, sometimes called *De Zolt's axiom*, plays an important role:

1. *Two arbitrary polygons, one of which is [properly] contained in the other, are never equivalent by decomposition.*

---

[15] [Banach and Tarski [1924] 2014, translated in chapter 6, with background and summary in section 4.4.]

[16] Compare Amaldi [1900] 1914, §11, 161–172. [Tarski failed to mention the author, Ugo Amaldi. See also Dehn 1901–1902.]

[17] A disk and a polygon with equal areas are not equivalent in the sense of elementary geometry: compare Amaldi [1900] 1914, §§6–7, 151–157. [See the previous footnote. This problem, known as *Tarski's circle-squaring problem*, was published separately as Tarski 1925b. It has since been solved, affirmatively: see the discussion in section 4.3 of the present book. The wrong text was printed for Tarski 1925b in the *Collected Papers* volume Tarski 1986a; the original text is reproduced and translated in section 4.3.]

[18] [In the original, the summary was in French.]

## 5.4 Summary

Starting from this principle, one establishes the following theorem, which presents a necessary and sufficient condition for the equivalence of polygons:

2. *In order that two polygons should be equivalent by decomposition, it is necessary and sufficient that they should have equal areas.*

In my note I envisage the notion of equivalence in a sense more general than that of elementary geometry: two point sets (thus, in particular, two polygons or polyhedra) are termed *equivalent by decomposition* if they can be decomposed into the same finite number of respectively congruent *arbitrary* point sets that have no common points.

I prove that, *even admitting this definition of equivalence, theorems 1 and 2 remain valid.*

In demonstrating the cited theorems I rely on results obtained by Banach in measure theory (Banach 1923). Establishing theorem 2, I also make use of the following lemma:

*P being the interior of a polygon and Q the point set obtained from P by adding a finite number of segments, the sets P and Q are equivalent by decomposition.*

It is interesting to remark that in attributing to equivalence the sense established in this note, theorems 1 and 2 may not be extended to polyhedra. This results from the following theorem, which perhaps seems paradoxical:

*Two arbitrary polyhedra (with equal volumes or not) are equivalent by decomposition.*

This theorem is demonstrated in the note Banach and Tarski [1924] 2014.[19]

---

[19] [Summarized in section 4.4 of the present book and translated in full in chapter 6.]

# 6
# *On Decomposition of Point Sets into Respectively Congruent Parts (1924)*

This chapter contains an English translation of the paper *Sur la décomposition des ensembles de points en parties respectivement congruents*, [1924] 2014, by Stefan Banach and Alfred Tarski. It appeared in volume 6 of the journal *Fundamenta Mathematicae*. This is its first translation. Its best-known result is often called the *Banach–Tarski paradox*: any two balls with different radii can be decomposed into the same finite number of disjoint, respectively congruent parts. Background for the paper and a summary are provided in sections 4.1–4.4. Section 8.6 discusses its relationship with related later research by Tarski and others.

This translation adheres closely to the terse French of the original. A few set-theoretic symbols have been replaced by modern equivalents.[1] Some abbreviations have been spelled out, and some uses of alternative type styles for emphasis, enunciations, and personal names have been modified. Punctuation, paragraph breaks, and page layout have been streamlined or redesigned to conform to the style of the present book. Bibliographic references and some personal names have been adjusted to conform with the conventions used here. All [square] brackets in the translation enclose editorial comments. Those are inserted, usually as footnotes, to note further editorial changes and correction of errors in the original, and occasionally to provide additional information.

---

[1] The original symbols $\subset, \supset, +, \Sigma, \times, 0, (\ )$, and $'$ for subset, superset, binary and general union, intersection, empty and singleton sets, and interior have been replaced by $\subseteq, \supseteq, \cup, \bigcup, \cap, \phi, \{\ \}$, and $°$, respectively.

# FUNDAMENTA MATHEMATICAE

REDAKTOROWIE

STEFAN MAZURKIEWICZ i WACŁAW SIERPIŃSKI

KOMITET REDAKCYJNY

D$^R$ STANISŁAW LEŚNIEWSKI, D$^R$ JAN ŁUKASIEWICZ,
D$^R$ STEFAN MAZURKIEWICZ, D$^R$ WACŁAW SIERPIŃSKI
PROFESOROWIE UNIWERSYTETU WARSZAWSKIEGO.

TOM VI

Z SUBWENCJI WYDZIAŁU NAUKI MINISTERSTWA W. R. I O. P.

WARSZAWA 1924.
PARIS, LIBRAIRIE GAUTHIER-VILLARS ET C$^{ie}$,
QUAI DES GRANDS-AUGUSTINS, 55.

## TREŚĆ TOMU VI.

| | Str. |
|---|---|
| W. Sierpiński: Sur une propriété des ensembles ambigus | 1 |
| S. Kempisty: Sur les fonctions approximativement discontinues | 6 |
| A. Khintchine: Über einen Satz der Wahrscheinlichkeitsrechnung | 9 |
| S. Mazurkiewicz: Remarque sur un théorème de M. Mullikin | 37 |
| W. Sierpiński: Un exemple effectif d'un ensemble mesurable $(B)$ de classe $\alpha$ | 39 |
| A. Tarski: Sur les ensembles finis | 45 |
| H. Lebesgue: Sur le théorème de Schoenflies | 96 |
| R. L. Moore: Concerning the common boundary of two domains | 203 |
| R. L. Wilder: On the dispersion sets of connected point-sets | 214 |
| P. Urysohn: Über ein Problem von Herrn C. Carathéodory | 229 |
| S. Banach: Un théorème sur les transformations biunivoques | 236 |
| C. Kuratowski: Une propriété des correspondances biunivoques | 240 |
| S. Banach et A. Tarski: Sur la décomposition des ensembles de points en parties respectivement congruentes | 244 |

# On Decomposition of Point Sets into Respectively Congruent Parts

By

# Stefan BANACH (Lwów) and Alfred TARSKI (Warsaw)

In this note we study the notions of *equivalence of point sets by finite decomposition* and *by denumerable decomposition*. Two point sets situated in a metric space are called equivalent by finite or denumerable decomposition just in case they can be decomposed into one and the same finite number, or into a denumerable infinity, of disjoint, respectively congruent parts.

The principal results contained in the present article are the following:

*In a Euclidean space of at least three dimensions two arbitrary sets, bounded and containing some interior points (for example, two spheres with different radii), are equivalent by finite decomposition.*

*An analogous theorem holds for sets situated on the surface of a sphere; but the corresponding statement concerning Euclidean spaces of one or two dimensions is false.*

On the other hand,

*In a Euclidean space of any dimension two arbitrary sets (bounded or not), containing interior points, are equivalent by denumerable decomposition.*

The demonstration of the preceding theorems rests on results of Felix Hausdorff, Giuseppe Vitali, and Stefan Banach[2] about the general problem of measure. Thus, they make use of Ernst Zermelo's *axiom of choice*. The role that this axiom plays in our reasoning seems to us to merit attention.

We have in mind, indeed, the two following theorems that result from our research.

  I. Two arbitrary polyhedra are equivalent by finite decomposition.
  II. Two different polygons, one of which is contained in the other, are never equivalent by finite decomposition.[3]

Now, we do not know how to demonstrate either of these two theorems without appealing to the axiom of choice: neither the first, which perhaps seems paradoxical, nor the second, which is in full accord with intuition. Moreover, analyzing their demonstrations, one can see that the axiom of choice intervenes in the demonstration of the first theorem in a form much more restricted than it does in the second.

---

[2] Hausdorff 1914a, 401, 469; Vitali 1905; Banach 1923, 30–31.

[3] This theorem can be regarded as a generalization of the theorem, familiar in elementary geometry, sometimes called *De Zolt's axiom*. See Tarski [1924] 2014b [translated in chapter 5].

## §1

## General Properties of Equivalence by Finite or Denumerable Decomposition

The arguments in this section are valid for point sets situated in an arbitrary space, on which is imposed the single hypothesis that to each pair $(a,b)$ of points there should correspond a real number $\rho(a,b)$ called the *distance* between points $a$ and $b$. Only corollaries 14–15' concern Euclidean space.

**Definition 1.** Point sets $A$ and $B$ are *congruent*,

$$A \cong B,$$

if there exists a function $\varphi$ that transforms $A$ to $B$ bijectively and satisfies this condition: for two arbitrary points $a_1$ and $a_2$ of the set $A$,

$$\rho(a_1, a_2) = \rho(\varphi(a_1), \varphi(a_2)).$$

In what follows we suppose that the elementary properties of the notion of congruence are known.

**Definition 2.** Point sets $A$ and $B$ are *equivalent by finite decomposition*,

$$A \underset{f}{=} B,$$

if there exist sets $A_1, A_2, \ldots, A_n$ and $B_1, B_2, \ldots, B_n$ that fulfill the following conditions:

I. $A = \bigcup_{k=1}^{n} A_k$ and $B = \bigcup_{k=1}^{n} B_k$,

II. $A_k \cap A_l = \phi = B_k \cap B_l$ when $1 \le k < l \le n$,

III. $A_k \cong B_k$ whenever $1 \le k \le n$.

**Definition 2'.** Point sets $A$ and $B$ are *equivalent by denumerable decomposition*,

$$A \underset{d}{=} B,$$

if there exist sets $A_1, A_2, \ldots, A_n, \ldots$ and $B_1, B_2, \ldots, B_n, \ldots$ that fulfill the following conditions:

I. $A = \bigcup_{k=1}^{\infty} A_k$ and $B = \bigcup_{k=1}^{\infty} B_k$,

II. $A_k \cap A_l = \phi = B_k \cap B_l$ when $k \ne l$,

III. $A_k \cong B_k$ for all natural numbers $k$.

In the theorems 1–15' that follow we establish the elementary properties of the notions just introduced, without limiting ourselves just to those that are useful to us later. To each theorem concerning equivalence by finite decomposition corresponds a theorem on

6.1 §1 General Properties of Equivalence

equivalence by denumerable decomposition; the demonstrations of the corresponding theorems being completely analogous, we limit ourselves to giving only one of them here.

**Theorem 1.** If $A = B$, or even $A \cong B$, then $A \overset{=}{_f} B$.

This is an immediate consequence of definition 2. Taking into account that $\phi \cong \phi$, one deduces further,

**Theorem 1'.** If $A = B$, $A \cong B$, or even $A \overset{=}{_f} B$, then $A \overset{=}{_d} B$.

**Theorem 2.** If $A \overset{=}{_f} B$, then $B \overset{=}{_f} A$.

**Theorem 2'.** If $A \overset{=}{_d} B$, then $B \overset{=}{_d} A$.

Those theorems result immediately from definitions 2 and 2'.

**Theorem 3.** If $A \overset{=}{_f} B$ and $B \overset{=}{_f} C$, then $A \overset{=}{_f} C$.

*Proof.* Let

$$A = \bigcup_{k=1}^{n} A_k \quad \text{and} \quad B = \bigcup_{k=1}^{n} B_k \tag{1}$$

$$B = \bigcup_{l=1}^{m} B'_l \quad \text{and} \quad C = \bigcup_{l=1}^{m} C_l \tag{2}$$

be the decompositions of sets $A$ and $B$, and of $B$ and $C$, respectively, that satisfy conditions I–III of definition 2. We set

$$B_{k,l} = B_k \cap B'_l \quad \text{when} \quad 1 \le k \le n \quad \text{and} \quad 1 \le l \le m; \tag{3}$$

from (1)–(3) it follows immediately that

$$B_k = \bigcup_{l=1}^{m} B_{k,l} \quad \text{whenever} \quad 1 \le k \le n, \tag{4}$$

$$B'_l = \bigcup_{k=1}^{n} B_{k,l} \quad \text{whenever} \quad 1 \le l \le m. \tag{5}$$

Moreover, following (3) and condition II of the cited definition one obtains

$$B_{k,l} \cap B_{k_1,l_1} = \phi \quad \text{when} \quad k \ne k_1 \quad \text{or} \quad l \ne l_1. \tag{6}$$

Sets $A_k$ and $B_k$ being congruent when $1 \le k \le n$, one deduces according to (4) and (6) the existence of sets $A_{k,1}, A_{k,2}, \ldots, A_{k,m}$ that satisfy formulas

$$A_k = \bigcup_{l=1}^{m} A_{k,l} \quad \text{whenever} \quad 1 \le k \le n, \tag{7}$$

$$A_{k,l} \cap A_{k_1,l_1} = \phi \quad \text{when} \quad k \ne k_1 \quad \text{or} \quad l \ne l_1, \tag{8}$$

$$A_{k,l} \cong B_{k,l} \quad \text{whenever} \quad 1 \le k \le n \quad \text{and} \quad 1 \le l \le m. \tag{9}$$

In the same way, when $1 \le l \le m$, congruence of the sets $B'_l$ and $C_l$ implies by virtue of (5) and (6) that there exist sets $C_{1,l}, C_{2,l}, \ldots, C_{n,l}$ such that

$$C_k = \bigcup_{k=1}^{n} C_{k,l} \quad \text{whenever} \quad 1 \le l \le m, \tag{10}$$

$$C_{k,l} \cap C_{k_1,l_1} = \phi \quad \text{when} \quad k \ne k_1 \text{ or } l \ne l_1, \tag{11}$$

$$B_{k,l} \cong C_{k,l} \quad \text{whenever} \quad 1 \le k \le n \text{ and } 1 \le l \le m. \tag{12}$$

By virtue of (1), (2), (7), and (10), one concludes that

$$A = \bigcup_{k=1}^{n} \bigcup_{l=1}^{m} A_{k,l}, \tag{13}$$

$$C = \bigcup_{l=1}^{m} \bigcup_{k=1}^{n} C_{k,l} = \bigcup_{k=1}^{n} \bigcup_{l=1}^{m} C_{k,l}; \tag{14}$$

from (9) and (12) one finally obtains

$$A_{k,l} \cong C_{k,l} \quad \text{whenever} \quad 1 \le k \le n \text{ and } 1 \le l \le m. \tag{15}$$

Formulas (13) and (14) provide a decomposition of sets $A$ and $C$ into a finite number (equal to $n \cdot m$) of parts; following (8), (11), and (15), this decomposition fulfills the conditions of definition 2. Thus one has

$$A \overset{=}{_f} C, \qquad \text{Q. E. D.}$$

In a completely analogous way one can demonstrate the following

**Theorem 3'.** If $A \overset{=}{_d} B$ and $B \overset{=}{_d} C$, then $A \overset{=}{_d} C$.

In conformity with theorems 1–3' the relations of equivalence by finite and by denumerable decomposition are *reflexive*, *symmetric* and *transitive*.

**Theorem 4.** If sets $A$ and $B$ can be decomposed into disjoint subsets,

$$A = \bigcup_{k=1}^{n} A_k, \quad B = \bigcup_{k=1}^{n} B_k,$$

in such a way that

$$A_k \overset{=}{_f} B_k \quad \text{whenever} \quad 1 \le k \le n,$$

then

$$A \overset{=}{_f} B.$$

*Proof.* The hypothesis of the theorem implies that, for each $k$ such that $1 \le k \le n$, there exists a decomposition of the sets $A_k$ and $B_k$,

$$A_k = \bigcup_{l=1}^{m_k} A_{k,l}, \quad B_k = \bigcup_{l=1}^{m_k} B_{k,l}, \tag{1}$$

that satisfies the conditions

$$A_{k,l} \cap A_{k_1,l_1} = \phi = B_{k,l} \cap B_{k_1,l_1} \quad \text{when} \quad k \ne k_1 \text{ or } l \ne l_1, \tag{2}$$

6.1 §1 General Properties of Equivalence

$$A_{k,l} \cong B_{k,l} \quad \text{whenever} \quad 1 \le k \le n \quad \text{and} \quad 1 \le l \le m. \tag{3}$$

From (1) one obtains

$$A = \bigcup_{k=1}^{n}\bigcup_{l=1}^{m_k} A_{k,l}, \quad B = \bigcup_{k=1}^{n}\bigcup_{l=1}^{m_k} B_{k,l}. \tag{4}$$

In conformity with definition 2, formulas (2)–(4) immediately yield

$$A \underset{f}{=} B, \qquad \text{Q. E. D.}$$

**Theorem 4'.** If sets $A$ and $B$ can be decomposed into disjoint subsets,

$$A = \bigcup_{k=1}^{\infty} A_k, \quad B = \bigcup_{k=1}^{\infty} B_k \quad \left(\text{respectively,} \quad A = \bigcup_{k=1}^{n} A_k, \quad B = \bigcup_{k=1}^{n} B_k\right),$$

in such a way that

$$A_k \underset{d}{=} B_k \quad \text{for each natural number } k \quad (\text{respectively, when } 1 \le k \le n),$$

then

$$A \underset{d}{=} B.$$

**Theorem 5.** If $A \underset{f}{=} B$, there exists a function $\varphi$, defined for all points of the set $A$, that fulfills these conditions:

    I.   the function $\varphi$ transforms $A$ into $B$ bijectively,

    II.  for every subset $C$ of $A$, one has $C \underset{f}{=} \varphi(C)$.[4]

*Proof.* Let

$$A = \bigcup_{k=1}^{n} A_k, \quad B = \bigcup_{k=1}^{n} B_k \tag{1}$$

be a decomposition of the sets $A$ and $B$ that satisfies the conditions of definition 2. The sets $A_k$ and $B_k$ being congruent for $1 \le k \le n$, definition 1 yields the existence of functions $\varphi_k$ that map $A_k$ to $B_k$ without changing the distances between the points being transformed. Set

$$\varphi(p) = \varphi_k(p) \quad \text{in case} \quad p \in A_k, \quad \text{where} \quad 1 \le k \le n. \tag{2}$$

With no trouble, by virtue of the properties of the decomposition (1), one concludes from this that

$$\text{the function } \varphi \text{ maps } A \text{ into } B \text{ bijectively.} \tag{3}$$

Now let

$$C \subseteq A \tag{4}$$

and set

---

[4] For every function $\varphi$ defined for all points of a set $A$, if $C \subseteq A$, then $\varphi(C)$ denotes the set of all elements $\varphi(p)$ for which $p \in C$.

$$C_k = C \cap A_k \quad \text{whenever} \quad 1 \leq k \leq n. \tag{5}$$

From (1), (4), and (5) one obtains immediately,

$$C = \bigcup_{k=1}^{n} C_k, \tag{6}$$

$$C_k \cap C_l = \phi \quad \text{when} \quad 1 \leq k < l \leq n, \tag{7}$$

$$C_k \subseteq A_k \quad \text{whenever} \quad 1 \leq k \leq n. \tag{8}$$

From (3), (4), (6), and (7), it follows that

$$\varphi(C) = \bigcup_{k=1}^{n} \varphi(C_k), \tag{9}$$

$$\varphi(C_k) \cap \varphi(C_l) = 0 \quad \text{when} \quad 1 \leq k < l \leq n. \tag{10}$$

Finally, it results from (2) and (8), in conformity with the indicated property of the functions $\varphi_k$, that

$$C_k \cong \varphi(C_k) \quad \text{whenever} \quad 1 \leq k \leq n. \tag{11}$$

Formulas (6) and (9) provide a decomposition of the sets $C$ and $\varphi(C)$ that fulfills, following (7), (10), and (11), all the conditions of definition 2. Thus one has

$$C \overset{=}{_f} \varphi(C). \tag{12}$$

Conditions (7) and (12) prove that $\varphi$ is the desired function.

**Theorem 5′.** If $A \overset{=}{_d} B$,[5] there exists a function $\varphi$ defined for all points of the set $A$ and fulfilling these conditions:

    I.  the function $\varphi$ maps $A$ to $B$ bijectively,

    II.  if $C$ should be an arbitrary subset of $A$, then $C \overset{=}{_d} \varphi(C)$.

Theorems 5 and 5′ directly imply the following corollaries.

**Corollary 6.** If $A \overset{=}{_f} B$ and there exists a decomposition of the set $A$ into disjoint subsets,

$$A = \bigcup_{k=1}^{n} A_k \quad \left(\text{respectively,} \quad A = \bigcup_{k=1}^{\infty} A_k\right),$$

then there also exists a decomposition of the set $B$ into disjoint subsets,

$$B = \bigcup_{k=1}^{n} B_k \quad \left(\text{respectively,} \quad B = \bigcup_{k=1}^{\infty} B_k\right),$$

such that

$$A_k \overset{=}{_f} B_k \quad \text{whenever} \quad 1 \leq k \leq n \quad (\text{respectively, for every natural number } k).$$

---

[5] [Banach and Tarski stated this hypothesis as $A \overset{=}{_f} B$, evidently in error.]

**Corollary 6′.** If $A \underset{d}{=} B$ and there exists a decomposition of the set $A$ into disjoint subsets,

$$A = \bigcup_{k=1}^{n} A_k \quad \left(\text{respectively,} \quad A = \bigcup_{k=1}^{\infty} A_k\right),$$

then there also exists a decomposition of the set $B$ into disjoint subsets,

$$B = \bigcup_{k=1}^{n} B_k \quad \left(\text{respectively,} \quad B = \bigcup_{k=1}^{\infty} B_k\right),$$

such that

$A_k \underset{d}{=} B_k$ whenever $1 \leq k \leq n$ (respectively, for every natural number $k$).

**Corollary 7.** If $A \underset{f}{=} B$, then to each subset $C$ of $A$ corresponds a subset $D$ of $B$ subject to conditions

I. $C \underset{f}{=} D$,
II. if $C \neq A$, then $D \neq B$.

**Corollary 7′.** If $A \underset{d}{=} B$, then to each subset $C$ of $A$ corresponds a subset $D$ of $B$ subject to conditions

I. $C \underset{d}{=} D$,
II. if $C \neq A$, then $D \neq B$.

Theorems 8 and 8′, which we shall establish now, will play an important role in the arguments of the following subsections.

**Theorem 8.** If $A_1 \subseteq A$, $B_1 \subseteq B$, $A \underset{f}{=} B_1$, and $B \underset{f}{=} A_1$, then $A \underset{f}{=} B$.

**Theorem 8′.** If $A_1 \subseteq A$, $B_1 \subseteq B$, $A \underset{d}{=} B_1$, and $B \underset{d}{=} A_1$, then $A \underset{d}{=} B$.

*Proof.* In conformity with theorems 4–5′, the two relations under consideration in this work, equivalence by finite and by denumerable decomposition, possess properties ($\alpha$) and ($\beta$) defined in Banach 1924, 236. Therefore theorems 8 and 8′ are just immediate consequences of the theorem 3 established in that note, which concerns all relations possessing those two properties.

**Corollary 9.** If $A \supseteq B \supseteq C$ and $A \underset{f}{=} C$, then $A \underset{f}{=} B$ and $B \underset{f}{=} C$.

**Corollary 9′.** If $A \supseteq B \supseteq C$ and $A \underset{d}{=} C$, then $A \underset{d}{=} B$ and $B \underset{d}{=} C$.

Those corollaries follow directly from the preceding theorems if one replaces there $A_1$ by $B$ as well as $B_1$ by $C$ and applies in turn theorems 1 and 3, or 1′ and 3′, respectively.

**Theorem 10.** If $A \underset{f}{=} A \cup B_k$ whenever $1 \le k \le n$, then

$$A \underset{f}{=} A \cup \bigcup_{k=1}^{n} B_k.$$

*Proof.* We consider two cases.

(a) [The case in which] the sets $A, B_1, B_2, \ldots, B_n$ are disjoint. We shall proceed by induction. The theorem being evident for $n = 1$, we suppose it is true for $n = n'$ and prove that it holds as well for $n = n' + 1$. Thus one has

$$A \underset{f}{=} A \cup \bigcup_{k=1}^{n'} B_k, \tag{1}$$

$$A \underset{f}{=} A \cup B_{n'+1}, \tag{2}$$

$$A \cap B_{n'+1} = \phi = B_{n'+1} \cap \bigcup_{k=1}^{n'} B_k. \tag{3}$$

In conformity with theorem 4, one obtains from (1) and (3),

$$A \cup B_{n'+1} \underset{f}{=} A \cup \bigcup_{k=1}^{n'+1} B_k. \tag{4}$$

From (2) and (4) it follows immediately, by virtue of theorem 3, that

$$A \underset{f}{=} A \cup \bigcup_{k=1}^{n'+1} B_k, \qquad \text{Q. E. D.}$$

(b) The general case. We set

$$B_1' = B_1 - A, \quad B_k' = B_k - \left(A \cup \bigcup_{l=1}^{k-1} B_l\right) \quad \text{whenever} \quad 2 \le k \le n. \tag{5}$$

From this one concludes without pain,

$$A \cup \bigcup_{k=1}^{n} B_k' = A \cup \bigcup_{k=1}^{n} B_k, \tag{6}$$

$$A \cup B_k' \subseteq A \cup B_k \quad \text{whenever} \quad 1 \le k \le n. \tag{7}$$

Now we apply corollary 9, replacing there $A$ by $A \cup B_k$, $B$ by $A \cup B_k'$, and $C$ by $A$. One obtains from (7) and the hypothesis of the theorem,

$$A \underset{f}{=} A \cup B_k' \quad \text{whenever} \quad 1 \le k \le n. \tag{8}$$

By virtue of (5), the sets $A, B_1', B_2', \ldots, B_n'$ are disjoint. Case (a) already established, one deduces from (8) and (6) that

$$A \underset{f}{=} \bigcup_{k=1}^{n} B_k' = A \cup \bigcup_{k=1}^{n} B_k.$$

Theorem 10 is thus completely demonstrated.

## 6.1 §1 General Properties of Equivalence

**Theorem 10'.** If $A \overline{\overline{d}} A \cup B_k$ when $1 \leq k \leq n$, then
$$A \overline{\overline{d}} A \cup \bigcup_{k=1}^{n} B_k.$$

Theorem 10' can be extended to the case of a denumerable infinity of summands [sets $B_k$]; but that would require a special demonstration, rather more complicated.

The theorem to which we now pass is all-important in the arguments of §2C.

**Theorem 11.** If $A_1 \overline{\overline{f}} A_2$, $B_1 \overline{\overline{f}} B_2$, $A_1 \cup A_2 \overline{\overline{f}} B_1 \cup B_2$, and $A_1 \cap A_2 = \phi = B_1 \cap B_2$, then
$$A_1 \overline{\overline{f}} B_1.$$

*Proof.* Kazimierz Kuratowski has established a general theorem concerning reflexive, symmetric, and transitive relations that possess properties ($\alpha$) and ($\beta$).[6] In case $A_1 \cup A_2 = B_1 \cup B_2$, theorem 11 follows immediately from that. To pass to the general case, we note that by virtue of corollary 6, the set $A_1 \cup A_2$ splits into two disjoint subsets,
$$A_1 \cup A_2 = B_1' \cup B_2',$$
so that
$$B_1' \overline{\overline{f}} B_1, \quad B_2' \overline{\overline{f}} B_2. \tag{1}$$
Since $B_1 \overline{\overline{f}} B_2$, one obtains from (1), following theorem 3,
$$B_1' \overline{\overline{f}} B_2';$$
by virtue of the preceding case, one can thus conclude that
$$A_1 \overline{\overline{f}} B_1'. \tag{2}$$
Formulas (1) and (2) immediately yield
$$A_1 \overline{\overline{f}} B_1, \qquad \text{Q. E. D.}$$

**Theorem 11'.** If $A_1 \overline{\overline{d}} A_2$, $B_1 \overline{\overline{d}} B_2$, $A_1 \cup A_2 \overline{\overline{d}} B_1 \cup B_2$, and $A_1 \cap A_2 = \phi = B_1 \cap B_2$, then
$$A_1 \overline{\overline{d}} B_1.$$

In corollaries 12 and 12' we shall give an easy generalization of the preceding theorems.

**Corollary 12.** If $A_1, A_2, \ldots, A_{2^n}$ as well as $B_1, B_2, \ldots, B_{2^n}$ should be disjoint sets, and

 I. $A_1 \overline{\overline{f}} A_k$ and $B_1 \overline{\overline{f}} B_k$ when $1 \leq k \leq 2^n$,

 II. $\bigcup_{k=1}^{2^n} A_k \overline{\overline{f}} \bigcup_{k=1}^{2^n} B_k,$

then
$$A_1 \overline{\overline{f}} B_1.$$

---

[6] Kuratowski 1924, 243. See the demonstration for theorems 8 and 8'.

*Proof.* By virtue of theorem 11, the corollary is true in case $n = 1$. To apply the principle of induction, we suppose that the corollary holds for $n = n'$ and prove that it holds also for $n = n' + 1$. Thus one has[7]

$$A_1 \underset{f}{=} A_k \text{ and } B_1 \underset{f}{=} B_k \text{ when } 1 \leq k \leq 2^{n'+1}, \tag{1}$$

$$\bigcup_{k=1}^{2^{n'+1}} A_k \underset{f}{=} \bigcup_{k=1}^{2^{n'+1}} B_k. \tag{2}$$

We set

$$A' = \bigcup_{k=1}^{2^{n'}} A_k, \quad A'' = \bigcup_{k=2^{n'}+1}^{2^{n'+1}} A_k, \quad B' = \bigcup_{k=1}^{2^{n'}} B_k, \quad B'' = \bigcup_{k=2^{n'}+1}^{2^{n'+1}} B_k. \tag{3}$$

Following theorem 4, from (1)–(3) one concludes immediately,

$$A' \underset{f}{=} A'', \quad B' \underset{f}{=} B'', \quad A' \cup A'' \underset{f}{=} B' \cup B''. \tag{4}$$

Sets $A'$ and $A''$ as well as $B'$ and $B''$ being disjoint, one deduces from theorem 11 according to (4),

$$A' \underset{f}{=} B'. \tag{5}$$

Formulas (1), (3), and (5) prove that the sets $A_1, A_2, \ldots, A_{2^{n'}}$ and $B_1, B_2, \ldots, B_{2^{n'}}$ satisfy the hypothesis of the theorem. In conformity with our assumption one thus obtains

$$A_1 \underset{f}{=} B_1, \qquad \text{Q. E. D.}$$

**Corollary 12'.** If $A_1, A_2, \ldots, A_{2^n}$ as well as $B_1, B_2, \ldots, B_{2^n}$ should be disjoint sets, and

I. $A_1 \underset{d}{=} A_k$ and $B_1 \underset{d}{=} B_k$ when $1 \leq k \leq 2^n$,

II. $\bigcup_{k=1}^{2^n} A_k \underset{d}{=} \bigcup_{k=1}^{2^n} B_k,$

then

$$A_1 \underset{d}{=} B_1.$$

**Theorem 13.** If $A \underset{f}{=} B$ and $A$ belongs to a class $K$ of sets that satisfies conditions

I. when $X \in K$ and $Y \in K$ one has $X \cup Y \in K$,

II. when $X \in K$ and $Y \subseteq X$ one has $Y \in K$,

III. when $X \in K$ and $Y \cong X$ one has $Y \in K$,

then the set $B$ also belongs to the class $K$.

*Proof.* In conformity with definition 2, let

$$A = \bigcup_{k=1}^{n} A_k, \quad B = \bigcup_{k=1}^{n} B_k \tag{1}$$

---

[7] [Evidently in error, Banach and Tarski wrote $n_1$ for $n'$ after this in the proof. The translation uses $n'$.]

be a decomposition of sets $A$ and $B$ into disjoint, respectively congruent parts. According to (1), condition II of the hypothesis of the theorem implies

$$A_k \in \boldsymbol{K} \quad \text{whenever} \quad 1 \leq k \leq n.$$

From this, and from condition III, follows

$$B_k \in \boldsymbol{K} \quad \text{whenever} \quad 1 \leq k \leq n. \tag{2}$$

Since condition I can be extended by a simple induction to the case of a union of an arbitrary finite number of sets, one concludes by virtue of (1) and (2) that

$$B \in \boldsymbol{K}, \qquad \text{Q. E. D.}$$

**Theorem 13'.** If $A \underset{d}{=} B$ and $A$ belongs to a class $\boldsymbol{K}$ of sets that satisfies conditions

  I. when $X_n \in \boldsymbol{K}$ for every natural number $n$, one has $\bigcup_{n=1}^{\infty} X_n \in \boldsymbol{K}$,
  II. when $X \in \boldsymbol{K}$ and $Y \subseteq X$, one has $Y \in \boldsymbol{K}$,
  III. when $X \in \boldsymbol{K}$ and $Y \cong X$, one has $Y \in \boldsymbol{K}$,

then the set $B$ also belongs to the class $\boldsymbol{K}$.

The two preceding theorems directly imply the following corollaries: $A$ and $B$ being sets situated in a Euclidean space with an arbitrary number of dimensions, one has

**Corollary 14.** If $A \underset{f}{=} B$ and $A$ is nowhere dense,[8] then $B$ is also nowhere dense.

**Corollary 14'.** If $A \underset{d}{=} B$ and $A$ belongs to the first category in the sense of Baire, then $B$ also belongs to the first category.

**Corollary 15.** If $A \underset{f}{=} B$ and $A$ is a set measurable in the Peano–Jordan sense, with measure zero, then $B$ is also a set measurable in the same sense, with measure zero.

**Corollary 15'.** If $A \underset{d}{=} B$ and $A$ is a Lebesgue-measurable set with measure zero, then $B$ is also a Lebesgue-measurable set with measure zero.

In an analogous way one proves that if $A \underset{f}{=} B$ and $A$ is a bounded set, then $B$ is also bounded.

---

[8] A (point) set $A$ is called a *boundary* set if it contains no interior point. A set $A$ is called *nowhere dense* [*non-dense* in the original] if the set consisting of all the points of $A$ as well as the accumulation points of $A$ is a boundary set. A set $A$ is said to belong to the *first category in the sense of Baire* if it is the union of denumerably infinitely many nowhere dense sets.

## §2

## The Fundamental Theorems about Equivalence by Finite Decomposition

In the arguments of this section as well as the following one we consider point sets situated in a Euclidean space with a finite number of dimensions.

### A. Euclidean Space with One or Two Dimensions

The most important theorem of this subsection is theorem 16, which establishes a condition necessary for two Lebesgue-measurable linear or planar point sets to be equivalent by finite decomposition.

**Theorem 16.** If $A$ and $B$ should be bounded Lebesgue-measurable linear or planar sets and $A \underset{f}{=} B$, then $m(A) = m(B)$.[9]

*Proof.* By a theorem of Stefan Banach,[10] one can assign to each bounded set $A$ situated in Euclidean space with dimension one or two a nonnegative real number $f(A)$ in such a way that the following conditions should be satisfied:

  I. if $A \cong B$, then $f(A) = f(B)$,
 II. if $A \cap B = \phi$, then $f(A \cup B) = f(A) + f(B)$,
III. if $A$ is Lebesgue-measurable, then $f(A) = m(A)$.

In conformity with definition 2, the hypothesis of the theorem implies the existence of sets $A_1, A_2, \ldots, A_n$ and $B_1, B_2, \ldots, B_n$ that satisfy these formulas:

$$A = \bigcup_{k=1}^{n} A_k, \quad B = \bigcup_{k=1}^{n} B_k, \tag{1}$$

$$A_k \cong B_k \quad \text{whenever} \quad 1 \le k \le n, \tag{2}$$

$$A_k \cap A_l = \phi = B_k \cap B_l \quad \text{whenever} \quad 1 \le k < l \le n. \tag{3}$$

By virtue of condition I and formula (2) one obtains

$$f(A_k) = f(B_k) \quad \text{whenever} \quad 1 \le k \le n. \tag{4}$$

Condition II can be extended via an easy induction to the case of an arbitrary finite number of disjoint summands. One can thus conclude from (1) and (3),

$$f(A) = \bigcup_{k=1}^{n} f(A_k), \quad f(B) = \bigcup_{k=1}^{n} f(B_k). \tag{5}$$

Formulas (4) and (5) immediately yield

---

[9] $m(A)$ denotes the Lebesgue measure of a set $A$ (linear measure if one is considering a one-dimensional space, planar measure in the two-dimensional case, and so on).

[10] Banach 1923, 30–31 [discussed in section 4.2].

## 6.2 §2 Fundamental Theorems about Finite Decomposition

$$f(A) = f(B). \tag{6}$$

Following the hypothesis of the theorem, one finally deduces from III and (6) that

$$m(A) = m(B), \qquad \text{Q. E. D.}$$

The converse of the preceding theorem is not valid. This follows directly from corollary 14: two point sets $A$ and $B$, where the one is nowhere dense and the other is not, can have the same Lebesgue measure even though they are not equivalent by finite decomposition. Nevertheless, the converse does hold in a particularly simple case, notably in the case of polygons. To establish this, we prove first the following

**Lemma 17.** Should $A$ be a planar set that is not a boundary set,[11] and $B$ be the union of a finite number of segments, then $A \underset{f}{\neq} A \cup B$.

*Proof.* We consider two cases.

Case (a): $A \cap B = \phi$. Let $C$ be a disk satisfying the formula

$$C \subseteq A. \tag{1}$$

Evidently, the set $B$ can be decomposed into a finite number of (not necessarily disjoint) segments, each of which has length less than that of a radius of $C$:

$$B = \bigcup_{k=1}^{n} B_k. \tag{2}$$

We consider segment $B_k$ for an arbitrary $k$ such that $1 \le k \le n$. Let $D_1$ be a segment congruent to $B_k$ and situated on a radius of $C$, but not containing the center of the circle. We choose an angle $\alpha$ incommensurable with a right angle and, for each natural number $n$, designate by $D_{n+1}$ the segment obtained by turning segment $D_1$ through angle $n \cdot \alpha$ about the center of the disk (in a fixed sense).

We set[11.5]

$$E = \bigcup_{n=1}^{\infty} D_n, \tag{3}$$

$$F = \bigcup_{n=2}^{\infty} D_n, \tag{4}$$

$$G = A - E. \tag{5}$$

Since $E \subseteq C$, one obtains immediately from (1) and (3)–(5) the following decomposition of sets $A$ and $A \cup B_k$:

$$A = G \cup F \cup D_1, \quad A \cup B_k = G \cup E \cup B_k. \tag{6}$$

Evidently,

$$G \cong G, \quad F \cong E, \quad D_1 \cong B_k, \tag{7}$$

---

[11] See page 105, footnote 8.

[11.5] [The starting indices of the following two unions were incorrectly specified in the original paper.]

because $F$ is obtained from $E$ by rotation through angle $\alpha$. Finally, one deduces easily from (3)–(5) as well as from the indicated property of angle $\alpha$ (incommensurability with a right angle) that formulas (6) effect a decomposition of the sets $A$ and $A \cup B_k$ into disjoint subsets. Then, in conformity with (6), (7), and definition 2,

(8) $$A \overset{=}{_f} A \cup B_k.$$

The same reasoning being valid for each segment $B_k$ when $1 \leq k \leq n$, one can apply theorem 10. According to (2), one obtains

$$A \overset{=}{_f} A \cup B.$$

Case (b): the general case. Apparently, the set $A - B$ is not a boundary set. Since $(A - B) \cap B = \phi$, by virtue of (a) one concludes that

$$A - B \overset{=}{_f} (A - B) \cup B = A \cup B.$$

This formula and the evident inclusion

$$A \cup B \supseteq A \supseteq A - B$$

imply, following corollary 9, that

$$A \overset{=}{_f} A \cup B, \qquad \text{Q. E. D.}$$

Thanks to a remark of Adolf Lindenbaum, one can state a theorem for linear sets, analogous to the preceding lemma, by replacing the term *segment* by *point*.

**Theorem 19**. If polygons $A$ and $B$ have the same area, one has

$$A \overset{=}{_f} B.$$

*Proof.* As one knows, polygons $A$ and $B$ are equivalent by decomposition in the sense of elementary geometry. That is, one can decompose them into the same number of respectively congruent polygons without common *interior* points. Let $A_1, A_2, \ldots, A_n$ and $B_1, B_2, \ldots, B_n$ be the interiors of these polygonal parts.[12] Evidently,

$$\bigcup_{k=1}^{n} A_k \overset{=}{_f} \bigcup_{k=1}^{n} B_k. \qquad (1)$$

Since the sets

$$A - \bigcup_{k=1}^{n} A_k \quad \text{and} \quad B - \bigcup_{k=1}^{n} B_k$$

are composed of a finite number of segments, one concludes, by applying the preceding lemma, that

$$\bigcup_{k=1}^{n} A_k \overset{=}{_f} \bigcup_{k=1}^{n} A_k \cup \left( A - \bigcup_{k=1}^{n} A_k \right) = A \quad \text{and similarly} \quad \bigcup_{k=1}^{n} B_k \overset{=}{_f} B. \qquad (2)$$

Following theorem 3, one immediately obtains from (1) and (2),

$$A \overset{=}{_f} B, \qquad \text{Q. E. D.}$$

---

[12] [Banach and Tarski used here the term *polygones partiels*.]

Theorems 17 and 19 directly imply

**Corollary 20.** For two polygons to be equivalent by finite decomposition, it is necessary and sufficient that they should have the same area.

## B. Euclidean Space of Three (and More) Dimensions

The arguments of this part concern space of three dimensions. To extend the obtained results to a space of $n > 3$ dimensions, it would be necessary to consider, instead of spheres, the sets of all points $(x_1, x_2, \ldots, x_n)$, where these (rectangular) coordinates satisfy the conditions

$$(x_1 - a_1)^2 + (x_2 - a_2)^2 + (x_3 - a_3)^2 = \rho^2,$$
$$b \leq x_k \leq c \quad \text{whenever} \quad 3 < k \leq n,$$

where $a_1$, $a_2$, $a_3$, $\rho$, $b$, and $c$ are constants.

To establish the principal result of this work, theorem 24, we shall first demonstrate some lemmas.

**Lemma 21.** Every sphere $S$ contains two disjoint subsets $A_1$ and $A_2$ such that

$$S \overset{=}{_f} A_1 \quad \text{and} \quad S \overset{=}{_f} A_2.$$

*Proof.* According to the famous theorem known as *Hausdorff's paradox*,[13] one can decompose the surface of the sphere $S$ into four disjoint subsets $B'$, $C'$, $D'$, and $E'$, where $E'$ is a denumerable set and sets $B'$, $C'$, and $D'$ satisfy the formulas

$$B' \cong C' \cup D', \quad B' \cong C' \cong D'.$$

Let $p$ be the center of the sphere $S$. We designate by $B$, $C$, $D$, and $E$ the unions of all the radii of the sphere $S$, excluding the center $p$, whose endpoints belong to $B'$, $C'$, $D'$, and $E'$, respectively. In this way one evidently obtains a decomposition of the sphere $S$ into five disjoint parts,

$$S = B \cup C \cup D \cup E \cup \{p\}, \tag{1}$$

subject to conditions

$$B \cong C \cup D, \tag{2}$$
$$B \cong C \cong D. \tag{3}$$

As far as the set $E$ is concerned, we shall use here only the following property, already noted by Felix Hausdorff:

there is a proper subset $F$ of $B \cup C \cup D$ such that $E \cong F$. $\qquad(4)$

---

[13] Hausdorff 1914a, 469.

It is easy to convince oneself of this, by making the sphere $S$ turn suitably about one of its axes.

From (2) and (3), without effort, one obtains
$$B \underset{f}{\equiv} B \cup C, \quad B \cup C \underset{f}{\equiv} B \cup C \cup D,$$
from which [follows], by virtue of theorem 3,
$$B \underset{f}{\equiv} B \cup C \cup D. \tag{5}$$
We set
$$A_1 = B \cup E \cup \{p\}. \tag{6}$$
By theorem 4, formulas (1), (5), and (6) yield
$$S \underset{f}{\equiv} A_1. \tag{7}$$

On the other hand, it results immediately from (3) and (5) that
$$C \underset{f}{\equiv} B \cup C \cup D, \tag{8}$$
$$D \underset{f}{\equiv} B \cup C \cup D; \tag{9}$$
applying corollary 7, one deduces from (4) and (8) the existence of a set $G$ that satisfies the formulas
$$F \underset{f}{\equiv} G, \quad \text{which yields} \quad E \underset{f}{\equiv} G; \tag{10}$$
$$G \subseteq C \quad \text{and} \quad G \neq C. \tag{11}$$
Conforming with (11), let
$$q \in C - G, \tag{12}$$
and then let us set
$$A_2 = D \cup G \cup \{q\}. \tag{13}$$

Sets $C$ and $D$ being disjoint, one deduces from (11) and (12) that sets $D$, $G$, and $\{q\}$ are disjoint, too. Now, the sets $\{p\}$ and $\{q\}$ being evidently congruent, one concludes according to (1), (9), (10), and (13),
$$S \underset{f}{\equiv} A_2. \tag{14}$$
Finally, one easily obtains
$$A_1 \cup A_2 \subseteq S \quad \text{and} \quad A_1 \cap A_2 = \phi. \tag{15}$$

Formulas (7), (14), and (15) prove that $A_1$ and $A_2$ are sets such as those sought.

**Lemma 22.** If $S_1$ and $S_2$ should be congruent spheres, then
$$S_1 \underset{f}{\equiv} S_1 \cup S_2.$$

*Proof.* Conforming with the preceding lemma, let $A_1$ and $A_2$ be sets subject to the conditions

$$S_1 \overset{=}{\underset{f}{}} A_1, \quad S_1 \overset{=}{\underset{f}{}} A_2, \tag{1}$$

$$A_1 \cup A_2 \subseteq S_1 \quad \text{and} \quad A_1 \cap A_2 = \phi. \tag{2}$$

By virtue of (1) and the hypothesis of this lemma,

$$S_2 \overset{=}{\underset{f}{}} A_2.$$

Corollary 7 then implies the existence of a set $B$ such that

$$B \subseteq A_2, \tag{3}$$

$$B \overset{=}{\underset{f}{}} S_2 - S_1. \tag{4}$$

According to theorem 4, one concludes easily from (1)–(4) that

$$A_1 \cup B \overset{=}{\underset{f}{}} S_1 \cup (S_2 - S_1) = S_1 \cup S_2, \tag{5}$$

$$A_1 \cup B \subseteq S_1 \subseteq S_1 \cup S_2. \tag{6}$$

Formulas (5) and (6) yield immediately, by virtue of corollary 9,

$$S_1 \overset{=}{\underset{f}{}} S_1 \cup S_2, \qquad \text{Q. E. D.}$$

**Lemma 23.** If a bounded set $A$, situated in Euclidean space of three dimensions, contains a sphere $S$, then

$$A \overset{=}{\underset{f}{}} S.$$

*Proof.* Since $A$ is a bounded set, one can obviously decompose it into $n$ (not necessarily disjoint) subsets,

$$A = \bigcup_{k=1}^{n} B_k, \tag{1}$$

each of which, when $1 \leq k \leq n$, fulfills the condition

$B_k$ is contained in a sphere $S_k$ congruent to $S$. (2)

By virtue of the preceding lemma,

$$S \overset{=}{\underset{f}{}} S \cup S_k \quad \text{whenever} \quad 1 \leq k \leq n,$$

which, according to theorem 10, implies that

$$S \overset{=}{\underset{f}{}} S \cup \bigcup_{k=1}^{n} S_k. \tag{3}$$

On the other hand, one obtains from (1), (2), and the hypothesis of this lemma,

$$S \subseteq A \subseteq S \cup \bigcup_{k=1}^{n} S_k. \tag{4}$$

By reason of corollary 9, formulas (3) and (4) directly imply

$$A \overset{=}{\underset{f}{}} S, \qquad \text{Q. E. D.}$$

The lemma just demonstrated now permits us to establish

**Theorem 24**. If two arbitrary sets $A$ and $B$, situated in Euclidean space of three dimensions, are bounded and are not boundary sets, then
$$A \overset{=}{_f} B.$$

*Proof.* Let $S_1$ and $S_2$ be spheres contained in $A$ and $B$, respectively; evidently, one can assume that
$$S_1 \overset{=}{_f} S_2. \tag{1}$$
By virtue of lemma 23 one obtains
$$A \overset{=}{_f} S_1 \quad \text{and} \quad B \overset{=}{_f} S_2. \tag{2}$$
According to theorems 1 and 3, one concludes immediately from (1) and (2) that
$$A \overset{=}{_f} B, \qquad \qquad \text{Q. E. D.}$$

Thus one sees in particular that two spheres with different radii are equivalent by finite decomposition, whereas, as we have proved earlier, two circles are equivalent only when their radii are equal. This essential difference between spaces of two and three dimensions is intimately related to the fact that the problem of measure has a positive solution in the first case and negative in the second.

### C. The Surface of a Sphere

Theorem 31, which is fundamental in this part of our investigation, shows from the viewpoint of equivalence by finite decomposition that the surface of a sphere behaves in a way fully analogous to three-dimensional space.

**Lemma 25**. If $A$ and $B$ are sets situated on the surface of the same sphere, $A$ is not a boundary set (with respect to this sphere) and $B$ is composed of a finite number of arcs of great circles, then
$$A \overset{=}{_f} A \cup B.$$
The *proof* of this is completely analogous to that of lemma 17.

**Lemma 26**. If spherical polygons $A$ and $B$, situated on the surface of the same sphere, have equal areas, then
$$A \overset{=}{_f} B.$$

The *proof* is based on the preceding lemma and does not differ from that of theorem 19; one utilizes the familiar theorem according to which two spherical polygons, situated on the surface of the same sphere and possessing equal areas, are equivalent by finite decomposition in the sense of elementary geometry.[14]

---

[14] Gerwien 1833b. [This citation was incorrect in the original.]

## 6.2 §2 Fundamental Theorems about Finite Decomposition

**Lemma 27.** The surface $S$ of any sphere can be decomposed into two disjoint subsets $A_1$ and $A_2$ such that one has $S \overline{\overline{f}} A_1$ and $S \overline{\overline{f}} A_2$.

*Proof.* By reasoning as in the demonstration of lemma 21, one proves the existence of sets $A_1'$ and $A_2$ satisfying formulas

$$S \overline{\overline{f}} A_1', \quad S \overline{\overline{f}} A_2, \tag{1}$$

$$A_1' \cup A_2 \subseteq S \quad \text{and} \quad A_1' \cap A_2 = \phi. \tag{2}$$

We set

$$A_1 = S - A_2; \tag{3}$$

from (2) and (3) one obtains without pain,

$$S = A_1 \cup A_2, \quad A_1 \cap A_2 = \phi, \tag{4}$$

$$S \supseteq A_1 \supseteq A_1'.^{15} \tag{5}$$

By virtue of (1) and (5) one concludes again, by applying corollary 9, that

$$S \overline{\overline{f}} A_1 \quad \text{and} \quad S \overline{\overline{f}} A_2. \tag{6}$$

Formulas (4) and (6) show that $A_1$ and $A_2$ are sets such as those desired.

With the aid of corollary 6, this lemma is generalized by an easy induction in the following way:

**Lemma 28.** If $n$ is an arbitrary natural number, the surface $S$ of any sphere can be decomposed into $n$ disjoint subsets $A_1, A_2, \ldots, A_n$ such that $S \overline{\overline{f}} A_k$ whenever $1 \leq k \leq n$.

**Lemma 29.** If $n$ is an arbitrary natural number and the surface $S$ of a sphere is decomposed into $2^n$ congruent spherical polygons $B_1, B_2, \ldots, B_{2^n}$ without common interior points, then $S \overline{\overline{f}} B_1$.

*Proof.* We set

$$B_1' = B_1, \quad B_k' = B_k - \bigcup_{l=1}^{k-1} B_l \quad \text{whenever} \quad 2 \leq k \leq 2^n. \tag{1}$$

One evidently obtains a decomposition of $S$ into $2^n$ disjoint subsets:

$$S = \bigcup_{k=1}^{2^n} B_k'. \tag{2}$$

Since each set $B_k'$ with $1 \leq k \leq 2^n$ contains some interior points (with respect to the surface $S$) and the set $B_k - B_k'$ is composed of a finite number of arcs of great circles (which might be reduced to a single point), one concludes, according to lemma 25, that

$$B_k' \overline{\overline{f}} B_k' \cup (B_k - B_k') = B_k. \tag{3}$$

There results from this immediately, by virtue of the hypothesis of the theorem,

$$B_1' \overline{\overline{f}} B_k' \quad \text{whenever} \quad 1 \leq k \leq 2^n. \tag{4}$$

---

[15] [Banach and Tarski omitted the ' from $A_1'$ here.]

On the other hand, conforming with lemma 28, let
$$S = \bigcup_{k=1}^{2^n} A_k \tag{5}$$
be a decomposition of the sphere $S$ into some disjoint subsets such that
$$S \overset{=}{\underset{f}{}} A_k, \quad \text{and thus} \quad A_1 \overset{=}{\underset{f}{}} A_k, \quad \text{whenever} \quad 1 \le k \le n. \tag{6}$$
By (2) and (4)–(6) the sets $A_1, A_2, \ldots, A_{2^n}$ and $B_1', B_2', \ldots, B_{2^n}'$ fulfill all the conditions of corollary 12; thus one obtains
$$A_1 \overset{=}{\underset{f}{}} B_1'. \tag{7}$$
Formulas (1), (6), and (7) imply immediately that
$$S \overset{=}{\underset{f}{}} B_1, \tag*{Q. E. D.}$$

**Lemma 30.** If the set $A$, situated on the surface $S$ of a sphere, is not a boundary set (with respect to this sphere), then $A \overset{=}{\underset{f}{}} S$.

*Proof.* One proves easily that the set $A$ contains a spherical polygon $A_1$ whose area is $4\pi\rho^2/2^n$, where $\rho$ designates the length of a radius and $n$ is a sufficiently large natural number. We decompose $S$ into $2^n$ congruent polygons without common interior points:
$$S = \bigcup_{k=1}^{2^n} B_k.$$
According to the preceding lemma, one obtains
$$S \overset{=}{\underset{f}{}} B_1. \tag{1}$$
The spherical polygons $A_1$ and $B_1$ having the same area, one concludes by applying lemma 26 that
$$A_1 \overset{=}{\underset{f}{}} B_1. \tag{2}$$
From (1) and (2) this results immediately:
$$S \overset{=}{\underset{f}{}} A_1. \tag{3}$$
On the other hand,
$$S \supseteq A \supseteq A_1. \tag{4}$$
Conforming with corollary 9, formulas (3) and (4) yield
$$A \overset{=}{\underset{f}{}} S, \tag*{Q. E. D.}$$

Lemma 30 established, the proof of the fundamental theorem 31 is evident:

**Theorem 31.** If point sets $A$ and $B$, situated on the surface of the same sphere, are not boundary sets (with respect to this surface), one has
$$A \overset{=}{\underset{f}{}} B.$$

## §3

## The Fundamental Theorems about Equivalence by Denumerable Decomposition

The arguments of this section concern Euclidean spaces with an arbitrary number $n$ of dimensions; but to fix ideas, we shall operate in the space of $n = 1$ or $n = 2$ dimensions.

**Lemma 32.** If $A_1, A_2, \ldots, A_m, \ldots$ are disjoint $n$-dimensional intervals[16] congruent to a single interval $A$, then

$$A \underset{d}{=} \bigcup_{m=1}^{\infty} A_m.$$

*Proof.* Consider the case of dimension $n = 1$. As Felix Hausdorff has shown[17] (using an idea of Giuseppe Vitali), one can decompose every segment into a denumerable infinity of disjoint subsets equivalent in pairs by finite decomposition. Let these be the decompositions of the segments $A, A_1, \ldots, A_m, \ldots$:

$$A = \bigcup_{k=1}^{\infty} B_k, \tag{1}$$

$$A_m = \bigcup_{k=1}^{\infty} B_{m,k} \quad \text{for each natural number} \quad m. \tag{2}$$

All these segments being congruent, one can evidently suppose that

$$B_k \cong B_{m,k} \quad \text{for all natural numbers} \quad k \text{ and } m,$$

from which follows

$$B_k \underset{f}{=} B_{m,l} \quad \text{for all natural numbers} \quad k, l, \text{ and } m. \tag{3}$$

One concludes from (2) that

$$\bigcup_{m=1}^{\infty} A_m = \bigcup_{m=1}^{\infty} \bigcup_{k=1}^{\infty} B_{m,k}. \tag{4}$$

Since any double series can be transformed by the method of diagonals into a simple series, formulas (1) and (4) furnish decompositions of the sets

$$A \quad \text{and} \quad \bigcup_{m=1}^{\infty} A_m$$

into denumerable infinities[18] of disjoint parts, which according to (3) are respectively equivalent by finite decomposition. By virtue of theorem 4' one deduces from this that

---

[16] A point set is called an $n$-dimensional interval if it is composed of all points $(x_1, x_2, \ldots, x_n)$ subject to the condition $a \leq x_k \leq b$ whenever $1 \leq k \leq n$, where $a$ and $b$ are constants.

[17] Hausdorff 1914a, 401. Strictly speaking, Hausdorff decomposes not the whole segment, but the segment without one endpoint. However, this objection, which one can avoid, has only an insignificant influence on the arguments that will follow.

[18] [The original was singular here: "(1) et (4) fournissent une décomposition...en une infinité...."]

$$A \overset{=}{_d} \bigcup_{m=1}^{\infty} A_m,$$

which proves the theorem for the case of linear space.

To pass to the case of $n$-dimensional space for $n > 1$, it suffices to replace the points of each decomposed segment by $(n-1)$-dimensional intervals perpendicular to it.

**Lemma 33.** If $A_1, A_2, \ldots, A_m, \ldots$ are disjoint $n$-dimensional intervals congruent to each other, and $E$ denotes the entire $n$-dimensional space, then

$$E \overset{=}{_d} \bigcup_{m=1}^{\infty} A_m.$$

*Proof.* Consider the case $n = 2$. The plane $E$ can be decomposed easily into a denumerable infinity of (not necessarily disjoint) squares,

$$E = \bigcup_{m=1}^{\infty} B_m, \tag{1}$$

such that

$$A_m \cong B_m \quad \text{for every natural number } m. \tag{2}$$

Set

$$C_1 = B_1, \quad C_m = B_m - \bigcup_{k=1}^{m-1} B_k \quad \text{for each } m \geq 2. \tag{3}$$

According to (1) and (3) one easily obtains

$$E = \bigcup_{m=1}^{\infty} C_m, \tag{4}$$

$$B_m \supseteq C_m \quad \text{for each natural number } m. \tag{5}$$

Formulas (2) and (5) evidently imply the existence of sets $D_1, D_2, \ldots, D_m, \ldots$ satisfying the formulas

$$C_m \cong D_m, \tag{6}$$

$$A_m \supseteq D_m \quad \text{for each natural number } m. \tag{7}$$

Sets $C_1, C_2, \ldots, C_m, \ldots$ as well as $D_1, D_2, \ldots, D_m, \ldots$ being disjoint, one concludes from (4) and (6) in conformity with definition 2' that

$$E \overset{=}{_d} \bigcup_{m=1}^{\infty} D_m. \tag{8}$$

From (7) one also deduces

$$E \supseteq \bigcup_{m=1}^{\infty} A_m \supseteq \bigcup_{m=1}^{\infty} D_m. \tag{9}$$

By virtue of corollary 9', formulas (8) and (9) immediately yield

$$E \underset{d}{=} \bigcup_{m=1}^{\infty} A_m, \qquad \text{Q. E. D.}$$

**Lemma 34.** If the set $A$, situated in $n$-dimensional space $E$, is not a boundary set, then $A \underset{d}{=} E$.

*Proof.* Suppose as before that $n = 2$. The set $A$ evidently contains a square $A'$. Let $A_1, A_2, \ldots, A_m, \ldots$ be disjoint squares, congruent to $A'$ and situated in the same plane $E$: the existence of such squares in the plane is manifest. By applying the two preceding lemmas, one immediately obtains

$$A' \underset{d}{=} \bigcup_{m=1}^{\infty} A_m \quad \text{and} \quad E \underset{d}{=} \bigcup_{m=1}^{\infty} A_m,$$

from which, by virtue of theorem 3', follows

$$A' \underset{d}{=} E.$$

Since at the same time,

$$E \supseteq A \supseteq A',$$

one concludes, conforming to corollary 9', that

$$A \underset{d}{=} E, \qquad \text{Q. E. D.}$$

Lemma 34 established, one immediately deduces from it the fundamental theorem of this section, namely

**Theorem 35.**[19] If sets $A$ and $B$, situated in a Euclidean space of an arbitrary number of dimensions, are not boundary sets, one has

$$A \underset{d}{=} B.$$

Now we shall generalize the notion of equivalence by denumerable decomposition, by introducing the following definition:

**Definition 3.** Point sets $A$ and $B$ are *almost equivalent by denumerable decomposition*,

$$A \underset{p}{=} B,$$

if there exist sets $A_1$, $A_2$, $B_1$, and $B_2$ satisfying the following conditions:

 I. $A = A_1 \cup A_2$, $B = B_1 \cup B_2$, $A_1 \cap A_2 = \phi = B_1 \cap B_2$,
 II. $A_1 \underset{d}{=} B_1$,
 III. $A_2$ and $B_2$ are Lebesgue-measurable[20] and $m(A_2) = m(B_2) = \phi$.

---

[19] A particular case of this theorem has been announced in Sierpiński 1918, 142. [There, $A$ and $B$ were squares of different sizes.]

[20] In the arguments that will follow, we suppose that the notion of measure has been extended to unbounded sets. See Hausdorff 1914a, 416.

The relation just defined is evidently *reflexive* and *symmetric*; we shall prove that it is also *transitive*.

**Theorem 36.** If $A \underset{p}{=} B$ and $B \underset{p}{=} C$, then $A \underset{p}{=} C$.

*Proof.* Let
$$A = A_1 \cup A_2, \quad B = B_1 \cup B_2, \tag{1}$$
$$B = B_1' \cup B_2', \quad C = C_1 \cup C_2, \tag{2}$$

respectively, be decompositions of sets $A$ and $B$, and of $B$ and $C$, satisfying the conditions of definition 3. From (1) and (2) stem immediately the following formulas:
$$B_1 = (B_1 \cap B_1') \cup (B_1 \cap B_2'), \quad B_1' = (B_1' \cap B_1) \cup (B_1' \cap B_2).$$

Since $A_1 \underset{d}{=} B_1$ and $B_1' \underset{d}{=} C_1$, one then concludes from corollary 6 that sets $A_1$ and $C_1$ can be decomposed into disjoint parts,
$$A_1 = A' \cup A_3, \quad C_1 = C' \cup C_3, \tag{3}$$

such that
$$A' \underset{d}{=} B_1 \cap B_1', \quad C' \underset{d}{=} B_1' \cap B_1, \tag{4}$$
$$A_3 \underset{d}{=} B_1 \cap B_2', \quad C_3 \underset{d}{=} B_1' \cap B_2. \tag{5}$$

Set
$$A'' = A_2 \cup A_3, \quad C'' = C_2 \cup C_3; \tag{6}$$

according to (1)–(3) and (6), one obtains
$$A = A' \cup A'', \quad C = C' \cup C'', \tag{7}$$

and from this one can easily be convinced that sets $A'$ and $A''$ as well as $C'$ and $C''$ are disjoint. From (4) one deduces further,
$$A' \underset{d}{=} C'. \tag{8}$$

Finally, one can prove that sets $A''$ and $C''$ have measure zero. Conforming to the properties of decompositions (1) and (2) one has, in effect,
$$m(A_2) = m(B_2) = m(B_2') = m(C_2) = 0. \tag{9}$$

Since $B_1 \cap B_2' \subseteq B_2'$ and $B_1' \cap B_2 \subseteq B_2$, it follows from this that
$$m(B_1 \cap B_2') = m(B_1' \cap B_2) = 0. \tag{10}$$

Applying corollary 15', according to (5) and (10) one concludes that
$$m(A_3) = m(C_3) = 0, \tag{11}$$

and from (6), (9), and (11) one finally obtains
$$m(A'') = m(C'') = 0. \tag{12}$$

According to definition 3, formulas (6), (8), and (12) imply that
$$A \underset{p}{=} C, \qquad\qquad \text{Q. E. D.}$$

## 6.3 §3 Fundamental Theorems about Denumerable Decomposition

The fundamental theorem on denumerable equivalence manifestly entails the following consequence:

*If sets $A$ and $B$, situated in a Euclidean space, are not boundary sets, then $A \underset{p}{=} B$.*

We propose to give in theorem 41 a generalization of this proposition.

**Lemma 37.** If $A$ and $B$ are sets situated in an $n$-dimensional Euclidean space, $A$ is Lebesgue-measurable, $B$ is an open set, and $m(A) = m(B)$, then to each positive real number $\delta$ correspond two closed sets $A_1$ and $B_1$ such that

I. $A_1 \subseteq A$ and $B_1 \subseteq B$,
II. $A_1 \underset{f}{=} B_1$,
III. $m(A_1) = m(B_1) > m(A) - \delta$.

*Proof* [for the case] $n = 2$. According to a familiar theorem in measure theory,[21] there certainly exists a closed bounded set $A'$ satisfying the formulas

$$A' \subseteq A \quad \text{and} \quad m(A) > m(A') > m(A) - \delta. \tag{1}$$

Since $m(A') < m(B)$, one can prove the existence of squares $C_1, C_2, \ldots, C_m$ and $D_1, D_2, \ldots, D_m$ that satisfy the following conditions ($C_k^\circ$ and $D_k^\circ$ denote the interiors of squares $C_k$ and $D_k$, respectively):

$$\text{the sets} \quad C_1^\circ, C_2^\circ, \ldots, C_m^\circ \quad \text{as well as} \quad D_1^\circ, D_2^\circ, \ldots, D_m^\circ \quad \text{are disjoint,} \tag{2}$$

$$C_k \cong D_k \quad (\text{which implies} \quad C_k^\circ \cong D_k^\circ) \quad \text{whenever} \quad 1 \le k \le m, \tag{3}$$

$$A' \subseteq \bigcup_{k=1}^{m} C_k \quad \text{and} \quad \bigcup_{k=1}^{m} D_k \subseteq B. \tag{4}$$

From (1), (2), and (3) follows the existence of a closed set $A_1$ such that one has

$$A_1 \subseteq A' \subseteq A, \tag{5}$$

$$m(A_1) > m(A) - \delta, \tag{6}$$

$$A_1 \subseteq \bigcup_{k=1}^{m} C_k^\circ, \quad \text{and thus} \quad A_1 = \bigcup_{k=1}^{m} \left( A_1 \cap C_k^\circ \right). \tag{7}$$

Conforming to (3) and (7), let $E_1, E_2, \ldots, E_m$ be sets enjoying the following properties:

$$E_k \cong A_1 \cap C_k^\circ \quad \text{whenever} \quad 1 \le k \le m, \tag{8}$$

$$E_k \subseteq D_k^\circ \quad \text{whenever} \quad 1 \le k \le m, \tag{9}$$

and set

$$B_1 = \bigcup_{k=1}^{m} E_k. \tag{10}$$

According to (4), (9), and (10) one evidently has

---

[21] [The *regularity theorem*: see Hausdorff 1914a, 408–413, or footnote 43 in section 4.4 of the present book.]

$$B_1 \subseteq B. \tag{11}$$

By virtue of (2), (7), (9), and (10), one obtains

$$A_1 \underset{f}{=} B_1. \tag{12}$$

Finally, the sets $A_1 \cap C_1^\circ, A_1 \cap C_2^\circ, \ldots, A_1 \cap C_m^\circ$ being closed (as intersections of closed sets: $A_1 \cap C_k^\circ = A_1 \cap C_k$), one concludes following (8) and (10) that the sets $E_1, E_2, \ldots, E_m$ and $B_1$ are also closed, and that

$$m(A_1) = m(B_1). \tag{13}$$

Formulas (5), (6), and (11)–(13) prove that $A_1$ and $B_1$ are the desired sets.

**Lemma 38.** *If $A$ and $B$ are point sets situated in a Euclidean space, $A$ is Lebesgue-measurable, $B$ is open, and $m(A) = m(B)$, then $A \underset{p}{=} B$.*

*Proof.* We shall define two infinite sequences $\{A_n\}$ and $\{B_n\}$ of sets recursively in the following way:

I. $A_1$ and $B_1$ are closed sets contained in $A$ and $B$, respectively, and satisfying conditions

$$A_1 \underset{f}{=} B_1, \tag{1}$$

$$m(A_1) = m(B_1) \geq \tfrac{1}{2} m(A). \tag{2}$$

II. If $n$ is an arbitrary natural number, then $A_{n+1}$ and $B_{n+1}$ are closed sets such that

$$A_{n+1} \subseteq A - \bigcup_{k=1}^{n} A_k \quad \text{and} \quad B_{n+1} \subseteq B - \bigcup_{k=1}^{n} B_k,$$

and satisfying the formulas

$$A_{n+1} \underset{f}{=} B_{n+1}, \tag{3}$$

$$m(A_{n+1}) = m(B_{n+1}) \geq \tfrac{1}{2} m\left(A - \bigcup_{k=1}^{n} A_k\right). \tag{4}$$

Based on lemma 37 one proves[22] by an easy induction the existence of all terms of the sequences $\{A_n\}$ and $\{B_n\}$. That is indeed evident for $n = 1$; and if sets $A_1, A_2, \ldots, A_n$ and $B_1, B_2, \ldots, B_n$ exist, one obtains easily from (2) and (4),

$$A = \bigcup_{k=1}^{n} A_k \text{ is Lebesgue-measurable}, \quad B - \bigcup_{k=1}^{n} B_k \text{ is open}, \tag{5}$$

$$m\left(A - \bigcup_{k=1}^{n} A_k\right) = m\left(B - \bigcup_{k=1}^{n} B_k\right). \tag{6}$$

By virtue of the lemma mentioned, conditions (5) and (6) immediately imply the existence of closed sets $A_{n+1}$ and $B_{n+1}$ satisfying (3) and (4).

---

[22] [Evidently in error, Banach and Tarski cited lemma 36.]

Therefore, one can conclude

$$\lim_{n\to\infty} m(A_n) = \lim_{n\to\infty} m(B_n) = 0,$$

from which, by reason of (4) and (6),

$$\lim_{n\to\infty} m\left(A - \bigcup_{k=1}^{n} A_k\right) = \lim_{n\to\infty} m\left(B - \bigcup_{k=1}^{n} B_k\right) = 0. \tag{7}$$

Set

$$A' = \bigcup_{k=1}^{\infty} A_k, \quad A'' = A - A', \tag{8}$$

$$B' = \bigcup_{k=1}^{\infty} B_k, \quad B'' = B - B'. \tag{9}$$

Evidently, one has

$$A = A' \cup A'', \quad B = B' \cup B'', \quad A' \cap A'' = B' \cap B'' = \phi. \tag{10}$$

From (7), (8), and (9), one deduces directly,

$$m(A') = m(A), \quad m(B') = m(B),$$

from which follows

$$m(A'') = m(B'') = 0. \tag{11}$$

By virtue of (4), (8), and (9), one obtains finally, by applying theorem 4',

$$A' \underset{d}{=} B'. \tag{12}$$

In conformity with definition 3, formulas (10)–(12) immediately yield

$$A \underset{p}{=} B, \qquad \text{Q. E. D.}$$

**Lemma 39**. If a bounded set $A$ situated in a Euclidean space $E$ is Lebesgue-measurable and has positive measure, then there exists in the same space a set $C$ of measure zero such that

$$A \cup C \underset{d}{=} E.$$

*Proof.* Since $m(A) > 0$, there evidently exists a bounded open bound set $B$ whose measure is equal to that of $A$. Conforming to the preceding lemma, one can conclude that

$$A \underset{p}{=} B.[23]$$

Thus, let $A_1$, $A_2$, $B_1$, and $B_2$ be sets fulfilling the conditions of definition 3:

$$A = A_1 \cup A_2, \quad B = B_1 \cup B_2, \quad A_1 \cap A_2 = B_1 \cap B_2 = \phi, \tag{1}$$

$$A_1 \underset{d}{=} B_1, \tag{2}$$

$$m(A_2) = m(B_2) = 0. \tag{3}$$

---

[23] [In error, Banach and Tarski wrote $E$ in place of $B$ here.]

Further, let $C$ and $D$ be bounded sets satisfying the formulas

$$C \cong B_2 \quad \text{and} \quad D \cong A_2, \tag{4}$$
$$(C \cup D) \cap (A \cup B) = \phi; \tag{5}$$

sets $A$ and $B$ being bounded, the existence of sets $C$ and $D$ is evident.

By virtue of (1), (2), (4), and (5) one obtains easily, by applying theorem 4′,

$$A \cup C = A_1 \cup A_2 \cup C \overset{=}{_d} B_1 \cup B_2 \cup D = B \cup D. \tag{6}$$

The set $B \cup D$ not being a boundary set, one deduces from lemma 34 that

$$B \cup D \overset{=}{_d} E. \tag{7}$$

By reason of theorem 3′, formulas (6) and (7) immediately yield

$$A \cup C \overset{=}{_d} E.$$

Since in addition the set $C$, following (3) and (4), has measure zero, lemma 39 is completely demonstrated.

**Lemma 40.** If a set $A$ situated in Euclidean space has positive Lebesgue inner measure (finite or not), then $A \overset{=}{_p} E$.

*Proof.* The set $A$ evidently contains a bounded Lebesgue-measurable subset $A'$ with positive measure. Let $C$ be a set with measure zero fulfilling the conditions of lemma 39 with respect to $A'$. Therefore,

$$A' \cup C \overset{=}{_d} E, \tag{1}$$
$$m(C) = 0. \tag{2}$$

Since

$$A' \cup C \subseteq A \cup C \subseteq E,$$

one concludes from (1), by virtue of corollary 9′, that

$$A \cup C \overset{=}{_d} E.$$

From this, following corollary 6′, it follows that the space $E$ can be decomposed into two disjoint sets,

$$E = E_1 \cup E_2 \tag{3}$$

such that

$$E_1 \overset{=}{_d} A, \quad E_2 \overset{=}{_d} C - A. \tag{4}$$

From (2) and (4) one deduces easily, applying corollary 15′, that

$$m(E_2) = 0. \tag{5}$$

Set

$$A_1 = A, \quad A_2 = \phi, \tag{6}$$

which implies

## 6.3 §3 Fundamental Theorems about Denumerable Decomposition

$$A = A_1 \cup A_2. \tag{7}$$

Formulas (3) and (7) provide a decomposition of the sets $A$ and $E$ that, one proves quickly by virtue of (4)–(6), satisfies the conditions of definition 3. Therefore, one has

$$A \underset{p}{=} E, \qquad\qquad \text{Q. E. D.}$$

Theorem 36 and the lemma just demonstrated immediately imply the following

**Theorem 41.** If sets $A$ and $B$, situated in a Euclidean space with an arbitrary number of dimensions, have positive Lebesgue inner measures (finite or not), then

$$A \underset{p}{=} B.$$

It should be remarked that in the decompositions furnished by the fundamental theorems of this work some Lebesgue-nonmeasurable sets must necessarily occur. Indeed, one sees that two sets are equivalent by decomposition (finite or denumerable) into Lebesgue-measurable sets only under the condition that they should have the same measure. As far as almost equivalent sets are concerned, one has the following

**Theorem 42.** For two point sets situated in a Euclidean space with an arbitrary number of dimensions to be almost equivalent by denumerable decomposition into Lebesgue-measurable (or even closed) sets, it is necessary and sufficient that they should have the same measure.[24]

This theorem is deduced easily from lemma 38 and theorem 36, by analyzing their demonstrations.

---

[24] In the same order of ideas one can establish the following theorem: if $A$ and $B$ should be sets situated in a Euclidean space with an arbitrary number of dimensions, if $A$ is Lebesgue-measurable, $B$ is open, and $m(A) < m(B)$, then the set $A$ is equivalent to a subset of $B$ by denumerable decomposition into Lebesgue-measurable sets.

# 7
# Degree of Equivalence of Polygons

The elementary approach to the theory of area, described in section 4.1, includes definitions of *polygonal region*, of *equivalence* of a pair $V, W$ of regions under decomposition into the same finite number $n$ of respectively congruent subregions with disjoint interiors, and of the *area* of a polygonal region. It includes the Bolyai–Gerwien theorem: polygonal regions are equivalent if and only if they have the same area. The 1924 papers of Alfred Tarski and Stefan Banach, translated in chapters 5 and 6, extended that theory to include analogous results with a different, set-theoretic, definition of equivalence.

Tarski returned to the elementary theory in the early 1930s with two papers devoted to the *degree of equivalence* of a pair $V, W$ with the same area: the smallest possible number $n$ of pairs of congruent subregions in such a decomposition. The papers appeared in journals addressed to gimnazjum teachers and students. The papers' background involves Tarski's employment as a schoolteacher and in training teachers, which will be discussed later, in chapter 9. Tarski's first paper posed a question that was immediately answered and published by a reader, gimnazjum teacher Henryk Moese; his work is essential for understanding Tarski's second paper. Section 7.1 of the present chapter is concerned with publication details and with the mathematics in these three papers.

Draft translations, by Izaak Wirszup, of those three papers were published informally in 1952 by the University of Chicago. Beyond that, these papers have attracted little professionial attention. But Tarski presented this material in wonderful lectures to the public and to high-school students. One of the present editors, James T. Smith, attended those in Berkeley during the 1960s and in Regina, Saskatchewan, in 1970. Because this mathematics is so elegant and was only barely accessible, and because Tarski was evidently reluctant to disseminate those drafts further, the Tarski and Moese papers are translated anew here, in sections 7.2 to 7.4.[1]

---

[1] The drafts constituted the booklet Tarski and Moese 1952. The new translations are designated Tarski [1931] 2014a, Moese [1932] 2014, and Tarski [1932] 2014d.

## 7.1 Introduction

The three papers by Alfred Tarski and Henryk Moese that are translated in chapter 7 appeared in two different journals:

| | | | |
|---|---|---|---|
| 7.2 | Tarski | On the Degree of Equivalence of Polygons | *Młody matematyk* 1 (1931) |
| 7.3 | Moese | Contribution to A. Tarski's Problem "On the Degree of Equivalence of Polygons" | *Parametr* 2 (1931–1932) |
| 7.4 | Tarski | Further Remarks about the Degree of Equivalence of Polygons | *Parametr* 2 (1931–1932) |

Distributed together, the journals formed part of a project to enhance the quality of mathematics instruction in Polish schools, described in section 9.7. The main journal, *Parametr*, was addressed to gimnazjum teachers; its supplement, *Młody matematyk* (*Young Mathematician*), targeted their particularly interested students.

A Hebrew translation of the first paper, by Dov Jarden, was published in 1951 in the Israeli journal *Riveon lematematika*. The next year, the University of Chicago published a draft English translation of all three papers, by Izaak Wirszup.[2] The first paper was reprinted in Polish in 1975 in the Polish journal *Delta*.

Tarski began his paper *On the Degree of Equivalence of Polygons* (section 7.2) by defining the notions of polygonal region and equivalence by finite decomposition and presenting nontrivial examples. His style in addressing his student audience was less formal than that in his research papers: for example, he left the requirement of disjoint interiors to readers' interpretation of the verb *decompose*.[3] For polygonal regions $W, V$ that are equivalent— $W \equiv V$ —Tarski defined the *degree* $\sigma(W, V)$ *of equivalence* to be the *smallest* number of subregions into which each can be decomposed, so that all corresponding pairs of subregions are congruent.[4]

Tarski listed four elementary properties of this function $\sigma$. He omitted proofs, but suggested that the most involved of these,

$$W \equiv U \ \& \ V \equiv U \ \Rightarrow \ \sigma(W, V) \leq \sigma(W, U) \cdot \sigma(V, U), \qquad \text{(property 4)}$$

could be proved by imitating the proof of the transitive law for $\equiv$.[5] Next, Tarski defined the *diameter* $\delta(P)$ of a polygonal region $P$ as the maximum distance between two points

---

[2] The Hebrew translation is Tarski [1931] 1951. The English drafts constituted the booklet Tarski and Moese 1952, which was reproduced from typescript. The present editors do not know the circumstances of those publications. For information about Jarden and Wirszup, see the box on page 130.

[3] In Polish, *podzielić*.

[4] Two doctoral students of Felix Bernstein at the University of Halle–Wittenberg had started an investigation of this subject. Hans Brandes (1907) considered an analogous notion of degree of equivalence in which only triangular subregions are allowed; he used it to analyze the complexity of proofs of the Pythagorean theorem. Paul Mahlo (1908) derived some more-general results and applied them to the same problem. Kitizi Yanigahara (1927) extended this approach and reported related work of Wilhelm Süss. Tarski did not mention any of this earlier research. For more information, see Pambuccian 2004.

[5] See theorem 3 in section 5.1.

## 7.1 Introduction

of $P$. He gave a detailed proof of the conditional inequality

$$W \equiv V \ \& \ W \text{ convex} \Rightarrow \sigma(W,V) \geq \frac{\delta(W)}{\delta(V)}, \qquad \text{(property 5)}$$

then left to his readers its generalization to pathwise connected regions $W$. Tarski noted that such inequalities can be used to establish upper and lower bounds of $\sigma(W,V)$, and that any example decomposition also provides an upper bound. For several cases Tarski used these bounds to completely determine $\sigma(W,V)$. In others, his upper bound was larger than his lower bound, and thus the exact value of $\sigma(W,V)$ remained unknown.

The remainder of the paper was devoted to theorems about a new function $\tau$, defined for each positive real number $x$ as follows: $\tau(x) = \sigma(W,V)$, where $W$ is a unit square and $V$ is a rectangle with edge lengths $x$ and $1/x$.

Tarski encouraged his readers to establish for themselves the first major result,

$$\tau(x) \leq 2 + \lceil \sqrt{x^2-1} \rceil \text{ for every } x \geq 1.^6 \qquad \text{(theorem III)}$$

He suggested examining the demonstrations of two related theorems in elementary textbooks: for example, the well-known text *Outline of Elementary Geometry for Use in Secondary Schools*, by Władysław Wojtowicz.[7] Analyzing those proofs would, Tarski claimed, lead students quickly to remarkable new mathematics. As a further enticement, he indicated slyly that this study had led him to the puzzling decompositions in figure 2 of his paper (page 136). His arguments are presented in the following three paragraphs. Their sophistication suggests the level of involvement that Tarski expected of his student readers.

Tarski's first hint was Wojtowicz's figure 148, adapted below. It shows that two parallelograms $ABCD$ and $ABC'D'$ with the same base $AB$ and altitude $h$ can each be decomposed into $n$ subregions such that corresponding subregions are congruent. Denote by $g$ the altitude, with respect to the same base, of the intersection of those two parallelograms, shaded in the figure; then $n = 1 + \lceil h/g \rceil$. The case shown, in which $n = 5$, is typical for the situation in which $h > g$ and $h/g$ is not an integer. When $g$ does divide $h$, the details at the top of the figure become simpler, but the same formula holds. When $h \leq g$, the edges opposite the common base overlap, and the formula yields $n = 2$, the correct value.[8]

---

[6] For every real number $x$, $\lceil x \rceil$ denotes the *ceiling* of $x$: the smallest integer greater than or equal to $x$.

[7] Wojtowicz [1919] 1926, §§177, 191, 192. For information about Wojtowicz, see a box in section 9.3.

[8] The Archimedean axiom justifies replicating the shaded triangle upward by steps until a copy overlaps the line $CD$. Most authors, instead, note that $ABCD$ and $ABC'D'$ would be obviously equidecomposable if they overlapped, then move $CD$ rightward by steps, maintaining an overlap each time, until it collides with $C'D'$. Wojtowicz's method facilitates counting the subregions in the decomposition.

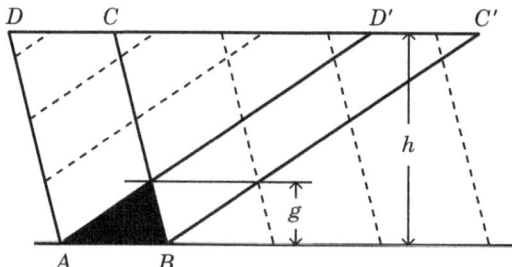

*Parallelograms with the Same Base and Altitude*

The second hint referred to a construction in Wojtowicz's §§191–192, which prepared for his proof of the Pythagorean theorem. The next figure is an adaptation for the situation at hand. Given a unit square $W = BCC'B'$ and a number $x > 1$, extend $CC'$ past $C$ to point $A$ so that $ABC$ is a right triangle with leg $BC$ of length 1 and hypotenuse $AB$ of length $x$. Locate point $A'$ on $AC'$ so that $ABB'A'$ is a parallelogram, and point $X$ where lines $BC$ and $A'B'$ intersect. (The figure shows the case in which $A'$ lies between $A$ and $C$ and, consequently, $X$ between $B$ and $C$; a similar argument is required when $A'$ lies between $C$ and $C'$.) Finally, locate points $D$ and $E$ on the line $A'B'$ so that $V = ABDE$ is a rectangle. By straightforward computation,

$BD$ has length $1/x$, $\quad BX$ has length $\dfrac{1}{\sqrt{x^2-1}}$.

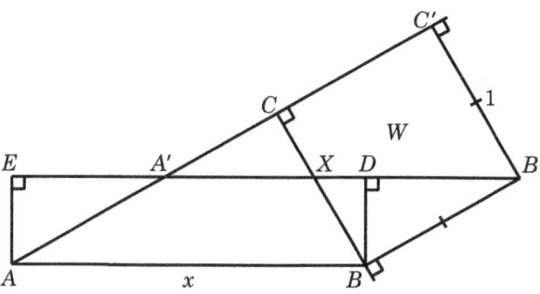

*Adaptation of Wojtowicz's Argument*

Square $W$ and parallelogram $ABB'A'$ have the same base $BB'$ and altitude 1. By the argument of the previous paragraph, they can each be decomposed into $n$ subregions such that corresponding subregions are congruent, where $n = 1 + \lceil h/g \rceil$ and $h$ and $g$ are the lengths of $BC$ and $BX$. Moreover, their triangular intersection $BXB'$ is a component of each decomposition. Split that into two right triangles $BXD$ and $BDB'$, and notice that the latter is congruent to triangle $AEA'$. Thus, square $W$ and rectangle $V$ are each decomposed into $2 + \lceil h/g \rceil$ subregions, so that corresponding subregions are congruent. Finally, as desired,

## 7.1 Introduction

$$\tau(x) = \sigma(W,V) \le 2 + \lceil h/g \rceil = 2 + \left\lceil \sqrt{x^2-1}\, \right\rceil.$$

The next results were much easier. The simplest is an improvement over the previous inequality for the case in which $x$ is an integer $n$:

$\tau(n) \le n$ for all integers $n > 0$. (theorem IV)

To see this, note that in this case,[9] each of $W$ and $V$ can be decomposed into $n$ rectangles with edge lengths 1 and $1/n$, hence $\sigma(W,V) \le n$. Further, for every $x > 0$, integer or not,

$$\delta(W) = \sqrt{2}, \ \delta(V) = \sqrt{x^2 + \frac{1}{x^2}},$$

and by property (5),

$$\tau(x) = \sigma(W,V) \ge \frac{\delta(W)}{\delta(V)} = \sqrt{\frac{x^4+1}{2x^2}}. \qquad \text{(theorem V)}$$

Theorems IV and V yield $1.4 \le \tau(2) \le 2$, and thus $\tau(2) = 2$; similarly, $\tau(3) = 3$. However, these results do not alone determine $\tau(4)$. In the problem section of the same journal issue, Tarski formally proposed finding a proof that $\tau(4) = 4$, a known result.[10]

Tarski concluded this paper with more exercises and two conjectures. Two exercises are simple applications of theorem III. A third, $\tau(2\frac{1}{4}) \le 4$, would require a new example, which Moese would describe in the next paper. Tarski's first conjecture, that $\tau(n) = n$ for every positive integer $n$, was suggested by its known instances for $n \le 4$. It would soon be confirmed by Moese. The second conjecture—that $\tau(x) \ge 3$ for every positive $x \ne 1/2, 1, 2$ —would in fact be *refuted* by Tarski himself in the third of these papers.

Moese began his paper, *A Contribution to A. Tarski's Problem "On the Degree of Equivalence of Polygons,"* by confirming Tarski's conjecture that $\tau(n) = n$ for all positive integers $n$.[11] As just noted, Tarski had already shown that $\tau(n) \le n$. To derive the opposite inequality, Moese introduced two ideas: strips of width $x$ between parallel lines, and $\sigma_W(x)$,[12] the minimum number of such strips whose union contains a specified bounded region $W$. This enabled him to substitute for the detailed analysis of the geometry of the unit square $V$ that of the simpler geometry of its inscribed disk and the corresponding hemisphere. It is easy to see that $\tau(n) \ge \sigma_V(x) = \sigma_K(x)$, where $x = 1/n$ and $K$ is a disk with radius $1/2$. Noting that the surface area of the portion of the hemisphere over $K$ that lies over one of these strips depends only on $x$, he showed that $\sigma_K(x) = \lceil 1/x \rceil$. It follows that $\tau(n) \ge n$.

---

[9] Generalize Tarski's figure 7 in section 7.2, page 141, where $n = 3$.
[10] This problem is Tarski [1931] 2014e: see section 12.10.
[11] For more information about Moese, see the box on page 130. His paper is translated in section 7.3.
[12] Moese's slightly different notation for this expression did not make the dependence on $W$ explicit.

**Dov Jarden** was born in 1911 in Motol, then in the Russian Empire (now Motal', in Belarus). He studied in Jewish schools there and in Warsaw, earned a rabbinical degree, and emigrated to Israel in 1934. At the Hebrew University of Jerusalem he earned master's and doctoral degrees in mathematics and Hebrew studies, respectively. In 1946 he founded the journal *Riveon lematematika*. Jarden authored, translated, and edited many works on geometry, number theory, the Hebrew language, and medieval Jewish history. He taught in secondary schools and universities in Israel and the United States. Jarden died in Jerusalem in 1986.*

**Izaak Wirszup** was born in 1915 in Vilnius, then under occupation by the German Army. He began university studies there during the 1930s, particularly with the mathematician Antoni Zygmund. After World War II, he rejoined Zygmund at the University of Chicago, earned the doctorate there in 1955, became a faculty member, and was appointed full professor there in 1965. Wirszup was influential in mathematics education both locally and nationwide, and remained active until his death in 2008.†

**Henryk Moese** was born in 1886 in the Małopolska region of southern Poland. He began a teaching career around 1910, probably at a private gimnazjum in Kolbuszowa, a small town in southeastern Poland, then in the Austrian Empire. In 1919 he moved from there to Śrem, a small city in western Poland, to become a teacher of mathematics, physics, and geography at the public General Józef Wybicki Boys' Gimnazjum. By 1929 he had become its assistant director. His workload consisted of twenty-four hours per week in classes, and twelve hours of other duties. Moese's namesake, probably his son, entered the first (youngest) class of that school for the year 1927/1928, and continued there through the fourth class, in 1930/1931.

In 1930 Moese was appointed director of the Liceum Ogólnokształcące, in Kępno, a small city in the southwest, at that time near the German border. The younger Henryk enrolled there a year later. A publicly funded gimnazjum with a classical emphasis, it enrolled about 275 students—about one-fourth, young women. Moese was responsible for a faculty of about twelve, and taught mathematics himself for six hours a week to the upper three classes. The teaching loads of the other faculty averaged about twenty-seven hours per week. The school is pictured on page 131; Moese served there until 1933. During that time he was also involved in scouting. That year, Moese evidently moved on. In September 1935 he was serving as director of the Mikołaj Kopernik Gimnazjum in Toruń, the historic city on the Vistula halfway between Warsaw and the Baltic. The portrait on page 131 shows him in his study there.

In a 1929 commemorative school report, Moese published a study of an intriguing arithmetic puzzle, the "seven sevens problem." During the early 1930s he contributed materials for publication in the journals *Parametr* and *Młody matematyk*, which were aimed at Polish gimnazjum teachers and their most interested students (see section 9.7). These included the paper described in this section and translated in section 7.3, and many smaller items.

The present editors have not yet traced the elder Moese's life before or after the period covered in this sketch. The younger Henryk was probably the Polish teacher and scholar born in Wysoka in southeastern Poland in 1917, who completed gimnazjum studies in Toruń in 1935 and earned a doctorate in philosophy from the university there in 1955.‡

---

*Yarden 2006.    † Chicago 2008.

‡ Zagórowski 1924–1926, volume 1, 275, and volume 2, 231. Śrem 1928, 7, 11, 13; 1929, 88; 1930, 61; 1931, 12, 58. Kępno 1931, 30, 33; 1932, 31–32, 73; 1933, 46, 65. Kurzawa and Nawrocki 1978, 129 ("Moese" is misspelled there). L. Sobociński 1935. Moese 1929. Rusiecki and Straszewicz 1932, 154. Toruń 1995, 478. For more information about Polish secondary schools of this era, see chapter 9.

*Gimnazjum in Kępno
Where Henryk Moese Taught
from 1930 to 1933*

*Henryk Moese
in His Study
in Toruń, 1935*

In his second paper, Tarski reported that Zenon Waraszkiewicz had been the first to prove that $\tau(n) = n$ for all positive integers $n$, but with an argument too complex for publication in *Parametr*. Moese proved it independently with the simpler argument just described.

---

**Zenon Waraszkiewicz** was born in 1909 in Warsaw, then part of the Russian Empire. His parents were schoolteachers. The family took refuge in Odessa during World War I, then returned to Warsaw. Zenon completed gimnazjum studies there in 1926, and entered the University of Warsaw to study mathematics. He earned a master's degree in 1930, and the doctorate in 1932, supervised by Stefan Mazurkiewicz. Waraszkiewicz continued research in point-set topology and analysis, and earned the *venia legendi* in 1937. Until World War II he taught in Warsaw secondary schools, served as assistant at the Warsaw Polytechnic University, and as dozent at the University of Warsaw. During the German occupation, Waraszkiewicz taught in the Polish clandestine schools. In 1945 he became a professor at the new University of Łódź, but died there that same year.*

*Derkowska 2001; Tatarkiewicz 2003.

---

Moese's next result, $\tau(x) \leq 1 + \lceil x \rceil$ for all real $x \geq 1$, strengthened the upper bound $2 + \lceil \sqrt{x^2 - 1} \rceil$ in Tarski's earlier paper.[13] That required only an example decomposition of the unit square and an $x$ by $1/x$ rectangle into that number of components. Moese merely presented a diagram[14] for the case $\lceil x \rceil = 4$; readers may discover his reasoning by reconstructing that figure. The paper continued with some further results about values of $\tau$, which can be used to solve exercises posed in Tarski's earlier paper. Moreover, a major result in Tarski's next paper would rest on Moese's final result: for any positive integers $n$ and $p$,

$$\tau\left(n + \frac{1}{p}\right) \leq n + 1. \qquad \text{(theorem 4)}$$

Moese concluded with two conjectures, the first of which concerned that theorem: $\tau(x) = n + 2$ for every positive integer $n$ and every $x$ such that $n < x \leq n + 1$, *except* when $x = n + 1/p$ for some positive integer $p$, in which case $\tau(x) = n + 1$.

---

[13] See theorem III on page 142; it is straightforward to show that the new bound is always less than or equal to the earlier one.

[14] Figure 4 on page 149.

## 7.1 Introduction

In his paper *Further Remarks about the Degree of Equivalence of Polygons* (section 7.4), Tarski defined the *width* $\omega(F)$ of a plane figure $F$ to be the smallest width of any strip that contains $F$, provided there is one.[15] He then proved that if $F$ is contained in some strip, contains a disk of the same width, and is the union of subsets $C_1, \ldots, C_n$, then

$$\omega(F) \le \sum_{i=1}^{n} \omega(C_i). \qquad \text{(theorem A)}$$

For the preliminary case in which $F$ is a disk with diameter $d = \omega(F)$, he suggested that readers imitate a proof in Moese's preceding paper. Indeed, each $C_i$ is contained in some strip with width $\omega(C_i)$; the portion of the hemisphere $H$ over $F$ that lies over this strip has surface area $\frac{1}{2}\pi d \omega(C_i)$; and the sum of these surface areas cannot be less than that of the entire hemisphere, which is $\frac{1}{2}\pi d^2 = \frac{1}{2}\pi d \omega(F)$. Tarski provided the remaining details of the proof of theorem A, then quickly derived his next result: if $W$ and $V$ are equivalent polygonal regions and $W$ satisfies the condition of theorem A, then

$$\sigma(W, V) \ge \frac{\omega(W)}{\omega(V)}. \qquad \text{(theorem B)}$$

Tarski noted that when applicable, theorem B can provide a larger lower bound for $\sigma(W, V)$ than property 5 in his previous article.[16] Thus, it would be useful to generalize it, requiring that $W$ be merely convex or connected rather than contain a disk of the same width. Such a generalization would follow from an analogous one of theorem A. The present editors are not aware of any such generalizations in the literature.

In Tarski's earlier paper, property 5 entailed theorem V, a lower bound for the function values $\tau(x)$. He showed that the discussion in his second paper leads easily to a much better lower bound: $\tau(x) \ge \lceil x \rceil$ whenever $x \ge 1$. Combining that with Moese's upper bound, Tarski concluded that for every $x \ge 1$,

$$\tau(x) = \lceil x \rceil \text{ or } 1 + \lceil x \rceil. \qquad \text{(theorem E)}$$

Consequently, for any positive integers $n$ and $p$,

$$\tau\left(n + \frac{1}{p}\right) = n + 1. \qquad \text{(theorem F)}$$

Indeed, this value of $\tau$ is $n + 1$ or $n + 2$ by theorem E, and Moese had shown already[17] that it cannot exceed $n + 1$. Moese had conjectured that $\tau(x) = \lceil x \rceil$ *only* when $x = n + 1/p$ for some positive integers $n$ and $p$.

---

[15] Tarski used the notation $\pi(F)$ for the width of $F$; in the present book, $\omega(F)$ is used to avoid confusion with the mathematical constant $\pi$.

[16] See page 127 in the present section.

[17] See page 132 in the present section.

In his final theorem H, Tarski confirmed this for the case $1 < x \leq 2$: that is, if $1 < x \leq 2$, then a unit square $Q$ can be decomposed into connected polygonal subregions $A$ and $B$ that can be rearranged to form a nonsquare rectangle $R_x$ with edges of length $x$ and $1/x$ *only when* $x = 1 + 1/p$ for some positive integer $p$. Tarski attributed this result to Adolf Lindenbaum and Zenon Waraszkiewicz, but claimed that the proof was too subtle to include in this paper. The figure on the next page shows that a decomposition is possible for $x = 1 + 1/p$. Tarski suggested that their proof entailed a result stronger than theorem H: the only way to decompose $Q$ into two subregions $A$ and $B$ that can be rearranged to form $R_x$ for any $x$ is to make both $A$ and $B$ congruent to a region $A_p$, for some positive integer $p$, as shown in that figure. This confirms the conjecture of Henryk Moese that concluded his [1932] 2014 paper, translated in section 7.3.[18]

Tarski noted that if the restriction on $x$ in theorem H could be removed, study of the function $\tau$ would be rather complete. But he did mention several different directions in which this subject could be developed further. Later, Tarski himself studied the effect of limiting the number of edges of the polygons allowed in the dissections—his work made essential use of theorem H.[19] The present editors are not aware of any other studies. Tarski's paper concluded with a selection of exercises, of various levels of difficulty.

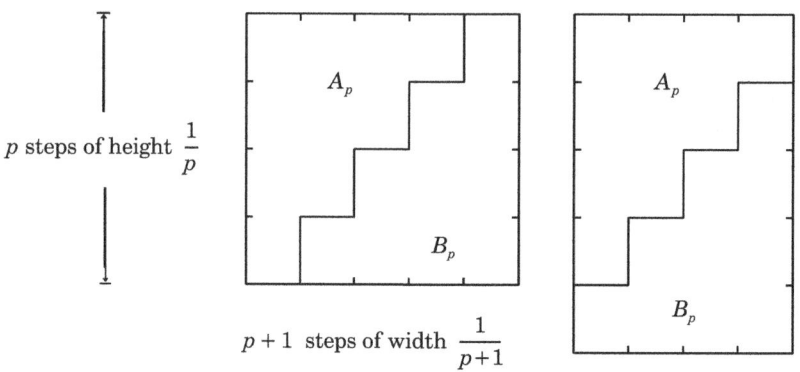

*Dividing a Square into Two Polygonal Regions*
*That Form a $(p+1)/p$ by $p/(p+1)$ Rectangle ($p = 4$)*

---

[18] See also the abstract, Lindenbaum [1937] 1938. The present editors have seen a promising sketch of an entirely elementary proof by J. Shilleto (2013).

[19] See section 8.5.

## 7.2 The Degree of Equivalence of Polygons (1931)

This section contains an English translation of Alfred Tarski's paper *O stopniu równoważności wielokątów*, [1931] 2014a, which appeared in volume 1 of the journal *Młody matematyk*. Aimed at particularly interested gimnazjum students, this journal was distributed with another one, *Parametr*, addressed to their teachers.[20] A draft English translation by Izaak Wirszup was published informally by the University of Chicago as Tarski and Moese 1952. The present translation was carried out independently, but used Wirszup's for confirmation.

This translation is meant to be as faithful as possible to Tarski's original text. Its only intentional modernizations are punctuation, and some changes in symbols, where Tarski's conflict with others used throughout this book. A bibliographic reference has been adjusted to conform with the conventions of the present book. All [square] brackets in the translation enclose editorial comments inserted for clarification, often as footnotes. As an aspect of adjusting punctuation, the editors increased white space to enhance visual organization.

DR. ALFRED TARSKI (Warsaw)

# On the Degree of Equivalence of Polygons

In this article I want to discuss some concepts, belonging entirely to the realm of elementary geometry, which until now have been investigated hardly at all.

As is well known, we call two polygons $W$ and $V$ *equivalent*, expressing this with the formula $W \equiv V$, if they can be divided into the same number of respectively congruent polygons.[21] This subdivision of equivalent polygons into congruent parts is not unique: two equivalent polygons can be divided into congruent parts in various ways, with respect to the number as well as the form of these parts.

We explain this with an example: figure 1, and figure 2 as well, show that a square with edge $a$ and a rectangle with edges $5/4\, a$ and $4/5\, a$ are equivalent to each other, but their subdivisions in the two figures are quite distinct.

In connection with this observation, a question arises in a natural way: what is the *least* number of respectively congruent parts into which two given equivalent polygons can be divided? We want to touch upon a problem of exactly this type in *Parametr*.

---

[20] For further information about these journals, consult sections 7.1 and 9.7.

[21] [This notion of equivalence is discussed in section 4.1. It is different from that used in chapters 5 and 6. The two notions are compared succinctly on page 59.]

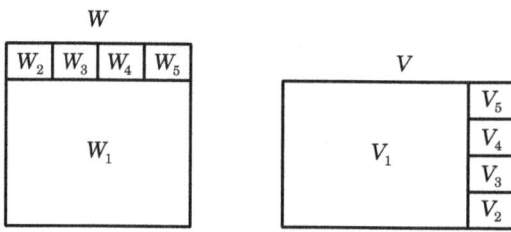

**Figure 1**

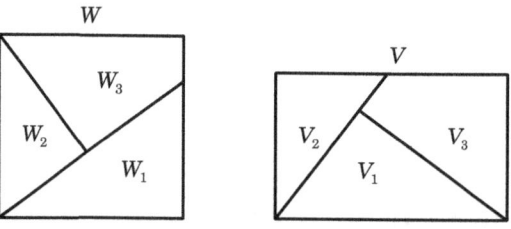

**Figure 2**

**Figure 3**   **Figure 4**

## 7.2 The Degree of Equivalence of Polygons (1931)

With this goal, we adopt the following definition. The *degree of equivalence*[22] of two equivalent polygons $W$ and $V$ shall be the *least* natural number $n$ satisfying this condition: each of the polygons $W$ and $V$ can be subdivided into $n$ polygons in such a way that the polygons obtained by subdivision of $W$ are respectively congruent to the polygons obtained by subdivision of $V$. We shall denote the degree of equivalence of polygons $W$ and $V$ by the symbol $\sigma(W,V)$.

A certain remark ought to be made here. It is convenient in the present considerations to give the word "polygon" a meaning broader than that which is used at the beginning of instruction in elementary geometry. Specifically, here we call a *polygon in the broader sense* a plane figure that is the *composition* of a finite number of polygons in the ordinary sense of that word. For example, the figure composed of rectangles $W_2$ and $W_4$ in figure 1 is a polygon in the broader sense, and so is the figure composed of both of those rectangles and the quadrilateral $W_3$ in figure 2 as well. We note in passing that the extension of the concept of polygon is extremely useful in the whole theory of the equivalence of polygons. Without this extension, many arguments from this theory [that are] encountered in elementary textbooks err from a lack of precision.

Applying the broader meaning to "polygon" presents a certain difficulty in properly defining the notion of congruence. We restrict ourselves here to the following visual explanation: two polygons in the broader sense—like geometric figures of any kind—are congruent if one of them can be "laid upon" the other (without changing the relative positions of the component parts of either of them) in such a way that they "coincide." For example, the polygon shown in figure 3 is not congruent to the polygon shown in figure 4, but is equivalent to it.

Until now we knew very little about the degree of equivalence of polygons. We give here as examples a few elementary properties of this concept.

1. *For any equivalent polygons $W$ and $V$,*
$$\sigma(W,V) = \sigma(V,W).$$

2. *In order for $\sigma(W,V) = 1$, it is necessary and sufficient that polygons $W$ and $V$ be congruent; in particular, for any polygon $W$, we have*
$$\sigma(W,W) = 1.$$

3. *If the polygon $W$ can be divided into polygons $W_1$ and $W_2$ and polygon $V$ into polygons $V_1$ and $V_2$ in such a way that $W_1 \equiv V_1$ and $W_2 \equiv V_2$, then*
$$\sigma(W,V) \leq \sigma(W_1,V_1) + \sigma(W_2,V_2).$$

---

[22] As far as we [Tarski] know, this concept was introduced by Dr. Adolf Lindenbaum (Warsaw) who together with the author of this article established some of the properties of this concept.

4. *If $W \equiv U$ and $V \equiv U$, then*

$$\sigma(W,V) \leq \sigma(W,U) \cdot \sigma(V,U).$$

Properties 1–3 are self-evident. Also, justification of property 4 does not present difficulties to those among the Readers who are aware of the application of the so-called *double-subdivision-network* method in the proof of the theorem that two polygons equivalent to a third are equivalent to each other.

For the formulation of the next property the concept of the diameter of a polygon is necessary. The *diameter* of a polygon $W$, designated by $\delta(W)$, shall be a *longest* one of the segments joining two points of the polygon $W$. It is easy to show that every polygon has a diameter; there may be many congruent diameters.

5. *If $W$ and $V$ are equivalent polygons, and in addition $W$ is a convex polygon, then*

$$\sigma(W,V) \geq \frac{\delta(W)}{\delta(V)}.$$

*Proof.* Let us apply reasoning by contradiction.[23] Specifically, suppose that contrary to the conclusion of [property] 5,[24]

$$\sigma(W,V) = n < \frac{\delta(W)}{\delta(V)}. \tag{1}$$

It follows immediately that

$$\delta(V) < \frac{\delta(W)}{n}. \tag{2}$$

In accordance with the definition of diameter there can be found in polygon $W$ two points $A_0$ and $A_n$ that are the endpoints of the diameter $\delta(W)$. Divide [segment] $A_0 A_n$ into $n$ congruent subsegments[25] and let $A_1, A_2, \ldots, A_{n-1}$ be the subdivision points. Each of the segments $A_k A_{k+1}$ (where $0 \leq k < n$) is congruent to an $n$th part of the diameter $\delta(W)$; therefore, in view of (2) we have $A_k A_{k+1} > \delta(V)$. From this we conclude, moreover, that

$$A_k A_l > \delta(V) \tag{3}$$

for any distinct natural numbers $k$ and $l$ lying between 0 and $n$ [inclusive].

According to the definition of degree of equivalence, from (1) it follows that polygons $W$ and $V$ can each be subdivided into $n$ [polygonal] parts that are respectively congruent. Let $W_1, W_2, \ldots, W_n$ be the polygons obtained by subdivision of $W$, and $V_1, V_2, \ldots, V_n$ be polygons respectively congruent to them, obtained from subdivision of $V$.

---

[23] [Tarski's term was *rozumowanie apagogiczne*: apagogic reasoning.]

[24] [Here and twice afterward, Tarski referred to property 5 as "theorem 5."]

[25] [Tarski's term was *części*: parts.]

## 7.2 The Degree of Equivalence of Polygons (1931)

As we know, points $A_0$ and $A_n$ belong to the polygon $W$. Since $W$ is by assumption a convex polygon, all points of the segment $A_0A_n$, particularly $A_1, A_2, ..., A_{n-1}$, belong to $W$. In this way, we distinguished $n+1$ points $A_0, A_1, ..., A_n$ in polygon $W$ and at the same time we subdivided this polygon into $n$ parts $W_1, W_2, ..., W_n$. We infer, therefore, that at least two of the indicated points belong to the same part: for instance, points $A_k$ and $A_l$, [with] $k \neq l$, belong to the part $W_m$.

Since the polygons $W_m$ and $V_m$ are congruent, we can certainly find in polygon $V_m$ two points $B_k$ and $B_l$ such that segment $B_kB_l$ is congruent to segment $A_kA_l$. Points $B_k$ and $B_l$, belonging to $V_m$, must also belong to $V$; consequently, segment $B_kB_l$ cannot exceed the diameter of polygon $V$. Replacing $B_kB_l$ with the congruent segment $A_kA_l$, we obtain the formula

$$A_kA_l \leq \delta(V), \qquad (4)$$

where $k$ and $l$ are two distinct natural numbers lying between 0 and $n$ [inclusive]. In view of the evident contradiction between (3) and (4), we must reject hypothesis (1) and accept [property] 5 as proved.

[Property] 5 may be generalized, replacing the condition "$W$ is a convex polygon" by the condition "$W$ is a connected polygon" (that is, a polygon, any two points of which can be joined by a broken line, all of whose points belong to this polygon). The proof of this generalized theorem, which requires a slight modification of the original proof, we leave to the Reader.[26]

Using the definition of degree of equivalence and the properties of this notion given above, it is possible to investigate the degree of equivalence in reference to different concrete pairs of equivalent polygons. In general, we are able to find only certain upper and lower bounds for the degree of equivalence of each particular pair of polygons.

We obtain an upper bound immediately in those cases in which we have a drawing establishing the equivalence of polygons $W$ and $V$ by decomposition into respectively congruent parts: if in each of those the number of parts is $n$, then by virtue of the definition of degree of equivalence, we will have

$$\sigma(W, V) \leq n.$$

Also, in establishing an upper bound, we can sometimes use properties 3 and 4.

For a lower bound we have first of all the trivial bound $\sigma(W, V) \geq 2$, which by virtue of property 2 follows for any pair of polygons $W$ and $V$ that are equivalent but not congruent. It is significantly more difficult to obtain a stronger lower bound; for the moment, we have available only property 5.

Only in a few cases has it been possible to obtain a lower bound coinciding with an upper bound, and thus to determine exactly the degree of equivalence of the polygons.

---

[26] [Let $p$ be a broken line in $W$ from $A$ to $A_n$. Let $B_0 = A$, $B_n = A_n$, and when $0 < i < n$, let $B_i$ be an intersection of $p$ with the perpendicular to the diameter $\delta(W)$ at $A_i$. Then $B_kB_l \geq A_kA_l \geq \delta(W)/n$ whenever $k \neq l$, and Tarski's argument with $B_0, ..., B_n$ in place of $A_0, ..., A_n$ yields the result.]

We give here a few examples.

A. Let $W$ and $V$ be the square and rectangle, respectively, in figure 1 or in figure 2. Figure 1 yields $\sigma(W,V) \leq 5$. On the other hand, from figure 2 we obtain a stronger bound: $\sigma(W,V) \leq 3$. According to property 2 we have $\sigma(W,V) \geq 2$; in this case, property 5 does not give a better bound. Thus, finally,

$$2 \leq \sigma(W,V) \leq 3.$$

The question which of the numbers 2 and 3 is the value for $\sigma(W,V)$ remains open.

B. Let $V$ be a rectangle with edges $5/4\,a$ and $4/5\,a$ (as in the previous example) and let $U$ be a rectangle with edges $5/2\,a$ and $2/5\,a$. Figure 5 gives $\sigma(U,V) \leq 2$. Since, on the other hand, we have $\sigma(U,V) \geq 2$ according to property 2 or 5, we finally obtain

$$\sigma(U,V) = 2.$$

C. Let $W$, $V$, $U$ be the figures described in examples A and B. According to property 4 we have

$$\sigma(W,U) \leq \sigma(W,V) \cdot \sigma(U,V).$$

As we showed in [examples] A and B, $\sigma(W,V) \leq 3$ and $\sigma(U,V) = 2$; therefore, $\sigma(W,U) \leq 6$. In this case, however, instead of applying property 4, it is better to base the argument directly on figure 6, which yields $\sigma(W,U) \leq 4$. By property 2, finally, we have

$$2 \leq \sigma(W,U) \leq 4.$$

The question of finding the exact value of $\sigma(W,U)$ again remains open.

D. Let $W$ be a square with edge $a$ and let $V$ be a rectangle with edges $3a$ and $1/3\,a$. From figure 7 it can be seen that $\sigma(W,V) \leq 3$. On the other hand, applying property 5, we obtain

$$\sigma(W,V) \geq \frac{\delta(W)}{\delta(V)}.$$

It is easy to see that the diameters of the rectangles are their diagonals; in view of this,

$$\delta(V) = a\sqrt{82/9}, \quad \delta(W) = a\sqrt{2}, \quad \text{and therefore} \quad \sigma(W,V) \geq \sqrt{41/9} > 2.$$

Since $\sigma(W,V)$ is a natural number, the inequality $\sigma(W,V) > 2$ can be replaced by the inequality $\sigma(W,V) \geq 3$. The upper and lower bounds coincide, and consequently we have

$$\sigma(W,V) = 3.$$

We content ourselves with these examples.

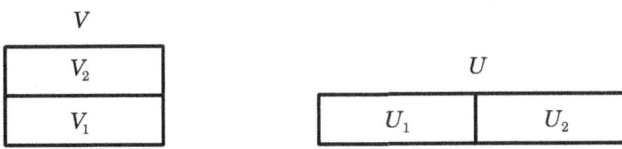

**Figure 5**

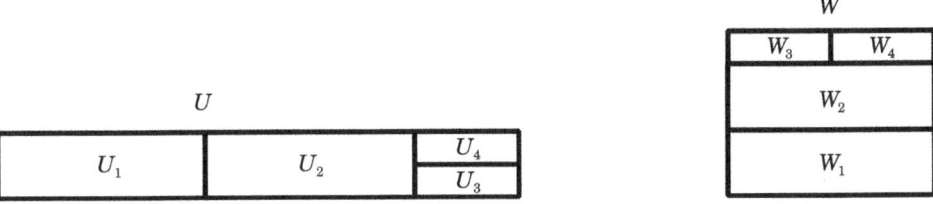

**Figure 6**

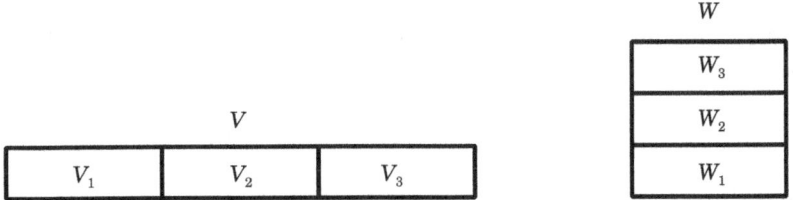

**Figure 7**

Referring to the first sentence of the present article, we repeat once more that in this part of elementary geometry nearly everything is left to do. A whole series of attractive topics to develop arises here, from which we extract the following.

Let $Q$ be a square with edge $a$, while $P$ should be a rectangle with edges $x \cdot a$ and $1/x \cdot a$, where $x$ is any positive real number. Polygons $Q$ and $P$ are clearly equivalent. It is easy to see that their degree of equivalence is a function of $x$; we shall denote it by the symbol $\tau(x)$. Thus, we set

$$\sigma(Q,P) = \tau(x).$$

The topic whose development we would keenly recommend would be a precise investigation of the function $\tau(x)$.

In the following theorems, we present some of the properties known to us.

I. *The function $\tau(x)$ is defined for all positive numbers and takes on as its values only positive whole numbers.*

II. $\tau(x) = \tau(1/x)$ *for every* $x > 0$.

These are direct consequences of the definition of the function $\tau(x)$ and of the definition of the degree of equivalence.

Moreover, denoting by the symbol $\lceil x \rceil$ the least integer $n$ not less than a given real number $x$ (and therefore satisfying the formula $n - 1 < x \leq n$), we have

III. $\tau(x) \leq 2 + \lceil \sqrt{x^2 - 1} \rceil$ *for any* $x \geq 1$.

We shall not give a proof of the above theorem. We remark only that this proof can be obtained by analyzing the proofs of two known theorems from the theory of equivalence of polygons, namely (1) the theorem of equivalence of two parallelograms whose bases and altitudes are respectively congruent, [and] (2) the theorem that the square constructed on a leg of an arbitrary right triangle is equivalent to the rectangle constructed from the hypotenuse and from the projection, on the hypotenuse, of the leg under consideration.[27] We recommend that the Reader give a complete proof for the theorem discussed, or at least for a few of its special cases, such as[28]

$$\tau(1\tfrac{1}{3}) \leq 3 \qquad \tau(2\tfrac{1}{4}) \leq 4 \qquad \tau(\sqrt{10}) \leq 5.$$

Theorem III establishes a certain upper bound for the function $\tau(x)$. In some cases in which $x$ is a rational number, it is possible to obtain other, often stronger, bounds for the function in question. Moreover, in establishing these bounds it is not necessary to

---

[27] Compare the proofs of both these theorems in the textbook Wojtowicz [1919] 1926, §177 and §§191–192. (By an analysis of the proof of the second of these theorems, we obtained figure 2, given earlier.)

[28] [Section 7.1 includes this proof of theorem III. Henryk Moese soon improved this result, in the paper Moese [1932] 2014 translated in section 7.3. The first and last special cases are straightforward applications, but the middle one required a new example subdivision, also provided by Moese.]

## 7.2 The Degree of Equivalence of Polygons (1931)

resort to subdividing the rectangles into figures that are not rectangles (in contrast with theorem III).

Omitting here the general case, we give the following easy theorem:

IV.  $\tau(n) \leq n$ *for every positive integer* $n$.

This follows from the following observation: a square with edge $a$ can be subdivided into $n$ congruent rectangles with edges $a$ and $1/n \cdot a$, from which a rectangle with edges $n \cdot a$ and $1/n \cdot a$ can then be constructed (compare figure 7 for $n = 3$).

An easy proof of [the following] theorem is based on property 5 of the degree of equivalence (compare example D given above):

V.  $\tau(x) \geq \sqrt{\dfrac{x^4 + 1}{2x^2}}$ *for every* $x > 0$.

We are unable at this point to strengthen in significant measure the lower bound for the function $\tau(x)$ established in the previous theorem.[29] However, a certain strengthening, devoid of greater significance, is presented by the following theorem, which we give here without proof:

$$\tau(x) \geq \sqrt{\dfrac{x^4}{2x^2 - 1}} \quad \text{for } x \geq 1.$$

Finally, we note a direct conclusion from theorem V:

VI.  $\tau(x) \to +\infty$ as $x \to +\infty$.

With the help of the above theorems, we can calculate the values of the function $\tau(x)$ for certain values of the argument—a few, anyway. Thus, we have $\tau(1) = 1$, $\tau(2) = \tau(1/2) = 2$, and $\tau(3) = \tau(1/3) = 3$ (concerning that last formula, compare example D). By a slightly different method it may be shown that $\tau(4) = \tau(1/4) = 4$: the justification for this formula we leave to the Reader.[30]

On the other hand, establishing the value of the function $\tau(x)$ for other values of $x$, even for integer values $x \geq 5$, still presents difficulties. In particular, we are unable at this point to prove the following theorem,[31] which seems highly probable:

$\tau(n) = n$ *for any positive integer* $n$.

---

[29] [A proof is included in section 7.1. Tarski did improve this result significantly with theorem C of the later paper Tarski [1932] 2014d translated in section 7.4.]

[30] [In the problem section of the same issue of *Młody matematyk*, Tarski posed the proof of $\tau(4) = 4$. This is an immediate consequence of a general theorem in the paper Tarski [1932] 2014d translated in section 7.4, but Tarski suggested another proof. See section 12.10.]

[31] [This conjecture was soon proved by Moese and others: see section 7.3.]

As another example of a theorem[32] unproved at this point, but probable, we quote the following statement:

$\tau(x) \geq 3$ *for every positive* $x$ *different from* $1/2$, $1$, *and* $2$.

This statement, together with theorem III, would permit calculation of values of the function $\tau(x)$ for infinitely many values of the argument. In fact, we would have $\tau(x) = 3$ for every $x$ satisfying the inequalities $1/\sqrt{2} \leq x \leq \sqrt{2}$ and $x \neq 1$.

From the remarks above it follows clearly that we are at the present moment still very far from precise knowledge of the course of values of the function $\tau(x)$.

---

*The editors encourage Readers to send in results of further investigations of topics touched on in the above article.*

## 7.3 Moese's Contribution (1932)

This section contains an English translation of Henryk Moese's paper *Przyczynek do problemu A. Tarskiego: "O stopniu równoważności wielokątów,"* [1932] 2014, which appeared in volume 2 of the journal *Parametr*. The journal's target audience was gimnazjum teachers and their most serious students. Moese was a gimnazjum teacher in Kępno, a small city in southwestern Poland.[33] In this paper he solved a problem posed in the [1931] 2014a paper of Alfred Tarski mentioned in the title, which is translated in section 7.2. Moese's work was fundamental for Tarski's subsequent [1932] 2014d paper, translated in section 7.4. For that reason, and because of its previous inaccessibility, it is included here in this volume devoted to Tarski. A draft English translation by Izaak Wirszup was published informally by the University of Chicago as Tarski and Moese 1952. The present translation was carried out independently, but used Wirszup's for confirmation.

This translation is meant to be as faithful as possible to Moese's original text. Its only intentional modernizations are punctuation, and some changes in symbols, where Moese's conflict with others used throughout this book. Some uses of alternative type styles for emphasis have been modified. Bibliographic references have been adjusted to conform with conventions of the present book. All [square] brackets in the translation enclose editorial comments inserted for clarification, often as footnotes. As an aspect of adjusting punctuation, the editors increased white space to enhance visual organization.

---

[32] [Tarski did write *theorem* (*twierdzenie*), but he refuted this conjecture in his later paper Tarski [1932] 2014d, translated in section 7.4.]

[33] For further information about *Parametr*, consult sections 7.1 and 9.7. For information about Moese, see the box on page 130.

HENRYK MOESE (Kępno)

# A Contribution to the Problem of A. Tarski, "On the Degree of Equivalence of Polygons"

In connection with the article *On the Degree of Equivalence of Polygons* by Dr. Alfred Tarski,[34] I would like to prove here some theorems concerning the function $\tau(x)$. On questions of notation, I refer Readers to the cited article by Tarski.

**Theorem 1.** $\tau(n) = n$ for every natural[35] number $n$.

*Proof* [which continues to page 148]. The part of the plane contained between two parallel lines at mutual distance $x$, we call simply an *x-strip*.[36] Figure 1, with its hatched part, sufficiently explains the definition of *x-strip*.

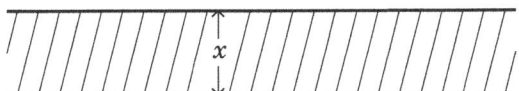

**Figure 1**

We define now a certain auxiliary function $\sigma_w(x)$. Let $W$ be an arbitrary bounded plane figure (that is, [one that] can be covered by a polygon). To cover this entire figure with $x$-strips, a certain number of strips must be used. The number of strips needed to cover the figure depends on the way of placing the strips on the figure. It can be shown easily that there is a least number of strips that can cover the entire figure. This minimum number of strips we denote precisely by the symbol $\sigma_w(x)$.[37]

**Lemma I.** If $W$ is a disk with radius $r$, then

$$\sigma_w(x) = \left\lceil \frac{2r}{x} \right\rceil \quad \text{for} \quad 0 < x \leq 2r.$$

(Here, $\lceil 2r/x \rceil$ denotes the least integer $n$ not less than the real number $2r/x$.)

---

[34] [Tarski [1931] 2014a, translated in section 7.2.]

[35] [For Moese, the natural numbers were evidently $1, 2, 3, \ldots$.]

[36] [Moese's term was *pas x*.]

[37] [The subscript in $\sigma_w$ is not a reference to the polygon in question—in this case $W$—but rather the first letter of the Polish word *wielokąt—polygon*.]

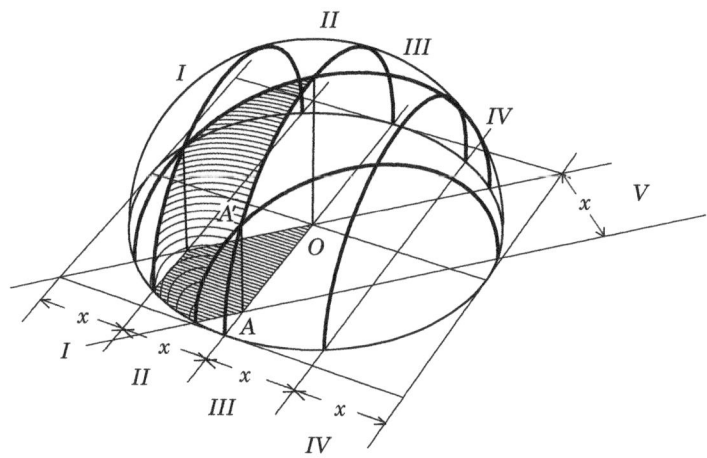

**Figure 2**

[Hemisphere over a disk with center $O$ and radius $r$]

*Proof.* Let us construct a hemisphere on the given disk (see figure 2).[38] Point $O$ is the common center of the disk and hemisphere; the radii of the disk and hemisphere equal $r$.

It is possible to establish the following correspondence between points belonging to the given disk and those lying on the surface of the hemisphere: let us draw from an arbitrary point of the disk $A$ a line perpendicular to the plane of the disk; this line will intersect the surface of the hemisphere at point $A'$, and we call points $A$ and $A'$ corresponding points.

Each point of the disk corresponds to just one point lying on the surface of the hemisphere, and vice versa. Points of the disk lying in the interior of an $x$-strip will correspond to points of the surface of the hemisphere lying in the interior of the spherical $x$-strip. However many $x$-strips are used to cover the disk, that many spherical $x$-strips cover the corresponding surface of the hemisphere.

In figure 2, the disk was covered by four $x$-strips $(x = r/2)$ running parallel, adjacent to one another and not overlapping each other, just as the four spherical $x$-strips $(x = r/2)$ running parallel, adjacent to one another but not overlapping each other, cover the surface of the hemisphere. The $x$-strips $V$ and $II$ have common points interior to the disk; similarly, the spherical $x$-strips $V$ and $II$ have common points. These common parts are hatched in figure 2.

---

[38] [Moese's verb was *opiszmy—let us describe*. His term *koło—circle—*has been translated here as *disk*. Figure 2 has been slightly simplified.]

## 7.3 Moese's Contribution

Now, instead of counting the least number of $x$-strips necessary to cover the whole disk, we will do the analogous task for the surface of the hemisphere. In both cases, on the basis of the correspondence established above, the result of the calculation must be the same. The solution of this problem for the hemisphere, however, is much easier than that for the disk.

The area of a spherical $x$-strip (on the hemisphere) is $\pi r x$. For a given hemisphere, this area depends solely on the altitude of the spherical $x$-strip, and not on the location of the strip on the hemisphere. Therefore, all spherical $x$-strips of the hemisphere have the same area. It follows from this that the least number of spherical $x$-strips that cover the entire hemisphere cannot exceed

$$\left\lceil \frac{2\pi r^2}{\pi r x} \right\rceil \quad \text{—that is,} \quad \left\lceil \frac{2r}{x} \right\rceil .{}^{39}$$

Nor can it be less than $\lceil 2r/x \rceil$. When $2r/x = n$ (where $n$ is an integer and positive), the least number of spherical $x$-strips that can cover the entire hemisphere must be $n$. Moreover, these strips cannot overlap each other, and thus they must run parallel to each other and adjoin each other. [On the other hand] if $2r/x$ is not an integer, then in covering the hemisphere with spherical $x$- strips, we have more freedom; the strips will overlap each other, but fewer than $\lceil 2r/x \rceil$ strips are not enough to cover the hemisphere completely.

Therefore, from the correspondence just established, it follows that

$$\sigma_w(x) = \left\lceil \frac{2r}{x} \right\rceil \quad \text{Q. E. D.}$$

For the function $\sigma_w(x)$ let us adopt the following notation:

$$\sigma_{kw}(x) \text{ for a square,} \quad \sigma_k(x) \text{ for a disk.}{}^{40}$$

Let us suppose that the edge of the square and the diameter of the disk are equal to $a$ (and thus the radius of the disk is $r = a/2$).

**Lemma II.** $\sigma_{kw}(x) = \sigma_k(x) = \lceil a/x \rceil$ when $0 < x \le a$.

*Proof.* From lemma I it follows that to cover a disk with the least number of $x$-strips it suffices to arrange these strips parallel to each other so that [they are] adjacent to each other, starting from the circumference of the disk. (See figure 3.) We arrange the $x$-strips parallel to the side $a$ of the square. Then it is apparent that the same least number of strips that covers the whole disk covers the square as well; and thus it must be [the case that] $\sigma_{kw}(x) = \sigma_k(x)$, [Q. E. D.]

---

[39] [This part of Moese's original paper was unclear, probably an editorial fault. The area of the hemisphere is $2\pi r^2$. Covering the hemisphere with $n = \lceil 2r/x \rceil$ parallel adjoining nonoverlapping strips confirms the displayed claim. The displayed material originally included the first sentence of the next paragraph.]

[40] [These subscripts are the first letters of the Polish words *kwadrat—square—*and *koło—circle*.]

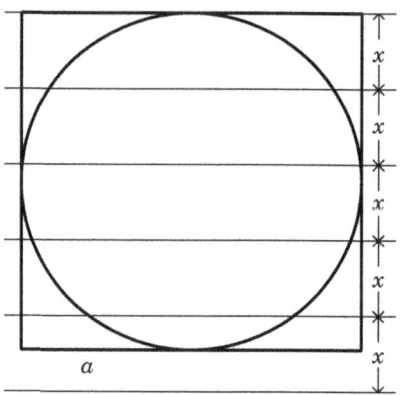

**Figure 3**

**Lemma III.** $\tau(x) \geq \sigma_{kw}(x \cdot a)$ when $0 < x \leq 1$.

*Proof.* It is clear that [when] $0 < x \leq 1$, each of the $\tau(x)$ respectively congruent polygons into which a square with side $a$ and an equivalent rectangle with sides $x \cdot a$ and $a/x$ may be subdivided must be wholly covered by an $(x \cdot a)$-strip. It follows from this that it cannot be [the case that] $\tau(x) < \sigma_{kw}(x \cdot a)$, Q. E. D.

Let us return now to the *proof* of theorem 1. Let us suppose that $1/x = n$, where $n$ is a positive integer; then on the basis of the lemmas, it must be [the case] that $\tau(n) \geq n$. Since on the other hand we have $\tau(n) \leq n$, therefore at last,

$$\tau(n) = n \qquad \text{Q. E. D.}$$

**Theorem 2.** $\tau(x) \leq n + 2$ when $n < x \leq n + 1$: in other words,

$$\tau(x) \leq 1 + \lceil x \rceil \quad \text{when} \quad x \geq 1.$$

This theorem strengthens the upper bound of the function $\tau(x)$ given in theorem III of the cited article Tarski [1931] 2014a.[41]

The *proof* follows directly from figure 4, drawn for $\lceil x \rceil = 4$.

**Theorem 3.** $\tau\left(\dfrac{n+1}{n}\right) = \tau\left(\dfrac{n}{n+1}\right) = 2$ for every natural number $n$.

Again, sufficient *proof* of this theorem is given by figure 5, where $n = 4$.

---

[41] [Translated in section 7.2.]

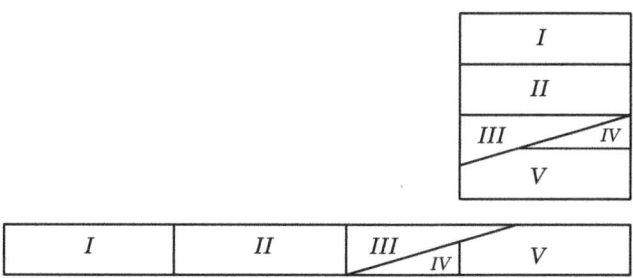

**Figure 4**

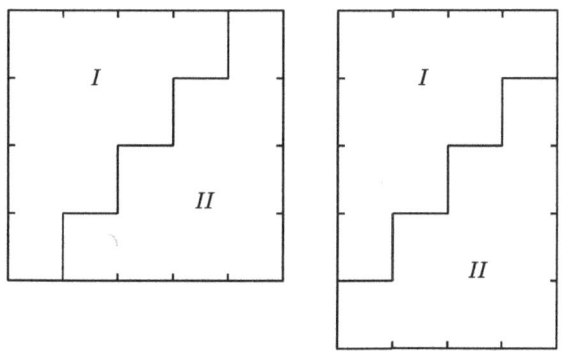

**Figure 5**

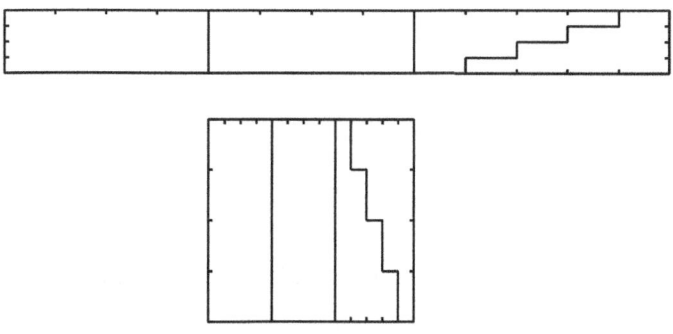

**Figure 6**

**Theorem 4.** $\tau\left(n+\dfrac{1}{k}\right) \leq n+1$, where $n, k$ are any positive integers.

This theorem is a generalization of theorem 3.

[*Proof.*] The truth of this theorem is made clear by figure 6, drawn for $n = 3$, $k = 4$. Figure 6 shows that $\tau(3\frac{1}{4}) \leq 4$.

And [figure 7 illustrates] yet another trick: polygon *II* should be understood in the broader sense (see the remark in Tarski's [1931] 2014a article[42]). Figure 7 yields $\tau(1\frac{2}{5}) \leq 3$. More such tricks could be given; they all become insignificant, however, insofar as they do not give a stronger bound for the function $\tau(x)$ than theorem 2. For $x = 1\frac{2}{5}$, theorem 1 gives the same bound as figure 7.

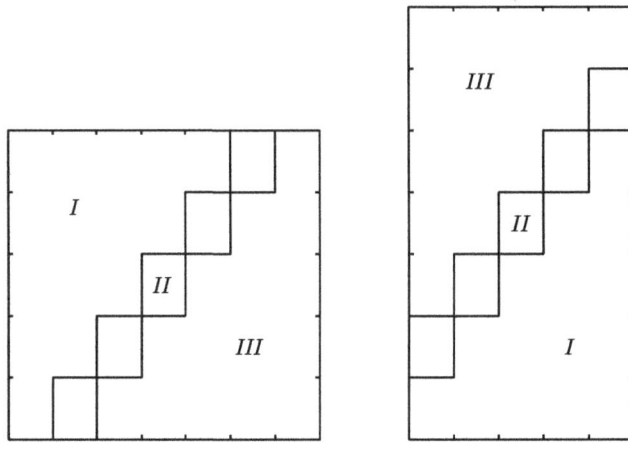

**Figure 7**

The exercises suggested by Tarski [in connection with his theorem III] now would have the following solutions:

$$\tau(1\tfrac{1}{3}) = 2, \quad \tau(2\tfrac{1}{4}) \leq 3, \quad \tau(\sqrt{10}) \leq 5, \quad \text{and moreover,} \quad \tau(\sqrt{15}) \leq 5.$$

In view of the above, I dare to mention the following probable theorem—

$$\tau(x) = n + 2 \quad \text{when} \quad n < x \leq n+1$$

—with the following exception:

---

[42] [Translated in section 7.2.]

$$\tau\left(n+\frac{1}{k}\right) = n+1,$$

where $n$ and $k$ are any natural numbers. The question of proof remains open.[43] Perhaps some Reader might prove the above theorem for $n = 1$, or even find a trick, like the previous ones, that would disprove the theorem in question.

For $n = 2$,[44] the following theorem seems to me very likely:

> *In case that the square can be divided into just two polygons, so that these polygons in a different arrangement can form an equivalent rectangle, both of these parts of the subdivision of the square must be right-angled polygons (interior angles 90° or 270°) congruent to each other.*

## 7.4 Remarks on the Degree of Equivalence (1932)

This section contains an English translation of Alfred Tarski's paper *Uwagi o stopniu równoważności wielokątów*, [1932] 2014d, which appeared in volume 2 of the journal *Parametr*.[45] The journal's target audience was gimnazjum teachers and their most serious students. The paper continued the discussion begun in Tarski [1931] 2014a and Moese [1932] 2014, translated in sections 7.2 and 7.3. A draft English translation by Izaak Wirszup was published informally by the University of Chicago as Tarski and Moese 1952.[46] The present translation was carried out independently, but used Wirszup's for confirmation.

The translation is meant to be as faithful as possible to Tarski's original text. Its only intentional modernizations are punctuation, and some changes in symbols, where Tarski's conflict with others used throughout this book. Some uses of alternative type styles for emphasis, enunciations, and personal names have been modified. Bibliographic references and personal names have been adjusted to conform with conventions of the present book. All [square] brackets in the translation enclose editorial comments inserted for clarification, often as footnotes. As an aspect of adjusting punctuation, the editors increased white space to enhance visual organization.

---

[43] [In his [1932] 2014d paper, translated in section 7.4, Tarski claimed that the proof for the case $n = 1$ was known, but too involved to include there, and that all other cases remained open. See the discussion at the end of 7.1.]

[44] [Moese probably meant $n = 1$ here.]

[45] For further information about *Parametr*, consult sections 7.1 and 9.7.

[46] The original paper and that draft were reprinted in Tarski's 1986a *Collected Papers*, volume 1, 595–602. Some serious typographical errors were introduced in those editions.

DR. ALFRED TARSKI (Warsaw)

# Remarks on the Degree of Equivalence of Polygons

In an article published in *Młody matematyk*,[47] I posed several problems concerning the degree of equivalence of polygons. Evidently, the article was written with a "lucky hand": the topic that I touched upon aroused the interest of several mathematicians. Thanks to their investigations, various hypotheses that I suggested were either established or disproved. And as to the main problem proposed in the cited article, about a complete investigation of the function $\tau(x)$, at the present moment not much is still lacking for a definitive solution.

In the following remarks, I wish to set down all the results obtained up to the present moment concerning the function $\tau(x)$, referring to the preceding article by Henryk Moese in this issue of *Parametr* and to some investigations by Adolf Lindenbaum and Zenon Waraszkiewicz[48] with which I am acquainted but which have not yet been published; and in addition, to extract from them certain facts of a more general nature. Moreover, I intend to propose a few further problems in the same field.

Recall that we call the *degree of equivalence of two equivalent polygons* $W$ and $V$, symbolically $\sigma(W,V)$, the least natural number $n$ satisfying the following condition: polygons $W$ and $V$ can [each] be divided into $n$ respectively congruent polygons. In particular, the degree of equivalence of a square with edge $a$ and a rectangle with edges $x \cdot a$ and $1/x \cdot a$ is denoted by the symbol $\tau(x)$.

We shall need the concept of width for a given geometric figure. As is well known, that part of a plane bounded by two parallel lines carries the name *strip*; the distance between these lines we call the *width of the strip*. The width of the narrowest strip covering a plane figure $F$ we shall call the *width of figure $F$*, and we shall denote it by the symbol $\omega(F)$.[49] It is not hard to show that every figure $F$ that can be covered by any strip at all possesses a definite width. For example, the width of a rectangle is equal to the length of the smaller of its sides; the width of an equilateral triangle is its altitude; and of a circle, its diameter. It is also obvious that if figure $F$ is part of figure $G$, then $\omega(F) \leq \omega(G)$; if figures $F$ and $G$ are congruent, then $\omega(F) = \omega(G)$.

---

[47] Tarski [1931] 2014a [translated in section 7.2].

[48] [For more information on Moese and Waraszkiewicz and on Lindenbaum see the boxes on pages 130 and 132 and in section 14.3. With their names Tarski used the honorifics *pan*, *pan magister*, and *pan doktor*, respectively. *Pan* means *Mr.*; *magister* indicated that Waraszkiewicz had a master's degree but no doctorate yet.]

[49] [Tarski used $\pi$ for width; this translation uses $\omega$ to avoid confusion with the numerical constant $\pi$.]

## 7.4 Remarks on the Degree of Equivalence (1932)

It is possible to justify the following theorem:

**A.** *If a figure $F$ contains in itself, as a part, a disk with diameter equal to the width of the figure* (for example, if figure $F$ is a disk or a parallelogram) *and if, moreover, we subdivide this figure into any $n$ parts $C_1, C_2, \ldots, C_n$, then*

$$\omega(F) \leq \omega(C_1) + \omega(C_2) + \cdots + \omega(C_n).$$

*Proof.* For the case in which figure $F$ is a disk, the proof is an almost word-for-word repetition of the proof of lemma I in the cited article by Moese.

For the general case, we argue in this way. In figure $F$ inscribe a disk $K$ of the same width. Some of the parts $C_1, C_2, \ldots, C_n$ of figure $F$ must have points in common with the disk $K$: for simplicity, let these be $C_1, C_2, \ldots, C_p$, where $p \leq n$. Further, for $k = 1, 2, \ldots, p$, let $C'_k$ be the common part of figure $C_k$ and the disk $K$. It is easy to see that the disk $K$ was subdivided into parts $C'_1, C'_2, \ldots, C'_p$; therefore, as we already know,

$$\omega(K) \leq \omega(C'_1) + \omega(C'_2) + \cdots + \omega(C'_n).$$

On the other hand, $\omega(C'_k) \leq \omega(C_k)$ for $k = 1, 2, \ldots, p$; therefore,

$$\omega(K) \leq \omega(C_1) + \omega(C_2) + \cdots + \omega(C_p)$$
$$\leq \omega(C_1) + \omega(C_2) + \cdots + \omega(C_n), \qquad \text{Q. E. D.}$$

The proof that all parallelograms, in particular, satisfy the assumptions of the above theorem does not present the Reader any difficulties.[50]

From theorem A we derive an important consequence concerning the degree of equivalence:

**B.** *If $W$ is a polygon containing in itself a disk as a part, with diameter equal to the width of the polygon* (for example, if $W$ is a parallelogram), *and $V$ [is] any polygon equivalent to it, then*

$$\sigma(W, V) \geq \frac{\omega(W)}{\omega(V)}.$$

*Proof.* According to the definition of degree of equivalence, polygons $W$ and $V$ can be subdivided into $\sigma(W, V) = n$ respectively congruent polygons: $W$ into $W_1, W_2, \ldots, W_n$, and $V$ into $V_1, V_2, \ldots, V_n$. According to theorem A,

$$\omega(W) \leq \omega(W_1) + \omega(W_2) + \cdots + \omega(W_n).$$

Moreover, obvious relations hold:

---

[50] [The proof requires a rather tedious school-geometry argument. Tarski proposed it as part of exercise 4 at the end of this paper.]

$$\omega(W_1) = \omega(V_1) \le \omega(V), \quad \omega(W_2) = \omega(V_2) \le \omega(V),$$
$$\ldots, \omega(W_n) = \omega(V_n) \le \omega(V).$$

Therefore, $\omega(W) \le \omega(V) \cdot n = \omega(V) \cdot \sigma(W, V)$, and thus

$$\sigma(W, V) \ge \frac{\omega(W)}{\omega(V)}, \qquad \text{Q. E. D.}$$

It is worthwhile to compare the theorem proved a moment ago with theorem 5 in my previous article.[51] Both of these theorems establish certain lower bounds for the degree of equivalence:

$$\sigma(W, V) \ge \frac{\delta(W)}{\delta(V)} \quad \text{and} \quad \sigma(W, V) \ge \frac{\omega(W)}{\omega(V)}.$$

In many instances the second inequality gives a much better result. For example, if $W$ is a rectangle with edges 16 and 12, and $V$, a rectangle with edges 192 and 1, then from theorem 5 we obtain[52]

$$\sigma(V, W) \ge \sqrt{\frac{192^2 + 1^2}{16^2 + 12^2}} = \sqrt{92.1625},$$

and thus $\sigma(W, V) \ge 10$; however, theorem B gives $\sigma(W, V) \ge 12$. On the other hand, one ought to notice that theorem 5 has a significantly wider scope of applicability than theorem B: the latter, we can at the present time prove only for certain special polygons (not even for triangles), while in theorem 5 the assumption that $W$ be a convex or even connected polygon is sufficient. For this purpose it would be worthwhile to return to theorem A: should it be possible to extend that theorem to arbitrary convex figures, we would immediately obtain the desired strengthening of theorem B.

In theorem V of my article,[53] I gave a certain lower bound for the function $\tau(x)$; thanks to theorem B this bound can be considerably strengthened:

**C.** $\tau(x) \ge \lceil x \rceil$ *for every number* $x \ge 1$.

*Proof.* Let $W$ be a square with edge $a$, while $V$ [should be] a rectangle with edges $x \cdot a$ and $1/x \cdot a$, [where] $x \ge 1$. Then $\sigma(W, V) = \tau(x)$, $\omega(W) = a$, and $\omega(V) = 1/x \cdot a$; and thus, according to theorem B, $\tau(x) \ge a/(1/x \cdot a) = x$. Moreover, since $\tau(x)$ is a whole number, and the symbol $\lceil x \rceil$ denotes the least whole number $\ge x$, then $\tau(x) \ge \lceil x \rceil$, Q. E. D.

---

[51] [Tarski [1931] 2014a, translated in section 7.2. There, that result is refered to as *property* 5.]

[52] [In the original, the first term of the following inequality was misprinted as $\tau(V, W)$.]

[53] [Theorem V is different from theorem (or property) 5.]

An upper bound that I had established in theorem III [of that article] has also been strengthened. This is theorem 2 in Moese's article:[54]

**D.**  $\tau(x) \le 1 + \lceil x \rceil$ *for every number* $x \ge 1$.

Recalling that $\tau(x)$ is a whole number, we derive immediately from theorems C and D the following conclusion:

**E.** *For every number* $x \ge 1$, *either* $\tau(x) = \lceil x \rceil$ *or* $\tau(x) = 1 + \lceil x \rceil$.

In view of the above theorem, for every value $x \ge 1$ we are able to establish the value of $\tau(x)$ within accuracy $1$.[55] It remains only to investigate for which values $x$ the function $\tau(x)$ takes on each of its two possible values. A partial answer to this question we find in theorem

**F.** *If* $x$ *is a number of the form* $x = n + 1/p$ *where* $n$ *and* $p$ *are natural numbers* $\ge 1$, *then* $\tau(x) = \lceil x \rceil = n + 1$.

Theorem F disproves one of my conjectures, according to which $\tau(x)$ should be $\ge 3$ for all values $x$ except $1/2$, $1$, and $2$. On the other hand, as a particular case of this theorem, a confirmation of another hypothesis, also suggested by me, can be obtained:

**G.** $\tau(n) = n$ *for every natural number* $n$.

It seems probable that numbers $x$ of the form $n + 1/p$ are the only numbers satisfying the condition $\tau(n) = \lceil x \rceil$. At this point, we are able to prove this conjecture only for the case $x \le 2$:

**H.** *If* $1 < x \le 2$, *then in order that* $\tau(x) = \lceil x \rceil$, *it is necessary and sufficient that* $x$ *be of the form* $x = 1 + 1/p$, *where* $p$ *is any natural number*.

I shall not include the proof here. It is somewhat complicated and requires some subtle methods of reasoning. From the proof of this follows, among other things, that in the case in which a square can be subdivided into two parts from which a rectangle can be constructed, the only possible method of subdivision is that described by Moese in the proof of theorem 3 in his article.[56]

If it should be possible to remove the condition $x \le 2$ from the assumptions of the above theorem, the problem of the function $\tau(x)$ would finally be solved. However,

---

[54] [Moese [1932] 2014, translated in section 2.5 of the present book. Tarski referred here to a figure 1, but that seems incorrect.]

[55] [In the original this last phrase was *z dokładnością do 1*.]

[56] [This conjecture was proposed formally at the end of Moese's article. Tarski referred here to a figure 2, but that seems incorrect. For more about theorem H and its consequences see section 8.5.]

reflecting on the challenges thus far and on the proof of theorem H [suggests that] this problem does not belong among the completely easy ones.

I must now devote a few words to the question of "author's priority" with respect to the results presented here. The question is somewhat complicated, as is usual in these situations, when one and the same group of problems interests several people simultaneously, and in addition some of them collaborate with each other. Entering into play were a few Warsaw mathematicians; Moese investigated these problems completely independently. The earliest result chronologically is theorem G. Waraszkiewicz proved it first; the same result was obtained later, although completely independently, by Moese. The ideas in both proofs are fairly closely related. Waraszkiewicz's reasoning is somewhat more complicated, and for this reason it will not appear in *Parametr*. By analyzing the proofs of theorem G, I came to the more general theorems A, B, and C. (By the way, theorem C, even if not clearly formulated by Moese, is contained in his lemmas II and III.) Theorems D and F come from Moese: F in a somewhat weaker formulation with the symbol ≤ instead of = in the conclusion. However, they were obtained independently in Warsaw as well: Lindenbaum proved theorem F, and D [was proved by] Bronisław Knaster and the author of these remarks. The results obtained bring to light the last hypothesis concerning the course of values of the function $\tau(x)$, which I wrote about here. A certain special case of this hypothesis was confirmed by Lindenbaum and Waraszkiewicz, to whom we owe theorem H.

As Lindenbaum noticed, *all the results so far obtained for the function $\tau(x)$ still hold when, in the definition of this function, the square is replaced by any rectangle.*

In conclusion, I propose a few more problems, closely connected with the concept of degree of equivalence.

Besides the function $\tau(x)$, several analogous functions may be defined, whose courses of values are not yet exactly known. For example, two parallelograms with equal bases and altitudes present one of the simplest examples of equivalent polygons. The degree of equivalence of such parallelograms depends, among other things, upon the magnitude of their angles. Let us suppose for simplicity that one parallelogram is a square and the second has an acute angle $\varphi$, and denote the degree of their equivalence by the symbol $T(\varphi)$. On the one hand using theorem B and on the other hand by analyzing the proof of the theorem on the equivalence of parallelograms given in elementary geometry textbooks,[57] it can be shown that

$$\lceil \csc \varphi \rceil \leq T(\varphi) \leq 1 + \lceil \cot \varphi \rceil.$$

It is also not hard to prove that in some cases both of the indicated bounds for the function $T(\varphi)$ coincide, and the value of the function is thus uniquely determined; in others, however, these bounds differ by 1. One is then concerned with establishing the value of the function $T(\varphi)$ in these situations, in which $\lceil \csc \varphi \rceil < 1 + \lceil \cot \varphi \rceil$.

---

[57] See Wojtowicz [1919] 1926, §177, [and page 128 in section 7.1 of the present book].

The functions $\tau(x)$ and $T(\varphi)$ remain in a close relationship to a certain more general concept, namely the degree of irregularity of a polygon. If we agree to consider a square as the most "regular" polygon, then the degree of equivalence of a polygon $W$ and a square equivalent to it might be called the *degree of irregularity of this polygon*, symbolically $\rho(W)$: the greater $\rho(W)$, the less regular the polygon $W$. If in particular, $W$ is a rectangle with sides $a$ and $b$, where $a \geq b$, then

$$\rho(W) = \tau\left(\sqrt{a/b}\right).$$

If $W$ is a parallelogram with base and altitude equal to $a$ and an acute angle $\varphi$, then $\rho(W) = T(\varphi)$. It would be an interesting thing to know the general properties of the new concept.

Finally, there are many problems concerning the *number of methods for [achieving] the best subdivision of equivalent polygons*. Let $W$ and $V$ be equivalent polygons and let $\sigma(W,V) = n$. The question is, how many methods can be used to subdivide these polygons into $n$ respectively congruent polygons? (We regard two subdivision methods as distinct if the subdivision nets are not congruent to each other.) We know situations in which there is only one method of subdivision. For example, as I mentioned in connection with theorem G, this happens when $W$ is a square with side 1 and $V$ [is] a rectangle with sides $(p+1)/p$ and $p/(p+1)$, where $p$ is any natural number. Again, in other cases there are many methods of subdivision. For example, if $W$ is a square with edge 1 and $V$ [is] a rectangle with edges $7/5$ and $5/7$, then we know at the present time three different methods. Two of those are mentioned in the article by Moese,[58] and I mentioned a third in my previous article.[59] It would be an interesting thing to establish which methods of subdivision are possible in this or other special cases.

## Exercises

1. What is the width $\omega(W)$ and the diameter $\delta(W)$ in the case that polygon $W$ is (1) a triangle, (2) a trapezoid, (3) a kite?[60]

2. Determine the width $\omega(F)$ of a figure $F$ that is a sector of a disk with radius $r$ and central angle $\varphi$. Investigate the change in $\omega(F)$ with respect to $\varphi$ and sketch its graph $(0° \leq \varphi \leq 360°)$.

3. [Suppose] $A$, $B$, and $C$ are the three vertices of an equilateral triangle with edge $a$. With point $A$ as center, draw the arc $BC$ that is less than a semicircle, and do the same with points $B$ and $C$. Let $W$ be the "circular polygon" bounded by these

---

[58] Remarks in connection with figure 5 in Moese [1932] 2014 [section 7.3, page 149].
[59] Tarski [1931] 2010a, figure 2 with its dimensions altered [section 7.2, page 136].
[60] [A *kite* is a quadrilateral with two pairs of adjacent congruent edges; Tarski's term for it was *deltoid*.]

three arcs. Calculate the width $\omega(W)$ and diameter $\delta(W)$. Repeat the exercise with any $n$-sided regular polygon, where $n$ is [an arbitrary] odd natural number.

4. Show that every parallelogram $R$ contains within itself, as a part, a disk with diameter $\omega(R)$. Check whether this theorem is [also] valid for a kite, a triangle, and a trapezoid.

5. Show that the area $s(W)$ of any polygon $W$ satisfies the inequality
$$s(W) < \delta(W) \cdot \omega(W).$$

6. If $W$ is a convex figure contained in the plane then
$$s(W) \geq 1/2 \cdot \delta(W) \cdot \omega(W)$$
($s$ —area of the figure, $\delta$ —diameter, $\omega$ —width).

Moreover, in the above formula we will have equality if and only if $W$ is a triangle.[61]

7. [Suppose] $W$ is a square with edge $a$, while $V$ [is] a rectangle with edges $21/16$ and $16/21$. Give at least three distinct methods for subdividing these two polygons into three respectively congruent parts. Investigate all possible methods for such a subdivision.

8. [Suppose] $W$ is a square with edge $a$, while $V$ [is] a parallelogram with base and altitude equal to $a$ and an acute angle $\varphi$. Show that
$$\lceil \csc \varphi \rceil \leq \sigma(W,V) \leq 1 + \lceil \cot \varphi \rceil.$$

Show that the numbers $\lceil \csc \varphi \rceil$ and $1 + \lceil \cot \varphi \rceil$ differ by at most 1, and investigate for which values of $\varphi$ they are equal.

9. [Suppose] $W$ is a parallelogram with edges 100 and 87 and acute angle 88°, and $V$ [is] an equivalent parallelogram with base 100 and acute angle 10°. Find $\sigma(W,V)$.

---

[61] Exercise 6 was suggested by A. L. from Warsaw [probably Lindenbaum].

# 8
# Research Threads

Previous chapters have included translations of research papers published by Alfred Tarski and others in 1921, 1924, and 1931–1932. They were accompanied by extensive background discussions of their topics and of Tarski himself and others involved in his career. The present chapter will give the reader a sense of the impact of that research, and it will provide references to closely related subsequent developments by Tarski and by others who continued his investigations.

The emphasis is on work in Poland that has not been widely discussed elsewhere. This chapter mentions only briefly some of Tarski's research in logic, which brought him great fame but is not emphasized in this book because it is already readily accessible. That discussion is connected with some short translations included in the supplementary chapter 15.

In 1999 Steven R. Givant published *Unifying Threads in Alfred Tarski's Work*, a wonderful biographical study of Tarski based on their collaboration and personal association during 1973–1983. Givant showed how Tarski, throughout his career, would return again and again to various research problem areas, each time advancing them, producing longer and longer threads of results. Some of these threads are composed of very tightly intertwined filaments. That pattern provides an organization for the study of Tarski's research. Some of the threads continued long after Tarski relocated to the United States in 1939. Others started anew there, sometimes splitting off and becoming independent from threads started earlier in Poland.

The sections of this chapter will follow threads and filaments of Tarski's research that began during 1920–1931:

| | |
|---|---|
| *Two Threads in Logic* | 1920– |
| *Thread: Well-Ordering and Finiteness* | 1921– |
| *Thread: Cardinal Arithmetic and the Axiom of Choice* | 1924– |
| *Thread: Application of Set Theory to Geometry* | (with four filaments) |
|    *Equidecomposability in Elementary Geometry* | 1931–1932 |
|    *Equidecomposability in Set Theory* | 1924– |
|    *The Measure Problem* | 1930– |
|    *Generalizing Cantor–Bernstein* | 1927– |

## 8.1 Two Threads in Logic

One thread of Alfred Tarski's research in logic began with his 1923 doctoral dissertation, supervised by Stanisław Leśniewski. Tarski showed that material equivalence and universal quantification over propositional variables could constitute the primitive notions of Leśniewski's system of logic. It is described in more detail in chapter 3, page 39. Leśniewski acknowledged the importance of that work and of Tarski's less formal contributions, but Tarski did not pursue them further. For information on developments influenced by this thread of Tarski's research, readers should consult works on Leśniewski's logic.

Partly in collaboration with his former teacher Jan Łukasiewicz, Tarski developed a second thread, a framework for research in logic and for its presentation, which has since become a standard in the mathematical world. Already in 1920, as a Warsaw student, Tarski presented a talk that included the deduction theorem, which would be fundamental for that framework. (See chapter 3, page 37.) Two small collaborations of Łukasiewicz and Tarski are described and translated in sections 15.1 and 15.4: Tarski's discussion of Łukasiewicz's 1925 paper *On a Certain Way of Understanding the Theory of Deduction*, and his contribution to Łukasiewicz's 1928–1929 paper *On Definitions in the Theory of Deduction*.[1] As a university dozent, Tarski worked closely with Łukasiewicz's research seminar, and in 1929 officially became Łukasiewicz's assistant, supervising student research. From this activity stemmed several publications on the fundamental ideas of logic, notably Tarski's [1930] 1983b paper *On Some Fundamental Concepts of Metamathematics*. Tarski developed his celebrated definition of *true sentence* as part of this program. His first publication on that subject, *On the Concept of Truth in Reference to Formalized Deductive Sciences* ([1930–1931] 2014), is discussed and translated in section 15.6.[2] Several key ideas in Tarski's approach stemmed also from his association with Leśniewski.

During the 1930s Tarski published expanded versions of those two works, which became the core of his framework for logic. They have become particularly well known through their translations in Tarski's [1956] 1983 book *Logic, Semantics, Metamathematics: Papers from 1923 to 1938*. A great expansion of research in model theory and formal semantics after World War II was based in large part on that framework. Chapter 18 of the present book lists surveys of Tarski's work and influence in these areas.

Occasionally Tarski published comments on subjects in philosophical logic. Sections 15.2, 15.3, 15.8, and 15.10 contain discussions and translations of his 1927 comments on the theories of action and causality, and his 1936 comments on idealism and metaphysics.

---

[1] Tarski [1925] 2014 and Łukasiewicz 1925; Łukasiewicz [1928–1929] 2014.

[2] A smaller contribution on this subject, Tarski's commentary on the paper Kokoszyńska 1936a, is translated and discussed in section 15.11.

## 8.2 Thread: Well-Ordering and Finiteness

Another thread, on the notions of well-ordering and finiteness, also began while Alfred Tarski was a student. His 1921 [2014] paper *A Contribution to the Axiomatics of Well-Ordered Sets*, translated in chapter 2, stemmed from the seminar of Stanisław Leśniewski, but Wacław Sierpiński seems to have had greater influence on this thread of Tarski's work. Tarski's 1924c abstract *Sur les principes de l'arithmétique des nombres ordinaux (transfinis)* summarized a presentation to the Warsaw section of the Polish Mathematical Society on axioms for the ordinal number system; it has a similar flavor. So does Tarski's major 1924b paper *Sur les ensembles finis*, which is concerned with various possible definitions of the notion of finiteness and the relationship between their equivalence and the axiom of choice. This highly readable paper, Tarski's *Habilitationsschrift*, earned for him the *venia legendi*—qualification to teach at a university.[3]

During the late 1930s, Tarski and his student Andrzej Mostowski returned to develop this thread much further, within the framework that Tarski had developed for logical research. World War II and the Cold War prolonged the delay. In 1949 Mostowski and Tarski published an abstract about their work, *Arithmetical Classes and Types of Well-Ordered Systems*. A major study by John E. Doner, Mostowski, and Tarski finally appeared in 1978, after Mostowski's death: *The Elementary Theory of Well-Ordering—A Metamathematical Study*. Doner had earned the doctorate in 1968 under Tarski's supervision in Berkeley.

## 8.3 Thread: Cardinal Arithmetic and the Axiom of Choice

As noted in chapter 3, the axiom of choice was a major interest of Alfred Tarski's teacher Wacław Sierpiński, and it came to play a major role in some of Tarski's own research threads. During the mid-1920s, Tarski published three papers on theorems related to cardinal arithmetic and to the axiom:

- *Sur quelques théorèmes qui équivalent à l'axiome du choix*
- *Quelques théorèmes sur les alephs*
- *Communication sur les recherches de la théorie des ensembles*[4]

The last was a collaboration with Adolf Lindenbaum. The publications in this thread are accessible, and thus beyond the scope of the present book. For further information on them, consult the 1988 survey by Azriel Lévy.

---

[3] Jadacki 2003a, 144–145.

[4] Tarski 1924e, Tarski 1925b, and Lindenbaum and Tarski 1926.

## 8.4 Thread: Application of Set Theory to Geometry

This thread would connect Alfred Tarski's research in set theory and logic to his interest in school mathematics. It includes his [1924] 2014b paper *On the Equivalence of Polygons* and the famous [1924] 2014 Banach–Tarski paper, which are described in sections 4.3 and 4.4, and translated in full in chapters 5 and 6. This thread also includes the more elementary articles on the degree of equivalence of polygons, featured in chapter 7.

Another thread of Tarski's geometric studies, axiomatics, began around 1927 and continued into the 1950s.[5] Readily accessible now, that work is not covered in this book. For information about that thread, see the surveys by Lesław W. Szczerba (1986) and by Tarski and Steven R. Givant (1999).

The following sections discuss four filaments of this applied-set-theory thread:

- equidecomposability in elementary geometry (1931–1932)
- equidecomposability in set theory (1924–)
- the measure problem (1930– )
- generalizing the Cantor–Bernstein theorem (1927– )

## 8.5 Equidecomposability in Elementary Geometry

This elementary-geometry filament probably took root during the mid-1920s, when Alfred Tarski began teaching the subject. But it surfaced only later, with his [1931] 2014a and [1932] 2014d papers on the degree of equivalence of polygonal regions $P$ and $Q$ with the same area: the smallest integer $n$ such that $P$ and $Q$ can each be subdivided into $n$ polygonal subregions with disjoint interiors such that pairs of corresponding subregions are congruent. A problem left open in the first of these papers was solved by Henryk Moese in the paper [1932] 2014, and that work was discussed and utilized in Tarski's second paper. In a [1937] 1938 oral presentation, Adolf Lindenbaum proved the final theorem H of Tarski's second paper, which had been attributed to him and Zenon Waraszkiewicz; but he did not publish details. Tarski's first paper on this filament was reprinted in 1975 and translated and published in Hebrew in Tarski [1931] 1951. Draft English translations of all three of these Polish articles constituted the obscure publication Tarski and Moese 1952. Beyond that, these works have had little impact.[6] New translations and a discussion of all three papers are included in chapter 7 and section 9.8, with the hope of bringing them to light.

---

[5] The geometric-axiomatics thread may have begun as early as 1924: see the first paragraph of Tarski [1924] 2014b, §1, translated in section 5.1.

[6] The present editors have found only one reference by another author to this material: Howard W. Eves (1963–1965, volume 1, chapter 5) mentioned inequality III of Tarski [1931] 2014a, crediting Tarski without citing any specific source or mentioning the strengthened version of the inequality in Tarski [1932] 2014d. Eves may have learned about inequality III orally, for he also credited Walter B. Carver and did not mention that the inequality was incorrectly typeset in the draft translation Tarski and Moese 1952. The present editors have not been able to confirm Carver's role.

Tarski himself continued this research to some extent. For example, some open conjectures about the degree of equivalence can be stated in the form, *if n is a natural number, then* $\Phi(n)$, where $\Phi(n)$ is a formula with a variable $n$, which would become a sentence of the elementary algebra of real numbers if $n$ were replaced by a constant representing a specific natural number. In the report *A Decision Method for Elementary Algebra and Geometry*, Tarski suggested that applying his method serially to many cases $\Phi(1), \Phi(2), \ldots$ would either refute such a conjecture or provide considerable intuitive evidence for its validity.[7]

Around 1970 one of the present editors, James T. Smith, more than once heard Tarski lecture on this subject. A principal object of consideration was the function $\tau(x)$, the degree of equivalence of the unit square $Q$ and an $x$ by $1/x$ rectangle $P$. Tarski also considered the analogous number $\tau_m(x)$ that results by limiting the subregions to polygons with at most $m$ edges. Clearly, $\tau_m(x) \geq \tau(x)$. Applying his own 1930b metamathematical result about the undefinability of the concept of natural number within elementary real arithmetic, Tarski showed that for each $m$ there must exist $x$ such that $\tau_m(x) > \tau(x)$: there is no uniform bound on the complexity of the polygons required to decompose $P$ and $Q$ each into a minimal number of subregions with correponding components congruent. His argument made essential use of the final theorem H of his [1932] 2014d paper.[8]

In 2011 Tom M. Apostol and Mamikon A. Mnatsakanian considered the extended problem of equidecomposability of polygonal regions with equal areas *and* perimeters, requiring boundary pieces to correspond as well as interior pieces. Their methodology does not overlap Tarski's.

## 8.6 Equidecomposability in Set Theory

Alfred Tarski's [1924] 2014b paper *On the Equivalence of Polygons*, and the famous 1924 paper *On Decomposition of Point Sets into Respectively Congruent Parts* by Stefan Banach and Tarski, began a filament of Tarski's research that tied together his interests in the axiom of choice and in elementary geometry. These works are concerned with a notion of decomposition different from the elementary one: decomposition of arbitrary point sets into arbitrary subregions, which are required to be entirely disjoint.

These papers have had enormous impact. During the next half century, Tarski and others returned repeatedly to questions raised there, developing whole new areas of mathematics. The intricate underlying constructions by Felix Hausdorff, the intricate reasoning employed by Banach and Tarski, and their startling conclusions have posed

---

[7] Tarski [1948] 1957, 4–5.

[8] Tarski [1930] 1983b, 134; see also [1948] 1957, 61. Theorem H is discussed in section 7.1. Recently, J. Shilleto (2012) noted that if Theorem H were stated to include more of what Tarski's suggested proof actually entailed, then it would imply Tarski's result about $\tau_m$ directly, without appealing to metamathematical results.

attractive challenges for expository writers—certainly for the editors of the present book. A notable early example was Karl Menger's 1934 lecture to a lay audience in Vienna. In the same vein, but much more detailed, is the book Wapner 2005. Sally and Sally 2007 is a sophisticated text aimed at students of various levels; its chapter 5 is an excellent introduction to this problem area. Stromberg 1979 is an elementary but extremely detailed mathematical account of the "paradoxical" Banach–Tarski theorem. For a more advanced approach, and to survey the large literature stemming from the Hausdorff and Banach–Tarski results, the best source is the monograph Wagon 1993. Other surveys, emphasizing Hausdorff's legacy, include Schreiber 1996 and Czyż 1994, chapter 1.

The 1990 solution by Miklós Laczkowicz of Tarski's circle-squaring problem, more than six decades after its statement in Tarski [1924] 2014b, has been discussed already at the end of section 4.3. Trevor Wilson (2005) discovered a notable strengthening of Laczkowicz's result and the Banach–Tarski paradox.

The Banach–Tarski paper sparked immediate questions about the generality of its results and the most effective ways to present them. Such questions, in turn, led to new research areas. Some further results are included in section 5 of Tarski's 1926 joint paper with Adolf Lindenbaum, already mentioned in section 8.3 in connection with Tarski's set-theoretic research. Others contributed too, particularly Johann (John) von Neumann. In 1929, Neumann recast the measure-theoretic arguments of Hausdorff and Banach to emphasize the role of the group of transformations under which the measures should be invariant. For any point set $E$, any $I \subseteq E$, and any group $\mathcal{G}$ of transformations of $E$, he defined an $[E, I, \mathcal{G}]$-*measure* to be an additive function $\mu$ from the power set of $E$ to the set of all nonnegative real numbers, invariant under all transformations in $\mathcal{G}$, such that $\mu(I) = 1$. He showed that an $[E, I, \mathcal{G}]$-measure exists if and only if a $[\mathcal{G}, \mathcal{G}, \mathcal{G}]$-measure exists. This result allows measure-theoretic questions to be shifted from the geometric context to the group-theoretic; sometimes that is a simplification. Today, a group $\mathcal{G}$ is called *amenable* when such a measure exists.[9] The theory of amenable groups underlies all subsequent work in this area. The first milestone in that theory was stated in Tarski's 1929–1930 abstract on additive functions:

> Adapting to the notions introduced by J. v. Neumann in his recent work...this result can be generalized in the following way:...in order for an $[E, I, \mathcal{G}]$-measure to exist it is necessary and sufficient that $I$ not be equivalent to its half under any finite decomposition relative to the group $\mathcal{G}$.

By *its half* Neumann had meant any subset $H \subseteq I$ that is equivalent to its complement, $I - H$. Thus, "paradoxical" decompositions of point sets arise when, and only when, the measure problem fails to have a solution.[10] Much later, this result played a pivotal role between the two main parts of Stan Wagon's monograph: he claimed, "Tarski's theorem

---

[9] Neumann 1929, 78–82. Neumann called such groups *measurable* (*messbar*). For more information, consult Wagon 1993, chapter 10, and de la Harpe 2004.

[10] Tarski 1929–1930, 117. That preliminary report was submitted to the Warsaw Society of Sciences and Letters on 2 May 1929 by Sierpiński. There are several apparent errors in statements of definitions and theorems, perhaps due to mistranslation into French.

is not as well known as it deserves to be."[11] Tarski 1929–1930 contained no proofs; the proof for this theorem would not appear in any form until nearly a decade had passed.

After relocating to the United States, Tarski maintained interest in questions about set-theoretic equidecomposability, and he inspired work by others. His [1946–1947] 2014 letters to Wacław Sierpiński, translated in full in section 15.12, explain some of that activity. For example, his new Berkeley colleague Raphael M. Robinson studied the Banach–Tarski decompositions of certain point sets $S$ into $k + l$ disjoint pieces, of which the first $k$ can be "reassembled" to constitute all of $S$, as can be the last $l$ pieces. In 1947, Robinson showed that for a solid sphere $S$, the minimum number $k + l$ of pieces is five $(k = 3$ and $l = 2)$, one of which may consist of a single point; for a spherical surface $S$, the minimum is four.

## 8.7 The Measure Problem

The remaining two filaments of Alfred Tarski's research under consideration here elaborated results of Stefan Banach that underlay their original papers on equidecomposability. One of these filaments has to do with the measure problem: to assign, to all bounded subsets of the line, nonnegative numbers called their measures, such that two congruent sets always have the same measure, the unit interval has measure 1, and the measure of the union of two disjoint sets is the sum of their individual measures. Analogous problems can be stated for higher-dimensional point sets. Banach's 1923 solution of the measure problem in dimensions one and two was described in detail in section 4.2. Felix Hausdorff's 1914 "paradox" had shown that the problem has no solution in higher dimensions. Their work started a new area of mathematics: existence and properties of additive set functions invariant under various transformation groups. Tarski's interest in this area emerged as a series of major works in the 1930s. In his 1929–1930 abstract *Sur les fonctions additives dans les classes abstraites et leur application au problème de la mesure*, Tarski announced that a modified problem, without the requirement of invariance under congruence, thus for the trivial transformation group, always has a positive solution: for each set $E$ there is a nonconstant additive function $\mu$ from the power set of $E$ to the two-element set $\{0,1\}$ such that $\mu(\{x\}) = 0$ for each $x \in E$. Tarski proved this in his 1930a paper *Une contribution à la théorie de la mesure* by constructing $\mu$ as the characteristic function of a prime filter in the power set.[12] The proof can be regarded as a streamlining of Banach's original argument, but to build the filter, Tarski still used well-ordering and recursion with ordinal numbers, rather than a maximal principle. This line of inquiry was explored in great detail in the two-part paper Tarski 1939–1945, *Ideale in vollständigen Mengenkörpern*.

In the presentation [1932] 2014a, *On Geometric Properties of Banach's Measure*, Tarski outlined a plan for recasting Banach's work into a geometric framework that would unfold

---

[11] Wagon 1993, 125, 144.

[12] Marshall H. Stone (1938, §5) explained how this idea was articulated independently at this time by Tarski and several others.

more naturally from the familiar theory of Peano–Jordan content. Previously published only in Polish, its abstract is translated in section 15.7. It includes no details at all. Tarski returned to this project again in his 1938b paper, *Über das absolute Mass linearer Punktmengen*; unfortunately, even there the exposition was meager.

During the late 1930s, Tarski collected much of his work in this area into publications that are really extended abstracts.[13] Section 4 of his proofless 1937c paper *Über additive und multiplikative Mengenkörper und Mengenfunktionen* is a list of results relating cardinality considerations to various versions of the measure problem. In his 1938a paper *Algebraische Fassung des Massproblems*, Tarski presented the results of his 1929–1930 abstract in much more detail, including his important theorem that "paradoxical" decompositions of point sets arise when, and only when, the measure problem fails to have a solution. The 1938a paper is difficult to read, for two reasons. First, its basic notion is not the algebra of subsets of a metric space, but an abstract version of the algebra of their equivalence classes under finite equidecomposability. This allowed Tarski to vary more freely the family of equivalence classes under consideration, and to omit disjointness provisions from additivity conditions, but it resulted in an uncomfortable setting for some otherwise familiar concepts. Second, Tarski included only minimal explanations and only hints of proofs. One hopes that some mathematician will rework with better exposition the results in this important paper.

Tarski's 1938b article presented a notion that he had outlined in his [1932] 2014a abstract. He termed a set $X$ of real numbers *absolutely measurable* if the supremum of the lengths of all segments equivalent to subsets of $X$ by finite decomposition should be equal to the infimum of the lengths of all segments that contain a subset equivalent to $X$. He presented many properties of this notion, and comparisons with more-standard concepts, but with little exposition and only hints of proofs. For example, a set $X$ is absolutely measurable just in case all measures satisfying the requirements of Banach's version of the measure problem assign it the same value. In that case, this value coincides with the supremum and infimum just mentioned: Tarski called it the *absolute measure* of $X$. Hugo Hadwiger developed these ideas further during the 1950s and made them the basis of section 3.5.3 of his 1957 monograph *Vorlesungen über Inhalt, Oberfläche und Isoperimetrie*.

Tarski returned to this area of inquiry after he became established at Berkeley. The broad 1948b study *Measures in Boolean Algebras*, coauthored with his junior colleague Alfred Horn, generalized and cleaned up some of the steps in Banach's solution of the measure problem (see section 4.2 of the present book) and clarified some of the results in Tarski 1938a. Chapters 13–16 of Tarski's 1949a book *Cardinal Algebras* also presented much material of this sort.[14] For further references about the work of others, consult Wagon 1993.

---

[13] Tarski complained bitterly about his lack of sufficient time during those years, due to his dependence on two jobs (see section 9.1). Perhaps that explains the lack of exposition in these papers.

[14] The abstract Horn and Tarski 1948a covers part of 1948b, but not this part.

## 8.8 Generalizing Cantor–Bernstein

In 1924 Stefan Banach published an elaboration of the Cantor–Bernstein theorem that was used heavily in the Banach–Tarski paper. His formulation (see the box on page 44) was not particularly graceful nor as useful as it might be. That probably spurred Tarski to reconsider it several times. In February and December 1927, Tarski and his close friend Bronisław Knaster presented three papers on this subject to the Warsaw section of the Polish Mathematical Society:

- Tarski 1927a  *Sur quelques propriétés caractéristiques des images d'ensembles*
- Tarski 1927b  *Quelques théorèmes généraux sur les images d'ensembles*
- Knaster 1927  *Un théorème sur les fonctions d'ensembles*

The first just introduced a setting simpler than Banach's; the others were successive generalizations of Banach's result. Tarski later described the third of these papers as containing joint work that led directly to his further research two decades later.[15] The next year, at the International Congress of Mathematicians in Bologna, Tarski presented a major survey of results on this subject: Tarski [1928] 1930, *Über Äquivalenz der Mengen in Bezug auf eine beliebige Klasse von Abbildungen*. He mentioned several other contributors, and also indicated that the 1924 theorem of Kazimierz Kuratowski that was used in the Banach–Tarski paper would fit into the new framework.

Soon after his 1939 voyage to the United States, Tarski presented material from this research filament at several conferences. He adapted some of these results for use in the 1949a book *Cardinal Algebras*, published a preliminary abstract 1949b, and finally the full 1955 paper, *A Lattice-Theoretical Fixpoint Theorem and Its Applications*. Its main result, the core of the Cantor–Bernstein theorem, can be stated succinctly: every increasing function on a complete lattice has a fixed point. In more recent years, largely through the work of Dana S. Scott, Tarski's fixpoint theorem has become a foundational block in the theory of computing. In his 2006 survey of Tarski's impact on that field, Solomon Feferman noted,

> ...the influence of Tarski on the semantics of programming languages is so pervasive that to detail it would require an entire presentation in itself.... [The publication] most cited in the computer science literature, namely his lattice-theoretic fixpoint theorem...is an elegant abstract formulation of the essential characteristic of definition by recursion.

For further information, consult Feferman 2006 and the literature cited there.[16]

As the present book went to press, an extraordinarily detailed book-length study of the mathematics and logic related to this section appeared: Hinkis 2013.

---

[15] Tarski 1955, 286. These papers were abstracts with considerable detail. The latter two were presented on the same day. Knaster was Tarski's next-door neighbor, according to the roster Polskie Towarzystwo Matematyczne 1927, 137, 139, and to Golińska-Pilarek, Porębska-Srebrna, and Srebrny 2009a, A29.

[16] Tarski 1949a, chapters 11, 16; 1955, 286. In 1951 Tarski's doctoral student Anne C. Davis (later Anne C. Morel) proved a converse: all lattices with this fixed-point property are complete (see Davis 1955). Feferman 2006, 7.

# Part Three
# Teaching

Part Three of this book is devoted to Alfred Tarski's work as a secondary-school mathematics teacher and teacher-trainer in Poland during 1924–1939. It contains translations from and commentary on all his publications in that field. The biographical and background information in chapter 9 concentrates on Tarski's involvement with secondary education and on his family.

During 1924–1935, Tarski became involved in efforts to improve mathematics instruction at the secondary level in Poland. Chapter 10 is a translation of his report to secondary teachers about the First Congress of Mathematicians of Slavic Countries, held in Warsaw in 1929. Tarski concluded that report by noting that such a congress did not serve the needs of secondary education in mathematics; a separate organization and activities should be established. Responding to those needs, the journals *Parametr* and *Młody matematyk* were founded during the next two years by Antoni M. Rusiecki and Stefan Straszewicz. A description of the journals is presented in section 9.7. An article by Tarski in *Parametr* on teaching about the circumference of a circle is translated in chapter 11. Tarski was a major contributor of exercises posed in these journals for consideration by teachers and students. Those are all translated and discussed in detail in chapter 12.

Tarski published two further articles in *Parametr* and *Młody matematyk*, about the degree of equivalence of polygons with equal areas. These were already included in Part Two of this book because they can be interpreted as research papers closely related to other material featured there, as well as enrichment for secondary teachers and students.

In 1935, with coauthors Zygmunt Chwiałkowski and Wacław Schayer, Tarski published the text *Geometrja* for use in the third year of secondary school. That book is described in section 9.9; chapter 13 consists of translations of representative excerpts.

Tarski's contribution to the enhancement of mathematics instruction in Poland was limited by his everyday teaching duties, by his preoccupation with the research into logic and foundations of mathematics that made him famous, and by adverse economic conditions and the onset of World War II, which impaired dissemination of his ideas. On the other hand, the considerable influence of his secondary-teaching experience on his mathematical and logical research is confirmed by the numerous intellectual filaments that connect ideas in Part Three of this book to those that dominated the previous Part Two on geometry, and to his pioneering work on logic.

Part Three concludes with the brief chapter 14, which ties Tarski's early career as an educator more closely to his research activity, particularly in logic, and recounts his fortunate relocation to the United States at the very outbreak of World War II.

The material gathered in Part Three explains some of the relationships of Tarski's early work to the intellectual, political, and social milieu of Poland between the world wars. The present editors hope that it will spur broader investigation into the connection between mathematical research and mathematics education during that era. This hope is a major reason for including works of such contrasting mathematical sophistication in a single volume of selected translations.

Tarski's life in Poland unfolded amid a chaotic vortex of political, social, economic, and scientific developments. Accounts of events, people, and ideas were merged from several dimensions to form the linear sequence of pages of this book. This is reflected in Part Three by the use of singly outlined boxes interspersed in the main narrative. They contain biographical sketches of some persons associated with Tarski, and informational essays about some other topics. Each box can be read independently: readers are not expected to visit them in sequence. Cross-references refer to them from the main narrative.

Doubly outlined boxes are used in chapter 12 to distinguish the exercises, written by Tarski, from their discussions, written by the present editors.

# 9
# Career and Family

This chapter describes Alfred Tarski's work as a secondary-school mathematics teacher and teacher-trainer in Poland during 1924–1935. It also provides background for that activity, and for all of Tarski's publications that stemmed from it. The first of those is translated in chapter 10: his [1929] 2014a report to secondary teachers about the First Congress of Mathematicians of Slavic Countries, held in 1929 in Warsaw. Tarski noted that such a congress did not serve the needs of secondary education in mathematics; a separate organization and activities should be established. The present chapter includes a description of the journals *Parametr* and *Młody matematyk*, which were founded soon after to meet some of those needs. This provides background for the exercises Tarski posed to stimulate both teachers and students, for his article on teaching about the circumference of a circle, and for his two research papers on the degree of equivalence of polygons with equal area.[1] They were all published in those journals, and are translated in chapters 12, 11, and 7, respectively. Finally, the present chapter provides an overview of the school geometry text[2] that Tarski coauthored in 1935; chapter 13 includes translations of representative excerpts.

## 9.1 Employment and Marriage

The account in chapter 3 of Alfred Tarski's life through age 24 included his attaining some financial independence through employment as an instructor in logic courses for teachers and as a mathematics teacher in Warsaw secondary schools.

Warsaw was then midway through a period of furious expansion of education. The city was growing rapidly:

| year | 1914 | 1921 | 1931 |
|---|---|---|---|
| population[3] | 884,500 | 936,700 | 1,170,800 |

Moreover, under Russian oppression until 1915, only about one-third of Warsaw children in the group aged seven through thirteen had been attending school; for most, schooling lasted only one or two years. Warsaw's literacy rate in 1915 was only about 53%. This situation began to improve under the 1915–1918 German occupation, and the 1921 Polish

---

[1] Tarski [1932] 2014e, [1931] 2014a, and [1932] 2014d.
[2] Chwiałkowski, Schayer, and Tarski [1935] 1946.
[3] Wynot 1983, 93, 107.

constitution instituted compulsory schooling for that age group. By 1925, about 63% of those children were attending, and the literacy rate had risen to 85%. By 1931, these proportions were 93.5% and 90%. World War I and its aftermath actually caused a decline in population of this age group during the 1920s; nevertheless, Warsaw had to expend great effort to achieve that improvement:[4]

| year | 1918/19 | 1923/24 | 1929/30 |
|---|---|---|---|
| population, age 7–13 | 150,000 | 132,000 | 110,000 |
| number in school | 50,000 | 83,160 | 102,850 |

In 1918 the new Polish school system was a chaos inherited from three fallen empires, incorporating their institutions and a great variety of private schools. Reforms were undertaken immediately, but were required only in public schools, and began to take effect in Warsaw only in 1922–1923. Elementary schools might offer as many as seven years of instruction. Secondary schools overlapped them: the first secondary year generally paralleled the fifth elementary year. Qualified pupils could switch then to a *gimnazjum*: a Polish secondary school that prepared students for university studies. The full gimnazjum course of study lasted eight years, usually through age eighteen.[5] Unlike elementary-school enrollment, that in secondary schools did not grow during the 1920s, but hovered around 30,000. Thus, it was probably not easy for anyone to find a good teaching job at that level during the 1920s. Whereas elementary schools were populated mostly by students from working-class families, the last years of secondary school were dominated by those from white-collar and small-business families. A large majority of secondary-school students attended private schools, whose high tuition made them inaccessible to working-class families.[6,7]

Poland reformed its school system in 1932: elementary schools would offer seven years of instruction. Qualified students could transfer after the sixth year to a gimnazjum. Gimnazjums would offer four years of instruction, covering general education—including Latin—and some practical subjects. New institutions called *liceums* would offer a final two years of secondary school.[8]

---

[4] Konarski 1971, 222. Wynot 1983, 235–237. Many details in this section stem from sparse Polish government and school records. Stanisław Konarski (1971, 215; 1973, 179) reported that most of the archives were destoyed during World War II and what remains is of little use.

[5] Kasperowiczowa 1969, 188; Konarski1971, 221; Sadowska 2001, 71. Although her 2001 book is much more comprehensive than those two earlier studies, Joanna Sadowska did not mention them.

[6] Wynot 1983, 241–244; private secondary-school tuition in 1930 was about 700 zlotys per year.

[7] According to Joseph Marcus (1983, 44, 184), the average annual income for a Polish nonfarm worker in 1929 was 2034 zlotys. He claimed that sixty percent of urban households could not afford even the average annual rent for minimal accommodation, one room and kitchen: 480 zlotys.

[8] Kasperowiczowa 1969, 194; Sadowska 2001, 71; Poland 1932. The reforms were gradually phased in. During 1932–1934, students did not transfer into the first two classes of secondary schools. In fall 1934, the secondary classes changed to the new numbering system. Starting in 1937–1938, three types of liceums were instituted, emphasizing classics, natural science, or mathematics and physics. The liceums graduated only one class before World War II began.

## 9.1 Employment and Marriage

In fall 1925 Tarski began employment as a mathematics teacher at the Third Boys' Gimnazjum of the Trade Union of Polish Secondary-School Teachers, financed by the Society of Friends of Polish Secondary Schools in Warsaw.[9] It was located at 29 Nowolipski Street, a building that it shared with some apartments. The school had been founded in 1922 under other auspices, with only a hundred students. At first it covered only four years—I and IV–VI—and relied on other schools to provide facilities that it could share during afternoons. The students' parents soon arranged that it be taken over by the Union. By fall 1924, the Gimnazjum had acquired its own premises and switched to the conventional morning schedule. By fall 1925, it covered all eight years and enrolled three hundred students. In 1932 the school was reconstituted as the Boys' Gimnazjum of the Society of Friends of Polish Secondary Schools, under the same director, and relocated at 150 Marszałkowska Street, with entrance on Rysia Street.[10] (See the figure below.) In 1933, the school was renamed in honor of the noted Polish writer Stefan Żeromski, who had died in 1925.[11] According to the philosopher Karol Martel, who had been a student there around 1930, the Gimnazjum

> ... had high academic standards and a liberal-minded faculty, many of whom were assimilated Jews and socialists. The director, Teofil Wojeński, was the author of a book about Żeromski and an active member of the Polish Socialist Party. Tarski immediately found the atmosphere intellectually, politically, and culturally congenial.... The students were good and the teachers serious....[12]

*Warsaw, 1925*
*140–150 Marszałkowska Street*

Rysia Street enters Marszałkowska midway back on the side opposite the viewer. The Żeromski Gimnazjum was evidently behind No. 150, the tall building most distant from the viewer, with two turrets.

---

[9] The Polish names of these organizations were *III Gimnazjum Męskie Związku Zawodowego Nauczycielstwa Polskich Szkół Średnich* and *Towarzystwo Przyjaciół Polskiej Szkoły Średniej w Warszawie*.

[10] The Polish word for *reconstituted* was *zlikwidowane*. The school had evidently lost its connection with the trade union. The two school locations, Tarski's family home, and the university were all in central Warsaw, at most three kilometers distant from each other.

[11] The school was destroyed during World War II; it is not related to a present-day Warsaw school with the same name. This account of the school stems from Jadacki 2003a, 144–145; Konarski 1973; Petrozolin-Skowrońska et al. 1994; Zagórowski 1924–1926, volume 1, 135, and volume 2, 248; and especially the 1926 article by Henryk Raabe. For more information about Raabe, see the box on page 177.

[12] Feferman and Feferman 2004, 56. Wojeński served during 1922–1939, except for one year.

Tarski met his bride, Maria Witkowska, at that gimnazjum. She had come there a year later than he, as counselor to the first-year students. Her sister Józefa was a secretary there.[13] The wedding took place in the chancery of the Roman Catholic All Saints Church in Warsaw, on 24 June 1929. The marriage record lists Alfred's and Maria's positions as *docent* at the university and as *official* (in Polish, *urzędniczka*). Maria's brother Antoni, then an army lieutenant stationed in Vilnius,[14] attended as witness. Tarski's father and mother were listed, but as "Ignacy and Róża, the Tarski couple," with no mention of their surname Teitelbaum.[15]

The marriage record indicates that Maria was from Minsk,[16] and twenty-seven: born in 1902, one year younger than her groom. Her parents, deceased, were Wincenty Witkowski and Maria Janczewska. In their biography of Tarski, Anita B. and Solomon Feferman wrote,

> The gods were good to Tarski when they sent him Maria Witkowska as a life partner.

They have provided a detailed and sympathetic account of an interesting, complicated, heroic woman whom they knew, respected, and admired. They noted,[17]

> Many of Tarski's acquaintances had the impression that Maria's family was gentry.... However, the facts according to the Tarski children are that Maria's parents owned a butcher shop where they both worked long hours; they may also have had some small land holdings... but they weren't rich.

In his biographical sketch of Tarski, Jacek Juliusz Jadacki provided much more detailed information about Maria's background, with an emphasis that at first seems incompatible with that quotation. The following three paragraphs are based mostly on his account.[18]

Maria's maternal grandfather Julian Janiczewski owned a large estate near Minsk. He was killed in the German attack on that city during the final stages of World War I. Maria's paternal grandfather Antoni Witkowski and her father also owned property near Minsk. For advocating Polish independence, they had once been exiled to Siberia by the tsarist regime. Besides her older sister Józefa and younger brother Antoni, born in 1900 and 1905, Maria had sisters Agnieszka, Jadwiga, and Helena, born in 1904, 1909, and 1911. Their mother died at age thirty-eight, soon after the birth of her sixth child in eleven years. At this time, Maria's father was seventy years old. He then married his

---

[13] Feferman and Feferman 2004, 65; Jadacki 2003a, 148; Raabe 1926. First-year students were about twelve years old.

[14] Long part of part of Russian-held Poland, Vilnius (in Polish, *Wilno*) was occupied at various times during 1915–1922 by German, Polish, Russian, and Lithuanian forces. It was part of Poland until 1939, when it was annexed by the Soviet Union. Since 1991 it has been the capital of Lithuania.

[15] Jadacki 2003a, 133. Parafia Wszystkich Świętych [1929] 1940.

[16] Minsk (in Polish, *Mińsk*) also had been part of Russian Poland. During 1915–1921 it changed hands several times between Russian, German, and Polish control, suffering grievously. Afterward it became part of the Soviet Union. It has been the capital of Belarus since 1991.

[17] Feferman and Feferman 2004, 64.

[18] Jadacki 2003a, 133–137, 148–149. Jadacki's sketch seems to have been prepared from informal notes with little editing. It relied on personal contacts with surviving members of Maria's parents' families.

## 9.1 Employment and Marriage

first wife's younger sister Antonina, who became stepmother to Maria and her siblings. These details are not incompatible with the impressions that the Fefermans reported: Maria's forbears were landowning families, but at least one suffered years of ill fortune.

During the 1919–1920 Polish–Soviet War, in front of his family, at age seventy-eight, Maria's father, Wincenty, was murdered by Soviet troops. The children were moved to an orphanage on a collectivized farm. The three eldest sisters quickly joined the Polish Military Organization—East.[19] They safeguarded and transmitted communications, helped prisoners escape, and transported weapons. Agnieszka was a nurse in a hospital for Polish prisoners; Maria was treasurer for her unit and assembled explosives. Józefa became deputy commander of the women's section, and made several daring and dangerous trips across enemy lines to Moscow, in times of chaos and blockade, always without identification or with false papers. The photograph below shows the three sisters with

*Maria Tarska in the 1919–1920 Polish–Soviet War*

From left, the women are Maria, Józefa, and Agnieszka Witkowska. The man is Józefa's fiancé Bolesław Zahorski.

---

[19] The name in Polish is *Polska Organizacja Wojskowa–Wschód*. Organized by Józef Piłsudski during World War I for diversionary and intelligence operations, this secret body persisted in the east for several years afterward (N. Davies 1982, volume 2, 381). The following description of the sisters' exploits is based on Ziemiański 1933, 94–99, 196–198. Jadacki (2003a, 148) cited this account of the efforts of women in the organization, but his report differs in some respects.

fellow soldier Bolesław Zahorski, who would become Józefa's husband.[20] After the war, the three sisters could not return to Minsk, in the Soviet Union; they went to Warsaw. All three were awarded the Cross of Valor medal (Krzyżem Walecznych) for heroism. During 1924–1927, Maria studied law at the University of Warsaw. She and Józefa found employment at the Third Boys' Gimnazjum of the Trade Union of Polish Secondary-School Teachers, where Tarski taught. Maria became the counselor for its first-year students (aged about twelve), and Józefa became the school secretary.

Soon after their marriage, the Tarski couple moved to an apartment on Sułkowski Street in the Żoliborz district.[21] (See the photograph on page 178.)

> Inspired by the Bauhaus... principles of design, this area of stylish new apartments, flats, and houses surrounded by greenbelts was considered a choice place to live... out of the hustle and bustle of the center of town yet... only a ten-minute trolley ride to the university.

Tarski made frequent recreational visits to the Tatra mountains in southern Poland, particularly to the resort town Zakopane. Maria sometimes accompanied him and his professional associates, although she did not travel elsewhere with him. His excursions there and abroad included affairs with some female colleagues, evidently with Maria's knowledge.

> Her way of dealing with her husband's... affairs was to accept them as graciously as she could. The ethos of the culture... was that husbands would have their mistresses and wives would look the other way; moreover, in their circle of friends, monogamy was seen as an unnatural bourgeois convention that scarcely anyone adhered to.

Maria continued university studies during 1930–1935, in education. She also kept up her martial skills and nerve: in 1931 she took honors in a two-position pistol-shooting match for Warsaw teachers.[22] Sharing two incomes, she and Alfred were enjoying "relatively good times... surrounded by friends and like-minded people both at home and at work." Nevertheless, Alfred's

> ...biggest problem was finding enough time to do research and write papers; his solution, when the pressure became too great and he absolutely needed uninterrupted time to bring a piece of work to conclusion, was to resort to... calling in sick.

Tarski's gimnazjum substitute during his leaves was Roman Hampel.[23]

---

[20] For information about Zahorski, see the box on page 177.

[21] This paragraph and the next are based mostly on Feferman and Feferman 2004. The quotations are from pages 66 and 92. They specified Tarski's address as No. 8, but Jadacki (2003a, 153) said that until late 1934, the address was No. 2, apartment 5. Perhaps the Tarski family moved during that time. See also Nowicki 1992, 154.

[22] *Ogniwo* 1931, 112. This article identified her employer as *Warszawa III Gimnazjum Związku Zawodowego Nauczycielstwa Polskich Szkół Średnich*; it had not yet changed its name to Żeromski Gimnazjum.

[23] Feferman and Feferman 2004, 66; Jadacki 2003a, 145. Concerning Hampel, see the box on page 177.

## 9.1 Employment and Marriage

---

**Roman Hampel** was born in 1907 and schooled in Warsaw, when it was in the Russian Empire. During 1924–1929 he studied mathematics at the university there, obtaining a master's degree. During the 1930s he taught in secondary schools and obtained another master's at Cracow in education and psychology. In 1951 he became chair of management mathematics and then vice-dean of the electrical engineering faculty at the Warsaw Polytechnic. Hampel died in 1963.*

Born in Warsaw in 1882 when it was in the Russian Empire, **Henryk Raabe** became active in socialist politics around 1902. He earned a doctorate in biology at the University of Cracow in 1915 and started teaching in secondary schools. Raabe helped found the Trade Union of Polish Secondary-School Teachers[†] in 1919 and served as its president during 1922–1927. In 1922 the union took over the school that became the Żeromski Gimnazjum. Raabe taught biology there, and Alfred Tarski joined the faculty in 1925. As president of the union, Raabe edited its journal *Ogniwo*, which published his 1926 account of the school and Tarski's [1929] 2014 report, translated in chapter 10. Raabe worked as a union organizer through the 1930s. During the 1939–1941 Soviet occupation, he became professor at the University of Lwów. After World War II, he participated in the establishment of the Communist regime in Poland, serving as ambassador to the Soviet Union, rector of the University of Lublin, and member of parliament. A successful popularizer of science, Raabe died in 1951.[‡]

**Bolesław Zahorski** was born in 1887. A journalist active in the Polish independence movement, he published under the pseudonym Bolesław Zygmunt Lubicz, and was imprisoned in Siberia. In 1920 he authored a volume of poetry about service in the Polish–Soviet war. After that war he married his fellow soldier Józefa Witkowska, who would become Alfred Tarski's sister-in-law. In 1922, soon after returning to civilian life, Zahorski died in an accident.[§]

---

\* Jakubowski 1964.   [†] *Związek Zawodowy Nauczycielstwa Polskich Szkół Średnich*   [‡] Brzęk 1983
[§] Roliński 1996, 482.

---

In summer 1934, Maria was pregnant with the Tarskis' first child. Alfred spent 30 August through 7 September at conferences in Prague, and was preparing for a prolonged visit abroad during the next year.[24] They gave up their Żoliborz apartment, and evidently sublet a room in Alfred's parents' apartment. Maria bore their son Janusz Andrzej on 11 December 1934.[25] Soon, Alfred left for a research visit to Vienna: a Rockefeller grant supported him there from January through June 1935. He returned home briefly, then spent the week of 15 September at the First International Congress for the Unity of Science in Paris.

---

[24] See section 14.2. Jadacki, probably in error, suggested (2003a, 153) that Tarski may also have studied in Vienna during 1934.

[25] On passport documents in the Archiwum Akt Nowych (Central Archives of Modern Records) in Warsaw, dated August 1934, and on the 29 November letter Tarski [1934] 2014, Alfred used his parents' address. His loan application, Tarski [1935] 2014, indicated that the Tarski family was living in one room *podnajętym przy większej rodzinie—sublet (from or with) (a or the) larger family*. It is not clear whether Tarski was emphasizing or downplaying his reliance on his parents. These letters are translated in section 16.3. See also Jadacki 2003a, 153–158. The Fefermans gave Janusz's birth date incorrectly (2004, 66).

*51 Koszykowa Street
Tarski's Parents' Residence*

*4 Sułkowski Street
Tarski's Residence, 1936–1939*

## 9.2 Teaching Geometry

Afterward, the couple, now with a child, realized that their quarters with Alfred's parents were impossibly crowded. Because rents were exhorbitant, Alfred applied to the Ministry of Religious Denominations and Public Education[26] for an advance equal to his university salary for six months to cover a down payment of 3500 zlotys toward purchasing a new apartment in Żoliborz. He would pay it back in installments over three years. After a short time, Tarski withdrew the application, noting that he had obtained a loan from another source. From early 1936, the Tarski family lived at an apartment they owned, 4 Sułkowski Street, apartment 18. It consisted of two and a half rooms, kitchen, bathroom, and basement.[27] This building and Tarski's parents' building are contrasted in the figures on page 178. Some statistics permit comparing the Tarskis' life with that of Warsaw residents in general: around 1930, only 38% of the Warsaw population were housed in quarters of more than two rooms; and repaying that down-payment loan would have cost the Tarskis more than half of the average annual salary, 2034 zlotys, of a single Polish nonfarm worker.[28]

### 9.2 Teaching Geometry

The context in which Alfred Tarski began teaching geometry at the Third Boys' Gimnazjum of the Trade Union of Polish Secondary-School Teachers included its faculty and curriculum and the tradition of geometry instruction in which Tarski would craft his own approach. This section considers these aspects in turn, and emphasizes two geometry texts on which he relied.

When Tarski arrived at the Gimnazjum in 1925 to teach mathematics, it covered the full eight years of secondary school, including those that overlapped elementary school. Its faculty numbered twenty-one, including the headmaster and a priest. Besides Tarski, only one other teacher, the art instructor, was younger than thirty. There were three women, who taught languages, social studies, and biology. Two men who had earned doctorates taught science. The one other mathematics teacher had an engineering degree. All faculty members were identified as Christian—fourteen, including Tarski, as Roman Catholic. The Gimnazjum offered instruction in the following subjects:[29]

| Subjects | Number of Teachers |
|---|---|
| Arts and Gymnastics | 4 |
| Biology, Chemistry, and Physics | 3½ |
| Latin | 3 |
| Polish | 2½ |
| French and German | 2½ |
| Geography and History | 2 |
| Philosophy and Religion | 2 |
| Mathematics | 1½ |

---

[26] Ministerstwo Wyznań Religijnych i Oświecenia Publicznego, often abbreviated MWRiOP.

[27] The Fefermans suggested that the loan may have been facilitated by the Teachers' Union. The new apartment was described in the letter Zakład 1935 from the Social Insurance Institution, which had just developed a large complex of new apartments in Żoliborz. That organization still exists.

[28] Marcus 1983, 44, 186. It is reasonable to assume, but not verified, that the required 3500 zlotys amounted to about six months' university salary for Tarski.

[29] Zagórowski 1924–1926 volume 2, iv–vi, 248–249. Teachers with two subjects are counted half in each.

There was no instruction in Russian! One of the biology teachers, Henryk Raabe, reported that the school was remodeled in 1926 and "completely met the needs of a modern school, as far as is possible under the conditions of being located in a residential building." His account showed that the faculty remained stable for that year; the philosophy teacher retired and was replaced by the already well-known philosopher Kazimierz Ajdukiewicz.[30]

Evidently, instruction at the Gimnazjum included mathematics each year—both algebra and geometry. The national standard recommended four hours per week for years IV–VI and five hours for the final years VII–VIII.[31] Tarski taught only mathematics, but the other mathematics instructor also taught physics. Tarski's teaching load was probably all five of these classes: twenty-two hours per week total.[32]

What comprehensive geometry texts were available for use in a gimnazjum in 1925? Surveying library catalogs and advertisements in Polish journals for teachers, the present editors discovered seven such texts, three of which were written originally in Polish:

| *Originally Polish* | *Translated* |
|---|---|
| Łomnicki 1923 | Enriques and Amaldi [1903] 1916 (from Italian) |
| Wojtowicz [1919] 1926 | Kiselev 1917 (from Russian) |
| Zydler 1925 | Močnik 1896 (from German) |
| | Zupančič 1918–1921 (from German)[33] |

Perhaps Tarski based his first gimnazjum classes on one of those texts. Some of them are notable. Andrei Petrovich Kiselev's book was originally published in Russian in 1892. An English edition appeared in 2006–2008; the publisher claimed, "It is by far the most famous Russian textbook, in all subjects, ever." Tarski himself probably learned from a Russian edition when he was in school. Somewhat modernized versions of the texts by Władysław Wojtowicz and Jan Zydler are currently available, too. Tarski referred to Enriques and Amaldi [1903] 1916 to establish the elementary foundation required for his own [1924] 2014b research work. In his [1931] 2014a article on equidecomposability of polygons, Tarski used specific arguments from a later edition of Wojtowicz's text. These two texts will now be described and compared.[34]

---

[30] Raabe 1926. For more information on Raabe and Ajdukiewicz, see the boxes on page 177 and chapter 3.

[31] Poland 1931. The present authors do not know what was recommended for years I–III.

[32] According to his Berkeley colleague John W. Addison (1983), Tarski sometimes taught as many as twenty-nine hours per week. In section 9.3 it is shown that during 1930–1939, Tarski's university teaching load averaged 5.75 hours per week. See also Givant 1999, 50.

[33] Books by Rihard Zupančič are often listed with the German spelling, Richard Suppantschitsch. A few specialized books were available for trigonometry, analytic geometry, and descriptive geometry, which were covered to some extent in the recommended curriculum. Kalicun-Chodowicki 1925 is an example.

[34] Tarski's references to these texts are on pages 79 and 142 of the present book. For more information on Wojtowicz, see the box on page 189.

**Ugo Amaldi** was born in 1875 in Verona. He earned the laureate from the University of Bologna in 1898 under supervision of Salvatore Pincherle. Amaldi won a professorship at Cagliari in 1903, then transferred to Modena in 1906, Padova in 1919, and Rome in 1924. There he taught mathematics in the architecture faculty until 1942, when he joined the science faculty. Amaldi's main research lay in the determination and classification of all continuous groups of transformations that depend on finitely or even infinitely many parameters. In 1901 Amaldi began extending Sophus Lie's pioneering work, and he eventually published a complete, detailed exposition for dimensions one through four. In 1918 Amaldi received a gold medal for this work from the Società Italiana delle Scienze, detta dei XL, and was elected to the Reale Accademia dei Lincei. Amaldi was best known, however, for his vital collaborations with Federigo Enriques and Tullio Levi-Civita in producing many highly readable expositions of elementary mathematics, starting in 1903, and mathematical physics, starting in 1920. He was sincerely but unostentatiously religious, with a reputation for calmness and impartiality. His son and daughter both became scientists. Amaldi died in Rome in 1957.[*]

**Federigo Enriques** was born in 1871 in Leghorn, to a prosperous family. They soon moved to Pisa, where he attended secondary school and the Scuola Reale Normale Superiore. He earned the laureate there in 1891 under supervision of Enrico Betti, then spent five years in postgraduate study and temporary positions. Enriques won the competition for a permanent chair in Bologna in 1896; he remained there until 1923. His best-known mathematical work was with Guido Castelnuovo on the birational classification of surfaces. Enriques also investigated questions in foundations of geometry and in differential geometry. In 1907 he wrote a major article on foundations for the *Encyklopädie der mathematischen Wissenschaften*. Later he collaborated with Castelnuovo on two more such articles, about algebraic surfaces and birational transformations. Enriques developed close relationships with scholars of many other disciplines. From 1900 on, his work included elementary texts and books for mathematics teachers, many in collaboration with Ugo Amaldi, and books on history and philosophy of mathematics and science. Polish translations of some of these works played a major role in the geometry discussed in the present book. In 1906 Enriques was elected to the Reale Accademia dei Lincei. During 1907–1913 he served as president of the Italian Philosophical Society. He organized and presided over the 1911 International Congress of Philosophy held in Bologna. In 1923 Enriques was appointed to the chair of higher geometry at the University of Rome. There he founded the National Institute for the History of Science. He remained in Rome until 1938, when he was dismissed because he was Jewish. Enriques spent most of his remaining years in hiding. He died in 1946.[†]

---

[*] Terracini 1958, Viola 1957.

[†] Castelnuovo 1947, Eisele 1971. Castelnuovo and Enriques 1908, 1914. The present account of Enriques's last years is from Castelnuovo 1947. According to Eisele 1971, however, Enriques "retired from teaching" during those years.

*Ugo Amaldi  
about 1910*

*Federigo Enriques  
in 1914*

## 9.2 Teaching Geometry

Of the texts just listed, *Elements of Geometry for Use in Higher Secondary Schools* by Federigo Enriques and Ugo Amaldi probably played the most important role in Tarski's early geometry teaching. It was first published in Bologna in 1903. Both authors were noted research mathematicians, early in their careers, specializing in connections between algebra and geometry.[35] At that time in Italy it was common for professors to spend serious effort simultaneously on research and on improving instruction in the schools. This book appeared in many editions, tailored to various audiences. In 1916, Wojtowicz translated it into Polish, presumably for use in Polish schools.

Enriques and Amaldi followed an axiomatic approach, but tempered it by describing various physical experiences before codifying them as geometric postulates. These were phrased in terms of the undefined notions *point*, *line*, *plane*, *incidence* of these objects, *order* of collinear points, and *congruence* of segments and of angles, using the notion of *set* when appropriate. The authors gave great care to formulation and examples of the basic concepts, and presented proofs of theorems in detail. There are ample illustrations and hundreds of exercises. A French reviewer wrote,

> The authors (through perhaps excessive imitation of the ancient geometers) exclude the idea of number, or at least postpone its introduction as long as possible, and reason with figures and geometric magnitudes. Such a method seems somewhat artificial today.... In sum, the authors have made logical rigor paramount... but sometimes that seems to be at the expense of the natural order and simplicity.[36]

The authors assumed that their student readers could tolerate and absorb the content of long, meticulous discussions. They made no allowance for those whose reading skills had not yet attained the level of highly educated adults. The result was a book that is mathematically correct and delightful reading for a sophisticated, motivated, independent learner, such as a professional mathematician.

Wojtowicz evidently perceived this disconnect between the book's style and its *intended* readers: students like those whom Tarski would teach. In 1919 Wojtowicz published his own text that covered most of the same material and added some new topics. He spent considerable effort to help readers, by clarifying emphasis and simplifying the language. According to Stefan Kwietniewski's brief 1923 review in the *Handbook for Self-Education*,

> Of all the *geometry* textbooks originally written in Polish, Wojtowicz's book stands out for its accuracy and [abundance of] content.

Nevertheless, Wojtowicz's style was not different enough to make it a substantial improvement over Enriques and Amaldi [1903] 1916. It still assumed of its readers a level of sophistication beyond that of any but exceptionally talented secondary-school students.

This survey of geometry books published for use in Polish-language secondary schools reveals that from 1900 on, administrators and mathematics teachers could select from a variety of approaches and textbooks. By 1925, the Polish government had begun to standardize the curriculum. That process led to new publications and a greater number

---

[35] Enriques and Amaldi [1903] 1913. For more information on those authors, see the box on page 181.

[36] *Revue de métaphysiques* 1904.

from which to choose. For example, Bazyli Kalicun-Chodowicki, the director of a boys' gimnazjum in Lwów, published in 1925 a textbook with the full title

*Descriptive Geometry for the Higher Classes of Secondary Schools (A Systematic Course Following the Program of the Ministry of Religious Denominations and Public Education).*

Stricter curriculum standards probably improved the overall quality of mathematics instruction in Poland, but also curtailed the variety of approaches available[37] and may have made it more difficult for the most knowledgeable teachers to experiment with alternative approaches, such as Tarski's own axiom system for geometry, discussed in section 9.4.

## 9.3 Teaching Teachers

*To prepare students for participation in a rapidly evolving society, teachers should not merely present their subjects as they were themselves taught.* In 1914, Stefan Kwietniewski and Władysław Wojtowicz emphasized application of that principle to the teaching of mathematics in Poland:

> One of the most outstanding attributes of modern mathematics is the work in elucidating its foundations.... These results... are so momentous that they must exert a profound influence on the exposition of the whole of mathematics, from the lowest to the highest levels of teaching. Today teacher[s] may no longer take the precritical position, *allez en avant et la foi vous viendra*. [They] must not only thoroughly understand the newest research in the foundations of mathematics, but also restructure their lessons from the ground up, in order to make them most straightforward and in accordance with modern science. This reconciliation... can only be established through many trials and the concerted efforts of a whole multitude of people —scholars as well as educators.[38]

Through his research, Alfred Tarski had become part of that forward movement. This section will argue that this principle played an essential role in his university teaching, both in elementary courses intended for general audiences and in courses offered specifically for training teachers. Other noted mathematicians had contributed major works emphasizing the connection between foundational research and mathematics education. Tarski referred to them in his own research reports, and presumably used them in preparing his own classes in schools and those at the National Pedagogical Institute, for preparing other teachers (see chapter 3).

Three of those works are described next: David Hilbert's *Foundations of Geometry*, Mario Pieri's *Elementary Geometry Based on the Notions of Point and Sphere*, and the collection *Questions Regarding Elementary Geometry* edited by Federigo Enriques.[39] These descriptions are followed by a detailed account of Tarski's efforts in the training of school teachers.

---

[37] A study is needed of the approaches to secondary mathematics instruction prevalent in the early 1900s, contrasted with the Polish reforms of the interwar period; but that is beyond the scope of the present book.

[38] Kwietniewski and Wojtowicz 1914. The French clause (italicized by the present editors) means *go on, and it will come to you*. For more information on Kwietniewski and Wojtowicz, see the box on page 189.

[39] Hilbert [1899] 1971, Pieri [1908] 1915, and Enriques [1900] 1914–1917.

*David Hilbert
in 1900*

*Hilbert's
Foundations of Geometry
Title Page*

In 1899 Hilbert had published *Foundations of Geometry*. For at least half a century it would be the standard rigorous presentation of Euclidean geometry.[40] Hilbert's undefined terms were *point*, *line*, and *plane*; *incidence* of those objects; *betweenness* of points; and *congruence* of segments and of angles. (Enriques and Ugo Amaldi adopted those primitive terms for their [1903] 1913 text.) Hilbert developed a large portion of Euclidean geometry in a few pages. He adhered to the style conventional for mathematical research: rigorous, but without extreme detail. His postulates stated well-known properties of geometric objects familiar to all. Many were quite succinct. But some were formulated in terms of undefined properties of defined notions, such as angle congruence, and became logically complicated when the latter were replaced by their definitions.

Hilbert introduced early the use of points on a selected line as scalar coordinates, and showed how various geometric properties of points correspond to algebraic properties of scalars. For example, he showed how versions of Desargues's theorem are connected with the scalar addition and multiplication operations. Although that theorem involves only incidence of points and lines, all in a single plane, and in three dimensions is derivable from incidence axioms alone, he showed that additional postulates—his congruence axioms—are required to prove it in plane geometry.

Hilbert's striking new results, the scope of the geometry he considered, his effective style, and the placement of his work in the mainstream of the mathematics of the day made *Foundations of Geometry* instantly famous. A French translation, [1899] 1900, was published the next year. Hilbert's work became identified with the birth of the modern axiomatic method in mathematics, even though that had been employed already for some years by Italian mathematicians. Tarski relied on Hilbert's book to provide the necessary foundation for his own [1924] 2014b research work and for the [1931] 2014f exercise that he posed for gimnazjum students and teachers.[41]

The second foundational work on which Tarski relied for the development of his approach to geometry was Mario Pieri's [1908] 2007 memoir *Elementary Geometry Based on the Notions of Point and Sphere*.[42] An alternative to Hilbert's foundation, it also presented its subject axiomatically, but with all its notions and postulates defined and formulated in terms of the notion *point* and the relation that holds between points $a, b$, and $c$ just when $a$ and $b$ are *equidistant* from $c$. The paper's title reflects Pieri's extensive use of elementary set theory in developing geometry from his postulates: he defined the sphere through $b$ centered at $c$ as the set of all points $a$ such that $a$ and $b$ are equidistant from $c$. Pieri's second aim was to foster more extensive use of properties of spheres in presenting elementary geometry, even in school courses. A third goal, which Pieri had already pursued for a decade, was to promote the use of transformations in elementary geometry. Pieri introduced various geometric transformations early through definitions, and employed them extensively throughout the memoir.

---

[40] Hilbert [1899] 1971: see the bibliography for details about its editions. This paragraph and the two following were adapted from Marchisotto and Smith 2007, subsection 3.10.2, with the publisher's permission.

[41] Tarski's references to Hilbert's text are on pages 79 and 264 of the present book.

[42] For information about Mario Pieri's life and mathematics, consult Marchisotto and Smith 2007.

*Mario Pieri
around 1908*

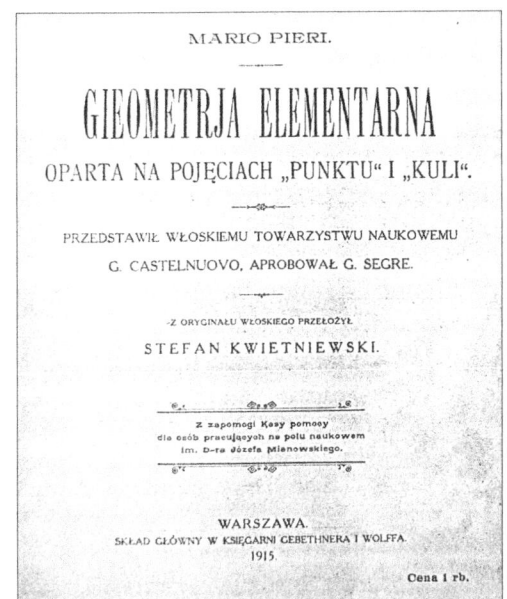

*Pieri's
Point and Sphere Memoir
Polish Translation*

Pieri's presentation of elementary geometry differed greatly from Hilbert's. Pieri's postulates employed, besides his two undefined terms, only the simplest defined notions, such as *sphere*. Many of his postulates were not at all succinct, but when phrased solely in terms of the primitive notions *point* and *equidistance*, their logical structures were simple. Pieri's exposition departed from the conventional mathematical research style: he provided virtually every detail of every proof! Hilbert's exposition was much more readable, but Pieri's was easier to analyze.

In 1915, Pieri's *Point and Sphere* memoir was translated into Polish by Stefan Kwietniewski, one of the original members of the mathematics faculty at the University of Warsaw. The Mianowski Fund supported its publication. Tarski studied it intensely. His later colleague Steven R. Givant reported,

> Tarski was critical of Hilbert's axiom system from a logical perspective ... [He] preferred Pieri's system, where the logical structure and the complexity of the axioms were more transparent.[43]

Although the formats of Hilbert [1899] 1922 and Pieri [1908] 2007 differed greatly, each presented much the same material as a traditional secondary-school geometry course based on Euclid's centuries-old approach. Hilbert did adopt an advanced viewpoint to discuss the reasoning behind his approach, but provided only exceedingly brief explanations. In contrast, the third foundational work on which Tarski relied was Enriques's [1900] 1914–1917 book *Questions Regarding Elementary Geometry*. This was a collection of articles by thirteen Italian authors, who described various aspects of elementary geometry in considerable detail from an advanced viewpoint. Publication of its translation was also supported by the Mianowski Fund. The authors of its articles ranged from secondary-school teachers to leaders in mathematical research. According to its translators Kwietniewski and Wojtowicz, it

> can be considered an extensive scientific and didactic commentary on a modern course in elementary geometry.... But in addition to this (despite certain shortfalls unavoidable in a collective work), it is a straightforward and clear introduction to the new research in axiomatics, written especially with the needs of teachers in mind.[44]

Three of the articles were devoted to philosophical and pedagogical questions, five to fundamental geometric concepts, and seven to the theory of Euclidean constructions and constructibility. Two very long articles were devoted to non-Euclidean geometry and to isoperimetric problems. Several of the authors were major figures in Italian mathematics, and some of these papers even today provide excellent background for theoretical questions about elementary geometry.

Having described the foundational works on which Tarski relied for developing his approach to geometry, this section now surveys the contexts in which he presented it: his university courses and three "vacation courses" for in-service teachers in Poland.

---

[43] Pieri [1908] 1915. Givant 1999, 50. See also Szczerba 1986, 908.

[44] Kwietniewski and Wojtowicz 1914. For more information about these men and the Mianowski Fund, see the box on page 189.

**Stefan Kwietniewski** was born in 1874 in Warsaw, which was then in the Russian Empire. His father, Władysław (1840–1902), a mathematician, taught in secondary schools and sometimes as docent at the University of Warsaw. He also translated mathematical texts, and served in academic administrative roles, one with the Mianowski Fund. Stefan's brother became professor of zoology at Lwów. Stefan studied in Munich and Göttingen, then earned the doctorate from the University of Zurich in 1902 with a dissertation on surfaces in a space of four dimensions, supervised by Heinrich Burkhardt. Afterward, he supported himself by lecturing on and writing about mathematics and translating mathematics texts. From 1915 to 1939, he regularly gave a geometry course at the University of Warsaw, but to few students. Withdrawn and solitary, he did not participate in its rich mathematical life, and never became a professor. His lectures eventually constituted almost the only evidence of his existence. An avid outdoorsman, he moved to a rural environment in 1939, but died suddenly in 1940. With support from the Mianowski Fund, Kwietniewski translated several texts that are important to the mathematics in the present book, and contributed significantly to volumes 1 and 3 of the Heflich and Michalski 1915–1932 *Handbook for Self-Education*, which introduced many Poles to advanced mathematics.*

**Władysław Wojtowicz** was born in 1874 in Wiśniowiec and schooled in Żytomierz—then in the Russian Empire, those places are now in western Ukraine. He studied law at the University of Kiev, then mathematics and philosophy in Berlin and Zurich. From 1907 on, Wojtowicz taught in and served as director of several secondary schools. During 1919–1922 he also served in the Ministry of Religious Denominations and Public Education. Wojtowicz was editor of *Wektor* and, with Stefan Straszewicz, of *Przegląd matematyczno-fizyczny*—journals devoted to secondary-school mathematics. The latter published Alfred Tarski's [1924] 2014b paper *On the Equivalence of Polygons*. Wojtowicz translated into Polish several well-known mathematical works, including those of Federigo Enriques and Ugo Amaldi, which were important sources for the mathematics described in the present book. Wojtowicz himself authored a volume of mathematical tables and several popular secondary-school textbooks. In research reports, Tarski referred specifically to Wojtowicz's [1919] 1926 geometry text. Wojtowicz died in 1942 in Warsaw.†

**Mianowski Fund**. Under Russian rule, education in the Polish language and for the advancement of the Polish people was heavily suppressed. In 1869 the Russian government closed the only university so devoted. On the death of its last rector, Józef Mianowski, members of its former community obtained permission to organize a foundation named for him, ostensibly to pay for printing treatises. The government required it to be supported solely by private funds, and forbade it to initiate other activities in science and education. The fund was remarkably successful in raising support, including ownership of an oil field in the Caucasus. It became Poland's largest institution supporting science. The fund financed a series of Polish guidebooks for self-instruction in university-level subjects and numerous translations of monographs in advanced mathematics and physics. It published the research journal *Prace matematyczno-fizyczne*, the journals *Przegląd pedagogiczny* and *Wektor*, which served teachers and supported mathematics instruction, and the journal *Nauka polska*, devoted to scientific news. These publications substituted for instruction not provided to Polish students under the Russian regime, and lay well beyond the scope of the Russians' intention for the foundation. Several of them were particularly important for the geometry in the present book. During the first decade of Polish independence, the fund supported seventy percent of the scholars receiving doctoral degrees in Poland. It was supporting the journal *Przegląd filozoficzny* when Alfred Tarski's first paper, Tarski 1921 [2014], was published there. There is no evidence that it supported Tarski further, but it did fund some of the activity of his colleague Stefan Banach and teachers Tadeusz Kotarbiński and Stanisław Leśniewski. The Mianowski Fund survives today.‡

---

*Dobrzycki 1971; Marczewski 1971.

†W. Piotrowski 2003. Tarski's references to Wojtowicz [1919] 1926 are on pages 142 and 156 of the present book.

‡Kuzawa 1968, chapter 3; Hübner, Piskurewicz, and Zasztowt 1992.

In 1925, Tarski had been granted the *venia legendi*, or right to teach at the university, probably based on his 1924b paper about definitions of finiteness. From fall 1925 on, the University of Warsaw hired him each year, always on a temporary part-time basis, to teach various courses concerned with logic research and with elementary mathematics from an advanced viewpoint. Many of these were aimed at some combination of general audiences and future teachers. Tarski began working closely with Jan Łukasiewicz and the students associated with his research program in logic. In most trimesters Tarski offered a problem seminar (proseminar) and supervised some of those students' research. In autumn 1929 he officially became assistant to Łukasiewicz.[45]

Tarski's university teaching assignments are detailed in the table on page 191.[46] In 1919 the university had adopted an academic year of three trimesters:

> autumn ... October–December,
> winter .... January–March,
> summer ... April–June.

Most of Tarski's courses lasted all year; exceptions are noted in the table. Except in 1925/1926, each of his classes met once a week; the number of hours is shown. Some items in the table reflect notes in Jadacki 2003a indicating that scheduled courses were canceled. During 1925–1929, Tarski's average university teaching load was 2.6 hours per week. His appointment as Łukasiewicz's assistant brought an increase: during 1930–1939 his average was 5.75 hours. With difficulty, Tarski managed that on top of about 22 hours per week teaching at the Żeromski Gimnazjum. (See section 9.2.)

Tarski's courses listed in the table with names in roman type were probably offered principally for students beginning research in logic. Those with names in *italics* were probably attended by general audiences, including future and in-service teachers. For example, Warsaw elementary-school teacher Wacław Schayer enrolled in Tarski's courses *Topics in Methodology* and *Topics in Elementary Geometry* during 1930/1931 and 1931/1932, respectively. He earned a master's degree, became a school director, and coauthored with Tarski the secondary-school textbook *Geometrja*, described and translated in section 9.9 and chapter 13.[47] The borderline between those types of courses was probably not clear,[48] but even with that allowance, it is apparent that nearly half of Tarski's classroom work for the University of Warsaw supported its role in teacher preparation.

---

[45] Jadacki 2003a, 144–149.

[46] The data are from Warsaw University 1925–1939. Jacek Juliusz Jadacki (2003a) has noted that there were occasional differences between course advertisements and what actually took place. Evidently, the university adjusted the schedule according to student demand and availability of instructors and funds.

[47] For more information about Schayer, see the box in section 9.9, page 221.

[48] For example, Mojżesz Presburger attended Tarski's courses of both types during 1925–1929 (Zygmunt 1991, 213).

## Tarski's Teaching Load at the University of Warsaw

| Year | Hr | Courses | Remarks |
|---|---|---|---|
| 1925/26 | 2 | Theory of cardinal numbers | Fall and winter only. |
| 1926/27 | 2 | *Elementary mathematics* | Origin of Tarski's axioms for geometry. Canceled: 2-hr logic research course. |
| 1927/28 | 2 | Logic research seminar | Canceled: 4-hr *elementary mathematics* course |
| 1928/29 | 2<br>1<br>2 | *Arithmetic of real numbers*<br>*Algebra in secondary schools*<br>Logic research seminar | Tarski recorded (1933, 56; [1933] 1983, 205) that the 1927/28 and 1928/29 seminars included reports on the completeness of various theories. |
| 1929/30 | 2 | Logic research seminar | Canceled: 2-hr *elementary geometry* course. |
| 1930/31 | 2<br>2<br>2 | *Arithmetic of natural numbers*<br>*Topics in methodology*<br>Exercises and proseminar in logic | With Łukasiewicz (Ł). *Exercises=ćwiczenia.* |
| 1931/32 | 2<br>2<br>2 | *Topics in elementary geometry*<br>*Methodology of deductive sciences*<br>Exercises and proseminar in logic | (Ł) |
| 1932/33 | 2<br>2<br>1<br>2 | *Topics in elementary geometry*<br>*Methodology of deductive sciences*<br>Logic exercises for scientists<br>Logic exercises, proseminar | Stressing area, volume. Only 1 hr in winter.<br><br>(Ł)<br>(Ł) Only 1 hr in winter. |
| 1933/34 | 2<br>2<br>1?<br>1 | *Theoretical arithmetic*<br>*Methodology of deductive sciences*<br>Logic exerc. for mathematicians<br>Logic discussion, proseminar | <br><br>(Ł). Fall only.<br>(Ł). *discussion=konwersatorjum.* 2 hr, spring. |
| 1934/35 | 3<br>2<br>2 | *Theory of school algebra*<br>*Methodology of deductive sciences*<br>Logic and methodology exercises | *Theory=podstawy teoretyczne.*<br>These courses were all scheduled for the full year but Tarski spent the winter and spring in Vienna. |
| 1935/36 | 2<br>1<br>2<br>1 | *Theory of school algebra*<br>*Exercises on the theory of algebra*<br>*Methodology of deductive sciences*<br>Logic and methodology exercises | |
| 1936/37 | 2<br>2<br>1 | *Theory of arithmetic*<br>*Methodology of deductive sciences*<br>Logic and methodology exercises | Described in Hiż 1971, 234. |
| 1937/38 | 2<br>2<br>1 | *Teaching elementary mathematics*<br>Methodology of mathematics<br>Logic and methodology exercises | The course title has changed slightly. |
| 1938/39 | 2<br>1<br>2<br>1 | *Teaching mathematics*<br>*Exercises in elementary math.*<br>Methodology of mathematics<br>Logic and methodology exercises | |

During 1930–1934 Tarski's research seminars were described as conducted under Łukasiewicz's supervision. In January 1934, Tarski was appointed *adjunkt* in the philosophy faculty, and those notations were discontinued. Tarski's presentations of courses here named *seminar*, *exercises*, or *discussion* probably did not precisely reflect the distinction in course names, which would have been assigned administratively. They certainly involved discussion and individual guidance, probably at several levels—from fundamental exercises in elementary logic courses to problems under investigation by doctoral-research students. The latter were called *travaux dirigés* (*directed studies*) in Jadacki 2003a.

Tarski offered the course titled *Zarys metodologji nauk dedukcyjnych* for six years. He adapted this title for the English translation of his logic text: *Introduction to Logic and to the Methodology of the Deductive Sciences* ([1936] 1995). That translated title is used in the teaching-load table. Tarski probably began each year lecturing at the elementary level of this marvelously clear text, appealing to a general audience. But after three trimesters, he was introducing his newly sophisticated students to recent research of the Polish school of logicians. According to one student, not then at research level, who attended Tarski's 1937 general *Methodology of Deductive Sciences* course,

> Almost everything that he said during those lectures was his own, new and surprising. Today it belongs to the classic material that forms the foundations of metamathematics.[49]

Evidently, Tarski's guideline for that course, and probably for his entire career in Poland, was the principle articulated at the start of the present section: current work in foundations of mathematics is so momentous that it "must exert a profound influence on the exposition of the whole of mathematics, from the lowest to the highest levels of teaching."[50] For Tarski, there was no clear border between guiding research in logic and mathematics and offering new results for consideration and use by his community.

During the summers of 1929, 1930, and 1931, Tarski combined teaching and recreation: he gave lectures during vacation courses for qualified future and in-service secondary-school mathematics teachers. Sponsored by the Ministry of Religious Denominations and Public Education, the courses took place, respectively, in Warsaw; in the Baltic resort city of Puck; and in Nowy Targ, a small city in mountainous southern Poland. Organized by Stefan Straszewicz, they lasted approximately two weeks in July, and featured many lectures by leading Polish mathematicians. Their stated goal was

> to acquaint students with the modern scientific views of basic mathematical concepts related to the teaching of school subjects.

The 1930 course in Puck, about which the most detail is available, seems typical. The following account reveals the high intellectual level of these courses, and reflects the seriousness with which the Polish government and academic establishment regarded their task of incorporating recent scientific advances into public education and awareness.[51]

---

[49] Hiż 1971, 234. Henryk Hiż later became a noted professor of linguistics.

[50] Kwietniewski and Wojtowicz 1914.

[51] Rusiecki and Straszewicz 1930–1931; the previous quotation is from page 111.

## 9.3 Teaching Teachers

The course at Puck extended for three or four weeks during July 1930:

| Speaker | Hours | Subject |
|---|---|---|
| Wacław Sierpiński | 14 | Real analysis |
| Stanisław Ruziewicz | 14 | Algebra |
| Otton Nikodym | 20 | Mathematics teaching |
| Zygmunt Szulczyński | 6 | Educational psychology |
| Alfred Tarski | 14 | Arithmetic or geometry (?) |
| Stefan Straszewicz | 20 | Geometry |
| Stefan Straszewicz | 8 | Discussions |

The lectures were described as covering elements of a subject from an advanced viewpoint, or as covering selected topics, and they included time for exercises. Sierpiński, the leading Polish mathematician, is an important figure in the present book. Ruziewicz, a professor at Lwów, had been his student before World War I. They lectured for two hours a day during the first and second weeks, respectively. A noted researcher in analysis, Nikodym published that same year a large work on teaching. Szulczyński was a ministry official. Tarski was added to the program after its first announcement; his topic is not known, but in the other two summer courses he lectured on arithmetic and on geometry. Those other courses totaled 105 and 88 hours; this one would total 96 if Tarski also lectured for 14 hours. There were forty students, including six women. Some students and speakers brought their families. One participant reported that there were ample opportunities for seaside recreation, a weekend guided tour of the nearby free city of Danzig with its monumental brick architecture dating from Hanseatic times, and a tour of an enormous construction project, the new Polish port at Gdynia.[52]

The previous discussion has concentrated on Tarski's effort in training teachers. As just noted, this effort was intertwined with that for his research-oriented courses. The latter had a profound effect on the worlds of philosophy and mathematical research. That aspect of Tarski's career is not the focus of this book, because his work in logic and mathematics has been relatively accessible and has become well known. However, entirely omitting an account of the research aspect of Tarski's teaching at the University of Warsaw would be awkward and misleading. Therefore, the following section 9.4 will sketch this aspect of Tarski's career. Section 9.5 will return to his work in mathematics education.

---

[52] Nikodym 1930–1937, volume 1. Rusiecki and Straszewicz 1930–1931, volume 1(6), 224.

## 9.4 Research Seminars

Background for Alfred Tarski's early research as a student at the University of Warsaw was described in chapters 1 and 3. Section 9.3 emphasized his work in presenting mathematics and logic to general audiences, including future and in-service teachers, at the university and during summers. But about half of the courses that he taught at the university were devoted to specialized research. Many of those were connected with his position as research and teaching assistant to Jan Łukasiewicz. As noted in section 9.3, it is difficult to distinguish his activities with those two types of audience, and Tarski himself probably did not do so sharply. He introduced general students to new results that clarified the subjects they were studying. Moreover, his own research often stemmed from questions encountered in elementary teaching. In a sketch of Tarski's life, Jan Zygmunt wrote,

> It is obvious that all researchers and intellects of distinction use teaching to announce, above all, their own theories, results, and opinions. But it is not always the case that teaching is a source of research inspiration.... Tarski was inspired by his teaching and his students. When he taught elementary mathematics, he undertook problems from the foundations of school geometry and arithmetic in his research; this yielded results on the equivalence of geometric figures ... and the ... completeness of the elementary arithmetic of real numbers.[53]

Tarski's University of Warsaw teaching assignments were detailed in the table on page 191. Courses whose names are not italicized there were offered for students beginning research in logic. Tarski's first course, in 1925/1926 on cardinal numbers, may not have been closely related to his other teaching, but he would certainly have presented some of his recently published results on that subject (see Tarski 1924e and 1925b).

In 1926/1927 Tarski presented a course on elementary mathematics from an advanced viewpoint, in which he introduced his own axiom system for elementary Euclidean geometry. He needed a modern system to provide a rigorous basis for Euclid's classical results. Tarski objected to the famous system that David Hilbert had introduced in 1899, because the logical interrelationship of its primitive concepts and postulates was insufficiently clear. Instead, Tarski adapted a system of Oswald Veblen based on just three primitive terms: *point*, a point's lying *between* two others, and *congruence* of the segments determined by two pairs of points. Using only those terms Tarski phrased a set of very simple postulates. He then adopted the approach used by Mario Pieri to derive from these the entire theory, providing all details of the proofs. Tarski used this system himself from then on, in research and presumably in teaching. But for various reasons, it was not published in any detail until three decades later. Then it became a standard for his own and others' extensive further study of foundations of geometry. Detailed proofs were not published until the year Tarski died.[54]

---

[53] Zygmunt 1995, xvi.

[54] Tarski and Givant 1999, 175. Hilbert [1899] 1971, Veblen 1911, Pieri [1908] 2007: for a comparison of those three systems, see section 9.3 and Marchisotto and Smith 2007, section 3.10. Tarski featured his distinctive version of the parallel axiom in the [1931] 2014f exercise that he posed for consideration by students and teachers: see section 12.11. Tarski [1957] 1959. Schwabhäuser, Szmielew, and Tarski 1983.

## 9.4 Research Seminars

While teaching the 1926/1927 elementary-mathematics course, Tarski was also working on an entirely different axiomatization of Euclidean geometry based on the undefined notions *individual*, *part of*, and *sphere* (in the sense of *ball*). That system incorporated aspects of *mereology*, Stanisław Leśniewski's part–whole theory; of general topology, also under intense study at Warsaw; and of Mario Pieri's axiomatization of elementary geometry, mentioned in section 9.3. Tarski announced his results on this subject at the First Polish Mathematical Congress in Lwów in September 1927; Tarski [1927] 1983 is an enlarged English translation of his French article in the congress proceedings. Although closely related to the material in this book, that work is not discussed further here because it is already easily accessible.[55]

During the next year, 1927/1928, Tarski took charge of the seminar on problems in logic, which included supervision of some of the students' research. Łukasiewicz had founded the seminar a year earlier. Its exciting agenda has been described by Jan Zygmunt in his 1991 study of one of the students, Mojżesz Presburger. The seminar participants immediately began producing significant results, and under Tarski's guidance continued to do so.[56]

The results from this seminar that are most significant for this book have to do with elimination of quantifiers. The following example of this technique, applied to an open sentence in elementary real arithmetic, is familiar from the quadratic formula:

$$\exists x\, (x^2 + bx + c = 0) \Leftrightarrow b^2 \geq 4c.$$

The existential quantification $\exists x$ has been eliminated from the open sentence on the left: it does not appear in the equivalent condition on the right. For an elementary theory $T$, such as this arithmetic or Tarski's system for elementary geometry,[57] it is sometimes possible to find a larger theory $T'$ such that every open sentence of $T$ is equivalent to an open sentence of $T'$ with no quantifiers. Theories $T$ with this property have many other important features. Tarski liked to formulate exercises in elementary arithmetic or geometry that amounted to more complicated examples analogous to the one given above. He evidently felt that by solving these, students would gain some basic familiarity with a technique that would later play a major role in mathematics and its applications.

---

[55] In 1948, Łukasiewicz's doctoral student Stanisław Jaśkowski published a simplification of Tarski's presentation, possibly influenced by Tarski's research seminar two decades earlier. See also Jaśkowski 1949. For further development of this subject, consult Gruszczyński and Pietruszczak 2008.

[56] Łukasiewicz and Tarski [1930] 1983, 38. Research reports by Łukasiewicz and Tarski stirred up much controversy at the 1927 Second Polish Philosophical Congress: see section 14.2. For another glimpse of the milieu of logic in Warsaw at this time, see Woleński 1995a, 379.

[57] An *elementary* language is one that can be specified using only individual variables and constants, finitary operations on and relations between them, Boolean connectives, and the universal and existential quantifiers ∀ and ∃. A *sentence* is an expression of the language that, given an interpretation of its constants, operators, and relations, is either true or false: all its variables are bound by quantifiers. An *open sentence* is an expression that is a sentence or would become one if its unbound variables were bound by initial quantifiers. (Open sentences are often called *propositional functions*.) A *logical consequence* of a set $S$ of sentences is a sentence that is true in any interpretation in which all members of $S$ are true. A set $T$ of sentences is a *theory* if it contains the logical consequences of each of its subsets $S$.

*Mojżesz Presburger in 1925*

*Jan Łukasiewicz in 1935*

The most memorable result that stemmed directly from Tarski's 1927/1928 seminar is Presburger's application of quantifier elimination to the elementary arithmetic of natural numbers with addition but without multiplication. This led to an algorithm for deciding the truth or falsity of all sentences in that theory, and that algorithm has found applications in both theoretical computer science and in software engineering. Presburger presented his work in Warsaw at the First Congress of Mathematicians of Slavic Countries, 23–27 September 1929.[58,59] By 1930, Tarski himself was able to use the technique to devise algorithms for deciding the truth or falsity of all sentences in the richer elementary theories of real arithmetic and geometry. Many other significant new results, on other areas of logic, originated in the 1927/1928 seminar: for example, the one now known as the upward Skolem–Löwenheim theorem. Also this year, Tarski began presenting publicly his organization of the fundamental concepts of metamathematics.[60]

---

[58] Presburger [1929] 1991. The Congress is described in the following section, and in Tarski's [1929] 2014a article, which is translated in chapter 10.

[59] **Mojżesz Presburger** was born in Warsaw in 1904; his father soon died and he was brought up by his mother. Mojżesz entered the University of Warsaw in 1923, and attended most of Tarski's lectures and seminars during 1925–1930. He prepared the monographs Ajdukiewicz 1928 and Łukasiewicz [1929] 1963 for publication. Having earned a master's degree from the university in 1930, he left academia for the insurance industry. Presburger and his wife were murdered by the Nazis. For further information about him, consult Zygmunt 1991.

[60] For various reasons, Tarski's decision procedures were published only after many years' delay, in [1940] 1967 and [1948] 1957. For additional information about the 1927/1928 seminar, see *Fundamenta Mathematicae* 1934 and Mancosu, Zach, and Badesa 2009, 132–134. See also Zygmunt 2010.

**Jan Łukasiewicz** was born in 1878 in Lwów, which was then part of the Austrian Empire. His father was an Army officer and his mother, the daughter of a civil servant. He was reared Roman Catholic. Graduating from gimnazjum in 1897, Łukasiewicz entered the University of Lwów to study mathematics and philosophy. There he earned the doctorate in 1902 under supervision of Kazimierz Twardowski. After three years' service as tutor and library clerk and one, studying at Berlin and Louvain, he earned the *venia legendi* and began teaching in the University of Lwów, offering its first course in mathematical logic. In 1910 Łukasiewicz published a noted book on Aristotelian logic, a subject to which he returned many times. In 1911 he was appointed professor.

By 1915 Łukasiewicz had published about forty books and papers. That year, he became one of the first professors of the newly reopened Polish University of Warsaw. He held adminstrative positions there from 1916 to 1918, when he became head of the division of higher education in the Ministry of Religious Denominations and Public Education. In his address on that occasion he unveiled the idea of three-valued logic, which had evolved from his earlier work on probability theory. For the year 1919 Łukasiewicz served as that minister in the cabinet of Ignacy Paderewski.

That same year, Stanisław Leśniewski, also a student of Twardowski, was appointed to a second professorship in logic at Warsaw. Together, Łukasiewicz and Leśniewski founded the Warsaw school of logic. Their most illustrious student, Alfred Tarski, earned the doctorate there in 1924. Łukasiewicz pursued very fundamental questions. He invented prefix notation for logical operators that year, and his [1929] 1963 introductory logic text served as a standard for many years. In 1926 he organized a celebrated research seminar, for which Tarski soon became partially responsible; it continued until 1939 and gave rise to much of the logical methodology used worldwide during the next decades.

Łukasiewicz served as university rector twice, during 1922–1923 and 1931–1932. In 1928 he married Regina Barwińska, who stemmed from an aristocratic background.* In 1932 he invited Heinrich Scholz, the only professor of logic in Germany, to give lectures in Warsaw; this led to fortunate association during the next years. On Scholz's recommendation, Łukasiewicz received an honorary doctorate from the University of Münster. The German ambassador presented it at a ceremony in Warsaw in December 1938. In February 1939, Łukasiewicz traveled to Münster to give a celebratory lecture. In September 1939, the Germans invaded Poland, bombed Warsaw, and destroyed everything Łukasiewicz owned.

During the German occupation, Łukasiewicz worked as a translator in a government archive, and taught clandestinely. For several years, Scholz provided financial and bureaucratic support for the Łukasiewicz couple at considerable danger to himself. Due to their backgrounds, their opposition to Bolshevism, their association with the Polish government, and their German connection, the couple realized that the coming Soviet invasion and conquest would threaten their lives. Scholz arranged for them to move to Münster in July 1944 just before it, too, was destroyed, by Allied bombing.

After the armistice, Łukasiewicz taught logic in a displaced-persons camp and in Brussels. In March 1946 he accepted an appointment as lecturer at the Royal Irish Academy in Dublin. Łukasiewicz lived there until he died in 1956. He continued research in logic, finishing the last fifteen of his hundred-odd publications, including his second book, [1951] 1957, on Aristotelian logic.†

---

* In some publications, the marriage date is given incorrectly as 1939.

† For further information, consult Łukasiewicz [1953] 1994, Schmidt am Busch and Wehmeier 2007, Schreiber 1999, and B. Sobociński 1956.

*Andrzej Mostowski around 1931*

*Wanda Szmielew in 1938*

As shown in the table on page 191, Tarski continued directing the seminar and student research until he left Warsaw in 1939. He was officially appointed assistant to Łukasiewicz in 1929. During the 1930s, at least six Warsaw students were conducting research in logic under direction of Łukasiewicz, with support from Tarski.[61] In 1934 Tarski was appointed *adjunkt* in the philosophy department; his courses were then listed independently of Łukasiewicz. Two students soon began logic research directly with Tarski: Andrzej Mostowski and Wanda Szmielew. Mostowski's 1938 doctorate was officially supervised by Kazimierz Kuratowski, because Tarski was not a professor. Under Tarski's guidance Szmielew achieved her first publishable result, [1938] 1947; she finished her doctorate with him at Berkeley after World War II. Both Mostowski and Szmielew became world leaders in logic research. For further information about them, see the box on page 199. Other aspects of Tarski's research program in the 1930s are discussed briefly in chapters 8 and 14.

During the period 1927–1939, Tarski himself published papers on nearly fifty different research projects in logic and mathematics; many of these must have been discussed in detail in the seminar. Tarski's lectures and seminar certainly played a major role in the worldwide development of mathematical logic during those years. It is a tragedy that virtually no detailed day-to-day record remains of them.

---

[61] Jadacki 2003a. Scholz 1957. According to Rickey 2011, Łukasiewicz was responsible for the doctorates of Mordchaj Wajsberg (1931), Zygmunt Kobrzyński, Stanisław Jaśkowski (1932), Bolesław Sobociński (1937), and Jerzy Słupecki (1938). Czesław Lejewski studied with Łukasiewicz but received the doctorate in London (1954).

**Andrzej Mostowski** was born in 1913 in Lwów, which was then in the Austrian Empire. His father, a medical doctor, died of typhus the next year. The family then moved to Zakopane, and in 1920 to Warsaw, where Andrzej's mother worked for a bank. Andrzej entered the University of Warsaw after graduating from gimnazjum in 1931. He studied mathematics and logic with Alfred Tarski and other noted faculty, began research closely related to Tarski's work in set theory from a decade earlier, and earned a master's degree in 1936. A wealthy uncle supported his study in Vienna and Zurich for the next two years. Returning to Warsaw, Mostowski resumed research with Tarski, extending a method for proving independence of various principles from the axioms of set theory with *Urelemente* (individuals that are not sets). Mostowski's thesis was published in 1938, and the degree awarded in 1939, under official supervision of Kazimierz Kuratowski, since Tarski was not a professor.

During World War II, Mostowski worked as a clerk and did clandestine teaching. In 1944 he married a logician, Maria Matuszewska; and in late 1945, he was appointed professor at the University of Warsaw. Mostowski quickly became a leader there and eventually a world leader in mathematical logic. He produced many major research papers and monographs, often closely connected with Tarski's work. Mostowski died suddenly and prematurely in 1975.[*]

**Wanda Szmielew** was born in Warsaw in 1918 to Dawid and Bronisława Montlak. She was schooled in Warsaw and married Borys Szmielew soon after graduation; they were divorced in 1954. She began studies at the University of Warsaw in 1935, particularly with Alfred Tarski and Adolf Lindenbaum. Her early research was connected with theirs in set theory from a decade earlier. Her first result ([1938] 1947) came quickly, but its publication was delayed by World War II. During that time Szmielew was employed as a surveyor, taught clandestinely, and continued her research. In 1945 she obtained employment at the University and Technical Institute in Łódź. She completed her master's degree at Warsaw in 1947 and obtained a position as assistant there. Two years later she traveled to Berkeley, where she completed research for the doctorate in 1950, the fifth one supervised by Tarski. In her dissertation, published in 1955, she demonstrated the decidability of the elementary theory of Abelian groups. Szmielew returned to a professorship at Warsaw, rose through the ranks, and became the leading figure in foundations of geometry in Poland. Her later research connected that field with set theory and the foundations of algebra. She died prematurely of a tumor, in 1976.[†]

---

[*] For further information, consult Krajewski and Srebrny 2008.

[†] Łódź 1946, Domoradzki 2011. For further information, consult Kordos, Moczyńska, and Szczerba 1977.

## 9.5 Warsaw/Poznań Congress, 1929

Alfred Tarski's student Mojżesz Presburger presented a now famous result (see the preceding section) at the First Congress of Mathematicians of Slavic Countries, held in Warsaw and Poznań, 23–27 September, 1929. The Congress was quite large: most of the Polish mathematicians with roles in the present book attended. There were about 180 participants, including 35 from outside the country. Some participants are pictured on page 201. Attendance would have been larger, had the Soviet Academy of Sciences not declined to participate. In his opening address, at the Warsaw Polytechnic University, Wacław Sierpiński said,

> [This is] a regional congress. There is no political substratum nor objective. It is only about facilitating the cooperation of mathematicians living in countries not very distant from each other, and speaking related languages.

Sierpiński's understanding of *political* was naively restrictive: any such meeting would have political import, particularly so at that time in Eastern Europe, as the Soviets realized.[62]

The eight plenary sessions of the congress featured invited talks by highly distinguished speakers, half from Poland and half from abroad, on various subjects:

| | | |
|---|---|---|
| Kazimierz Kuratowski | | topology |
| Stefan Mazurkiewicz | | topology |
| Otton Nikodym | | mathematics and society |
| Wacław Sierpiński | | topology |
| Abraham A. Fraenkel | Germany | Georg Cantor's legacy |
| Leon Lichtenstein | Germany | fluid mechanics |
| Karl Menger | Austria | dimension theory |
| Kyrille Popoff | Bulgaria | integration and mechanics |

At least a dozen of the contributed papers were closely related to the mathematics discussed in this book.

Tarski presented two papers:

> *Remarks on Some Basic Concepts of Metamathematics,*
> *Cantor's Hypothesis about the Alephs.*[63]

The first had the same title as his German paper [1930] 1983, completed in March 1930. There, Tarski indicated that this material stemmed from Tarski 1928, the report of a talk given in Warsaw a year previously. Tarski's congress talk was probably adapted from an intermediate draft. His second talk was probably similar to one on the same subject that he presented in Vienna in February 1930 and published as Tarski 1931.

---

[62] Sierpiński [1929] 1930, 10. The complete proceedings of the congress are included in Leja 1930. Academician Nikolai N. Luzin and two other Russians were on the committee that organized the congress. But its promise was eclipsed by events at home: during early 1929 the Soviet Academy had come under intense political pressure and investigation for counterrevolutionary bias and activities; a purge began in August, and by the end of the year eleven percent of its permanent staff had been dismissed (Levin 1988).

[63] In Polish, *Uwagi o kilku podstawowych pojęciach metamatematyki* and *Hypoteza Cantora o alefach*.

## 9.5 Warsaw/Poznań Congress, 1929

*First Congress of Mathematicians of Slavic Countries
Warsaw Polytechnic University, September 1929*[64]

The Congress offered at least seven scheduled social events. For example,

> On 24 September the participants were received at 17:00 by the prime minister, Dr. Kazimierz Świtalski, the honorary president of the congress, at the palace of the Prime Minister; afterward, at 20:00, they attended an evening performance at the opera, with a performance of the Polish ballet *Pan Twardowski*.[65]

The ballet, written during 1919–1920 by Ludomir Różycki, was based on a Polish folktale reminiscent of the Faust legend. On the last day of the congress in Warsaw, the participants attended a reception at the city hall, and many departed at 23:00 on a sleeper train for Poznań. The next day, after a closing ceremony at the university, they visited the General Polish Exposition, which had been open in Poznań since May.[66]

---

[64] The present editors have tentatively identified several in the front row:

| | | |
|---|---|---|
| 1 Janina Hosiassonówna | 4 Stefan Mazurkiewicz | 7 Jan Łukasiewicz | 10 Alfred Tarski |
| 2 Antoni M. Rusiecki | 5 Kyrille Popoff | 8 Leon Lichtenstein | 11 Mojżesz Presburger |
| 3 Samuel Dickstein | 6 Wacław Sierpiński | 9 Kazimierz Kuratowski | |

For other photographs of the same group, some participants changed positions. On one of those, in the Tarski Archive at the Bancroft Library in Berkeley, Maria Tarska identified Alfred and Sierpiński.

[65] Leja 1930, 28. The ballet had no connection with the philosopher Kazimierz Twardowski who figures in the present book.

[66] The journey to Poznań would have taken about five hours. See the poster and photograph on the facing page: Jastrzębowski 1929; Powszechna Wystawa Krajowa 1929, 3.

*General Polish Exposition, Poznań*
*May–September 1929*

Readers are invited to compare Tarski's impressions of these events in chapter 10 and to note his judgment that a conference of this sort could not meet the needs of secondary teachers, nor those whose main concern was training them:

> including in the program of one and the same assembly issues of the exact sciences and of didactics puts [didactics] at a disadvantage.

Tarski proposed as an alternative a joint conference of teachers and researchers, "devoted exclusively to the problems of mathematics in the framework of the educational system."[67]

---

[67] Tarski [1929] 2014a, translated in chapter 10.

## 9.6 Organizations and Journals for Teachers

In Alfred Tarski's 1929 report translated in chapter 10 of this book, he argued that conferences oriented toward mathematical research would not meet the needs of those teachers and scholars interested mainly in the pedagogy of mathematics. He proposed a joint effort by both groups, targeted exclusively at pedagogical issues. Was any such activity underway during 1925–1929?

In several other countries, societies had existed for years to foster interaction between the research and educational communities in mathematics, for example:

| Country | Society, year founded | Journal |
|---|---|---|
| Hungary | Bolyai János Matematikai Társulat, 1891 | *Matematikai és physikai lapok* |
| Italy | Mathesis, 1895 | *Periodico di matematica* |
| United Kingdom | The Mathematical Association, 1871 | *Mathematical Gazette* |
| United States | Mathematical Association of America, 1915 | *American Mathematical Monthly* |

Each of these published a journal aimed at both communites.

A 1927 Polish teachers' almanac listed only two organizations serving secondary-school teachers in general throughout Poland. Neither one addressed mathematics teachers specifically.[68] The larger, with 7000 members, was the Towarzystwo Nauczycieli Szkół Średnich i Wyższych, which published the journal *Przegląd pedagogiczny*.[69] Each year during 1921–1939 it organized a congress of Polish secondary-school teachers. The smaller organization, with 2000 members, was the Związek Zawodowy Nauczycielstwa Polskich Szkół Średnich, the organization that supported the gimnazjum where Tarski taught. It published the journal *Ogniwo*, which included Tarski's [1929] 2014 report.[70] In July 1930, this second organization merged with the Związek Polskiego Nauczycielstwa Szkół Powszechnych, an organization of 36,000 teachers in Polish schools at all levels, to form the Związek Nauczycielstwa Polskiego, which still exists. This was the organization that invested in the housing development where the Tarski family rented and purchased their homes from 1930–1939. These latter organizations published the journals *Głos nauczycielski* and *Ruch pedagogiczny*.[71] In addition to the journals named here, several smaller ones were published for various groups of teachers. During 1925–1929 all these journals included brief book reviews, announcements, and news about scientific congresses and courses. These included items about mathematics, but the journals presented little specifically about secondary-level instruction in mathematics.

---

[68] Tomczak 1927, 247–248.

[69] Society of Teachers in Secondary Schools and Universities. The journal name means *Educational Review*.

[70] Polish Union of Secondary-School Teachers. *Ogniwo* means *The Link*.

[71] Polish Teachers' Association. The journal names mean *The Voice of Teachers* and *Educational Trends*.

## 9.7 *Parametr* and *Młody Matematyk*

The previous section noted that unlike some other countries that also led the development of mathematics in the late 1920s, Poland had no organization specifically for mathematics teachers. While there were some journals targeted at teachers in general, none was devoted specifically to mathematics education. The journal *Przegląd matematyczno-fizyczny*, described in section 4.3, had attempted to address both the mathematics research and education communities. Edited by Władysław Wojtowicz and Stefan Straszewicz, it published Alfred Tarski's [1924] 2014b paper on equidecomposability of plane polygons, which is translated in chapter 5. But that journal continued for only three years.

In 1930, Straszewicz and Antoni M. Rusiecki, an official in the Ministry of Religious Denominations and Public Education, tried again to fill the need for a journal for mathematics teachers. They persuaded the Poznań firm Księgarnia Św. Wojciecha (St. Wojciech's Bookstore) to underwrite publication of a new journal, *Parametr*, which the two would edit. On its first page, the publisher stated its purpose:

> The genius of Polish mathematics has in recent years brought [it] to a leading position in world science. But there are widespread complaints about the unsatisfactory results of mathematics teaching in Polish schools. We...recognize the need for repairs in this area of Polish life, ...[and] believe that *Parametr* will fill the existing gap in Polish educational publishing and will contribute to raising the level of mathematics education in Poland. The journal will discuss issues related to the teaching of mathematics in...schools. It will also contain a section for older students.

The editors explained,

> The publisher, St. Wojciech, considers issuing *Parametr*...a public service for the Polish schools, and does not treat [it] as a source of income....

Although Rusiecki was principally employed by the ministry, it evidently played no formal role in publishing the journal.[72]

Subscriptions cost fifteen zlotys per year for ten issues: about the cost of an expensive book. The first year's volume constituted four hundred pages. Each issue contained four or five articles, some book reviews—often quite extensive—and lists of new books. There were sections for professional news, miscellaneous notes, problems for readers to solve, and solutions. Each issue but one concluded with a summary in Lingua Peano, an artificial language developed for international scientific discourse around 1900 by the mathematician and linguist Giuseppe Peano.[73] The ministry was impressed and encouraged by the journal. Late in the first year of publication, it described *Parametr* as

---

[72] Księgarnia Św. Wojciecha 1930; Rusiecki and Straszewicz 1930, 323. For further information about those editors, consult the boxes on pages 208 and 210. St. Wojciech is often referred to as St. Adalbert.

[73] Rusiecki 1931 is a brief introduction to Lingua Peano for readers of Polish. Peano's 1903 term for the language was *Latino sine Flexione*. It is very easy to learn!

> a journal devoted to mathematics education in schools .... It is maintained at a high level, and contains valuable methodological advice. It can be of service to teachers of mathematics in primary schools and secondary schools and teacher training. It should be found in every library for school teachers.[74]

The editors were concerned about the balance between material that would interest only teachers and that which might attract both teachers and older students. In five of the nine issues of volume 1, some articles of the latter type were segregated into a *section for youth*.[75] But some of the articles for teachers, mostly about pedagogy, did contain examples that might engage students. About one-third of all the articles in the 1930 volume 1 would have interested just teachers.

Beginning with the 1931/1932 volume 2 of *Parametr*, the editors moved its student-oriented content, including most of the problem section, to a new journal, *Młody matematyk*, distributed with *Parametr* at no additional cost.[76] *Parametr* then contained the material intended just for teachers. The editors actually increased the amount of material in *Młody matematyk* that would interest advanced students as well as teachers. The total number of pages in the two volumes increased by twenty percent. The journals are remarkable for their inclusion of so much material that related mathematics to the society in which their readers lived.

During these first years, the two journals included about one hundred articles. Rusiecki himself wrote seventeen. Sixty-nine were contributed by authors who apparently had no university affiliation: most of them were probably teachers. Nineteen were written by associates of postsecondary institutions. Only six of these latter authors wrote more than one article; Tarski contributed three. Example mastheads and tables of contents are shown in the figures on page 206.

After the first year, the problem sections were apportioned rather strangely between the two journals. Tarski contributed fourteen exercises to *Parametr*, volume 2, and to *Młody matematyk*, volume 1: more than any other contributor except Rusiecki, the editor. The problem sections are described in detail in chapter 12, and Tarski's contributions are all translated and discussed there.

---

[74] Ministerstwo 1930.

[75] *Dział dla młodzieży*

[76] A student subscription to *Młody matematyk* alone cost 4 zlotys; group discounts were available (Rusiecki and Straszewicz 1930, 322).

*Parametr* 2(1931/1932)(2–3) supplement,
*Młody matematyk* 1(1931/1932)(3):

> Alfred Tarski.
> On the Degree of Equivalence of Polygons

*Parametr* 2(1931/1932)(8–10):

> Alfred Tarski,
> *The Theory of the Circumference of a Circle in the Secondary School,*
> *Remarks on the Degree of Equivalence of Polygons.*

> Henryk Moese,
> *A Contribution to Tarski's Problem "On the Degree of Equivalence of Polygons."*

These periodicals were very interesting, delightful to look at and read, and should have provided a significant contribution to mathematics education. But, according to their editors,

> *Młody matematyk* did not find a resonance as *Parametr* did. Although issues were circulated to all public and private schools, only a few schools have responded by ordering a subscription.[77]

The journals may have suffered from poor marketing. But production problems became unmanageable as well. Responsibility for publication evidently lay almost entirely on Rusiecki. He pointed this out several times in editorial notes. Two of those, late in the third year of publication, are most poignant:

> *Parameter* has survived a serious illness: it was not published for half a year. ... [This] is not the slightest fault of the publisher. The entire responsibility for the lapse in production falls on the editorial office. ... The Editor of *Parametr* serves in the Ministry of Religious Denominations and Public Education as a ministerial school instructor. At the beginning of the current school year, he buckled under the workload: in September instructional visits began as well as participation in the Second Congress of Polish Mathematicians in Vilnius; in October and November there occurred more than twenty days of travel each month, including organization of a two-week instructors' course in Lwów; and finally in December there began intensive work on a framework of mathematics programs in [public] schools ... in connection with the intended school reform. Meanwhile, reporting activites and progress on "Mathematics for the fifth class of elementary school" [is] to be completed.[78]

> For the delay in the issuance of *Młody matematyk* we sincerely apologize. The Publisher does not bear any guilt for this. The whole responsibility falls on the editor, who could not manage to cope with all the current obligations on his time. The editors must admit that despite the bitterness that they have suffered as a result of not presenting *Młody matematyk* [on time], there was also a drop of sweetness: readers' letters proved that *Młody matematyk* has friends.[79]

After the 1931–1932 volumes, the journals ceased publication for several years. *Parametr* reappeared in 1939 for a brief run, until the German invasion.[80]

Straszewicz had already joined several others to found a new journal, *Matematyka i szkoła*, published by the Towarzystwo Nauczycieli Szkół Średnich i Wyższych.[81] Pursuing much the same goals as *Parametr* and structured similarly, it published five issues during 1938–1939.

---

[77] Rusiecki and Straszewicz 1932b, 33.

[78] Rusiecki and Straszewicz 1932b, 145.

[79] Rusiecki and Straszewicz 1931–1932, 97. The last phrase has a double meaning: the journal title means *the young mathematician*.

[80] For further information about these journals, consult Dubiel 1975, and Dubiel 1990, section 2.2.1, 31–34. But that report of their publication history is incorrect.

[81] For further information about this journal, consult Dubiel 1990, section 2.2.2, 34–35. This organization, the Society of Teachers in Secondary Schools and Universities, also published the journal *Przegląd pedagogiczny* to serve teachers of all disciplines.

In 1948, after the war, Rusiecki and Straszewicz collaborated again to found the journal *Matematyka: Czasopismo dla nauczycieli*. It was published by the Polish Mathematical Society on behalf of the Ministry of Education.[82] The new journal's structure and flavor clearly descended from that of the earlier ones. Its 1949 second volume included some solutions of exercises that had been posed in *Parametr* a decade or more earlier, including one for Tarski's rather difficult [1930] 2014c exercise[83] about factoring polynomials of the form $ax^4 + b$. The same volume began with a paper by Stanisław Jaśkowski that interpreted, for an audience of teachers, Tarski's [1927] 1983 axiomatization of Euclidean geometry based on the notion of *sphere*. In an interesting way, the journal continued the tradition of *Parametr* and *Młody matematyk* in connecting mathematics with its social context: the editorial note Iwaszkiewicz 1949a that immediately followed Jaśkowski's paper provided a Marxist commentary![84]

---

**Antoni Marian Rusiecki** was born in 1892 in Bodzechów, a village near Kielce in southern Poland, then part of the Russian Empire. His father was a railroad stationmaster. Antoni was schooled in Warsaw. He attended the Polytechnic Institute in St. Petersburg, then the University of Kiev, from which he was graduated in 1916. Rusiecki taught in secondary schools there and in Białystok. In 1922 he became instructor of mathematics at the Ministry of Religious Denominations and Public Education in Warsaw. His work involved development and administration, as well as teacher-training visits throughout the country. The autobiography of the noted mathematician Mark Kac provides a glimpse of one of those visits, which resulted in Kac's first publication, as a student in 1931. For a more detailed account of Rusiecki's workload, see his lament on page 207. Simultaneously, Rusiecki taught in several Warsaw schools. During 1930–1931 he and Stefan Straszewicz founded the journals *Parametr* and *Młody matematyk*, devoted to teaching mathematics in the schools.* He himself provided a substantial portion of their content, which must have required stupendous effort. Editing these impressive journals rested too heavily on his shoulders. The journals quickly became irregular, and disappeared during 1932–1939. During World War II, Rusiecki ran a used-book store and taught in clandestine schools and in a rural elementary school. After returning to Warsaw, he served as editor at two major book publishers and was a founding editor of the journal *Matematyka: Czasopismo dla nauczycieli* (*Journal for Teachers*). In addition, Rusiecki taught methodology of teaching at the University of Warsaw and worked in support of the Polish Mathematical Olympiad. He was author or coauthor of numerous elementary-school mathematics texts and books on methods of teaching, often in collaboration with Wacław Schayer or Adam Zarzecki. Rusiecki died in Warsaw in 1956.†

---

* These two journals, described in detail in the present section, included the material translated in sections 7.2–7.4 and chapters 11 and 12.

† W. Piotrowski 2003; Królikowski 1991; Kac 1985, 4.

---

[82] The title means *Mathematics: Journal for Teachers*. See its editor's report, Iwaszkiewicz 1949b. This handsome journal is still in publication.

[83] Tarski's exercise is translated and discussed in detail in section 12.3.

[84] Jaśkowski 1949, Iwaszkiewicz 1949a. Tarski's axiomatization is briefly described in section 9.4.

*Antoni Marian Rusiecki
around 1935*

*Stefan Straszewicz
around 1935*

**Stefan Straszewicz** was born in 1889 in Warsaw, then part of the Russian Empire. His father was an army officer and his mother, a teacher. He finished secondary school in 1906 in Białystok then studied for a year with the Society of Scientific Courses* in Warsaw. He left in 1907 to study in Zurich, returned in 1911 for a year's military service, then taught in several Warsaw schools. In 1913 he resumed studies at the University of Zurich, where he earned the doctorate in 1914 with a dissertation on convex geometry, supervised by Ernst Zermelo. Straszewicz remained there as a researcher until 1919, and then served in the Polish military during the Polish–Soviet War.

For a few years starting in 1919, Straszewicz held multiple part-time teaching positions at secondary schools, the National Pedagogical Institute (Państwowy Instytut Pedagogiczny), the Free Polish University (Wolnej Wszechnicy Polskiej), and the University of Warsaw. At the latter three, he was a colleague of Alfred Tarski. At the university, Straszewicz taught a course in mathematics for natural scientists every year from 1921 to 1934. He earned the *venia legendi* there in 1926, began teaching differential geometry as well, and continued mathematical research. By 1935 he had published about ten papers on the theories of point sets and curves. Then he decreased his university activity, and taught only one more course there, elementary geometry, during 1936/1937.

Straszewicz had begun teaching at the Warsaw Polytechnic in 1920, was appointed professor in its School of Civil Engineering in 1928, and was soon serving that university in administrative roles.

Straszewicz edited *Przegląd matematyczno-fizyczny*, a journal devoted to instruction at the secondary level, during its short life, 1923–1926. This *Review* published Tarski's paper that is translated in chapter 5. In 1930 Straszewicz was a founding editor of the journals *Parametr* and *Młody matematyk*, which served both teachers and students. Described in detail in the present section, those journals published the works of Tarski translated in sections 7.2–7.4 and chapters 11 and 12. Unfortunately, they were issued only during 1930–1932 and 1938–1939. In 1937 Straszewicz started another, similar journal, *Matematyka i szkoła*.

From 1926 until 1939, Straszewicz chaired the committee on mathematics at the Ministry of Religious Denominations and Public Education. He became heavily involved nationwide in curriculum development, teacher supervision, and teacher training. He had become a mathematical statesman with nationwide influence.

Straszewicz coauthored numerous secondary-school texts, including a 1935 geometry book known for its emphasis on transformations. It competed directly with the [1935] 1946 text by Zygmunt Chwiałkowski, Wacław Schayer, and Tarski, parts of which are translated in chapter 13 of the present book.

During World War II, Straszewicz headed the clandestine activity of the Warsaw Polytechnic. After the war he helped rebuild it, served for three years as its vice-president, and also lectured at the University of Warsaw. He spearheaded the Polish Mathematical Olympiad during the 1950s and served as president of the Polish Mathematical Society. Straszewicz died in Warsaw in 1983, a mathematical statesman of international influence.[†]

---

*Towarzystwo Kursów Naukowych, a private organization that provided the only postsecondary scientific instruction available under the Russian oppression.

[†] Warsaw University 1921–1939; Piłatowicz 2006.

## 9.8 Tarski's Contributions to *Parametr* and *Młody Matematyk*

During 1930–1932, Alfred Tarski contributed three articles to the journals *Parametr* and *Młody matematyk* and spurred another one by Henryk Moese.[85] In addition, Tarski submitted fourteen exercises to the journals' problem sections for students and teachers to consider. These made him one of the most significant contributors. All of these articles and exercises are translated in sections 7.2 and 7.4 and chapters 11 and 12 of the present book. The preceding section gave an overview of the journals. This section briefly describes the articles, sometimes by referring to discussions elsewhere in this book. Detailed discussions of the journals' problem sections and Tarski's exercises are found in chapter 12.

Traditionally, students learned in secondary school that any two polygonal regions $W$ and $V$ with the same area can be subdivided into the same number of polygonal subregions with disjoint interiors such that all pairs of corresponding subregions of $W$ and $V$ are congruent. Tarski called the smallest number of subregions for which this is possible the *degree* $\sigma(W,V)$ *of equivalence of* $W$ *and* $V$. His paper *On the Degree of Equivalence of Polygons*[86] introduced this notion and some of its most general properties, and then considered some special cases in detail. In particular, for each $x > 0$, Tarski defined $\tau(x) = \sigma(W,V)$, where $W$ is a unit square and $V$ is a rectangle with edge lengths $x$ and $1/x$. He proceeded to compute some values of $\tau$, such as $\tau(1) = 1$, $\tau(2) = 2$, and $\tau(3) = 3$, and proved that $\tau(n) \leq n$ for every positive integer $n$. Tarski left as exercises a number of details and the computation of $\tau(4)$. He concluded by stating some problems that were yet unsolved, such as whether $\tau(n) = n$ for every positive integer $n$. The methodology of this mathematics is elementary, within the scope of the secondary-school curriculum. Tarski presented it to students, providing a glimpse of what formulating and researching a real mathematical problem is like.

Henryk Moese, a secondary teacher and reader of *Młody matematyk*, took up Tarski's challenging unsolved problem and proved that $\tau(n) = n$ for all positive integers $n$, using reasoning only slightly more complicated than Tarski's. He settled some other questions and stated a new conjecture that would determine by simple formulas the values $\tau(x)$ for all positive real numbers $x$. Moese's work was published in the next volume of *Młody matematyk*'s parent journal *Parametr*, which targeted mathematics teachers.[87]

In the same issue of *Parametr*, Tarski published a continuation of his first paper on the degree of equivalence.[88] He extended his own and Moese's work to show that $\tau(x) = \lceil x \rceil$ or $1 + \lceil x \rceil$ *for all real* $x \geq 1$. Moese had conjectured that the former value obtains *only* when $x = n + 1/p$ for some positive integers $n$ and $p$. Tarski reported that he

---

[85] Front matter of those journal issues is displayed on page 206 of the present book.
[86] Tarski [1931] 2014a, translated in section 7.2.
[87] Moese [1932] 2014, translated in section 7.3. For information about Moese, see a box and portrait in section 7.1.
[88] Tarski [1932] 2014d, translated in section 7.4.

had proved this for the case $1 < x \leq 2$, but claimed that the proof was too subtle for that journal.

Detailed discussions of these three papers on the degree of equivalence are found in sections 7.1 and 8.5 of the present book. In 1952 Izaak Wirszup drafted English translations of the papers, and the College of the University of Chicago published them as typescripts with hand-lettered symbols, only partially edited.[89] The present editors have not been able to determine why that occurred. For decades the original articles and those drafts were the only form in which this mathematics was accessible, and they were difficult to find. Complete new translations are included in sections 7.2–7.4 of the present book.

That same issue of *Parametr* also contained Tarski's paper *The Theory of the Circumference of a Circle in the Secondary School.*[90] To his target audience of secondary-school geometry teachers, he advocated an alternative to the standard definition of the circumference of a circle $C$ and derivation of its basic properties.

Tarski first presented a definition of the circumference of a circle $C$ that he termed *usual*.[91] Essentially, he proved that

(a) the sequence of perimeters $p_n$ of regular polygons with $n$ edges, inscribed in $C$, is increasing;

(b) that of the perimeters $q_n$ of the corresponding circumscribed polygons is decreasing;

(c) $p_n < q_n$ for every $n$; and

(d) $q_n - p_n$ approaches zero as $n$ increases.

By a result in the theory of limits, already covered at this stage of the curriculum, there is a unique number $c$ such that $p_n < c < q_n$ for every $n$. This number $c$ is defined to be the circumference of $C$. Tarski analyzed its dependence on the radius $r$ of $C$, defined the number $\pi$ so that $c = 2\pi r$ always, and showed how to use his intermediate results to approximate $\pi$ using trigonometric tables.

Tarski claimed that this standard argument, which he called the *method of limits*, has three disadvantages:

---

[89] Tarski and Moese 1952. For information about Wirszup, see a box in section 7.1.

[90] First published in *Parametr* 2 (1931–1932), 257–267; translated in chapter 11.

[91] The present editors have found no clear source for Tarski's *usual* definition. It can be extracted easily, however, from some elementary-geometry texts popular in earlier decades, supplemented by a precise treatment of the theory of limits. For example, see C. Davies 1890, Book V, 149–162.

## 9.8 Tarski's Contributions to *Parametr* and *Młody Matematyk*

- The proofs of (a) and (b) are cumbersome, involving complicated trigonometry and algebra.[92]
- The regular polygons play an unnaturally special role in this definition, so that this method does not suggest a general way to define the lengths of other curves.
- The definition seems to use more of the theory of limits than is necessary.

To meet those objections, Tarski presented an alternative approach, which he called the *method of cuts*, in the sense of Dedekind cuts:

(d) Properties of polygons already covered at this stage of the curriculum imply that the perimeter of any simple polygon inscribed in $C$ is smaller than the perimeter of any polygon circumscribed about $C$.

(e) By the continuity principle, some number $c$ lies between all those inscribed perimeters and all the circumscribed perimeters.

(f) The number $c$ will be unique if *any* sequences $p_n$ and $q_n$ of inscribed and circumscribed perimeters can be found such that $q_n - p_n$ approaches zero as $n$ increases. Instead of using the perimeters $p_n$ and $q_n$ of regular polygons with $n$ edges for all $n$ as in the standard approach, Tarski suggested using just those of the regular polygons with $2^n$ edges, for all $n$. He showed that the trigonometry and algebra required is much simpler, and can be used in the same way to approximate $\pi$.

Certainly this method of cuts satisfies the first objection in the previous paragraph and alleviates the second. However, as Tarski noted, both methods use about the same amount of the theory of limits, just in different order.

Tarski was not breaking new ground here. Steps (d) and (e) of his method of cuts were used in the text by Federigo Enriques and Ugo Amaldi, with which Tarski was familiar. Step (f)—doubling the number of edges of the inscribed and circumscribed polygons at each step of the sequences—had been employed by Archimedes two millennia earlier.[93]

---

[92] Tarski's method of proof included demonstration of the inequality

$$\frac{\sin \beta}{\sin \alpha} < \frac{\beta}{\alpha} < \frac{\tan \beta}{\tan \alpha},$$

for commensurable $\alpha, \beta$ with $0° < \alpha < \beta < 90°$. This kind of reasoning was familiar to ancient Greek geometers, and was handed down to modern times by historians. See Heath 1921, volume 2, 5, 276–283.

[93] Enriques and Amaldi [1903] 1916, chapter 7, §§522–532. Tarski cited that text as a recommended source of geometric lore on the first page of his [1924] 2014b paper, translated in chapter 5. See also Archimedes [1897] 2002a.

## 9.9 Tarski's Coauthored Textbook *Geometrja*

During the mid-1920s, as Alfred Tarski began his career as a geometry teacher at the Third Boys' Gimnazjum, there was little variety among the Polish textbooks available for his classes. Only three had originated in Poland. To fill the need, several new ones appeared during the next decade. In 1935 Tarski collaborated with two colleagues to publish a textbook, *Geometrja dla trzeciej klasy gimnazjalnej* (*Geometry for the Third Gimnazjum Class*), called simply *Geometrja* in this chapter. Excerpts are translated in chapter 13.[94]

The senior coauthor, Zygmunt Chwiałkowski, had earned a doctorate in mathematics. He was an experienced Warsaw gimnazjum teacher, active in teachers' organizations. By 1925 he had begun publishing mathematics textbooks for use in elementary and secondary schools. The junior coauthor, Wacław Schayer, began teaching in Warsaw elementary schools in 1927 and then earned a master's degree in mathematics from the University of Warsaw.[95] Schayer attended Tarski's courses *Topics in Methodology* and *Topics in Elementary Geometry* during 1930/1931 and 1931/1932, respectively. *Geometrja* was the only schoolbook that Tarski authored. But both Chwiałkowski and Schayer continued publishing textbooks for elementary and secondary schools, many in collaboration with each other and with other mathematicians mentioned in the present book. After World War II, Schayer became a nationally prominent politician. Little is known about Chwiałkowski; for example, the present editors do not know how or when he met his coauthors.[96]

*Geometrja* was published in Lwów by the firm Państwowe Wydawnictwo Książek Szkolnych (National Schoolbook Publisher). An advertisement[97] for it is depicted on the facing page. The text at the top says that the publisher recommended the following gimnazjum textbooks for the academic year 1937/1938. The other books advertised are for classes in history, French, German, Latin, and English. The prices are in zlotys. *Geometrja* was republished twice, during and just after the war—for further details, see the box on page 224.

*Geometrja* adhered to the official curriculum published in 1934 by the Ministry of Religious Denominations and Public Education. That document contained guidelines for implementing the major 1932 reform of the Polish school system: the curriculum for

---

[94] Chwiałkowski, Schayer, and Tarski [1935] 1946. Textbook availability in the 1920s was discussed in section 9.2 of the present book.

[95] Tarski's coauthor Wacław Schayer was a first cousin of **Stanisław Schayer** (1899–1941). The latter once taught mathematics with Chwiałkowski in the same Warsaw gimnazjum, was a member of the Warsaw Society of Sciences and Letters, and founded the Oriental Institute of the University of Warsaw (Zagórowski 1924–1926, volume 2, 250; Konarski 1994; Jerzmanowski 2013).

[96] Chodorowski et al. 2005, 31, 44. For more information on Chwiałkowski and Schayer, see the boxes on pages 219 and 221.

[97] Państwowe Wydawnictwo Książek Szkolnych 1937, from page 221 of the 1 September 1937 issue of the journal *Przegląd pedagogiczny*, published by the Society of Teachers in Secondary Schools and Universities, which was mentioned in section 9.6.

gimnazjum years I–IV. *Geometrja* conforms so closely with the specifications that one wonders whether its authors had a hand in formulating them. But the guidelines do not suggest the style of presentation or details of the exercise sets: those are the province of the textbook authors.[98] Three books competing with *Geometrja* were published the same year: Iwaszkiewicz 1935, Łomnicki 1935, and Straszewicz and Kulczycki [1935] 1948.

Unlike many geometry textbooks, *Geometrja* is not self-contained. Intended only for the third gimnazjum year, it relied on other textbooks for prerequisite material covered in the preceding years.[99] Its authors assumed that student readers of *Geometrja* had already become familiar with various techniques for reasoning, calculating, measuring, and drawing, and with some of the most basic ideas and theorems of deductive geometry. Moreover, it left many important topics in geometry to be presented as high points in the textbooks for the later years of gimnazjum.

*Advertisement in* Przegląd pedagogiczny, *1937*

---

[98] Poland 1934, 117–123. The reform is often referred to under the name of the minister: Janusz Jędrzejewicz. See page 172 for a broader description. Its recommendations for the liceum years V–VI would not be phased in until 1937. The present editors do not know which mathematicians delineated this curriculum.

[99] In 1934, the year before *Geometrja* appeared, a different publisher had published a similarly titled textbook for the preceding year's course, written by the Lwów mathematician Antoni Łomnicki. For more information about him, see a box in section 4.2.

Among the logical notions prerequisite for *Geometrja* were

- definition, axiom, use of the word *arbitrary*;
- proof, including use of exhaustive and of mutually exclusive cases;
- reduction to previously solved problems;
- converse and equivalent statements;
- informal use of inclusion, intersection, and union of sets.

*Geometrja* also assumed familiarity with these features of elementary geometry:

- points, segments, and lines;
- segment arithmetic, including the axiom of Archimedes;
- oriented angles, angle arithmetic, and degree measure;
- triangles of various sorts, their congruence and anticongruence, and the basic inequalities;
- erecting and dropping perpendiculars;
- parallels and their use in subdividing a segment into equal parts;
- polygons of various types and their basic properties;
- circles, and intersection of two circles.

No solid geometry is involved in *Geometrja*.[100]

The following paragraphs describe the contents of *Geometrja*: see its full translated table of contents in the introduction to chapter 13 of the present book. The thirty-five sections of *Geometrja* are divided into three parts:

|     |                                    |          |
|-----|------------------------------------|----------|
| I.  | *Geometric Loci* .................. | §1 –§13  |
| II. | *Angles in Relation to a Circle* . | §14–§23  |
| III.| *Measuring Segments and Areas* ... | §24–§35  |

Sections §1–§7 and §11–§13 in part I of *Geometrja* continue (and to some extent, review) the presentation of very elementary topics begun in a previous geometry course. These include

- lines and circles and their intersections, with an explicit axiom about them;
- perpendicular bisectors;
- parallel lines.

Three sections—§2, §4, and §7—are devoted to construction problems involving these notions. Logical principles are repeatedly discussed in geometrical context. Conditions for existence and uniqueness of solutions of construction problems are a recurring theme.

Section §3 contains another example, rather novel, of this incorporation of logic. Considering a line $g$ and a circle $C$ with given radius and center, it presents in great detail equivalent systems of exhaustive and mutually exclusive conditions for relating the number of intersections of $C$ with $g$ to the relationship of the radius and the distance from $g$ to the center. In *Geometrja*, applications of this technique are called *closed*

---

[100] The title of §3 of *Geometrja* does use the word *plane*, but the theory in the book would be valid even if it stipulated that there should be just one plane.

## 9.9 Tarski's Coauthored Textbook *Geometrja*

*systems of theorems*, and occur several times. Tarski also featured the technique in a general context under the name *Hauber's law* in his well-known logic textbook, written at the same time.[101]

The three remaining sections of part I of *Geometrja* and all of part II can be regarded as applications of the preceding elementary topics in specific, somewhat more involved, situations. Sections §8–§10 are devoted to angle bisectors and the circumcircle and incircle of a triangle; §14–§19, to arcs and chords of a circle. They present the standard theorems on their subjects. Sections §20 and §21 consist of related construction problems. Sections §22 and $23, on quadrilaterals inscribed in and circumscribed about a circle, and on regular polygons, provide distinctive climaxes for this part of *Geometrja*.

Throughout the first two parts of *Geometrja*, arithmetic is done with segments, not with their numerical lengths.[102] This approach is evidently part of the prerequisite material; the authors do not really emphasize it. Sections §1 and §2 are ambiguous in this regard: distances could be segments or numbers. But the first sentence of §3 (section 13.3) makes it clear:

> We know that the distance from a point $M$ to a line $PQ$ ... is the segment $MN$ constructed from point $M$ perpendicular to $PQ$.

On the other hand, the first four sections in part III are about segments *and* numbers. The first, §24, is a review of arithmetic with rational numbers and its relation to the manipulation of commensurable segments.

Section §25 shows that the diagonal of a square is incommensurable with its edge, so that if lengths are to be measured by numbers, irrational numbers must be used. Its method is remarkable: from the assumption that the ratio $r$ of diagonal to edge should be rational, the authors derive the equation $r^2 = 2$ without the usual resort to the Pythagorean theorem. Then they follow the traditional route to contradiction: no rational fraction that represents $r$ could be irreducible. This section and the next two describe in some detail, but informally, the relationships between irrational numbers and approximating rationals, and culminate with the following criterion for equality:

> If for each value of $m$ two irrational numbers $\gamma$ and $\delta$ have the same approximations
> $$\frac{l}{m} \quad \text{and} \quad \frac{l+1}{m}$$
> with precision $1/m$, then the numbers are equal to each other: $\gamma = \delta$.[103]

*Geometrja* makes no attempt to *define* irrational numbers, but regards them as given and introduces student readers to methods for handling them.

---

[101] Tarski [1936] 1995, §50. For Hauber's law see Hoormann 1971 and Hauber 1829, 265, §287.

[102] With segment arithmetic, an author regards certain segments as sums of two others, and develops the required properties of this addition without referring to numerical lengths of segments. An alternative approach assumes that the required properties of numbers have already been established, and postulates existence and properties of a correspondence between segments and their numerical lengths.

[103] Chwiałkowski, Schayer, and Tarski [1935] 1946, §27, property 2, translated in section 13.8 of the present book.

*Zygmunt Chwiałkowski in 1918*

← *Chwiałkowski*

*Teachers at the Jan Zamoyski Gimnazjum, around 1930*

> **Zygmunt Chwiałkowski**, the senior author of the 1935 secondary-school text *Geometrja* coauthored with Wacław Schayer and Alfred Tarski, was born in 1884. Chwiałkowski earned a doctorate, and in 1914 published a monograph in Russian on functional equations. In 1918, when the photograph on the facing page was taken, he was teaching mathematics at the Zofia Sierpiński girls' gimnazjum in Warsaw. Later, he transferred to the Stanisław Staszic boys' gimnazjum, which was directed by the mathematician Jan Zydler, author of a popular secondary-school geometry text. From 1923 to 1934, Chwiałkowski taught at the Jan Zamoyski boys' gimnazjum on Smolna Street. He was well liked there, and was described as merry, witty, and a passionate hunter. He was depicted affectionately by an anonymous student: the rightmost figure in the caricature reproduced on the facing page. Chwiałkowski was active in the mathematics section of the organization of specialized teachers' circles in Warsaw. By 1925, he had begun publishing mathematics textbooks for both elementary and secondary schools. Simultaneously with *Geometrja*, and with the same publisher, he coauthored with Schayer a textbook on elementary arithmetic and geometry. They produced at least twenty such books during 1935–1939. In fall 1949, Chwiałkowski served briefly on the committee that established Poland's secondary-school Mathematics Olympiad.*
>
> ---
>
> *Zagórowski 1924–1926, volume 1, 123, and volume 2, 249. Chodorowski et al. 2005, 31, 44. *Przegląd pedagogiczny* 1929, 1931. Dobrzyńska and Olszewska 1998. Schinzel 2000, 157.

The traditional treatment of similarity occupies §28–§32 of *Geometrja*. The first of these sections, entitled *A Theorem of Thales*, is not about the familiar Thales characterization of angles inscribed in semicircles, but about the proposition that parallel projection of one line $g$ onto another preserves ratios of lengths of segments of $g$. For commensurable segments, this follows from earlier material about parallels. For incommensurable segments, equality of the ratio of two segments and that of their projections is derived through use of the equality criterion just mentioned. This Thales theorem then leads in §29 to the familiar AA, SAS, and SSS similarity theorems for triangles. These sections and the next are replete with example applications and exercises. Section §32 presents a theorem that is required for the area theory that will conclude *Geometrja*:

> If through corresponding vertices in two similar convex polygons we pass all the diagonals, then they will divide the polygons into the same number of triangles, respectively similar and identically arranged.[104]

The concluding three sections of *Geometrja* are devoted to the area of polygons. Section §33 merely introduces the notion of sum of polygonal regions with disjoint interiors, and defines polygonal regions to be equivalent if they are sums of equal numbers of subregions, congruent in pairs. Having established some segment as having unit length, the next section informally assumes four area axioms:

---

[104] Chwiałkowski, Schayer, and Tarski [1935] 1946, §32.

- To each polygonal region corresponds a number, its area.
- The area of a square with unit edge is 1.
- Congruent polygonal regions have the same area.
- The area of a sum of polygonal regions with disjoint interiors is the sum of their areas.

On this basis, the area of an $a \times 1$ rectangle with rational $a$ is computed directly. For irrational $a$, it then follows from the equality criterion for irrational numbers mentioned earlier. The area of an $a \times b$ rectangle with rational $b$ is then computed directly, and for irrational $b$, it follows again from that criterion. The formulas for areas of other special polygons are then derived as usual. Finally, §35 shows that if $r$ is the ratio of corresponding edges of two similar triangles, then the ratio of their areas is $r^2$. The analogous result for similar polygons in general follows from the §32 result mentioned earlier.

*Wacław Schayer
in 1924*

**Wacław Maria Schayer** was born in Warsaw, then a part of the Russian Empire, in 1905. His father, a lawyer, died two years later. Later, his mother married a mining engineer. In 1918 the family moved to Sosnowiec, an industrial city in southern Poland. During the 1920 Polish–Soviet War, at age fifteen, Wacław served for three months in the Polish army. After his stepfather died in 1922, Wacław completed his schooling in Warsaw. He entered the university in 1924 to study mathematics and physical sciences.

Simultaneously with university studies, Schayer obtained employment as an elementary-school teacher in the Marymont district, joined the teachers' union, and became active in socialist politics. In 1927 he married Genowefa Zając. Their two daughters, Alicja and Krystyna, were born in 1931 and 1933. During 1930/1931 and 1931/1932, Schayer attended Alfred Tarski's courses *Topics in Methodology* and *Topics in Elementary Geometry*. In 1932 Schayer became director of the Bolesław Limanowski elementary school in the Żoliborz district, sponsored by the Workers' Society of Friends of Children (Robotnicze Towarzystwo Przyjaciół Dzieci); he also taught in the affiliated gimnazjum.

In 1935 Schayer coauthored the secondary-school textbook *Geometrja* with Zygmunt Chwiałkowski and Tarski. Excerpts are translated in chapter 13 of the present book. During 1935–1939 Schayer coauthored an extensive series of elementary-school texts with Chwiałkowski, and started another series with Antoni M. Rusiecki. Schayer continued his involvement with socialist and teachers' organizations, serving during 1938–1939 as vice-president of the "New Paths" Society for Democratic Education, founded by Czesław Wycech.

In September 1939 Schayer moved to Lwów, just days before the Soviet invasion of that region. He worked there briefly as a teacher, then as editor and translator. In 1941, after the Germans seized Soviet-occupied Poland, Schayer returned to Warsaw. By that time, the Department of Education and Culture of the Polish underground government had taken form under Wycech's leadership. Under the pseudonym Wiktor Jerzmanowski (the maiden name of his paternal grandmother), Schayer became its general secretary, and skillfully coordinated the work of many smaller organizations throughout occupied Poland. He continued teaching at the secretly functioning Limanowski School and helped organize the Polish People's Party (Polskie Stronnictwo Ludowe). Schayer's daughter Alicja, a thirteen-year-old messenger, was killed in the 1944 Warsaw Uprising. Immediately after the war, Wycech became minister of education in the provisional government, and Schayer served during 1945–1947 as vice-minister for schools. Schayer was awarded the Commander's Cross of the Order of Polish Rebirth. His son Andrzej Jerzmanowski was born in 1946.

During the postwar era, Schayer maintained a high level of political activity and editorship. He served as member of Parliament for the United People's Party in 1952–1956, during 1954–1957 as vice-minister of agriculture, and in 1958 as vice-minister for education. He died suddenly at age 54 in 1959. Since the war, he had resumed his collaboration with Rusiecki to publish school mathematics texts; their series continued long after Schayer's death.*

---

*For further information about Schayer and the underground government, consult Schayer 1924, Turkowski 1994, Karski 1944, and Redzik 2004. Turkowski reported that Schayer earned a master's degree in 1931, but Schayer's university records show that he was still attending classes in Spring 1932. Some of this information is from Jerzmanowski 2013. Andrzej Jerzmanowski is now professor of biochemistry and molecular biology at the University of Warsaw, and a member of the Polish Academy of Sciences.

Two parts of the plane-geometry curriculum very closely related to the content of *Geometrja* were left for inclusion in later geometry courses:

- the Pythagorean theorem;
- precise description of the use of real numbers, particularly to define the circumference $\pi$ of a unit circle and other irrational numbers needed for geometry.

The Pythagorean theorem is easily obtainable from the area theory at the end of *Geometrja*, and would have been an attractive culmination, but the authors opted neither to use nor to mention it! It appeared early in the official curriculum for the following year IV.[105] Tarski himself had sketched a compatible approach to the real number system in the article Tarski [1932] 2014e, translated in chapter 11 of the present book. That paper related to the curriculum for year V.[106]

Even so truncated, *Geometrja* is an interesting and impressive book. Already noted is the abundance of exercises in its earlier parts. That continues throughout: there are fifty-five in the sections §33–§35 on area. Second, its selection of 130 figures (for a book of 108 pages!) is impressively rich and effective. Third, its simplicity of language is startling, especially when contrasted with that of Enriques and Amaldi [1903] 1916. In spite of that simplicity, *Geometrja* presents details, avoids false statements, gives convincing arguments where they are feasible, and states assumptions clearly where they are not.

These qualities of a geometry textbook make a difference to some student readers. For example, the senior editor of the present book first studied geometry from a popular American textbook, more comprehensive and polished than *Geometrja*, but written at the same time for somewhat similar students. That exciting experience opened up the world of mathematics for him. But it left some bad tastes. For example, that book related the area of a region to counting squares on an underlying grid—a good idea—but did not bother to summarize the properties of this process on which it might base the area reckoning; nor did it discuss refining the grid. In its plan for a proof of the theorem of Thales mentioned earlier, the American textbook may even have misled students by suggesting that *any* two segments were commensurable! In fact, it never mentioned irrational numbers at all, leaving that present editor to hear about those only by rumor in a calculus class three years later.[107] Perhaps by chance, he took the resulting discomfort as a challenge to find further compensating triumphs in mathematics, rather than as a disincentive for continued exploration! The authors of *Geometrja* did not leave that to chance: they took painstaking care to be correct and clear.

---

[105] Poland 1934, 160–164.

[106] The curriculum for year V was to be phased into the new liceum schools in 1937. It was based on the penultimate year of the pre-reform curriculum: Poland 1931, 66–67.

[107] Stone and Mallory 1937, 343, 284.

## 9.9 Tarski's Coauthored Textbook *Geometrja*

No detailed study has been made of the genesis of the curriculum standard Poland 1934 and the extent to which it was enforced in textbooks like *Geometrja*. No record survives of the contributions to the content and style of that book by its individual coauthors. Tarski's influence seems most noticeable in three aspects. First is the frequent use and discussion of logical principles in geometrical context. Sometimes that is explicit, as in the discussion of closed systems of theorems in §3. Sometimes it is informal, as in the words *let us suppose* that introduce area axioms in §34. These features reflect Tarski's concurrent authorship of his well-known [1935] 1946 logic textbook. Second is the unusually deep and precise attention to the properties of the real number system in §25–§27, couched in simple language appropriate for its informal introduction to serve the need to measure all segments. Several years earlier, Tarski had sketched a compatible approach adequate for dealing with the circumference of a circle at a somewhat more advanced level. Finally, in *Geometrja* there is a flavor of persistent attention to detail, precision, and completeness that reflects the same qualities required for solution of the exercises for talented students that Tarski had proposed earlier in the journals *Parametr* and *Młody matematyk*.[108] Tarski did not leave the development of these talents, and understanding mathematics, to chance.

For translation in chapter 13 of the present book, the editors selected ten sections of *Geometrja* that are representative of its content and support these assessments:

§1 –§4 introduce the flavor of the text;
§25–§27 present its treatment of incommensurable segments and irrational numbers;
§33–§35 develop the theory of polygonal areas.

The material gathered in the present book increases the accessibility of Tarski's early work and explains some of its relationships to the intellectual, political, and social milieu of Poland between the world wars. The editors hope that it will spur broader investigation into the relationship of mathematics and its cultural setting during that era. That would be particularly welcome for the connection between mathematical research and mathematics education, as displayed by Tarski's research on geometry and his practice both in teaching secondary-school students and training secondary-school teachers. This hope is a major reason for including in a single volume of selected translations works of such contrasting mathematical sophistication as Tarski's research on geometric decomposition in Part Two and the secondary-school geometry lessons in chapter 13.

---

[108] See Tarski [1932] 2014e, translated in section 3.3 of the present book, and the exercises translated and discussed in 3.4.

**Later editions of Geometrja.** A second edition of the 1935 text *Geometrja* by Zygmunt Chwiałkowski, Wacław Schayer, and Alfred Tarski was published in 1944 in Jerusalem by the Ministerstwo Wyznań Religijnych i Oświaty Publicznej, Sekcji Wydawniczej Armii Polskiej na Wschodzie (Ministry of Religious Denominations and Public Education, Publishing Section for the Polish Army in the East). This branch of the Polish government-in-exile in Great Britain was set up to provide services for the army recruited during 1941–1942 by General Władysław Anders from Polish prisoners in the Soviet Union, and for the much larger group of civilians accompanying that army.

The Soviets invaded eastern Poland two weeks after the German invasion from the west on 1 September 1939. They deported about 1.5 million Poles from the conquered territory to prisons and labor camps deep in Siberia. They especially targeted middle-class Poles likely to oppose Soviet rule. During the following two years, about half died. When the Germans invaded the Soviet Union in June 1941, the Polish government-in-exile persuaded the British authorities to coerce the Soviets into releasing their remaining Polish prisoners to form an army against the Germans. Polish agents recruited about forty thousand men. The Soviets allowed them and about seventy-five thousand civilians to leave, but provided virtually no means for them to do so. After harrowing journeys through Central Asia to Iran, the Poles made their way to camps in Palestine and other Middle-Eastern lands. The Soviets discontinued even this cooperation with the Poles in April 1943. The Polish Army units formed by then played significant roles in the defeat of the Germans, which was concluded in May 1945. During this period a complex network of elementary, secondary, and trade schools was established for the refugee children at many sites worldwide. According to Polish-American sociologist Tadeusz Piotrowski (2004, 98), "The greatest impediment to education was the lack of textbooks. The problem was somewhat mitigated by having them printed in Palestine and later, in Iran."

After the surrender, about six million Poles were stranded outside the new frontiers of Poland—nearly 1.2 million in western Germany, most of whom had been enslaved there during the war. Most could not travel, could not return to areas incorporated into the Soviet Union, or would not return as Soviet rule descended over the rest of their country. To provide for displaced persons, more than seven hundred camps were established worldwide. Many of these refugees were able to repatriate or emigrate to another country within two years, but some remained in the camps for as long as nine years. Polish society was reconstituted in them, and by fall 1945, about thirty-five thousand children were attending Polish middle and secondary schools in the camps.*

The second edition of *Geometrja* was reprinted in 1946 in Hanover by Polski Związek Wychodźctwa Przymusowego (Polish Association of Forced Emigration) to serve those students.

---

*For further information about this history, consult N. Davies 1982, volume 2, 271–272 and chapter 20; Snyder 2010, chapter 4; T. Piotrowski 2004; and Jaroszyńska-Kirchmann 2002, and 2004, chapters 1 and 2.

# 10
# Congress of Mathematicians of Slavic Countries (1929)

This chapter contains an English translation of Alfred Tarski's [1929] 2014a article *Zjazd matematyków*, which appeared in the journal *Ogniwo*. The journal identified the author only as *A.T.* It is a brief account of the First Congress of Mathematicians of Slavic Countries, held in Warsaw and Poznań, 23–27 September 1929. The journal's full title means *The Link: Organ of the Trade Union of Polish Secondary-School Teachers and Newsletter of Its Central Administration*.[1] For further information about the Congress, the journal, and its milieu, consult section 9.5.

The translation is meant to be as faithful as possible to the original. Its only intentional modernization is punctuation. Personal names have been adjusted to conform with conventions of the present book. All [square] brackets in the translation enclose editorial comments inserted for clarification, sometimes as footnotes.

---

[1] The complete proceedings of the Congress are included in Leja 1930. The full title of the journal is *Ogniwo: organ Związku Zawodowego Nauczycielstwa Polskich Szkół Średnich i biuletyn zarządu głównego Z. Z. N. P. S. Śr.* The title of the article in the journal's table of contents was *Zjazd Matematyków Krajów Słowiańskich*; the added phrase means *of Slavic Countries*. Tarski claimed authorship in his 1965a bibliography.

A. T.

# AN ASSEMBLY OF MATHEMATICIANS

During the days 23–27 of the previous month [September 1929], the first Congress of Mathematicians of Slavic Countries took place in Warsaw. In the welcoming speech, delivered at the opening session of the Congress in the presence of the Minister of Religious Denominations and Public Education[2] and diplomatic representatives from Bulgaria, Czechoslovakia, and Yugoslavia, the chairman of the executive committee, Prof. Wacław Sierpiński, emphasized among other things that the Congress did not in any way have a political basis. Its goal was to establish closer intellectual relationships among researchers who work in the same branch of knowledge, who live in nearby territories and for whom collaboration is made somewhat easier by the kinship of their native languages. As a confirmation of his words Prof. Sierpiński pointed to the presence of several mathematicians who came from non-Slavic countries with the goal of participating in the work of the Congress.

Nearly one hundred participants took part in the proceedings of the Congress. Naturally, Polish researchers predominated—from Warsaw as well as from other university circles: from Lwów, Cracow, and Vilnius, and even from abroad, like Prof. Leon Lichtenstein from Leipzig. In addition, a number of mathematicians came from Czechoslovakia, Bulgaria, and Yugoslavia, among them the eminent researcher Prof. Kyrille Popoff from Sofia. Among the guests from non-Slavic countries should be named Prof. W. H. Young from London, President of the International Mathematical Union; Prof. Abraham A. Fraenkel from Kiel; Prof. Karl Menger from Vienna; Prof. Petre Sergescu from Cluj in Romania; and even two Japanese mathematicians, Prof. Akitsugu Kawaguchi and Prof. Kinjiro Kunugui. Several eminent Western European mathematicians, among them Prof. Jacques Hadamard from Paris and Prof. Leonida Tonelli from Bologna, sent their communiqués to be read at the Congress sessions, with the goal of making evident their affinity[3] with the attendees. Striking, however, was the complete lack of mathematicians from Soviet Russia, with whom Polish researchers remain in constant research contact, and who participated in numbers in the first Polish Mathematical Congress in Lwów in 1927 as well as in last year's International Congress in Bologna, but whom the Soviet government expressly prohibited from taking part in the present Congress.

---

[2] [Tarski wrote *p. Ministra W. R. i O. P.*, which stands for *pan Minister Wyznań Religijnych i Oświecienia Publicznego*. The Minister was Dr. Sławomir Czerwiński.]

[3] [Tarski's word was *łączność*.]

The proceedings of the Congress took place in general sessions and in five sections:
  I. Foundations of mathematics, history, didactics of mathematics
  II. Arithmetic, algebra, analysis
  III. Set theory, topology, and their applications
  IV. Geometry
  V. Mechanics and applied mathematics.

Altogether, about eighty talks and communiques were delivered at the Congress. The proceedings were carried out in Slavic as well as international languages, especially in German and French. During the Congress there took place several official receptions—for example, at the Presidium of the Council of Ministers—and social gatherings. During the evening of 26 September a significant number of participants went on a special train to Poznań, where the following day the Congress came to a close.[4]

Congress participants from abroad did not spare the organizers words of acknowledgment for efficient organization as well as the high standard of the proceedings. The Congress was one more proof of the outstanding position of Polish mathematics in the modern scientific world.

In general, the community of teachers played a very small role in the Congress. Among the presented talks only one, by Prof. Antoni Łomnicki from Lwów, carried a strict didactic quality. In general, the experience of the most recent mathematical assemblies reveals that including in the program of one and the same assembly issues of the exact sciences and of didactics puts [didactics] at a disadvantage. Incidentally, that can be foreseen a priori: for the most part, organizers of assemblies happen to be researchers—professors in institutions of higher learning, little interested in the affairs of the educational system of the primary and secondary schools. And in addition, even those mathematicians who have understanding of and interest in these affairs are inclined to take them up less at any time in the duration of the assembly, wishing to take advantage of a rare occasion to establish and deepen contacts with out-of-town professional colleagues. On the other hand, every teacher of mathematics undoubtedly feels a need to discuss in a broad group many of the problems from the realm of elementary mathematics and mathematical didactics. Specific conditions of our educational system, related to the frequent changes in the curriculum and the difficulties of realizing its individual sections, strengthen and intensify this need. For all these reasons, it would be greatly desirable to organize soon a special assembly, devoted exclusively to the problems of mathematics in the framework of the educational system, primary and secondary, in which theoreticians [and] researchers interested in these problems as well as practitioners—teachers active in the primary and secondary schools—could participate significantly.

---

[4] [In the original, the date of the trip to Poznań was incorrectly printed as 2 September.]

# 11
# Circumference of a Circle (1932)

This chapter contains an English translation of Alfred Tarski's [1932] 2014e article *Teorja długości okręgu w szkole średniej*, which appeared in the October–December 1932 issue of the journal *Parametr*.[1] The journal's target audience consisted of gimnazjum teachers and their most serious students. For further information about that publication, consult section 9.7 of the present book.

The translation is meant to be as faithful as possible to the original. Its only intentional modernizations are punctuation and some changes in symbols where Tarski's conflict with today's mathematical practice in English. In this paper, Tarski used the same symbol, for example $u_n$, for a sequence $u_1, u_2, u_3, \ldots$ of terms and for a typical term of that sequence; and the same symbol, for example $u(x)$, for a function $u$ and a typical value of that function. He also omitted notation such as $n \to \infty$ that conventionally appears under the limit symbol lim. The translation maintains his practices. Bibliographic references and personal names have been adjusted to conform with conventions of the present book. Some uses of alternative type styles for emphasis, enunciations, and personal names have been modified. The present editors used white space to enhance visual organization of the paper. All [square] brackets in the translation enclose editorial comments inserted for clarification, often as footnotes.

---

[1] This paper was reprinted in Tarski's 1986a *Collected Papers*; several typographical errors were introduced there. The Tarski 1965a and Givant 1986 bibliographies listed the paper's title as *The Theory of the Measure of the Circumference of a Circle for High School Teaching*.

DR. ALFRED TARSKI (Warsaw)

# The Theory of the Circumference of a Circle in the Secondary School

According to the curriculum requirements for secondary school,[2] calculating the circumference of a circle, the area of a disk, and the surface area and volume of basic solids of revolution should be covered in close connection with the theory of limits, as examples of applications of that theory. This recommendation is usually interpreted in such a way that the circumference of a circle and the area of a disk, and so on, are defined as limits of certain sequences; and from these definitions are derived the formulas used in the calculation of the quantities under consideration. This way of proceeding I will call the *method of limits*.

It is possible, however, to choose as a starting point such definitions that do not contain the concept of a limit at all, but on the other hand imply a close relationship between the quantities mentioned and the notion of continuity, characterizing them as certain cuts.[3] Nevertheless, in this case, in deriving from the given definitions the formulas for calculation, it is convenient to use facts from the theory of limits. This method, which I will call the *method of cuts* for short—although I will not use the term *cut* explicitly[4]—seems more suitable in many respects.

The arrangement of material in the required curriculum allows facts from trigonometry to be used while working out problems that interest us—to a wider or narrower extent depending on the type of school. I have a suspicion that in this way one can present a clearer and more easily accessible lecture.

In the present article I shall focus attention mainly on the theory of the circumference of a circle. Making use of trigonometry, I shall outline a lecture about this theory following [each of] the two aforementioned methods; and then I shall compare both methods from the point of view of their didactic value.

---

[2] [The reprint of this paper in Tarski 1986a referred here to the curriculum of the *seventh class* of secondary school. Before the 1932 reform, that was the penultimate year of Polish secondary schools, when students were about seventeen years old. For more information about the Polish school system, see section 9.1.]

[3] [Tarski used *cut* in the sense of *Dedekind cut*. His word was *przekrój*.]

[4] [Tarski used here the Latin adverb *explicite*.]

## § 1. The Method of Limits

I assume that the class knows about the theory of limits to the extent provided in the curriculum. The student has thus already grasped the notions of *sequences*, *increasing* and *decreasing* sequences, sequences *bounded above* and *below*, and finally, *convergent* sequences and *limits*. I also assume a familiarity with the following theorems:

1.  *Every convergent sequence has exactly one limit.*
2.  *Every increasing sequence bounded from above is convergent, and its limit is greater than all its terms; every decreasing sequence bounded from below is convergent, and its limit is less than all its terms.*
3.  *If all terms of a convergent sequence $u_n$ are $\leq$ (respectively, $\geq$) a given number $c$, then $\lim u_n \leq c$ (respectively, $\lim u_n \geq c$); if all terms of a sequence $u_n$ are $= c$, then $\lim u_n = c$.*
4.  *If $c$ is any number and $v_n$ is a convergent sequence, then $\lim(c \pm v_n) = c \pm \lim v_n$ and $\lim(c \cdot v_n) = c \cdot \lim v_n$.*
5.  *If $u_n$ and $v_n$ are convergent sequences, then $\lim(u_n \pm v_n) = \lim u_n \pm \lim v_n$ and $\lim(u_n \cdot v_n) = \lim u_n \cdot \lim v_n$.*

Later in the lecture it will be convenient to use the notion of *mutually convergent* sequences:

6.  *We call sequences $u_n$ and $v_n$ mutually convergent if*
    (1) *$u_n$ is an increasing sequence and $v_n$, decreasing;*
    (2) *the inequality $u_n < v_n$ always holds; and finally,*
    (3) *$\lim(v_n - u_n) = 0$.*[5]

From this definition and from the theorems given above, the following conclusions, among others, can be derived:

7.  *If the sequences $u_n$ and $v_n$ are mutually convergent, then the sequences are convergent [and] have a common limit, and this limit is the only number greater than all terms of the first sequence but smaller than all terms of the second.*
8.  *If $c > 0$ and sequences $u_n$ and $v_n$ are mutually convergent, then the sequences $c \cdot u_n$ and $c \cdot v_n$ are mutually convergent.*

Last, the following lemma will be necessary for us:

---

[5] Instead of *mutually convergent* sequences, one usually speaks of *convergent* sequences. However, because of the double meaning of the word *convergent*, such terminology can be a source of misunderstanding. This is also why in the present article I use the term *convergent* only to describe individual sequences having a limit, and not as an expression of dependency between two sequences. [Tarski's terms for *convergent* and *mutually convergent* were *zbieżny* and *współzbieżny*.]

9. *If $k < l$ and $u_n$ is an increasing (respectively, decreasing) sequence, then the arithmetic mean of the first $k$ terms of the sequence is less than (respectively, greater than) the arithmetic mean of the first $l$ terms.*

*Proof.* We have

$$l \cdot (u_1 + u_2 + \cdots + u_k) =$$
$$k \cdot (u_1 + u_2 + \cdots + u_k) + (l-k) \cdot (u_1 + u_2 + \cdots + u_k), \quad (1)$$
$$k \cdot (u_1 + u_2 + \cdots + u_k + u_{k+1} + \cdots + u_l) =$$
$$k \cdot (u_1 + u_2 + \cdots + u_k) + k \cdot (u_{k+1} + \cdots + u_l).$$

The product $(l-k) \cdot (u_1 + u_2 + \cdots + u_k)$ can be expressed as a sum of $k \cdot (l-k)$ parts, each of which is one of the terms $u_1, u_2, \ldots, u_k$:

$$(l-k) \cdot (u_1 + u_2 + \cdots + u_k) =$$
$$(u_1 + u_2 + \cdots + u_k) + (u_1 + u_2 + \cdots + u_k) + \cdots \qquad (l-k \text{ times})$$
$$+ (u_1 + u_2 + \cdots + u_k).$$

Similarly, the product $k \cdot (u_{k+1} + \cdots + u_l)$ can be expressed as a sum of equally many parts, each being one of the terms $u_{k+1}, \ldots, u_l$:

$$k \cdot (u_{k+1} + \cdots + u_l) =$$
$$(u_{k+1} + \cdots + u_l) + (u_{k+1} + \cdots + u_l) + \cdots \qquad (k \text{ times})$$
$$+ (u_{k+1} + \cdots + u_l).$$

Since the sequence $u_n$ is increasing, every part of the first sum is less than every part of the second sum; and since the number of terms in both sums is the same,

$$(l-k) \cdot (u_1 + u_2 + \cdots + u_k) < k \cdot (u_{k+1} + \cdots + u_l). \quad (2)$$

Adding to both sides of inequality (2) the number $k \cdot (u_1 + u_2 + \cdots + u_k)$, we obtain

$$k \cdot (u_1 + u_2 + \cdots + u_k) + (l-k) \cdot (u_1 + u_2 + \cdots + u_k) <$$
$$k \cdot (u_1 + u_2 + \cdots + u_k) + k \cdot (u_{k+1} + \cdots + u_l);$$

and hence, because of (1),

$$l \cdot (u_1 + u_2 + \cdots + u_k) < k \cdot (u_1 + u_2 + \cdots + u_l). \quad (3)$$

Finally, dividing both sides of inequality (3) by $k \cdot l$ then simplifying, we obtain the inequality

$$\frac{u_1 + u_2 + \cdots + u_k}{k} < \frac{u_1 + u_2 + \cdots + u_l}{l}, \qquad \text{Q.E.D.}$$

Moving on to trigonometry, I assume that the class already knows the theory of angle measurement (in degrees, but not in radians), the definitions and variation of the main trigonometric functions $\sin \alpha$, $\cos \alpha$, and $\tan \alpha$ —at least to the extent of the first quadrant—and the fundamental relationships between these functions. Moreover,

## 11.1 §1 Method of Limits

familiarity with the formulas for the functions of sums and differences of angles is desired, as well as with the transformations based on those formulas.

Based on the definitions of limit and of the function $\cos \alpha$, we show without any difficulty that

10.  $\lim \cos \alpha/n = 1$.

Further, we establish the following lemma, which incidentally merits attention by itself:

11.  *If* $0° < \alpha < \beta < 90°$, *then* $\dfrac{\sin \beta}{\sin \alpha} < \dfrac{\beta}{\alpha} < \dfrac{\tan \beta}{\tan \alpha}$.

It could be expressed in this way: *in the first quadrant, the sine increases more slowly than the angle, and the tangent, faster.*

*Proof.* We shall prove only a particular case of the theorem, which, incidentally, completely suffices for our purposes: specifically, [the case] in which angles $\alpha$ and $\beta$ are commensurable, so that there should exist natural numbers $k$ and $l$ and an angle $\gamma$ such that

$$k < l, \tag{1}$$
$$\alpha = k \cdot \gamma, \quad \text{and} \quad \beta = l \cdot \gamma. \tag{2}$$

For $n = 1, 2, \ldots, l$ we set

$$u_n = \sin n \cdot \gamma - \sin(n-1) \cdot \gamma. \tag{3}$$

Thus, in particular, $u_1 = \sin \gamma$, $u_2 = \sin 2\gamma - \sin \gamma$, and so on. We show that

*The sequence $u_n$ is decreasing.* (4)

In fact, in accordance with (3), when $n < l$ we have

$$u_n = \sin n \cdot \gamma - \sin(n-1) \cdot \gamma = 2 \sin \tfrac{1}{2}\gamma \cdot \cos(n - \tfrac{1}{2}) \cdot \gamma,$$
$$u_{n+1} = \sin(n+1) \cdot \gamma - \sin n \cdot \gamma = 2 \sin \tfrac{1}{2}\gamma \cdot \cos(n + \tfrac{1}{2}) \cdot \gamma.$$

Since the cosine decreases in the first quadrant,

$$\cos(n - \tfrac{1}{2}) \cdot \gamma > \cos(n + \tfrac{1}{2}) \cdot \gamma,$$

which yields

$$2 \sin \tfrac{1}{2}\gamma \cdot \cos(n - \tfrac{1}{2}) \cdot \gamma > 2 \sin \tfrac{1}{2}\gamma \cdot \cos(n + \tfrac{1}{2}) \cdot \gamma;$$

that is, $u_n > u_{n+1}$.

Should the class not yet know the trigonometric transformations used here, we can establish the necessary inequalities by following a path of straightforward geometric reasoning from the definition of sine.[6]

According to lemma 9, from (1) and (4) it follows that

$$\frac{u_1 + u_2 + \cdots + u_k}{k} > \frac{u_1 + u_2 + \cdots + u_l}{l}. \tag{5}$$

On the other hand, from (3) we conclude with ease that

$$u_1 + u_2 + \cdots + u_k = \sin\gamma + (\sin 2\gamma - \sin\gamma) + \cdots$$
$$(\sin k\cdot\gamma - \sin(k-1)\cdot\gamma) = \sin k\cdot\gamma$$

and similarly, $u_1 + u_2 + \cdots + u_l = \sin l\cdot\gamma$. Therefore, in view of (5),

$$\frac{\sin k\cdot\gamma}{k} > \frac{\sin l\cdot\gamma}{l}$$

and hence

$$\frac{\sin l\cdot\gamma}{\sin k\cdot\gamma} < \frac{l}{k}. \tag{6}$$

Statements (2) and (6) immediately yield

$$\frac{\sin\beta}{\sin\alpha} < \frac{\beta}{\alpha}.$$

We similarly justify the second required inequality,

$$\frac{\beta}{\alpha} < \frac{l}{k}.$$

Specifically, for $n = 1, 2, \ldots, l$ we set $v_n = \tan n\cdot\gamma - \tan(n-1)\cdot\gamma$, and we show that this sequence is increasing—either with the help of simple trigonometric transformations, or following a path of straightforward geometric reasoning,[7] then argue further as before. Thus, finally,

$$\frac{\sin\beta}{\sin\alpha} < \frac{\beta}{\alpha} < \frac{\tan\beta}{\tan\alpha}, \qquad \text{Q.E.D.}$$

With the help of the lemma above we establish theorem

---

[6] Such reasoning, along with a suitable drawing, is contained implicitly in the article Mihułowicz 1930, 220–221.

[7] Compare Mihułowicz 1930, 221–222.

12. *The sequences $u_n = n \cdot \sin 180°/n$ and $v_n = n \cdot \tan 180°/n$, for $n \geq 3$, are mutually convergent.*

*Proof.* Beginning with $n = 3$, angles
$$\alpha = \frac{180°}{n+1} \quad \text{and} \quad \beta = \frac{180°}{n}$$
are acute and commensurate; moreover,
$$\alpha < \beta \quad [\text{and}] \quad \frac{\beta}{\alpha} = \frac{n+1}{n}.$$

Therefore, according to lemma 11,
$$\frac{\sin \dfrac{180°}{n}}{\sin \dfrac{180°}{n+1}} < \frac{n+1}{n} < \frac{\tan \dfrac{180°}{n}}{\tan \dfrac{180°}{n+1}}.$$

Multiplying both sides of the first part of this statement by $n \cdot \sin \dfrac{180°}{n+1}$, we obtain
$$n \cdot \sin \frac{180°}{n} < (n+1) \cdot \sin \frac{180°}{n+1},$$
so that $u_n < u_{n+1}$. Similarly, the second part of the statement gives $v_n > v_{n+1}$. In this way,

$u_n$ is an increasing sequence, whereas $v_n$ [is] decreasing. (1)

Further, we have
$$u_n = n \cdot \sin 180°/n = n \cdot \tan 180°/n \cdot \cos 180°/n = v_n \cdot \cos 180°/n.$$
Since $\cos 180°/n < 1$, therefore, we always have
$$u_n < v_n. \tag{2}$$

According to theorem 2, the sequence $v_n$ must be convergent, since [it is] decreasing and bounded below (for example, by the number zero). It also follows from theorems 10 and 5 that the sequence $1 - \cos 180°/n$ is convergent; moreover,
$$\lim(1 - \cos 180°/n) = 1 - \lim \cos 180°/n = 1 - 1 = 0.$$

Thus, in accordance with theorem 4, the sequence $v_n \cdot (1 - \cos 180°/n)$ is also convergent. Moreover,
$$\lim(v_n \cdot (1 - \cos 180°/n)) = \lim v_n \cdot \lim(1 - \cos 180°/n) = \lim v_n \cdot 0 = 0.$$

On the other hand
$$v_n \cdot (1 - \cos 180°/n) = v_n - v_n \cdot \cos 180°/n =$$
$$v_n - n \cdot \tan 180°/n \cdot \cos 180°/n = v_n - n \cdot \sin 180°/n = v_n - u_n.$$

Thus, finally,
$$\lim(v_n - u_n) = 0. \tag{3}$$
In view of (1)–(3) and according to definition 6, sequences $u_n$ and $v_n$ are mutually convergent, Q. E. D.

It follows from theorems 7 and 12 that the sequences $u_n = n \cdot \sin 180°/n$ and $v_n = n \cdot \tan 180°/n$ have a common limit. To denote it we introduce a special symbol:

13. *The common limit of the mutually convergent sequences*
   $$u_n = n \cdot \sin 180°/n \quad and \quad v_n = n \cdot \tan 180°/n,$$
   *where* $n \geq 3$, *we call the number* $\pi$.

As a direct conclusion of theorem 8 and definition 13 we obtain

14. $n \cdot \sin 180°/n < \pi < n \cdot \tan 180°/n$ *for every natural number* $n \geq 3$.

This last statement gives a student the means to determine the number $\pi$ with fairly great precision with the aid of ordinary four-digit tables. (The question whether knowing approximate values for this number was necessary in the construction of the tables does not come into play here.)[8] For example, setting $n = 180$, we calculate
$$n \cdot \sin 180°/n > 180 \cdot 0.01745 = 3.141;$$
for $n = 90$ we obtain
$$n \cdot \tan 180°/n < 90 \cdot 0.03495 < 3.146.[9]$$
Thus,
$$3.141 < \pi < 3.146.$$

Already now, we can formulate a definition for the circumference of a circle and derive from it a formula to calculate this quantity.

15. *The circumference of a circle is the common limit of the sequence of perimeters of all regular polygons inscribed in the circle and the sequence of perimeters of all regular polygons circumscribed about the circle.*[10]

The circumference of a circle is often defined as the common limit of all sequences of perimeters of regular polygons inscribed in the circle or [of those] circumscribed about

---

[8] [Tarski's verb phrase was *nie wchodzi tu w grę*. That is, the question is not covered in this lecture.]

[9] [In the four-place table in Łomnicki 1930, 8, for example, the approximation for $\sin 1°$ is 0.0175, which indicates that $0.01745 \leq \sin 1° \leq 0.01755$. Were $\sin 1° = 0.01745$, this would have been rounded down to 0.0174 because 4 is even. Thus, one can conclude from the table that $0.01745 < \sin 1°$. For $\tan 2°$ the approximation is 0.0349, which indicates that $0.03485 \leq \tan 2° \leq 0.03495$. Were $\tan 2° = 0.03495$, this would have been rounded up to 0.0350 because 9 is odd. Thus, according to the table, $\tan 2° < 0.03495$. Analogous reasoning with $\tan 1°$ would not have yielded a strict inequality.]

[10] [The proof of theorem 16 shows that this common limit exists.]

the circle, [with the additional provisos that] the first polygon in the sequence have an arbitrary number of sides, and each subsequent one have twice as many sides as the previous one. This definition offends with its unnaturalness, and moreover does not display any practical or theoretical advantages in comparison with definition 15.[11]

The fundamental theorem in the theory of the circumference of the circle reads,

16. *Every circle has a precisely defined circumference $l$; if $r$ is the length of the radius of the circle, then $l$ is expressed by the formula $l = 2\pi \cdot r$.*

*Proof.* Let $u_n$ (respectively, $v_n$) be the perimeter of a regular $n$-gon[12] inscribed in the circle (respectively, circumscribed about the circle). With the help of a picture, it is easy to show that $u_n = 2r \cdot n \cdot \sin 180°/n$, while $v_n = 2r \cdot n \cdot \tan 180°/n$. Applying theorems 8 and 12, therefore, we conclude that sequences $u_n$ and $v_n$, [being] mutually convergent, have a common limit; moreover, in accordance with theorem 4 and definition 13,

$$\lim u_n = \lim v_n = 2r \cdot \lim(n \cdot \sin 180°/n) = 2r \cdot \lim(n \cdot \tan 180°/n) = 2\pi \cdot r.$$

According to definition 15, exactly that common limit is the circumference of the circle: $l = 2\pi \cdot r$, Q.E.D.

As a direct consequence of the above theorem we obtain

17. *The circumferences of circles are directly proportional to the lengths of their radii.*

The theory of the area of a circle does not show any essential differences. We define the area of a circle as the limit of the sequence of areas of all regular polygons inscribed in the circle or [of those] circumscribed about the circle. Denoting by the symbols $r$, $s$, $u_n$, and $v_n$ the length of the radius of the circle, the area of the circle, the area of the regular $n$-gon inscribed in the circle, and the area of the regular $n$-gon circumscribed about the circle, respectively, we demonstrate that the sequences

$$u_n = r^2 \cdot n \cdot \sin 180°/n \cdot \cos 180°/n \quad \text{and} \quad v_n = r^2 \cdot n \cdot \tan 180°/n$$

have a common limit: $\lim u_n = \lim v_n = \pi \cdot r^2$. Therefore, according to the definition, $s = \pi \cdot r^2$.

In a completely analogous manner we develop the theory of the surface area and volume for basic solids of revolution: the cylinder, cone, and sphere. Concerning the latter in particular, let us consider the solids that arise from rotating regular polygons with $2n$ sides, inscribed in a circle and circumscribed about the circle, about an axis passing through two opposite vertices of the polygon. We denote the length of the radius of the circle by $r$, the surface area of the solid resulting from rotating the inscribed (respectively, circumscribed) polygon by $u_n$ (respectively, $v_n$), and the volume of the

---

[11] Jerzy Mihułowicz (1930) has already mentioned this.
[12] [Tarski's term for closed polygon with $n$ vertices was *n-kąt*.]

solid by $u'_n$ (respectively, $v'_n$). Using the formulas for surface area and volume of the cylinder and cone, we show that

$$u_n = 4\pi \cdot r^2 \cdot \cos 90°/n, \qquad v_n = \frac{4\pi \cdot r^2}{\cos 90°/n},$$

$$u'_n = 4/3\,\pi \cdot r^3 \cdot \cos^2 90°/n, \qquad v'_n = \frac{4\pi \cdot r^3}{3\cos 90°/n}.$$

From this, by passing to the limit, we obtain the known formulas for surface area and volume of a sphere.

## § 2. The Method of Cuts

Facts from the theory of limits[13] are necessary here to the same extent as in the previous method; only lemma 9 drops out.

On the other hand, a reminder to the class about the *axiom of continuity* (due to Richard Dedekind) is imperative. This axiom played an essential role in the previous discussions, too: without its help we would not have been able to develop the theory of measuring segments and angles, to prove the existence of limits of mutually convergent sequences, nor even to inscribe in a circle regular polygons with an arbitrary number of sides. Now, however, we will apply it in a direct manner.

For a suitable formulation of the axiom, we shall first of all make the following two conventions:

18. *We say that a set $A$ of numbers precedes a set $B$ of numbers if every number in set $A$ is less than every number in set $B$.*

19. *We say that a number $c$ separates a set $A$ of numbers from a set $B$ of numbers if every number in set $A$ is $\leq c$ and every number in set $B$ is $\geq c$.*

The axiom takes the following form:

20. *If a set $A$ of numbers precedes a set $B$ of numbers, then there exists at least one number $c$ that separates the two sets.*

This statement is one of the axioms of algebra or arithmetic;[14] and similarly, we accept a completely analogous axiom in geometry, formulating it not for numbers, but for points and segments.

---

[13] [Tarski wrote *sequences*, not *limits*, here.]

[14] If the real numbers are defined as cuts in the set of rational numbers (and are not introduced axiomatically), then statement 20 forfeits its role as an axiom and becomes a theorem.

Reasoning based directly on the axiom of continuity (and on the notion of cut, [which is] implicitly inherent in it) is generally harder than the considerations in which results from the theory of limits are applied. To avoid these difficulties we establish the following theorem, which in some sense extends a bridge between both of the methods considered:

21. *If a set $A$ of numbers precedes a set $B$ of numbers, if moreover $u_n$ and $v_n$ are two mutually convergent sequences and all terms of the first sequence belong to set $A$ and [those of] the second to set $B$, then there exists exactly one number separating set $A$ from set $B$. This unique number, which in addition does not belong to either of the two sets, is the common limit of both sequences $u_n$ and $v_n$.*

*Proof.* According to axiom 20, there exists at least one number $c$ that separates set $A$ from set $B$. Considering definition 19, every such number must be greater than all terms of the sequence $u_n$ and less than all terms of the sequence $v_n$. Applying theorem 7, we come to the conclusion that the unique such number is the common limit of the sequences $u_n$ and $v_n$, which is exactly what we were supposed to prove.

As to trigonometry, the preparatory considerations within its scope admit a certain simplification for the method of cuts.

Instead of equation 10, we establish the following equation, whose proof does not present greater difficulty:

10′.  $\lim \cos \alpha/2^{n+1} = 1$.

Lemma 11, whose justification was quite troublesome, basically drops out altogether; we shall need one very special case of this lemma in which $\beta = 2\alpha$:

If $0 < \alpha < 45°$ then $\sin 2\alpha < 2 \sin \alpha$ and $\tan 2\alpha > 2 \tan \alpha$.

The proof is based either on the identities

$$\sin 2\alpha = 2 \sin \alpha \cos \alpha, \qquad \tan 2\alpha = \frac{2 \tan \alpha}{1 - \tan^2 \alpha},$$

or on straightforward geometric reasoning, which is very simple.

We modify theorem 12 in the following way:

12′. *The sequences $u_n = 2^{n+1} \cdot \sin 180°/2^{n+1}$ and $v_n = 2^{n+1} \cdot \tan 180°/2^{n+1}$ are mutually convergent.*

The proof of this differs from the proof of theorem 12 only in that instead of lemma 11 we use 11′.

Definition 13 changes analogously:

13′. *The common limit of the mutually convergent sequences*

$$u_n = 2^{n+1} \cdot \sin 180°/2^{n+1} \quad \text{and} \quad v_n = 2^{n+1} \cdot \tan 180°/2^{n+1}$$

*we call the number* $\pi$.

Coming to the theory proper of the measure of the circumference of a circle, we accept the following definition—already applied more than once, incidentally, in the elementary lecture:[15]

15′. *The* circumference of a circle *is the number greater than the length of every closed simple polygon inscribed in the circle, and less than the length of every closed polygon circumscribed about the circle.*[16]

The question of equivalence of definitions 15 and 15′, and of 13 and 13′, does not present any great difficulties, nor will it interest us further here.

Taking definition 15′ as a starting point, we prove theorem 16—that is, the main theorem of the whole theory—with the help of the following reasoning.

Let $A$ (respectively, $B$) be the set of all those numbers that are lengths of closed polygons inscribed in a circle (respectively, circumscribed about the circle). Set $A$ precedes set $B$ because, as is familiar from geometry, a convex polygon contained in the interior of another closed polygon is shorter than that polygon.[17]

We form two mutually convergent sequences such that all terms of the first sequence belong to set $A$, and of the second, to set $B$. We could use here those sequences that were mentioned in definition 15; then, however, we would have to base the proof on theorem 12, and thus indirectly on lemmas 9 and 11. To avoid this, we construct two other sequences, skipping infinitely many of the terms of each of the sequences mentioned just a moment ago. We take into consideration specifically the sequences of regular polygons inscribed in a circle (respectively, circumscribed about the circle) in which the first is a square and each subsequent one is obtained by doubling the number of edges of the previous one. It is easy to prove that the $n$th polygon in each of these sequences has $2^{n+1}$ edges.

---

[15] Compare Enriques and Amaldi [1903] 1916, 207–214 [chapter VII, §520].

[16] [Tarski's words were

Długość okręgu *jest to liczba, większa od długości każdej łamanej zamkniętej (niezwiązanej), wpisanej w okrąg, a mniejsza od długości każdej łamanej zamkniętej, opisanej na okręgu.*

He followed Enriques and Amaldi [1903] 1916, chapter VII, but modified its version of that definition. In that version, all polygons under consideration were closed and convex, and thus simple. Tarski lifted that restriction on the circumscribed polygons.]

[17] [The polygons in $A$ must be convex: see the preceding footnote. Tarski must have intended those in $B$ to be simple, else the notion of interior would be problematic. For the familiar result, consult Enriques and Amaldi [1903] 1916, 208. Tarski did not indicate how to extend that presentation to allow the polygons in $B$ to be nonconvex or nonsimple.]

Denoting by $u_n$ (respectively, $v_n$) the perimeter of the $n$th inscribed (respectively, circumscribed) polygon, we obtain

$$u_n = 2r \cdot 2^{n+1} \cdot \sin 180°/2^{n+1}, \quad v_n = 2r \cdot 2^{n+1} \cdot \tan 180°/2^{n+1}.$$

With the help of theorems 8 and 12', we conclude from here that sequences $u_n$ and $v_n$ are mutually convergent. In accordance with theorem 4 and definition 13' we have in addition $\lim u_n = \lim v_n = 2\pi \cdot r$.

Essentially, since all terms of the sequence $u_n$ belong to set $A$, and [those of] sequence $v_n$ to set $B$, it follows from theorem 21 that the common limit of both of these sequences is the unique number separating these two sets, and hence is the circumference $l$ of the circle,[18] in accordance with definition 15'. Therefore, $l = 2\pi \cdot r$. Q.E.D.

From the point of view of pure research, the difference between the two methods considered here is minimal. The situation is different from a didactic point of view. The following ideas ought then to be emphasized.

1. The supporting body of knowledge indispensable in the development of the theory is simpler for the second method than for the first, since two lemmas with somewhat complicated proofs drop out.

2. The basic definition of the circumference of a circle is logically simpler in the second case than in the first, since it does not depend on the notion of a limit. It is more natural, since the regular polygons do not play "privileged" roles in it. Finally, it is more instructive, since it extends without any changes to arbitrary convex curves.

3. The main theorem in the theory, which contains a formula for calculating the circumference of a circle, is easier to derive from the first definition than from the second; however, I do not suppose that it would be possible to speak here of a fundamental difference in the degree of difficulty.

In all these respects, the method of cuts possesses, in my opinion, greater didactic value in the secondary-school setting than the method of limits.

---

[18] [The previous $l$ was missing from the original.]

# 12
# Exercises Posed by Tarski

This chapter contains translations and discussions of the fourteen exercises that Alfred Tarski published for enhancing instruction in Polish gimnazjums. The translations appear in boxes in each of the numbered sections that follow. They are meant to be as faithful as possible to the originals, and follow the conventions used for the longer translations in other chapters. In particular, all [square] brackets in the translations enclose editorial comments inserted for clarification. Each individual exercise is accompanied by a discussion by the present editors: the material outside the box. These discussions include contextual information, descriptions of responses from students, teachers, and other mathematicians, and full or partial solutions by the present editors. The paragraphs immediately following this one describe general aspects of all of the exercises.

The journal *Parametr* was founded in 1930 to remedy some perceived inadequacies in Polish mathematics instruction at the *gimnazjum* level—at the schools that prepared students to enter university. It was edited by Antoni M. Rusiecki and Stefan Straszewicz. Rusiecki, a government official and trainer of teachers, was evidently the executive editor; Straszewicz was a professor at the Technical University of Warsaw. *Parametr* was intended to address both gimnazjum teachers of mathematics and their best students. That breadth proved difficult to manage; after a year, the student-oriented content was diverted to a second new journal, *Młody matematyk* (*Young Mathematician*). These two journals featured coordinated problem sections that contained exercises presented for readers' consideration, and solutions submitted by readers or crafted by the journal editors.[1]

In the present book, the word *problem* often indicates a significant mathematical question with no known solution, offered to challenge research mathematicians. The word *exercise* indicates a question offered to help instruct a class of students or to guide individual students to a higher level of achievement.[2] The phrase *problem section* is conventionally used to designate a collection of either sort of questions, often with discussion

---

[1] For a broader discussion of these journals see section 9.7. For more about Rusiecki and Straszewicz, consult boxes in section 9.7.

[2] Tarski used the word *ćwiczenia* for routine exercises commonly included in textbooks, such as those translated in chapter 13. He used *zadania* for both nonroutine exercises such as those translated in the present chapter, and for research problems.

or solutions. In his authoritative 1986 Tarski bibliography, Steven R. Givant included a list of problems and exercises posed by Tarski. Givant did not distinguish individually between the nine open research problems and the fourteen exercises intended for instruction. All of the latter were originally published in *Parametr* or *Młody matematyk*, and receive their first translations in this chapter.[3] For seven of Tarski's exercises the author was identified in the original only by the initials *A.T.* The journals did not include any table of such identifiers, but authorship was confirmed in Tarski's 1965a bibliography. All of Tarski's exercises were reprinted in Polish in his *Collected Papers*; some typographical errors were introduced there.[4]

*Parametr* and *Młody matematyk* included little explicit discussion to justify their problem sections. They simply extended a tradition already well established in other countries. The journal editors *may* have considered some or all of the following goals for the problem sections and their potential contributors:

(1) to exhibit mathematical bravado
(2) to display mathematical beauty
(3) to identify and stimulate the elite mathematics students
(4) to enhance mathematical instruction of all gimnazjum students.

Those are listed roughly in order of increasing importance, in the present editors' view. Goal (1) was of course not stressed, but is everpresent in mathematical publication. The problem sections did not specifically emphasize beauty—goal (2)—but these journals were *filled* with beautiful mathematics! The journal editors were serious about using substantial exercises to further instruction of good students—goal (3):

> ...sometimes there will be harder problems...for those undertaking a more serious study of mathematics. May the younger readers not grumble about such problems also being printed: many of the mathematics lovers will try to tackle [such] a problem, and many will probably be happy to have studied solutions of these problems.[5]

The editors' attention to goal (4) is revealed by their commenting repeatedly about using exercises to further instruction, without singling out the elite group. Their frustration with their lack of success with goals (3) and (4) is described in the following paragraphs. This may have been partially due to their own confusion about those goals, obscured by Rusiecki's overcommitment. They wrote,

> We draw attention to the "Problems from Advanced Teachers' Courses," which we regard as instructional exercises; we intend to give their solutions only in cases where readers encourage us to do so by sending solutions (taking into account the comments in the article Straszewicz 1930). We also present more difficult problems for the advanced readers.

---

[3] One of the research problems, Tarski 1925a, is translated in section 4.3, page 62. The others are easily accessible elsewhere in French or English.

[4] Tarski 1965a, 3; and 1986a, volume 4, 688–693.

[5] Rusiecki and Straszewicz 1931–1932, 2.

Thus, they intended to use some of their exercises explicitly as instruction, but only if readers were already willing to air their own solutions![6]

During 1930–1932, the journals published 218 numbered exercises and a few others. Rusiecki, evidently *the* editor of the problem sections, took personal credit for 76 of them, about 35%. Tarski contributed the second largest number: fourteen, about 6%. Three others—two teachers and one professor—each contributed more than four exercises. Rusiecki collected most of the rest from various examinations.[7]

Nearly half of these 218 exercises received point ratings according to their difficulty. No criterion for the ratings was published, and they seem somewhat haphazard. Some ratings were assigned, and some revised, in later issues. Of the rated exercises,

- 38% received ratings below five points
- 52% —five to fourteen
- 10% —fifteen or more.

The journal editors would award generous prizes to readers who submitted complete solutions for specific *contest* exercises or for various others whose ratings should total at least one hundred points. The prizes were vouchers for 25 zlotys or more, to be spent on books that the recipients could select. The large prize for Tarski's very elementary contest exercise (translated in section 12.1) and the journal editors' remarks about the desired completeness of solutions suggests that they were stressing the organization of solutions and the quality of writing.[8]

The journal editors repeatedly expressed their dissatisfaction with the response to the problem sections. For example,

> The editors are receiving too few solutions. This is a bad sign. In order to improve the problem section, readers are asked to send their criticism and desiderata regarding the section. Whoever is unhappy with the problem section may send in friendly criticism.[9]

Nevertheless, solutions were published for 84—about 39%—of the 218 numbered exercises posed during 1930–1932. Most solutions were detailed; some, even for elementary exercises, extended over several illustrated pages. In quite a few instances all or portions of the published solutions were attributed to individual solvers; in some, only the proposer was credited. The many remaining solutions were apparently constructed by the editor or were composites of contributed solutions. Solutions were published for only two

---

[6] Rusiecki and Straszewicz 1930. The article Straszewicz 1930 emphasized some specific criteria for completeness of a solution. For Rusiecki's workload, consult section 9.7.

[7] Those three contributors were Kazimierz Gilewicz (from Poznań), Samuel Steckel (Białystok), and Władysław Wojtowicz (Warsaw). For more about Wojtowicz, see a box in section 9.3. Exercises published after 1932 (that is, in 1939) were not analyzed for the present discussion, because there was no time for readers to respond to them.

[8] Rusiecki and Straszewicz 1930, 229, and 1931; Rusiecki 1931a. Contemporary issues of *Parametr* contained advertisements for recently translated novels by James Oliver Curwood, which cost about five zlotys, and for many Polish scientific books priced well below that.

[9] Rusiecki and Straszewicz 1930.

exercises rated ten or higher. During 1930 and 1931, fifty different readers submitted correct solutions:

- 20 gimnazjum students
- 13 gimnazjum faculty
- 17 others.[10]

The problem sections of *Parametr* and *Młody matematyk* were mainly the work of their editor, Rusiecki. He provided 35% of the exercises, collected many others from various examinations, contributed very many solutions, and probably assembled many others from readers' contributions. Even with a substantial backlog of material, his production of the 1930 and 1931–1932 volumes must have taken a stupendous effort. No one could have sustained that for long, on top of a full-time government position devoted to fostering and administering secondary education in mathematics countrywide. The journals ceased publication after 1932. *Parametr* was resurrected during 1939 until the German invasion.

During 1930–1932, Tarski submitted articles and exercises for this serious, exciting, and potentially influential journal, connected with both government and academia. He was a major contributor and his intent was serious. It is worthwhile to investigate his contributions to determine what that was. To what extent did he seem to be motivated by the goals described earlier? Tarski eventually developed a notably graceful style of presenting mathematics; in fact he achieved it at great effort.[11] But he was not known for writing or speaking about that in public. Some of his exercises stand as examples of elegance, but that seems not to have been a major goal for the whole set of exercises.

Tarski's intent could have been to *foster* the education of talented students, addressing both them and their teachers. He would have realized from experience that these groups' capabilities overlapped. Moreover, improvement in teaching could be stimulated by these students' perception and increased interest. Or, he could have intended to help *ferret out* the very topmost mathematics students in the gimnazjums, to recruit them to university studies and mathematical careers. Many of Tarski's contributions were advanced exercises, which received no further attention in the journals. That suggests that he and the journal editors did not intend them principally to foster the development of talented students in general, but as stimuli for the very topmost and a means of identifying them.

The present editors analyzed the difficulty of Tarski's exercises. They devised criteria as follows, according to their own experience as students, teachers, and mathematicians.

- Half of Tarski's exercises seem to involve only mathematics *at gimnazjum level*. The others seem slightly or well above that.

---

[10] Rusiecki and Straszewicz 1932a.

[11] See Givant 1991, 18, for an account of Tarski's private conversations about writing, decades later. Another colleague reported, "…the more one worked with Tarski, the result tended to look less and less laborious. In fact Tarski would work over a mathematical presentation until it achieved an elegance and simplicity which disguised the difficulties hidden beneath the surface" (Szczerba 1986, 910).

- Only two apply this mathematics in a way that seems to fit *clearly within the curriculum*. The subjects of the others would have been unfamiliar to some or all students.

- Most require *reasoning beyond the sophistication* of even the best students.

Four of Tarski's exercises scored low on all three grounds: the present editors designated them *elementary*. Six others scored highest on at least two of these criteria: they are *advanced*. The other four are *intermediate*. The journals themselves assigned point ratings to eleven of the exercises; most are consistent with the present editors' designations.[12] All these ratings are given with the discussions of the exercises in the following sections.

In Tarski's exercises there is a major emphasis on *geometry*: only three of the fourteen seem *not* to require any geometric reasoning. All of the geometry exercises involve *symmetry* in some way; in three, that could be called a main concern. There is an emphasis, too, on reasoning with *inequalities*: eight exercises involve that in some way, and in three of those it is the main concern. Three exercises—two elementary and one advanced—emphasize core areas of the curriculum: quadratics and logarithms. Three more involve subjects—integer arithmetic and the floor and ceiling functions—that may not have appeared significantly in the curriculum; one of those was elementary.

Organization of *case-ridden arguments* was emphasized in nine of Tarski's fourteen exercises. Some of those involve almost unmanageable proliferations of cases, demanding either bookkeeping bordering on the obsessive or intense concentration beyond the ability of most. Three more display that feature to a lesser extent. As scholar and university teacher, Tarski was advancing the application of formal logic, and this type of argument provided an excellent opportunity for that. But he rarely stated that goal explicitly. The techniques are familiar now, probably because of their widespread use in computer science.

Five of Tarski's exercises asked, under what conditions on some parameters $b, c, \ldots$ does a problem involving them and a variable $x$ have a solution $x$? For example, his first exercise, in section 12.1, involved the discriminant condition $b^2 - 4c \geq 0$ for solvability of a quadratic equation $x^2 + bx + c = 0$. In general, such a problem amounts to finding a condition $P(b, c, \ldots)$ equivalent to $\exists x \, Q(x, b, c, \ldots)$ for a given condition $Q(x, b, c, \ldots)$ —that is, *elimination of the quantifier* $\exists x$. That was also a major thread in Tarski's logical research, stemming from his 1927–1928 University of Warsaw research seminar. But its general discussion even now remains limited to advanced logic courses.

Only five of Tarski's exercises were stated with complete clarity. For seven others, the desired form of the solution was incompletely specified. While this is a way of assessing students' perspicacity and assertiveness, it is also a way of forcing solvers to learn elsewhere, from some in-group, what form a solution should take.

---

[12] The rectangle-cutting exercise in section 12.9 seems rated somewhat too high by the journal, and the section 12.11 exercise on the parallel postulate, too low.

Tarski's exercises received little attention after their 1930–1932 publication in *Parametr* and *Młody matematyk*. Although the journals published solutions for about 39% of the exercises posed during 1930–1932, only three solutions appeared for exercises by Tarski, about 14%. Two were elementary; solution of the third—see section 12.10—was provided by a general theorem in a research paper by Henryk Moese, also published in *Parametr*. The journal *Matematyka: Czasopismo dla nauczycieli* (*Mathematics: A Journal for Teachers*), began publication soon after World War II. Its editors included Rusiecki and Straszewicz, and its problem section clearly descended from those of the earlier journals. One of Tarski's advanced problems, from section 12.3, was reprinted and solved there. Two others, from 12.9 and 12.10, were reprinted in 1975 as exemplars of the mathematics in *Młody matematyk*.[13]

What accounts for this apparent lack of impact? Tarski may have intended most of his exercises not to enhance instruction but to identify the very top students and to stimulate them and some of their teachers. As such, they would not have required much follow-up. But even had the exercises succeeded according to that goal, the journals offered no feedback. On the other hand, Tarski and the journal editors may simply have underestimated the difficulty of his exercises. A common fault of mathematicians successful in research is to assume that good students in general learn the same way they did. Moreover, many of the exercises require organizational persistence beyond the patience of young students. That might have deterred readers from considering them, especially when they were posed alongside other exercises that were simpler and offered better opportunity for star students to shine.

Tarski's 1931 exercise in section 12.11 was the first publication of his version of Euclid's parallel postulate. Although Tarski used that form in his own research and teaching, he did not discuss it in print again until 1957. Its role in subsequent research on foundations of geometry is described in section 12.11.

Tarski's exercises confirm the connection between his research in mathematics and logic and his work as a teacher. They show that he was keenly aware of that connection, and made serious effort to ensure that each of these aspects of his activity influenced the other.

---

[13] Moese [1932] 2014, translated in section 7.3; for its impact, see sections 7.5 and 8.5. Iwaszkiewicz 1949b. Dubiel 1975.

## 12.1 [1930] 2014a, Exercise on Diluting Wine

Tarski's contribution *Zadanie o rozcieńczaniu wina* to the July–August 1930 issue of *Parametr* (volume 1, page 229) is translated and discussed here. Unlike the other exercises in this chapter, this one had no number, nor was it assigned a point value.

---

### CONTEST PROBLEM II
### Exercise on Diluting Wine

A winemaker had two barrels; one had a volume of $a$ liters, and the second, $b$ liters. The first barrel was filled with pure wine but the second was empty. The winemaker poured a certain amount of wine— $x$ liters—from the first barrel to the second, and filled up the second barrel with water. After mixing the wine and water, the winemaker poured from the second barrel to the first just enough so that the first barrel became full. It turned out that in the first barrel was the kind of wine that we would obtain if, into the barrel containing $a$ liters, we poured $c$ liters of pure wine and filled up the remaining [volume] with water.

Find $x$ and report for what values of the givens $a, b, c$ we have (1) two solutions, (2) one solution, (3) a problem with no solution.

*Dr. Alfred Tarski (Warsaw)*

---

*Solution*.[14] After mixing, the concentration of pure wine in the second barrel was $x/b$. After the first barrel was refilled with the mixture, it contained

$$(a - x) + (x/b)x = c$$

liters of pure wine. After rearranging this equation into standard quadratic form and considering the discriminant $d$, we see that the three conditions correspond to the cases (1) $d > 0$, (2) $d = 0$, and (3) $d < 0$. The quadratic formula yields the $x$ values.

---

[14] Readers are reminded that the solutions in this chapter are the present editors' unless otherwise noted.

*Discussion.* This exercise is *elementary* and devoted to a core subject. Its solution involves the quadratic formula, a very familiar example of quantifier elimination. The journal editors characterized it as *thought-provoking*.[15]

There were three other contest exercises that year.[16] The first challenged readers to plan the longest airplane flight from sunrise to sunset, entirely within Poland; in the present editors' view, it was too vague. The other two were stated as precisely as Tarski's: to create a certain type of geometric puzzle based on equidecomposable polygons, and to find a parameter $m$ such that exactly six pairs of positive integers $x, y$ satisfy the equation $5x + 3y = m$.

For the best two solutions of Tarski's contest problem, the journal editors offered substantial prizes. The size of the reward for solving this elementary problem indicates that the editors were stressing the organization of solutions and the quality of writing.[17] There seems to have been no provision in the journal for recognizing contributed solutions for this exercise.

## 12.2 [1930] 2014b, Exercise 83: Iteration of the Absolute Value Symbol

Tarski's exercise *Nr. 83: Równanie ze znakami bezwzględnej wartości* is translated and discussed in this section. It was published in the July–August 1930 issue of *Parametr* (volume 1, page 231).

---

Exercise 83. *Iteration of the absolute value symbol.* Prove that

$$||a| - |b|| = |a + b| + |a - b| - |a| - |b|.$$

[*Three points for solution.*]   A. T. (Warsaw)

---

*Discussion.* This exercise is *elementary* and emphasizes inequalities and symmetry. The equation is equivalent to

$$|a| + |b| - |a + b| = |a - b| - ||a| - |b||.$$

---

[15] "*Arcydyskusyjny*"—see Rusiecki and Straszewicz 1930, 229.
[16] Rusiecki 1930c, 1930d, and Kaptur 1930.
[17] See the discussion on page 245.

That is, the amount by which one member of the triangle inequality $|a| + |b| \geq |a+b|$ exceeds the other is the same as that for the related inequality $|a - b| \geq ||a| - |b||$, which is often employed in real-analysis texts.

The journal recognized contributed solutions from two students of a gimnazjum in Łańcut, a town in southeastern Poland. One received full credit. The next year, the journal published an unsigned solution, which treated the cases separately.[18]

*Solution.* The coordinate axes and diagonals divide the $a,b$-plane into the origin and eight closed unbounded wedge-shaped regions that overlap just along those lines. The equation obviously holds in the region where $a \geq b \geq 0$, and is invariant under the reflections across the boundary lines; therefore it is valid in general. While that concise argument is convincing for mathematicians, it might not be appropriate for a gimnazjum.

## 12.3 [1930] 2014c, Exercise 88: Decomposition into Factors

Tarski's exercise *Nr. 88: Rozkład na czynniki* is translated and discussed in this section. It was published in the September 1930 issue of *Parametr* (volume 1, page 277).

---

Exercise 88. *Decomposition into factors.* What conditions must integers $a$ and $b$ —both different from zero—satisfy so that these conditions are necessary and sufficient conditions to be able to express the binomial $ax^4 + b$ in the form of a product of polynomials of at least first degree with integer coefficients?

[*Ten points for solution.*]     Dr. Alfred Tarski (Warsaw)

---

*Discussion.* This exercise is *advanced*. It emphasizes a core subject—factoring— and another that may not have been encountered often in schools—integer arithmetic. The exercise was stated awkwardly and gave no clue to the form of the desired solution.[19] Two teachers each received full credit for solving this exercise: Michał Hornowski from Warsaw[20] and Władysław Stojda from Piotrków Trybunalski, a city in central Poland.

---

[18] The students were Józef Burda and Bronisław Szul. For the solution, see Rusiecki 1931b.

[19] The condition "to be able to express..." in the exercise itself satisfies the stated requirement, but that is not what Tarski intended.

[20] **Michał Hornowski** was born in 1893 in Russia and educated in Warsaw. From about 1920 to 1958 he taught mathematics in Polish institutes that trained teachers. He edited, coauthored, and reviewed many textbooks and books about mathematics. Hornowski died in Warsaw in 1966. (Piotrowski 2003, 76.)

*Solution.* Soon after World War II, the exercise was reprinted in the journal *Matematyka: Czasopismo dla nauczycieli*. It received two additional correct solutions and published one a year later: the condition that there should exist nonzero integers $k$, $\lambda$, and $\mu$ such that either

$$k\lambda^2 = a \quad \text{and} \quad k\mu^2 = -b, \tag{1}$$

or

$$k\lambda^2 = a \quad \text{and} \quad k\mu^2 = b, \tag{2}$$

$$\text{and} \quad 2\lambda\mu = \nu^2 \quad \text{for some integer } \nu. \tag{3}$$

The following proof sketch is adapted from that source.[21]

First use elementary algebra to verify that condition (1) implies that the given quartic is the product of two quadratics with integer coefficients, and that the conjunction (2) & (3) also implies this. Next, show that (1) is equivalent to the condition that

$$-a/b \text{ should be the square of a rational number.}^{22} \tag{1'}$$

Now suppose that the given quartic is the product of two quadratics with integer coefficients $\overline{\alpha}, \ldots, \overline{\eta}$:

$$ax^4 + b = (\overline{\alpha}x^2 + \overline{\beta}x + \overline{\gamma})(\overline{\delta}x^2 + \overline{\varepsilon}x + \overline{\eta}). \tag{4}$$

Let

$$\lambda = \gcd(\overline{\alpha}, \overline{\delta}), \quad k = \overline{\alpha}\overline{\delta}/\lambda^2, \quad \mu = \gcd(\overline{\gamma}, \overline{\eta}).$$

By comparing coefficients of the members of equation (4), one sees that $\overline{\beta} = 0$ if and only if $\overline{\varepsilon} = 0$: that is, either

$$\overline{\beta} = 0 = \overline{\varepsilon} \quad \text{or} \quad \overline{\beta}, \overline{\varepsilon} \neq 0. \tag{5}$$

By a brief argument, the first alternative implies condition (1), and thus also (1'). Similarly, the second alternative of (5) implies condition (2). Moreover, comparing coefficients also yields $0 = \overline{\beta}\overline{\delta} + \overline{\alpha}\overline{\varepsilon}$. The integers $\alpha = \overline{\alpha}/\lambda$ and $\delta = \overline{\delta}/\lambda$ are relatively prime; since $\lambda\alpha\overline{\varepsilon} = -\lambda\overline{\beta}\delta$ and $\lambda$ is nonzero, $\alpha$ divides $\overline{\beta}$. That is, there exists an integer $\nu$ such that $\overline{\beta} = \nu\alpha$. Another brief argument yields condition (3).[23]

---

[21] See Iwaszkiewicz 1949b. Stojda was given priority for the solution. The postwar solutions were by Tadeusz Czarliński and Mieczysław Warmus. Born in Breslau in 1918, **Mieczysław Warmus** became a noted professor of mathematics and computer science; he died in Australia in 2007.

[22] Moreover, the conjunction (2) & (3) is equivalent to the simpler condition that $4a/b$ should be the fourth power of some rational number; but that result is not used in this argument.

[23] Readers are invited to complete the "brief" arguments mentioned in this paragraph. The first two rest on showing that $\alpha$ and the ratio $\gamma = \overline{\gamma}/\mu$ divide each other, so that $\gamma = \pm\alpha$. The third uses elementary algebra to derive $\pm 2\lambda\mu = \nu^2$. The two $\pm$ signs must therefore be $+$.

Finally, suppose that the given quartic $Q(x)$ has a linear factor $cx + d$ with integer coefficients. Since $Q(x)$ is even, it has roots $\pm d/c$, and thus $Q(x) = q(x)(c^2x^2 - d^2)$ for some quadratic $q(x)$ with rational coefficients. Let $m$ be the least common multiple of their denominators, so that the quartic $mQ(x)$ is the product of two quadratics with integer coefficients, which must fall under the first alternative of (5). Therefore, $mQ(x)$ must satisfy condition (1'): $-(ma)/(mb)$ should be the square of a rational number. But that ratio is the same as $-a/b$, so (1'), and hence condition (1), holds for the given quartic $Q(x)$.

## 12.4 [1930] 2014d, Exercise 102: Interesting Identity

Tarski's exercise *Nr. 102: Ciekawa tożsamość* is translated and discussed in this section. It was published in the September 1930 issue of *Parametr* (volume 1, page 278).

---

Exercise 102. *Interesting Identity*. Show that for every natural number $n \geq 2$ this identity holds:

$$\sum_{g=2}^{n} \mathrm{E} \log_g n = \sum_{k=2}^{n} \mathrm{E} \sqrt[k]{n}.$$

*Remark.* For a real number $x$, we denote by the symbol $\mathrm{E}x$ the whole number $p$ satisfying the conditions $p \leq x < p + 1$. The symbol E is French shorthand for the word *entier*; we read the symbol $\mathrm{E}x$ as *the entirety of* $x$.

Dr. Alfred Tarski (Warsaw)

---

*Discussion.* This exercise is *intermediate* in level. While it emphasizes core subjects—logarithms, roots, and notation—it provides a taste of symmetry, inequalities, and counting, and serves to introduce the floor function.[24] Its elementary nature is disguised by the bravado of its notation. But Tarski slyly gave a subtle hint by using different indices with the sums. Two readers were recognized for contributing solutions: Perez Halamon, from Kalisz, a city in central Poland, and Władysław Stojda. No solution was published.

---

[24] In this exercise, $\mathrm{E}x$ stands for a value $\lfloor x \rfloor$ of the floor function.

*Solution.* The decreasing function $g \to \log_g n$, graphed below in figure 1, maps the interval $[\sqrt[n]{n}, n]$ to $[1, n]$. Its inverse is the decreasing function $k \to \sqrt[k]{n}$. The first and second sums in the exercise are the numbers of lattice points in the regions $G$ and $K$, hatched /// and \\\ in the figure, respectively, not including the points on the axes. There are exactly $n - 1$ lattice points in each of the regions $G - K$ and $K - G$, all on the lines with equations $k = 1$ and $g = 1$, respectively. Thus, $G$ and $K$ contain the same number of lattice points, and the sums are equal.

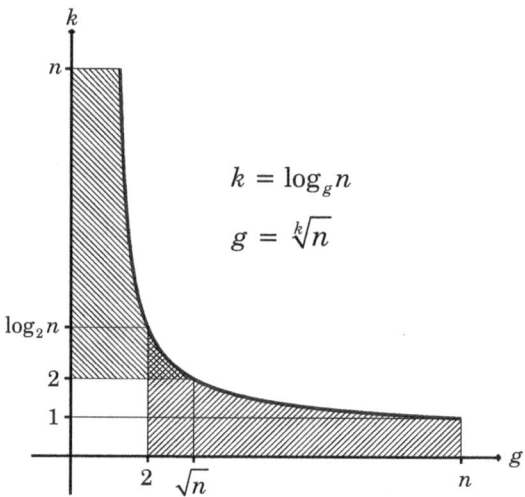

**Figure 1** (for Section 12.4)

## 12.5 [1930] 2014e, Exercise 118

Tarski's exercise *Nr. 118* is translated and discussed here. It was published in the October 1930 issue of *Parametr* (volume 1, page 318). Unlike the other exercises in this chapter, it had no descriptive title. It was included among "problems associated with the content of articles"—probably selected as related to the paper Cwojdziński 1930 on sequences, in the section of the same issue that targeted student readers.

> Exercise 118. A sequence $a_n$ was defined in the following way:
>
> if $n$ is an odd number, then $a_n = \dfrac{2}{n+1}$, and
>
> if $n$ is an even number, then $a_n = -\dfrac{2}{n}$.
>
> Construct a formula $a_n = F(n)$ that would give all the terms of the sequence.
>
> A. T. (Warsaw)

*Discussion.* This exercise is *elementary*, but touches on a subject perhaps less commonly encountered in schools—integer arithmetic. Teodor Hrycak, a teacher and engineer from Stanisławów, a district in southeastern Poland, contributed a solution, but none was published.

*Solution.* A glance at the previous exercise posed by Tarski, in section 12.4, would suggest using the floor or ceiling operator. Here are two sample solutions, one with and one without:

$$a_n = \frac{(-1)^{n+1}}{\lceil n/2 \rceil} \qquad a_n = \frac{4}{1-(-1)^n(2n+1)}.$$

## 12.6 [1930] 2014f, Exercise 136: Equidistant

Tarski's exercise *Nr. 136: Ekwidystanta* is translated and discussed in this section. It was published in the November–December 1930 issue of *Parametr* (volume 1, page 398).

---

Exercise 136. *Equidistant*.[25] Find the set of all points whose distance from the boundary of a given rectangle with edges $a$ and $b$ is equal to a given segment of length $r$.

[*Four points for solution.*]   A. T. (*Warsaw*)

---

*Discussion.* This exercise is *elementary*, with a taste of symmetry, inequalities, and argument by cases. The word *find* (*wyznaczyć*) in the exercise statement was unfortunately vague. No solution by a reader was reported, but the journal published a solution several years later.[26] The journal editor asked a nice question: what is the analogous result in three dimensions?

*Solution.* Illustrated below by figure 2 for the case $a < b$ and $r < a/2$, the desired curve consists of an internal rectangle and an external one whose corners have been replaced by quarter circles centered at the vertices of the original rectangle as indicated. Tedious arguments will show that the points on this locus are exactly those desired.

When $a < b$ and $r = a/2$, the inner rectangle becomes a line segment; when $a = b$ and $r = a/2$ it becomes a point. When $r > a/2$ the locus has no interior component.

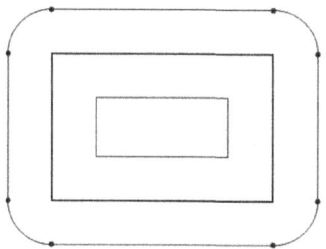

**Figure 2** (for Section 12.6)

*Equidistant Curve*

---

[25] In English the exercise title is an adjective used nominally to refer to the equidistant curve that the exercise describes. The Polish title is a noun.

[26] Rusiecki 1939, which was incorrectly labeled as a solution to a different exercise.

## 12.7 [1931] 2014b, Exercise 167: System of Inequalities

Tarski's exercise *Nr. 167: Układ nierówności* is translated and discussed in this section. It was published in the February 1931 issue of *Młody matematyk* (volume 1, page 46), distributed with *Parametr*, volume 2.

---

Exercise 167. *System of inequalities.* Investigate what conditions numbers $k$, $l$, $m$ must satisfy so that there should exist an angle $\varphi$ in the first quadrant ($0 \leq \varphi \leq 1/2\,\pi$) satisfying the following two inequalities simultaneously:

$$k \sin \varphi + l \cos \varphi \leq m$$
$$k \cos \varphi + l \sin \varphi \leq m.$$

*Ten points for solution.* A. Tarski (Warsaw)

---

*Discussion.* This exercise and its companion, exercise 168 in the next section, were identified as biproducts of Tarski's research on a question that he posed in his [1931] 2014a paper on the degree of equivalence of polygons, translated in section 7.2. That question was repeated in his exercise 170, translated in section 12.10. Shortly afterward, Tarski's original question was answered by an entirely different method in the paper Moese [1931] 2014, translated in section 7.3.

Thus, this exercise's interest lies mainly in its emphasis on types of reasoning that Tarski considered important for the next generation of mathematicians. It is *advanced*, and emphasizes symmetries and inequalities. Their relation to an asymmetric feature—specification of first quadrant—is particularly engaging. The exercise also makes the investigator confront the need to reformulate arguments to reduce the great number of cases involved. The word *conditions* in the statement of the exercise should probably have been explained, to make it clear that this is an exercise in quantifier elimination. No solutions by readers were reported, and no solution was published.

*Partial solution.* The following discussion assumes that $k$, $l$, $k \pm l$, and $m$ are nonzero. Figure 3, on the next page, uses a Cartesian $x, y$ coordinate system with origin $O$ to display lines with equations as follows:

|  |  | $x, y$ intercepts |
|---|---|---|
| $g$ : | $kx + ly = m$, | $m/k$, $m/l$, |
| $g'$ : | $lx + ky = m$, | $m/l$, $m/k$. |

The picture is symmetric with respect to the main diagonal and the lines intersect at point

$$P = \left\langle \frac{m}{k+l}, \frac{m}{k+l} \right\rangle \quad \text{with distance} \quad OP = \left| \frac{m}{k+l} \right| \sqrt{2}.$$

The figure was drawn for the case in which $P$ lies in the first quadrant: its coordinates are positive.

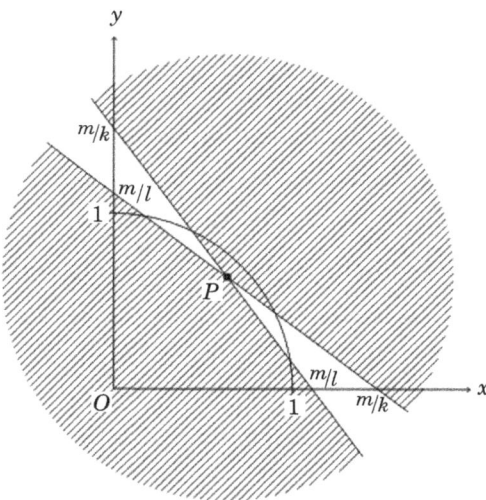

**Figure 3** (for Section 12.7)

The set of points $\langle x, y \rangle$ that satisfy the simultaneous inequalities

$$\begin{cases} kx + ly \leq m \\ lx + ky \leq m \end{cases}$$

is one of the two hatched regions in the figure. It is the one containing $O$ just in case $m > 0$. The inequalities in Tarski's exercise can be formulated as the two just shown, plus the condition that $\langle x, y \rangle$ lie on the first quadrant of the unit circle $\mathcal{U}$, as shown in the figure for a case in which $OP < 1$.

Should $P$ fall in the third quadrant, Tarski's inequalities will have a solution if and only if $m > 0$. Should it lie in the first, there will be a solution just when $m > 0$ and $P$ or the intercept nearer $O$ lies on or outside $\mathcal{U}$, or when $m < 0$ and $P$ lies on or inside $\mathcal{U}$. Therefore, under the listed assumptions, Tarski's inequalities have a solution if and only if

- $\dfrac{m}{k+l} < 0$ and $m > 0$; or

- $\dfrac{m}{k+l} > 0$ and $m > 0$, and $\dfrac{m}{k+l}\sqrt{2} \geq 1$ or $\dfrac{m}{k} \geq 1$ or $\dfrac{m}{l} \geq 1$; or

- $\dfrac{m}{k+l} > 0$ and $m < 0$, and $\dfrac{m}{k+l}\sqrt{2} \leq 1$.

These conditions can be reformulated as Boolean combinations of linear inequalities involving $k$, $l$, and $m$.

Readers of the present book may wish to complete this analysis to accommodate the cases in which one or more of $k$, $l$, $k \pm l$, and $m$ is zero, and to simplify these conditions.

## 12.8 [1931] 2014c, Exercise 168: Another System of Inequalities

Tarski's exercise *Nr. 168: Układ nierówności* is translated and discussed in this section. It was published in the February 1931 issue of *Młody matematyk* (volume 1, page 46), distributed with *Parametr*, volume 2.

---

Exercise 168. *System of inequalities.* Investigate what conditions numbers $k$ and $l$ must satisfy so that the following three inequalities should possess at least one common solution:

$$(k+1)x^2 - 2lx + (k-1) \geq 0$$
$$(l+1)x^2 - 2kx + (l-1) \geq 0$$
$$x^2 - x \leq 0.$$

Indicate the relationship between this exercise and the previous exercise.

*Ten points for solution.* A. Tarski (Warsaw)

---

*Discussion.* Background for this exercise was discussed with that of its companion, exercise 167 in the previous section.

This exercise, too, is *advanced*. It is ostensibly about inequalities and quadratics, and quickly evolves into reasoning about many cases. Further investigation into its relationship with the exercise in the preceding section might reveal a geometric connection. Like that one, its statement could have been clarified to make it clear that this is an exercise in quantifier elimination. No solutions by readers were reported, and no solution was published.

*Partial solution.* Tarski's system of three inequalities can be solved by straightforward but tedious algebraic manipulation, outlined as follows. It is assumed that constants $k, l \neq -1$.

The Cartesian graphs of the first two quadratics are parabolas that open upward or downward depending on whether the leading coefficients are positive or negative. Each has two roots, one, or none, depending on whether its discriminant $4(l^2 - k^2 + 1)$ or $4(k^2 - l^2 + 1)$ is positive, zero, or negative. The sum of the discriminants is 8. One or both of those quadratics must have two roots: otherwise, that sum would be negative or zero. If (a) one or both parabolas should open downward, the set $X$ of solutions of Tarski's first two inequalities will be empty, a single isolated point, two isolated points, or a finite interval, depending on the linear order of the roots. If (b) both open upward, $X$ is the union of two or three disjoint intervals, two of which are infinite, depending on the linear order of their roots. The isolated points and interval endpoints mentioned in the preceding two sentences are roots of the quadratics.

Tarski's third inequality merely describes the points $x$ in the closed unit interval $U$. Thus a solution of the system of all three inequalities exists just in case

    a.    one or both parabolas open downward, $X$ is nonempty, and $U$ intersects one of the intervals described, or

    b.    both open upward and $U$ intersects one of the intervals described.

The conditions describing the opening of the parabolas, whether $X$ is empty in case (a), and the order of the roots of the quadratics and endpoints of the unit interval, can all be expressed as inequalities involving $k$ and $l$. Thus, the existence of a solution of Tarski's system of three inequalities involving $k, l, x$ is equivalent to a Boolean combination of these inequalities involving $k$ and $l$.

Readers of the present book may wish to construct these inequalities, simplify them into manageable form, and then investigate how they can be reformulated, if necessary, to accommodate the additional cases in which $k$ or $l$ equals $-1$. Moreover, the outline just presented does not address the second part of Tarski's exercise: how is this set of inequalities related to those in the previous exercise 167 in section 12.7? That, too, is left for future consideration.

## 12.9 [1931] 2014d, Exercise 169: Cutting a Rectangle Out of a Square

Tarski's exercise *Nr. 169: Wycięcie prostokąta z kwadratu* is translated and discussed in this section. It was published in the February 1931 issue of *Młody matematyk* (volume 1, page 46), distributed with *Parametr*, volume 2.

> Exercise 169. *Cutting a rectangle out of a square.* With regard to exercise 1 posed in issue 1 of *Parametr* from 1930, and to the letter to the editor published in issue 5 of that year,[27] prove the following assertion:
>
> > So that from a square with edge $a$ it would be possible to cut out a rectangle with edges $b$ and $c$, it is necessary and sufficient that either $b \leq a$ and $c \leq a$, or that $b + c \leq a\sqrt{2}$.
>
> *Fifteen points for solution.*      A. Tarski (*Warsaw*)

*Discussion.* The assertion in this exercise was stated awkwardly in the original. Exercise 1 in issue 1 asked for a general method to cut as many rectangular pieces as possible, each with edges $b$ and $c$, from a given rectangular sheet of paper (at least for the case in which the edges of the rectangles are parallel to that of the sheet). In the cited letter to the editor, Włodzimierz Krysicki wrote,

> [Exercise 1] is a very interesting question, which initially seems very simple, but in reality is very complicated. It seems as if the problem escapes mathematical methods.

Although exercise 1 was rated fifteen points, the editors awarded Krysicki twenty points for his partial solution of a restricted case. A solution of Tarski's exercise 169 would provide a tool for solving the general case. This result and the method used for the solution presented in this section would provide useful tools for computer-assisted design. However, the present editors have not found any reference to it in the literature of that subject.[28] In his 1975 historical description of *Młody matematyk* described in section 9.7, Władysław Dubiel included this exercise among the six exemplars that he chose to reprint.

---

[27] Rusiecki 1930a; Krysicki 1930. **Włodzimierz Krysicki** was born in Warsaw in 1905 and educated there. After earning a master's degree in mathematics in 1928, he taught in a secondary school. Krysicki earned the doctorate from the University of Łódź in 1950 and taught there until his retirement in 1975. A noted statistician and writer on mathematics, he died in 2001 (Pawlikowska-Brożek 2003, 119).

[28] Slightly more general versions of Tarski's exercise were posed and solved in Garnett and Carver 1925 and in Ford and Carver 1957.

This exercise is *intermediate* in level. It emphasizes inequalities, but involves symmetry as well, and is also an example of quantifier elimination.[29] The assigned point value seems too high. No solutions by readers were reported, and no solution was published.

*Solution.* If $b \leq a$ and $c \leq a$, the rectangle can be placed in a corner of the square. Assuming $b + c \leq a\sqrt{2}$ on the other hand, consider the subsegment $AB$ of a diagonal of the square, centered, with length $b$, as shown in figure 4, below. The remainder of the diagonal consists of two segments, each of length $d = 1/2(a\sqrt{2} - b) \geq c/2$. The perpendiculars to the diagonal at $A$ and $B$ meet the edges of the square at four points that form a rectangle with edges of lengths $b$ and $2d \geq c$. The desired rectangle can be cut from this. Thus, Tarski's condition is sufficient.

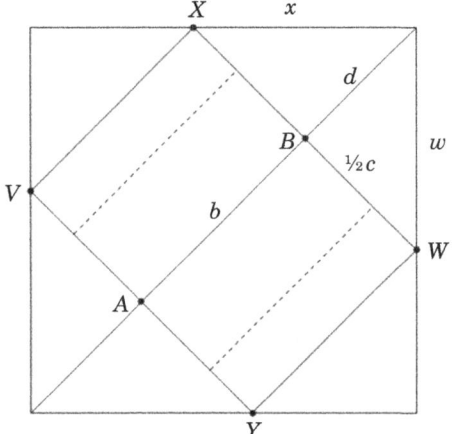 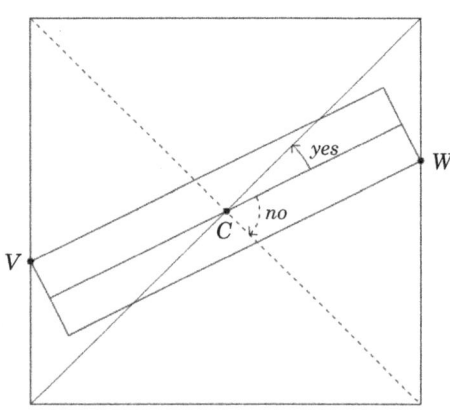

**Figure 4** (for Section 12.9)     **Figure 5** (for Section 12.9)

To prove that the condition is necessary, suppose the desired rectangle can be cut from the square in figure 5 but $a < b$ or $a < c$. If $b$ denotes the length of the longer edge, then $a < b$ in either case. Translate the rectangle horizontally then vertically within the square, so that its left and right vertices become equidistant from the left and right edges of the square, and its top and bottom vertices, equidistant from the top and bottom edges. The translated rectangle is congruent to the original one, and has the same center $C$ as the square. Attempt to rotate the translated rectangle about $C$, within the square, to make its major axis fall on one of the diagonals of the square. Should that fail, two opposite vertices $V, W$ of the rotated rectangle would have collided with opposite edges of the square, as shown by the dashed rotation in figure 5. If only one pair of opposite vertices collided with edges of the square, choose the other diagonal, which will intersect the quadrants containing those vertices; after rotation, the rectangle would be situated

---

[29] The word *possible* stands for an existential quantifier.

## 12.10 Exercise 170: On the Degree of Equivalence of Polygons

like the inner one in figure 4, and the previous reasoning will yield the desired inequality. If *both* pairs of opposite vertices collided with edges, the rectangle would be situated like $VWXY$ in figure 4, and the previous reasoning would imply $b + c = a\sqrt{2}$.

To see that the axes of the rotated rectangle do coincide with the diagonals of the square when its vertices all collide with edges, relax that requirement in figure 4. The diagonals of $VWXY$ must nevertheless be equal, and they have simple formulas in terms of $a$ and the indicated lengths $w$ and $x$. The resulting equation implies $w = x$.

### 12.10 [1931] 2014e, Exercise 170: On the Degree of Equivalence of Polygons

Tarski's exercise *Nr. 170: Twierdzenie o stopniu równoważności wielokątów* is translated and discussed in this section. It was published in the February 1931 issue of *Młody matematyk* (volume 1, page 46), distributed with *Parametr*, volume 2.

---

Exercise 170. *Theorem on the degree of equivalence of polygons*. Use the assertion [in section 12.9, Exercise 169] to prove the equation

$$\tau(4) = 4.$$

In the proof of this statement, it is permissible to refer to the results of the exercises titled *System of inequalities* [in sections 12.7 and 12.8].

*Ten points for solution.*  A. Tarski (Warsaw)

---

*Discussion.* The degree of equivalence and the function $\tau$ were described and the exercise posed as an unsolved problem in Tarski's paper [1931] 2014a, translated in section 7.2. It was reprinted in the present form later in the same issue of *Młody matematyk*. Its level is *advanced*.

Tarski's problem was solved the same year by Henryk Moese, as a corollary of a more general result: $\tau(n) = n$ for *every* positive integer $n$. Moese published this result in the next volume of *Parametr* in the paper Moese [1931] 2014, which is translated in section 7.3. Moese's methods had nothing to do with the suggestions that Tarski included in the exercise statement.[30] Director of a gimnazjum in Kępno, a town in southern Poland, Moese was a regular contributor to these journals.

---

[30] The present editors have not attempted to apply Tarski's suggestions. For more information about Moese, see a box in section 7.1.

## 12.11 [1931] 2014f, Exercise 183: Postulate about Parallels

Tarski's exercise *Nr. 183: Postulat o równoległych* is translated and discussed in this section. It was published in the February–March 1931 issue of *Parametr* (volume 2, page 78).

> Exercise 183. *Postulate about parallels*. Accepting the system of axioms for Euclidean geometry given by David Hilbert, or any system established in one of the Polish school texts on elementary geometry,[31] show that the axiom of parallelism can be replaced by the statement,
>
> > *For every point interior to a convex angle there exists at least one segment that passes through this point and has endpoints on the edges of the given angle.*
>
> *Five points for solution.*        A. T. (*Warsaw*)

*Discussion.* This exercise is *advanced*, not because of its mathematical content —part of a core subject—but because of the sophistication required to consider alternatives in the complicated axiom system of elementary geometry. It is somewhat delicate because its interpretation and solution depend on how the notions of convex angle and interior point are defined and how the parallel postulate is formulated.

Before introducing his version of the parallel postulate, Hilbert[32] presented postulates that, for a given line $g$ in a given plane $\varepsilon$, allow classification of all points in $\varepsilon$ as lying on $g$ or on exactly one of two half-planes called the *sides* of $g$ in $\varepsilon$. Two distinct lines $g$ and $h$ in $\varepsilon$ that intersect at a point $O$ thus partition $\varepsilon$ into nine parts: $O$, four open-ended rays emanating from $O$, and four regions, each of which is the intersection of two sides of different lines. See figure 6 on page 266 for an example. The union of $O$ and a pair $r$ and $s$ of such rays, where $r \subseteq g$ and $s \subseteq h$, is called a *convex angle*; the rays are called its *arms*, and their intersection $O$, its *vertex*. Each arm lies entirely on

---

[31] Tarski was familiar with Hilbert [1899] 1922. Section 9.2 contains a list of Polish geometry texts in use at the time.

[32] Hilbert [1899] 1922 or [1899] 1971, §§1–4; this present discussion applies to the 1903 second edition and all later ones, but Hilbert's treatment of the parallel postulate in preceding editions was different. For further information see Hilbert 2003, 419.

## 12.11 Exercise 183: Postulate about Parallels

the same side of the line containing the other. The intersection of the side of $g$ containing $s$ and the side of $h$ containing $r$ is called the *interior* of the angle.[33]

Although Euclid's original statement of the parallel postulate involved the notions of side and angle, axiomatic presentations of Euclidean geometry often employ a simpler version called *Playfair*'s postulate: *in a given plane $\varepsilon$, through any point $O$ not on a given line $l$ there passes at most one line that does not intersect $l$.* Such a line $g$ is called a *parallel* to $l$ —in symbols, $g \mathbin{/\mkern-5mu/} l$.[34] The other postulates in a presentation of Euclidean geometry usually entail existence of at least one parallel to $l$ in $\varepsilon$.[35]

The present editors examined various editions of Polish texts available in the 1920s: Enriques and Amaldi [1903] 1916, Kiselev 1917, Łomnicki 1923, Wojtowicz [1919] 1926, and Zydler 1916. Each derived the existence of a parallel from postulates stated earlier, and derived its uniqueness from Playfair's postulate. Some discussed Euclid's original parallel postulate.

The form of the parallel postulate that Tarski presented in this exercise was apparently first mentioned as such in Roberto Bonola's [1906] 1955 history of non-Euclidean geometry. Bonola and others have suggested that the principle had been used earlier in Lorenz 1791–1792, but that seems doubtful.[36]

*Solution.* To show that Playfair's postulate implies Tarski's, assume the former and consider the convex angle $\angle ROS$ formed by rays $\overrightarrow{OR}$ and $\overrightarrow{OS}$ in lines $g$ and $h$ respectively: see figure 6 on page 266. Let $k$ be the line $\overleftrightarrow{RS}$. Given any point $P$ in the interior of the angle formed by those rays, use the other postulates to find $l \mathbin{/\mkern-5mu/} k$ in $\varepsilon$ through $P$. By Playfair's postulate, $g$ must intersect $l$ because $l \mathbin{/\mkern-5mu/} k$ and $g \neq k$; similarly, $h$ must intersect $l$, and thus the conclusion of Tarski's postulate holds.

---

[33] The word *convex* was used here because some geometry texts consider things called *nonconvex angles*, with radian measure $\leq 0$ or $\geq \pi$. They are not used in the presentation that Tarski was following.

[34] Proposition 29 in Euclid's book I ([1908] 1956, volume 1, 312) directly implies Playfair's postulate. For a proof of Euclid's original postulate from Playfair's, see the discussion by Thomas L. Heath (ibid., 313–314).

[35] For example, drop a line $k \perp l$ from $O$, then erect in $\varepsilon$ a line $g \perp k$ at $O$.

[36] Bonola (§28, page 58) reported that in an attempt to justify the parallel postulate, A. M. Legendre had tacitly used this principle a century earlier in constructing from a given triangle a larger one whose defect (180° – its angle sum) would be twice that of the original triangle. Bonola suggested that Lorenz had done likewise. But in Euclid [1908] 1956, 220, the editor Heath reported that Johann Friedrich Lorenz had in fact used a different principle. Neither Bonola nor Heath referred to a specific page of Lorenz 1791–1792. In his Berkeley doctoral dissertation supervised by Tarski, Haragauri N. Gupta (1964, page 11a) referred to Lorenz, volume 1, 101–102. The present editors have been unable to examine the original edition of Lorenz. The 1798 second edition is inconsistent with all three citations. The 1804 third edition (§§27, 28, 73) is consistent only with Heath's citation. The principle that Heath mentioned occurred also in the 1851 fourth edition (§28, which evidently stemmed from 1820) but not as a justification of the parallel postulate.

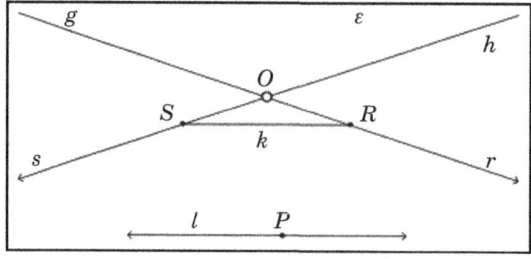

**Figure 6** (for Section 12.11)

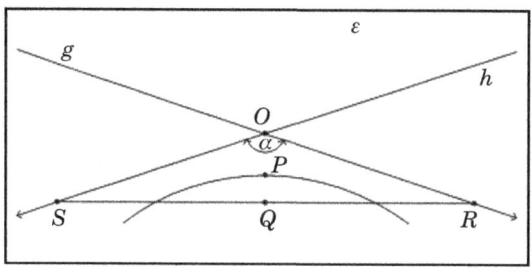

**Figure 7** (for Section 12.11)

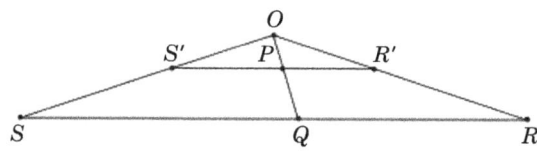

**Figure 8** (for Section 12.11)

*Tarski's Parallel Postulate, in Terms of Betweenness β Alone:*
$\beta OPQ$ & $\beta R'PS'$ & $O \neq P \Rightarrow \exists R, S(\beta OR'R$ & $\beta OS'S$ & $\beta SQR)$

## 12.11 Exercise 183: Postulate about Parallels

To show the converse, assume Tarski's postulate and (expecting a contradiction to ensue) suppose a point $O$ lay on distinct parallels $g$ and $h$ to a line $l$ that does not pass through $O$, as in figure 7 on page 266. Let $P$ be any point on $l$, so that $l$ would lie entirely in the interior of the angle $\alpha$ formed by the sides of $g$ and $h$ that contain $P$. On the line $\overleftrightarrow{OP}$ select any point $Q$ on the side of $l$ opposite $O$ and all other points on $g$ or $h$; then $Q$ would be interior to angle $\alpha$. By Tarski's postulate, there would exist a line $k$ through $Q$ that would intersect both $g$ and $h$ at some points $R$ and $S$, respectively, which would both have to lie on the side of $l$ opposite $Q$. Thus, segments $\overline{QR}$ and $\overline{QS}$ would both intersect $l$. Those intersections would have to be distinct, so that $k = l$, contrary to the hypotheses that $g, h \parallel l$. This contradiction shows that $O$ cannot lie on distinct lines $g, h \parallel l$, which is Playfair's postulate.

*Impact.* No reader submitted any solution for this exercise, and none was published. Tarski had begun using this form of the parallel postulate in his lectures on the axiomatics of Euclidean geometry at the University of Warsaw during 1926–1927.[37] The argument in the previous paragraph suggests that this form is closely entwined with the properties of the betweenness relation $\beta$ that are used to formulate reasoning about the sides of a line in a plane. Indeed, this form of the postulate can be expressed in terms of $\beta$ alone[38]:

$$\beta OPQ \ \& \ \beta R'PS' \ \& \ O \neq P \Rightarrow \exists R,S(\beta OR'R \ \& \ \beta OS'S \ \& \ \beta SQR).$$

This relationship is illustrated by figure 8 on page 266.

Tarski published his full system of postulates only after long delay, in the notable [1957] 1959 paper *What Is Elementary Geometry?* There, he discussed his system's organization in detail, but included no proofs. Nevertheless, that paper stimulated a considerable body of research by Tarski and others. For example, a version of the famous Pasch postulate, used to describe the betweenness relation in both Euclidean and non-Euclidean geometry, can be formulated in a form strikingly similar to Tarski's parallel postulate:

$$\beta OQP \ \& \ \beta R'PS' \Rightarrow \exists R,S(\beta ORR' \ \& \ \beta OSS' \ \& \ \beta SQR).$$

Readers of the present book may wish to consider the minor alteration of figure 8 required to illustrate this so-called *weak Pasch postulate*. This analogy has suggested some intriguing, still unsolved problems in foundations of geometry.[39]

For more detailed analyses of Tarski's system, see the papers *Tarski and Geometry* by Lesław W. Szczerba, and *Tarski's System of Geometry* by Tarski and Steven R. Givant. Givant constructed the latter publication from a long letter that he and Tarski had written to Wolfram Schwabhäuser in 1978 to assist in the preparation of the 1983 book *Metamathematische Methoden in der Geometrie* by Schwabhäuser, Wanda Szmielew, and Tarski. That work finally included the proofs that demonstrated the adequacy of Tarski's postulate system for elementary Euclidean geometry. His axiomatization is now often regarded as a standard for comparison of results in foundations of geometry.

---

[37] Tarski and Givant 1999, 175.

[38] Tarski and Givant 1999, 183. The symbol $\beta OPQ$ means that points $O$ and $P$ are equal, or $P$ lies between $O$ and point $Q$, or $P = Q$.

[39] Tarski and Givant 1999, 197–198.

## 12.12 [1931] 2014g, Exercise 186: Analytic Geometry of Space

Tarski's exercise *Nr. 186: Zadanie z geometrji analitycznej w przestrzeni* is translated and discussed in this section. It was published in the February–March 1931 issue of *Parametr* (volume 2, page 78).

> Exercise 183. *Analytic geometry of space.* Find the locus of points equidistant from three given lines. Give an exhaustive discussion of the possible cases.
>
> *Thirty points for solution.* A. T. (*Warsaw*)

*Discussion.* This exercise and the next seem out of order: see the following section 12.13. This exercise is well known, not usually attributed to Tarski, and probably not original to him. It is *advanced*, due to its requirement to consider very many cases, and to its use of solid analytic geometry, including the intersection of two hyperbolic paraboloids. It still plays a role in geometric research.[40] The word *find* (*znaleźć*) in its statement is unfortunately vague. No solutions by readers were reported, and no solution was published.

Discussion of this exercise is continued at the end of the next section, 12.13, because solution of the next exercise seems prerequisite to solution of this one.

## 12.13 [1932] 2014b, Exercise 213: Stereometry

Tarski's exercise *Nr. 213: Zadanie ze stereometrji* is translated and discussed in this section. It was published in the January–February 1932 issue of *Parametr* (volume 2, page 207).

---

[40] See Everett, Gillot, et al. 2009 and Everett, Lazard, et al. 2009.

## 12.13 Exercise 213: Stereometry

> Exercise 213. *Stereometry*. Find the locus of points in space equidistant from three given lines lying in one plane. Give an exhaustive discussion of all possible cases.
>
> *Five points for solution.*  A. T. (Warsaw)

*Discussion.* This exercise and the previous one, in section 12.12, seem out of order. Solution of this second one seems prerequisite to solution of the first. Following the solution of the present exercise is a partial solution of the earlier one. The level of the present exercise is *intermediate*, due to its requirement to consider very many cases; but it is devoted to a core subject in geometry. The word *find* (*podać*) in its statement is unfortunately vague. Aleksander Grzymała, a reader from Warsaw, contributed a solution and received full credit, but no solution was published.

*Solution.* First, let $\varepsilon$ be the given plane. Determining the locus $\lambda_\varepsilon$ of points in $\varepsilon$ equidistant from the given lines is almost a routine plane-geometry problem. If (1) all three coincide, then $\lambda_\varepsilon = \varepsilon$. If (2) just two of them are distinct and those intersect, then $\lambda_\varepsilon$ is the union of the two lines that bisect the resulting vertical angles; the bisectors are perpendicular to each other. If (3) just two of the given lines are distinct and those are parallel, then $\lambda_\varepsilon$ is the line in $\varepsilon$ parallel to and midway between them. If (4) the lines form a triangle, then $\lambda_\varepsilon$ consists of four distinct points: its incenter and excenters. If (5) the lines are distinct and concurrent, then $\lambda_\varepsilon$ is their common point. If (6) the lines are all distinct and parallel, then $\lambda_\varepsilon$ is empty.

The previous paragraph dealt with the routine planar cases. One planar case remains: (7) two of the lines, $g$ and $h$, are distinct and parallel and the third is a transversal intersecting them at points $V$ and $W$, respectively. The two pairs of bisectors of the vertical angles formed at $V$ and $W$ form a rectangle $VYWZ$ with center $X$, depicted with dashed lines in figure 9 on page 270. Its diagonals $VW$ and $YZ$ are congruent and bisect each other and form two pairs of congruent isosceles triangles. By the exterior-angle and isosceles-triangle theorems, $\angle WXZ$ is congruent to the angle between the transversal and $g$, on the side opposite $Z$. By the theorem about alternate interior angles, $Y$ and $Z$ fall on the line $m$ in $\varepsilon$ parallel to $g$ and $h$ and midway between them: thus $Z$, and $Y$ as well, belongs to the locus $\lambda_\varepsilon$ of points on $\varepsilon$ equidistant from all three given lines. Conversely, any point of $\lambda_\varepsilon$ must lie on one of the bisectors and on $m$, and hence must coincide with $Z$ or $Y$. In this nonroutine planar case, the locus $\lambda_\varepsilon$ consists of two points.

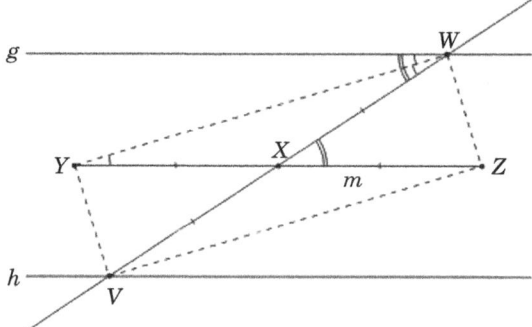

**Figure 9** (for Section 12.13)

Now it is simple to determine the desired locus $\lambda$ of all points in space that are equidistant from the three given lines in $\varepsilon$: it is the union of all perpendiculars to $\varepsilon$ through points in $\lambda_\varepsilon$. In the cases considered above, $\lambda$ consists of (1) all points in space, (2) the union of two intersecting planes perpendicular to $\varepsilon$, (3) a single plane perpendicular to $\varepsilon$, (4) four distinct lines perpendicular to $\varepsilon$, (5) a single line perpendicular to $\varepsilon$, (6) no points at all, or (7) the union of two distinct lines perpendicular to $\varepsilon$.

*Partial solution of exercise 183.* The exercise described in the previous section asked for the locus $\Lambda$ of all points in space that are equidistant from three given lines configured arbitrarily. Cases (1–3), discussed in the previous paragraph, cover the situations in which the three lines all coincide or just two are distinct but coplanar. Cases (4)–(7) cover the situations in which the three lines are all distinct but coplanar.

Several more cases can be analyzed relatively easily, in which the given lines are all distinct but two pairs are coplanar. For case (8), in which the three lines are all parallel, $\Lambda$ is the line parallel to them through the circumcenter of a triangle formed by their intersections with a common perpendicular plane. If (9) two of the given lines intersect and the third is parallel to one of them, then $\Lambda = \varepsilon \cap (\zeta \cup \eta)$, where $\varepsilon$ is the plane of points equidistant from the parallels, and $\zeta$ and $\eta$ are the two perpendicular planes of points equidistant from the intersecting lines; in this case, $\Lambda$ will consist of two intersecting lines. If (10) two pairs of given lines each intersect, then $\Lambda$ is the intersection of two unions of pairs of perpendicular planes; in general, $\Lambda$ will consist of four lines —two pairs of perpendiculars. In the concurrent case, these four lines will be concurrent as well. Readers may wish to puzzle about whether special configurations of the given lines in case (10) will yield a locus $\Lambda$ that is simpler than the general one just described.

The remaining cases involve nonlinear analytic geometry—the reason for the title of exercise 183. Of these, the simplest case (11) is the most important: just two of the given lines, $g$ and $h$, are distinct, but they are noncoplanar. The discussion here is limited to an example. Use a Cartesian coordinate system with origin $O$ and unit points $U$, $V$, and $W$. Let $g$ be the line through $U$ parallel to the axis through $V$, and let $h$ be the axis through *W*. The squares of the distances to *g* and *h* from a point $X = \langle x_1, x_2, x_3 \rangle$ are $(x_1 - 1)^2 + x_3^2$ and $x_1^2 + x_2^2$, respectively, and thus the locus $\Lambda$ has equation $(x_1 - 1)^2 + x_3^2 = x_1^2 + x_2^2$. This is equivalent to the equation

$$x_1 = \tfrac{1}{2}(x_3^2 - x_2^2) + \tfrac{1}{2},$$

which identifies the locus as a hyperbolic paraboloid. The general case is similar to this example in that two noncoplanar lines *g* and *h* always have a common perpendicular[41] analogous to the axis through *U*. But in general, *h* will not be perpendicular to the plane of *g* and that axis. More analytic geometry is required to find the distance from *X* to *h*, and the algebra leading to the equation of the locus $\Lambda$ becomes formidable. But it can nevertheless still be identified as a hyperbolic paraboloid.[42]

In the final cases, at most one pair of the given lines can be coplanar. If (12) those are parallel, then the locus $\Lambda$ is a conic section: the intersection of the plane of points equidistant from the parallels with the hyperbolic paraboloid of points equidistant from one of the other pairs. If (13) that given pair intersects, then $\Lambda$ is the intersection of a hyperbolic paraboloid with the union of two perpendicular planes. In general, such an intersection will be the union of two conics; readers may wish to investigate whether any special instance of this case might yield a simpler locus. Finally, if (14) no pair of the given lines is coplanar, the locus $\Lambda$ is the intersection of two hyperbolic paraboloids. Readers may wish to investigate the nature of that curve.[43]

## 12.14 [1932] 2014c, Exercise 214: Arranging Two Segments in a Plane

Tarski's contribution *Nr. 214: Układ dwóch odcinków na płaszczyźnie* is translated and discussed in this section. It was published in the January–February 1932 issue of *Parametr* (volume 2, page 207).

---

Exercise 214. *Arranging*[44] *two segments in a plane.* A plane figure consisting of two segments is given. Indicate all axes and centers [of symmetry] of this figure, lying in its plane. Give an exhaustive discussion of the possible cases.

*Five points for solution.*           A. T. (*Warsaw*)

---

[41] The common perpendicular to two noncoplanar lines is discussed by Thomas L. Heath in Euclid [1908] 1956, volume 3, 306–307, in his commentary on book XI, proposition 19. See also Enriques and Amaldi [1903] 1916, chapter 12, §780, 322.

[42] Readers of the present book may wish to ferret out interesting properties of this locus; see Forsythe 1969.

[43] See Everett, Gillot et al. 2009 and Everett, Lazard et al. 2009.

[44] Tarski's term for *arranging* was *układ*: *arrangement* or *system*. Givant (1986, 936) used *system*.

*Discussion.* The level of this exercise is *intermediate* although it requires no deep expertise in geometry. It forces students to consolidate familiarity with a broad scope of elementary geometric objects and their properties. Organization and exposition of a solution is its greatest challenge. Should a solution classify the methods of arranging the components of the figure, just the point set that results, or the possible sets of symmetries? The solution presented below takes the last approach, which seems to lead to the least redundancy. It also attempts to give enough detail that readers can reconstruct the sketches that guided the present editors. No solutions by readers were reported, and no solution was published.

*Solution.* Tarski did not specify whether his segments should contain their endpoints. This solution assumes that they do, and that points are segments with length zero. The degenerate case in which both given segments are the same single point $O$ exhibits the most symmetry: all rotations about $O$ and reflections across all axes through it.

Otherwise, the largest possible symmetry group is that of a square, which contains two quarter-turns and a half-turn about its center $O$ and reflections across the four axes that join $O$ with the vertices and edge midpoints. This symmetry occurs only when the two given segments are the diagonals of a square.

The next largest group is the symmetry group of a nonsquare rectangle: the *Vierergruppe*, which contains a half-turn about the center $O$ of the rectangle and reflections about the two axes that join opposite edge midpoints. This symmetry occurs when the two given segments are

- collinear and overlap, or
- collinear, disjoint, and congruent, or
- the diagonals or a pair of opposite edges of a nonsquare rectangle, or
- the diagonals of a nonsquare rhombus.

Some configurations are symmetric only by reflection about a single axis. This happens when the two given segments are

- collinear and disjoint but incongruent, or
- a pair of congruent edges of an isosceles triangle, or
- the parallel edges or the congruent edges or the diagonals of an isosceles trapezoid that is not a parallelogram, or
- the base of an isosceles triangle and a subsegment of its altitude, or
- the diagonals of a kite that is not a rhombus.

Two configurations are symmetric only by a single half-turn: the two given segments must be a pair of opposite edges or the diagonals of a parallelogram that is neither a rhombus nor a rectangle.

Other configurations of the given segments exhibit no symmetry.

# 13
# *Geometry for the Third Gimnazjum Class (1935)*

This chapter is devoted to a translation of parts of the school-geometry text *Geometrja dla trzeciej klasy gimnazjalnej* by Zygmunt Chwiałkowski, Wacław Schayer, and Alfred Tarski. The text was published in 1935 by Państwowe Wydawnictwo Książek Szkolnych, the National Schoolbooks Publisher.[1] Its title page is shown on page 276 of the present book. The note there under the publisher's logo says that the price of the book, including the cost of the stamp of the Society Supporting Construction of Common Public Schools, is 1.20 zlotys. That was comparable to the retail price of three kilograms of wheat flour, one kilogram of pork, five liters of milk, or thirteen eggs.[2]

This translation is even closer to the original than the other translations in the present book. Their intent is to convey content most accurately with changes in style only as necessary for that. However, a *main* reason for translating parts of the *Geometrja* book is to display its style,[3] to provide a better glimpse of the atmosphere of mathematics instruction in Poland during its time. All [square] brackets in the translation enclose editorial comments inserted for clarification; this includes all the footnotes.

At the beginning of *Geometrja*, lengths are regarded as segments. For example, consider the first sentence of §3 (page 281):

> We know that the distance from a point $M$ to a line $PQ$ ... is the segment $MN$ constructed from point $M$ perpendicular to $PQ$.

In the middle of *Geometrja* is a transition. The authors introduced numbers into geometry gradually. In §24, they used rational numbers as ratios of pairs of commensurable

---

[1] Chwiałkowski, Schayer, and Tarski [1935] 1946. For information about Chwiałkowski and Schayer and about later editions, see boxes in section 9.9.

[2] These data, from Poland 1936, 168, are for 1935 in Warsaw, where the cost of living was higher than in many other places in Poland.

[3] The original use of boldface type for emphasis has been maintained. In the original, uppercase Latin variables were not italicized; in the translation, they are. Minor changes in punctuation were sometimes made to adhere more closely to conventions of English, and sometimes to enhance clarity. Most paragraphs in the original, and in the translation, consist of single sentences.

segments. In §25 (page 292) they showed that the diagonal $BC$ of the unit square is incommensurable with its edge $AC$. Then (page 294) they gave numbers a greater role:

> We assume that the length $BC$ with respect to the unit $AC$ is a number. But it cannot be a whole or fractional number.... Therefore, this is a new kind of number. Previously we did not know about such numbers, and only now do we become familiar with them.

This changeover presented a problem in translation. At first, concepts referred to as segments, lengths, or distances are treated somewhat interchangeably. Congruent segments are called equal; so are certain segments and distances. Later, after introduction of irrational numbers, the distinction between geometric segments and numerical lengths or distances is clearer, but the authors' terminology is not fully consistent.

The figures were redrawn using *Mathematica* software: their relative dimensions are close approximations to those of the originals. The only systematic change was to use solid rather than hollow dots to indicate points. Placement of figures in relation to surrounding text is close to that of the original.

In the present book are translated the very beginning and very end of the *Geometrja* text, plus some material from the middle: its sections 1–4, 25–27, 34–35, and table of contents.[4] The table is shown on the facing page. Sections 34–35 are devoted to the theory of areas of polygonal regions; thus, they are closely related to Tarski's research reported in Part Two of the present book. Sections 25–27 of *Geometrja* introduce readers to irrational numbers. They contain some material necessary for the later sections, and are related to the approach to the real number system described in Tarski's [1932] 2014 article on the circumference of a circle, translated in chapter 11 of the present book. Brief editorial comments are inserted in boxes between the translations of sections 4 and 25, and between those of sections 27 and 34, to describe the sections of *Geometrja* that are not translated here.

For more detailed information about *Geometrja* and its background, see section 9.9 of the present book. As noted there, the editors hope that including in this single volume of selected translations works of such contrasting mathematical sophistication as Tarski's geometrical research in Part Two and the secondary-school geometry lessons in the present chapter will spur broader investigation into the relation between mathematical research and mathematics education, not just in Poland between the world wars, but in more general contexts.

---

[4] The book contained no significant front matter after the title page, and no back matter except the table of contents. The §28 title in the original table was wrong; it was corrected in later editions and here.

# TABLE OF CONTENTS

## PART ONE

### Geometric Loci

|   |   | page |
|---|---|---|
| §1. | Circle and Disk | 277 |
| §2. | Construction Problems | 279 |
| §3. | Relative Positions of a Circle and a Line Lying in a Plane | 281 |
| §4. | Construction Problems | 285 |

§5. Tangents to a Circle
§6. Axis of Symmetry of a Segment
§7. Construction Problems—Method of Geometric Loci
§8. Circumcircle of a Triangle
§9. Bisectors of Angles Formed from Two Intersecting Lines
§10. Incircle of a Triangle
§11. Parallel Lines
§12. Parallel Lines as Geometric Loci
§13. Geometric Locus of Points Equidistant from Two Parallel Lines

not translated

## PART TWO

### Angles in Relation to a Circle

§14. Central Angles
§15. Arcs and Chords
§16. Diameter of a Circle as Axis of Symmetry
§17. Chords and Their Distances from the Center of a Circle
§18. Angles between Chords and Tangents
§19. Inscribed Angles
§20. Arcs as Geometric Loci
§21. Construction Problems
§22. Quadrilaterals Inscribed in a Circle and Circumscribed about a Circle
§23. Regular Polygons

## PART THREE

### Measuring Segments and Areas

| | | |
|---|---|---|
| §24. | Commensurable Segments | |
| §25. | Incommensurable Segments—Irrational Numbers | 292 |
| §26. | Measuring Incommensurable Segments with a Unit | 296 |
| §27. | Rational Approximations of Irrational Numbers—Operations on Irrational Numbers | 297 |

§28. A Theorem of Thales
§29. On Similarity—Similar Triangles
§30. Method of Similarity in Construction Problems
§31. Practical Applications
§32. Similar Polygons

not translated

| | | |
|---|---|---|
| §33. | Equivalent Polygons | 302 |
| §34. | On Measuring Areas of Polygons | 304 |
| §35. | Areas of Similar Polygons | 316 |

Z. CHWIAŁKOWSKI — W. SCHAYER — A. TARSKI

# GEOMETRJA

## DLA TRZECIEJ KLASY GIMNAZJALNEJ

CENA
WRAZ ZE ZNACZKIEM NA
TOWARZYSTWO POPIERANIA
BUDOWY PUBLICZN. SZKÓŁ
POWSZECHNYCH WYNOSI
1·20 ZŁ.

PAŃSTWOWE WYDAWNICTWO KSIĄŻEK SZKOLNYCH
WARSZAWA — LWÓW
1935

# PART ONE

## Geometric Loci

### § 1. Circle and Disk

With point $O$ as center, we describe a circle with radius equal to a given segment $r$ (figure 1).

Every point of the circle, such as $A$, $B$, $C$, is one radius distant from the center of the circle and, conversely, each point of the plane whose distance from the center $O$ is equal to the radius lies on the circle.

If each point of a certain figure has a given property and, conversely, every point having this property belongs to the figure, we say that this figure is the geometric locus of points having the given property.

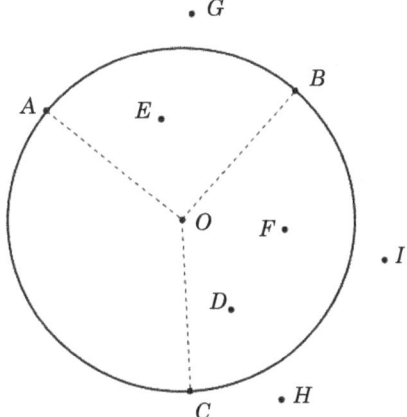

**Figure 1**

Thus we can provide the following definition of a circle.

**Definition. A circle is a geometric locus of points in a plane whose distances from a certain fixed point called the center are equal to each other.**

The points whose distances from the center $O$ are less than the radius, for example $D$, $E$, $F$, we call **interior** to the given circle, and points whose distances from the center $O$ are greater than the radius, for example $G$, $H$, $I$, we call **exterior**.

The points on a circle and points interior to the circle constitute a **disk.**

**Definition. A disk is a geometric locus of points in a plane whose distances from the center are less than or equal to the radius.**[5]

Recall from last year the definitions of chord, diameter, central angle, and arc, and write them out.

Now we recall the following axiom:

**Axiom. A segment or arc of a circle connecting a point interior to a circle with an exterior point intersects that circle at one and only one point.**

---

[5] [In the original the terms for *circle* and *disk* were *okrąg* and *koło*, respectively. The authors should have defined the term *radius* before *interior*.]

Based on on this axiom,[6] it is possible to prove the following theorem:

**Theorem. A line[7] passing through a point interior to the circle intersects the circle at two points.**

Let us also accept without proof this theorem:

**Theorem. Any two points interior to a circle can be connected with a segment or polygonal path having no point in common with the circle.[8]**

Consider whether the same can be said for any two points exterior to the circle.

**Exercises.** 1. Choose any point $A$ [and] draw the geometric locus of points distant from point $A$ by a segment [of length] (a) 2 cm, (b) 3 cm, (c) 5 cm.

2. Choose any point $A$ and shade the geometric locus of points distant from point $A$ by a segment [of length] (a) less than or equal to 2 cm, (b) greater than or equal to 3 cm.

3. Using any radius draw a circle and sketch the geometric locus of points distant from the center of the circle by a segment equal to (a) half of the radius of the given circle, (b) twice the radius of the given circle.

4. Draw any angle, not equal to a straight or full angle; choose point $A$ (a) interior to, (b) exterior to the angle.[9]
Find the points on the arms of the angle distant from point $A$ by a given segment.

5. With point $A$ as center, draw two circles, one with radius equal to 2 cm, the second with radius equal to 5 cm.
Indicate on the drawing the geometric locus of points whose distance from the center is (a) less than 2 cm, (b) equal to 2 cm, (c) greater than 2 cm but less than 5 cm, (d) equal to 5 cm, (e) greater than 5 cm.

---

[6] [Euclid's very first proposition ([1908] 1956, 241–242) rested on this principle but he failed to state it among his postulates. Now called the *circle* principle, it is often derived through analytic geometry from a much stronger geometric continuity axiom. *Geometrja* alludes to a continuity axiom informally in §26 (page 296). The circle principle was stated formally as an axiom in the Polish texts Enriques and Amaldi [1903] 1916 (§§191, 211, pages 56, 63, for a segment and an arc), Łomnicki 1934 (§§27, 28, pages 44, 47), and Wojtowicz [1919] 1926, 21. (Łomnicki 1934, a textbook for the preceding gimnazjum year, adhered to the same curriculum standards as *Geometrja*.) The text Zydler 1925 (22–23) considered the circle principle worth mentioning, but did not indicate that it must be introduced as an axiom or derived from a stronger one. See the discussion in Moise [1963] 1990, §§16.2 and 16.5.]

[7] [The word *prosta* in the original is translated here as *line*. It connotes *straight* line.]

[8] [In the original, the words for *segment* and *polygonal path* were *odcinek* and *łamana*, respectively. It is easy to prove this theorem using the exterior angle theorem and Euclid's propositions 18 and 19 ([1908] 1956, 279–284).]

[9] [What amounts to a definition of the phrase "interior of an angle with measure $m$ *such that* $0° \leq m \leq 360°$" is to be found in Łomnicki 1934, 22–23. The terms *pełny* and *półpełny* for *full* and *straight* angle (360° and 180°—*pół* means *half*) are explained there and used in this exercise in *Geometrja*.]

6. Draw a segment $AB$ [with length] equal to 7 cm; with point $A$ as center draw a circle with radius 5 cm; with $B$ [as center], a circle with radius 4 cm. Indicate the geometric locus of points whose

(a) distance from $A$ is less than 5 cm [and] from $B$, less than 4 cm;
(b) distance from $A$ is less than or equal to 5 cm [and] from $B$, greater than 4 cm;
(c) distance from $A$ is greater than 5 cm [and] from $B$, less than or equal to 4 cm;
(d) distance from $A$ is greater than 5 cm [and] from $B$, greater than 4 cm.

7. Choose any two segments $a$ and $b$ ($a > b$), and with an arbitrarily chosen point $A$ as center, draw concentric circles: one with radius $a$; a second with radius equal to the sum $a + b$; [and] a third with [radius equal to] the difference $a - b$. Where are the centers of the circles located, drawn with radius $b$, that with the first of these circles (a) do not have even one point in common, (b) have one common point, (c) have two common points?

8. With an arbitrarily chosen point as center, draw a circle with an arbitrary radius. Choose a point $A$ on this circle. Find points on the circle distant from point $A$ by a segment (a) equal to the radius of this circle, (b) less than the radius, (c) greater than the radius but less than the diameter, (d) equal to the diameter.

Are there any points on this circle distant from point $A$ by a segment greater than the diameter?

## § 2. Construction Problems

**Problem.** Construct a triangle, given two edges $a$ and $b$ and the median $m$ to edge $a$.

To find a way to solve this problem, we draw an arbitrary triangle $ABC$ (figure 2) and its median $AD$ and assume that it is precisely the triangle that we are going to construct and that $BC = a$, $AC = b$, $AD = m$.

Now we consider how to draw such a triangle.

We can immediately draw edge $CB$, equal to $a$: thus, we have two vertices of the triangle, $B$ and $C$ —[this] is only about finding the third vertex $A$.

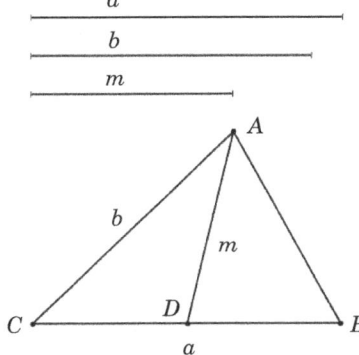

**Figure 2**

The vertex $A$ lies at a distance $m$ from point $D$, which is the midpoint of edge $BC$. Therefore $A$ lies on the geometric locus of points at distance $m$ from point $D$: that is, on the circle with center $D$ and radius $m$.

We can also say about the vertex $A$ that it is located at a distance $b$ from point $C$ [and] therefore lies on the geometric locus of points at distance $b$ from point $C$ —that is, on the circle drawn with radius equal to $b$ about point $C$ as center.

Therefore, point $A$ is a common point of these two circles.

We now know how to perform the construction of the triangle.

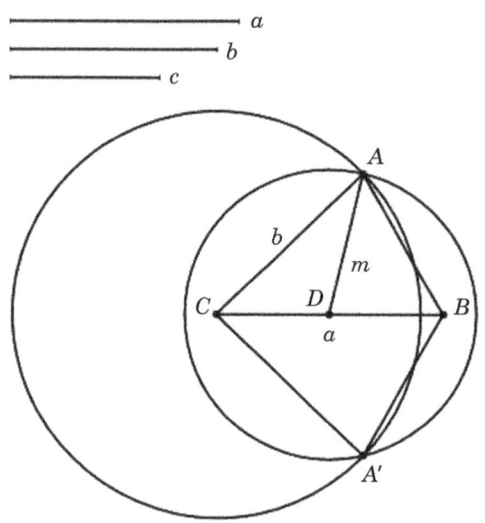

**Figure 3**

On a line (figure 3) we place the segment $BC = a$, [and] we find the point $D$ that is the midpoint of segment $BC$ ($CD = a/2$). About point $D$ we draw a circle with radius $m$; and about point $C$ we draw a circle with radius $b$. The points $A$ and $A'$ of intersection of these circles we connect with points $C$ and $B$. We have obtained two triangles: $ABC$ and $A'BC$.

Prove that each of these triangles has two edges $a$ and $b$ and median $m$.

Prove that $\triangle ABC$ and $\triangle A'BC$ are congruent.

But can you always perform this construction?

The following cases can arise: the circles drawn about points $C$ and $D$
   (1) have no points in common,
   (2) have only one point in common (are **tangent**),
   (3) have two points in common (they **intersect**).

If these circles did not intersect or were tangent, we could not construct the triangle.

Since we know what condition must be satisfied so that the two circles intersect, we can conclude that the desired triangle can be constructed if and only if the segment $CD = a/2$ is less than the sum of segments $b$ and $m$ and greater than their difference.

Many constructions amount to determining a point of intersection of two circles, then proceeding as in the previous problem.

**Exercises.** 1. Choose any two points $A$ and $B$ and find a point whose distance from point $A$ equals the segment $AB$, while the distance from point $B$ is equal to an arbitrarily chosen segment $a$.

Consider the cases in which (1) $a < 2AB$, (2) $a = 2AB$, (3) $a > 2AB$.

2. Choose in the plane any three points $A$, $B$, and $C$. Find a point whose distance from point $A$ is equal to the segment $AC$ and [whose] distance from point $B$ is equal to segment $BC$.

3. Construct a parallelogram, given an edge and the two diagonals.

4. Construct a rhombus, given an edge and a diagonal.

## § 3. Relative Positions of a Circle and a Line Lying in a Plane

We know that the distance from a point $M$ to a line $PQ$ (figure 4) is the segment $MN$ constructed from point $M$ perpendicular to $PQ$.

If we are given a circle and a line, one and only one of three cases must then occur (figure 5): the distance from the line to the center of the circle is (1) greater than the radius, (2) equal to the radius, (3) less than the radius.

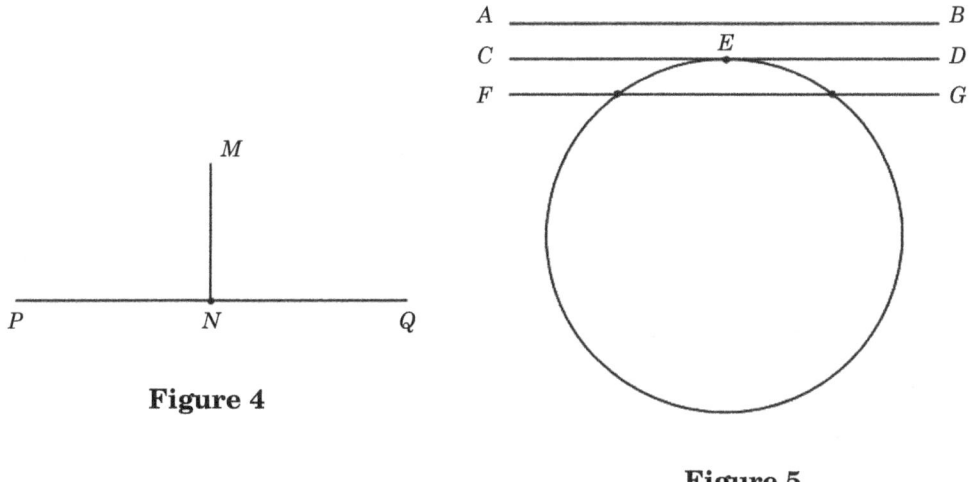

**Figure 4**

**Figure 5**

Now we prove the following three theorems. [Proofs follow the three statements.]

**Theorem 1. If the distance from a line to the center of a circle is greater than a radius, the line does not have any points in common with the circle.**

**Theorem 2. If the distance from a line to the center of a circle is equal to a radius, the line has one point in common with the circle.**

**Theorem 3. If the distance from a line to the center of a circle is less than the radius, the line has two points in common with the circle.**

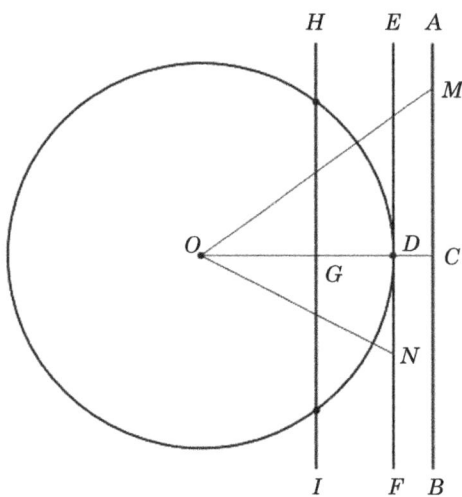

**Figure 6**

(1) Assumption: the distance from point $O$ to the line $AB$ is greater than the radius. We shall prove that every point of the line $AB$ lies exterior to the circle (figure 6).

Point $C$ lies exterior to the circle, since $OC > OD$. Any other point $M$ of the line $AB$ is farther from $O$ than $C$, because $OM > OC$ (the hypotenuse [of a right triangle] is greater than a leg); thus, all points of the line $AB$ are exterior to the circle, and therefore the line has no point in common with the circle.

(2) Assumption: the distance from point $O$ to the line $EF$ is equal to the radius $r$.

Therefore, when the perpendicular is dropped from point $O$ to $EF$, the segment $OD$ of this perpendicular is equal to $r$, and thus the point $D$ lies on the circle.

We shall prove that other than point $D$ the line $EF$ has no point in common with the circle. Let us take on line $EF$ any point $N$ different from $D$, [so that] $ON > OD$, [and] therefore $N$ lies outside the circle. Thus all points of the line $EF$ except $D$ lie in the exterior of the circle, and the line has only one point in common with the circle. In this case we say that the line $EF$ is **tangent** to the circle.

(3) Assumption: the distance from the line $HI$ to $O$ is less than the radius $r$. We construct $OG \perp HI$, so that $OG < r$; therefore $G$ lies in the interior of the circle.

We know from the preceding that a line passing through an interior point of a circle intersects the circle at two points; thus, $HI$ intersects the circle at two points.

We call a line that intersects a circle at two points a **secant**.

The question arises whether the theorems converse to the previous [ones] are true. We know that an assertion converse to a true one is not always also true. For example, the theorem "if two angles are vertical, then they are equal" is true, but the converse of this theorem, "if two angles are equal, then they are vertical," is false.

Let us examine whether the following statement is true:

**Assertion converse to theorem 1. If a line does not have any point in common with a circle, then the distance from the line to the center of the circle is greater than the radius.**

Assumption: the line does not have any points in common with the circle. [Desired] conclusion: the distance from the line to the center of the circle is greater than the radius.

There are only three possible cases concerning the distance from the line to the center of the circle: specifically, that this distance be (1) greater than the radius, (2) equal to the radius, (3) less than the radius. One and only one of these cases must occur.

Let us suppose that the distance from the line to the center of the circle were equal to the radius; then on the basis of theorem 2 the line would have one point in common with the circle, which is contrary to the assumption of our theorem. Similarly, if the distance from the line to the center of the circle were less than the radius, then on the basis of theorem 3 the circle would have two points in common with the line, which again contradicts the assumption of our theorem.

We have proved that the distance from the line to the center of the circle cannot be equal to the radius or less than the radius; therefore, it is greater than the radius. In this way the theorem has been proved.

Form the converses of the remaining two theorems, 2 and 3, and prove them by the method of reduction to absurdity.[10]

We notice that for the proof of each of the theorems converse to theorems 1, 2, and 3, it is sufficient to prove that the assumptions of theorems 1, 2, and 3 exhaust all possible cases—that is, one and only one of them must occur—and moreover the conclusions of theorems 1, 2, and 3 mutually exclude each other.

Thus, if we have several theorems whose assumptions exhaust all possible cases, and the conclusions following from them mutually exclude each other, then the assertions converse to the previous [ones] are true and can be proved by the method of reduction to absurdity.[11]

We call a system of theorems satisfying the above conditions a **closed system of theorems**.

The three previous theorems, which form a closed system of theorems, we can write in abbreviated form (we denote the distance from the line to the center of the circle by $d$, and the radius by $r$):

$$\text{If } \begin{cases} (1) \ d > r, \\ (2) \ d = r, \\ (3) \ d < r, \end{cases} \text{then the line and the circle } \begin{cases} (1) \ \text{have no points in common,} \\ (2) \ \text{have one point in common,} \\ (3) \ \text{have two points in common.} \end{cases}$$

---

[10] [The authors used Polish words meaning *reductio ad absurdum*.]

[11] [Tarski incorporated text similar to this in §50 of his well-known 1936 logic text. There, he called the principle stated in this paragraph *Hauber's law*. The assumptions of the several theorems need not be mutually exclusive as required here, but merely exhaustive.]

In this way, write down the converse theorems.

Last year[12] we encountered a closed system consisting of five theorems. Denoting by $d$ the distance between the centers of two circles, [and] by $r$ and $r'$ their radii ($r > r'$), we write these five theorems as follows:

If 
$\begin{cases} (1)\ d > r + r', \\ (2)\ d = r + r', \\ (3)\ r - r' < d < r + r', \\ (4)\ d = r - r', \\ (5)\ d < r - r', \end{cases}$ then the circles $\begin{cases} (1)\ \text{have no points in common—moreover, one lies in the exterior of the other;} \\ (2)\ \text{are externally tangent;} \\ (3)\ \text{have two points in common;} \\ (4)\ \text{are internally tangent;} \\ (5)\ \text{have no points in common—moreover, one lies in the interior of the other.} \end{cases}$

Formulate and write the converse theorems in this way.

In the future, when we prove that some theorems form a closed system, we will not prove the converse theorems separately.

From the three theorems previously proved follow [some] corollaries.

**Corollary 1.** The line $AB$ (figure 7) perpendicular to a radius $OC$ and passing through its endpoint $C$ is tangent to the circle. (The distance to $AB$ from $O$ is equal to the radius.)

This property gives us the ability to construct the tangent to a circle passing through a point on the circle.

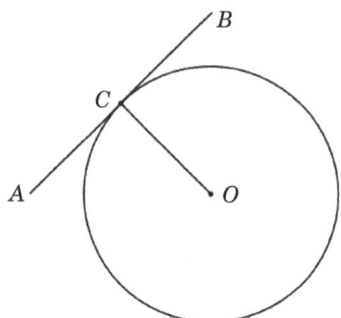

**Figure 7**

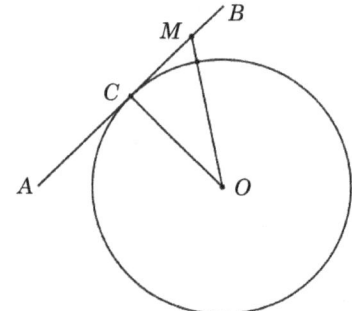

**Figure 8**

---

[12] [See Konarski 1934, §28, 44–49.]

**Corollary 2.** If through a point of contact $C$ (figure 8) of a line $AB$ with a circle we pass a radius $OC$, then that is perpendicular to $AB$.

*Proof.* If $OC$ were not perpendicular to $AB$, then we could drop $OM \perp AB$; since $AB$ is tangent, its distance from $O$ is equal to the radius $r$; therefore, $OM = r$. Point $M$ lies thus on the circle. Since $AB$ and the circle have only one point in common, point $M$ coincides with $C$, $OM$ coincides with $OC$, [and] segment $OC$ is perpendicular to $AB$.

**Exercises.** 1. Draw an arbitrary line and a point $A$ not lying on it. Choose any segment $a$ and locate a point on the line distant from point $A$ by the segment $a$. Does there exist such a point for all segments $a$, of any length?
What should the length of segment $a$ be so that (1) there does not exist such a point, (2) there exists only one such point, (3) there exist two such points?

2. Draw any line and choose a point $O$ not lying on it. Trace the circle[13] with center $O$ [and] tangent to that line.

3. On a line, choose any point $A$ and with radius equal to a given segment trace circles tangent to that line at point $A$.

4. Draw any line and a circle having no point in common with it. Pass a tangent to the circle (a) parallel, (b) perpendicular, to this line.

5. Draw a circle and a line, choose any angle, and pass a tangent to the given circle, forming with the given line an angle equal to the given [angle].

6. With a point $A$ as center, draw a circle tangent to a given circle (a) externally, (b) internally.

7. Trace a circle with any radius and a line having no point in common with it. On the circle choose any point $A$. Trace a circle with center on this line and tangent to the given circle at point $A$. For which locations of point $A$ is the problem impossible?

8. On any line, choose any point $A$. With point $A$ as center, draw a circle with radius 3 cm. Draw circles with 2-cm radii whose centers are on the line and that have one point in common with this circle. How many such circles can you draw?

## § 4. Construction Problems[14]

We have already solved several construction problems, using only **compass and straightedge**. In the future we will be performing geometric constructions using only these two tools. (For drawing parallels and perpendiculars, we can use set squares, since both of these constructions can also be performed using compass and straightedge.)

---

[13] [Here the original used the word *koło*, which usually meant *disk*.]
[14] [The original titles of sections 2 and 4 were the same.]

If besides compass and straightedge we are going to use still other tools, that will be specially discussed in the terms of the problems.

**Problem 1.** Construct a triangle, given two edges $a$ and $b$ and the angle $\beta$ lying opposite $b$, the smaller of them (figure 9). For what lengths of edge $b$ is the construction possible?

Let $ABC$ be the desired triangle (figure 9). We can draw the edge $BC$ and angle $\beta$ at once,[15] still seeking vertex $A$, which is located on line $BM$ (figure 10) and which is distant from point $C$ by segment $b$ [and] thus is a point common to line $BM$ and the circle with center $C$ and radius $b$.

From point $C$ we drop the line $CD$ perpendicular to $BM$.

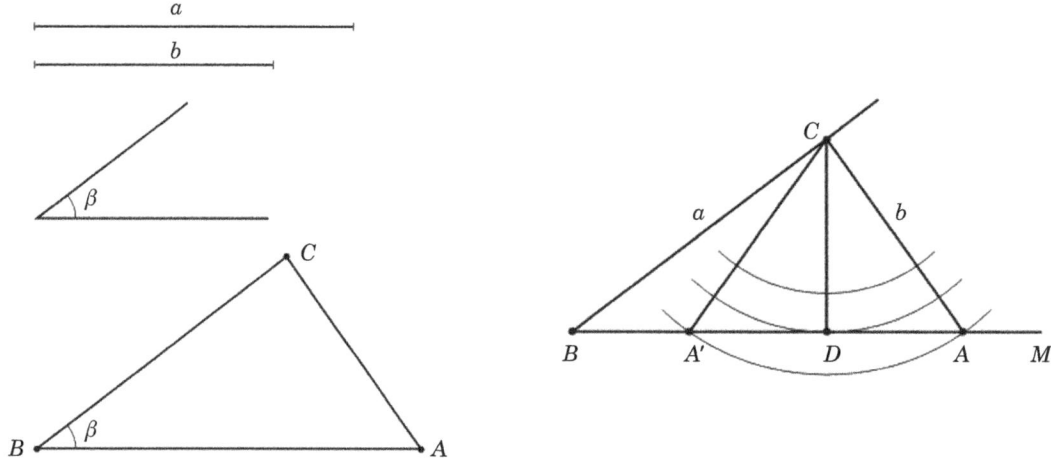

**Figure 9**            **Figure 10**

The following cases may occur.

(1) $b > CD$: then a circle with radius $b$ and center $C$ intersects $BM$ in points $A$ and $A'$. Therefore we have two triangles $ABC$ and $A'BC$ satisfying the conditions of the problem, and there are two solutions.

(2) $b = CD$: then the circle is tangent to $BM$, and we have only one triangle $BCD$.

(3) $b < CD$: the circle and line $BM$ have no point in common—the problem has no solution.

---

[15] [The original, in error, specified angle $B$ in this sentence.]

### 13.4 §4 Construction Problems

Solving a construction problem often comes down to determining the point of intersection of a given line $AB$ with a circle drawn about a given point $M$ as center.

Then there can occur three cases:

(1) the radius of the circle is less than the distance from point $M$ to $AB$, so that there are no points of intersection of the circle with $AB$;

(2) the radius of the circle is equal to the distance from $M$ to $AB$, so that the circle is tangent and there exists one point common to the circle and line $AB$;

(3) the radius of the circle is greater than the distance from $M$ to $AB$, so that there exist two points of intersection of the circle with $AB$.

If a construction problem is more complicated, then we proceed in the following way.

Assuming that the problem is solved, we draw by hand a figure similar to the one sought and suppose that it satisfies the requirements of the problem. Next, we try to find a relationship between the givens and the elements sought that allows us to reduce the solution of the given problem to [that of] another problem that can we can already solve. In addition, it is often useful to construct auxiliary circles or lines.

Such a procedure we call the **analysis** of the construction problem.

For example, we will solve the following problem.

**Problem 2.** Construct a triangle, given two edges $a$ and $b$ and the median $m$ with respect to the third edge.

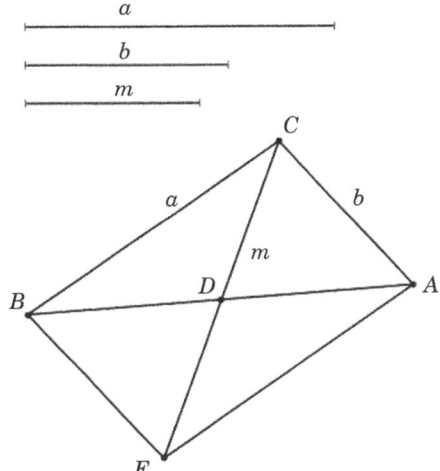

Figure 11

Let $\triangle ABC$ (figure 11) be the desired triangle and $CD$, the median to edge $AB$, [so that] $BD = DA$. We find that [by] extending $CD$, measuring off $DE = CD$, and joining successively the points $A$, $C$, $B$, $E$, $A$, we obtain the quadrilateral $ACBE$, which is a parallelogram because [its] diagonals divide each other into halves. Triangle $EBC$, which is half of the parallelogram, we can construct from three edges: $CB = a$, $BE = b$, $CE = 2m$. We can therefore construct the desired triangle $BCA$.

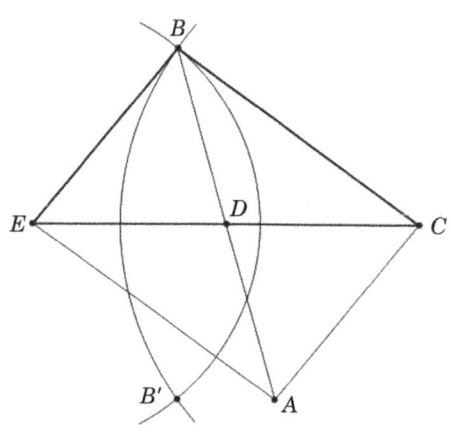

**Figure 12**

Now we perform the **construction** and describe it. We construct triangle $EBC$ with edges $a$, $b$, $2m$. With this as goal, we measure off (figure 12) $ED = DC = m$. About points $E$ and $C$ we draw circles with radii equal to $b$ and $a$, respectively.

These circles intersect if and only if

$$a - b < 2m < a + b.\ ^{[16]}$$

We have two intersection points $B$ and $B'$ that satisfy this condition, and construct two triangles $EBC$ and $EB'C$, "oppositely" equal—that is, congruent after reversing one of them. We draw only one of these triangles, $EBC$. We construct $BD$, extend $BD$, and on the extension we measure off $DA = BD$; joining $A$ with $C$, we obtain $\triangle ABC$, as desired.

Now it is necessary to **prove** that the figure as drawn meets the conditions of the problem. This part of the reasoning is called the **proof**. From the construction it follows that $BC = a$, $BD = DA$, $DC = m$.

Since $\triangle DBE = \triangle DAC$ ($DB = DA$, $DC = DE$, and angles $ADC$ and $BDE$ are equal), [it follows that]

$$AC = BE = b.$$

We see that $\triangle ABC$ satisfies the conditions of the problem. We still have to investigate whether the problem is always solvable; if not always, it is necessary to investigate the conditions under which solution is possible, and the number of solutions. This part of the reasoning we call the **determination** of the conditions for the possibility of solutions, and their number.

In our problem the construction of triangle $ABC$ is possible if and only if it is possible to construct the auxillary triangle $EBC$, and thus when the conditions (with $a > b$)

$$a - b < 2m < a + b$$

are satisfied.

---

[16] [According to figure 12 it is assumed that $a > b$. This result was part of the previous year's course: see Łomnicki 1934, §28, page 47.]

When these conditions are satisfied it is possible to construct two congruent triangles that satisfy the conditions of the problem.[17]

As we have explained in this example, a construction problem consists of the following parts: (1) analysis, (2) construction, (3) proof, (4) determination of conditions for possibility and of the number of solutions.

Often the analysis is so short that we omit it and proceed at once to construction.

**Exercises.** 1. Construct a right triangle given a leg $a$ and the hypotenuse $b$. Can the problem be solved for all $a$ and $b$?

2. Construct a triangle given two edges and the projection of the shorter of these onto the longer. Investigate the possibility of the conditions of the problem.

3. Construct a triangle given an edge, an angle, and the median to the one remaining edge that does not pass through [the vertex of] this angle.
Choose the given segments so that there exist (a) two solutions, (b) one, [and] (c) so that there should be no solution.

4. Construct a right trapezoid given one of the bases $a$, edge $b$ perpendicular to both bases, and edge $c$. Choose the second edge so that (a) there are two solutions, (b) there is one solution, (c) there is no solution at all.
Hint. The construction and the investigation of the existence and number of solutions is explained by figure 13.

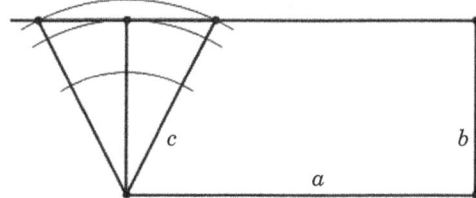

**Figure 13**

5. Construct a triangle given a median $a$ to an edge, and each of the angles that the median forms with the remaining edges.
Hint. Perform the analysis of the problem by extending the median as in a problem solved earlier.

6. In the interior of an arbitrary concave angle[18] choose any point $A$, and pass through $A$ the line whose segment contained between the arms of the angle is bisected by the point $A$.
Hint. The problem comes down to the construction of a parallelogram, two of whose adjacent edges lie on the arms of the given angle and whose diagonals intersect at point $A$.

---

[17] [This sentence is puzzling: infinitely many triangles, all congruent, satisfy the requirements. The authors' term for *congruent* was *przystający*. The meaning is unclear even if by that they meant *indirectly* congruent, referring to a solution obtained by choosing $B'$ in place of $B$, for it is possible to construct from any solution an indirectly congruent one.]

[18] [A *concave* (*wklęsły*) angle must be one whose measure $m$ satisfies $0° < m < 180°$.]

7. Choose any three segments $a$, $b$, and $c$. Draw three circles, one with radius $a$, the second, $b$, and the third, $c$, so that each of them should be externally tangent to the other two.

Is the construction possible for all $a$, $b$, and $c$?

8. Construct a right triangle given the hypotenuse and the sum of the legs.

Hint: extend a leg beyond the vertex of the right angle and measure off on the extension a segment equal to the second leg.

9. Draw a right triangle given an acute angle and (a) the sum of the legs, (b) the difference of the legs.

Hint. Perform an analysis and extend one of the legs beyond the vertex of the right angle by a distance equal to the other leg, [and] join the endpoint of this extension with the vertex of the right angle.[19] What angle results?

10. Construct a trapezoid given its two bases and the two diagonals.

Hint. Construct a triangle with a base equal to the sum of the trapezoid's bases and edges equal to the diagonals.

11. Construct a square given the sum of a diagonal and an edge.

Hint. Draw a square $ABCD$, construct a diagonal, for example $AC$, extend edge $AB$ beyond[20] point $A$ by the distance $AE = AC$, construct segment $EC$, [and] prove that $\sphericalangle ACE = \sphericalangle AEC = 22°30'$.

12. Draw two parallel lines $AB$ and $CD$, choose a point $E$, and on the line $CD$ mark a point equidistant from point $E$ and the line $AB$.[21]

Choose the point $E$ so that (a) there exist two such points, (b) one such point, (c) no such point.

13. Construct a parallelogram given the two diagonals and the angle included between them.

14. Construct a trapezoid given its four edges $a$, $b$, $c$, $d$ ($a$ and $c$ are the bases).

Hint. Construct a triangle one edge of which equals the difference between the bases of the trapezoid, the others being equal to $b$ and $d$.

---

[19] [The hint must be mistaken: try joining the endpoint with another vertex instead.]

[20] [Segment $AB$ was specified incorrectly as $AC$ in the original.]

[21] [This occurrence of $AB$ was printed incorrectly as $CD$ in the original.]

### 13.5 Summary of §5–§24

This box describes the sections of the *Geometrja* text that lie between the previous and next translated sections.

Some of these sections continue the review and presentation of very elementary topics begun in the previous geometry course and in §1–§4:

| | |
|---|---|
| §5 tangents to a circle | §11–§13 parallel lines |
| §6 perpendicular bisector | §24 rational arithmetic |
| §7 construction problems | |

In another group of sections these concepts are applied in somewhat more involved situations:

§8 –§10 circumcircle, angle bisectors, incircle
§14–§19 circular arcs and chords and related angles
§20–§21 construction problems involving these

Two sections can be regarded as intermediate goals for that development:

§22 inscribed and circumscribed quadrilaterals
§23 regular polygons

These intervening sections present the standard theorems on their subjects. The abundance of exercises is impressive. The simplicity of language is startling, especially when contrasted with that of Enriques and Amaldi [1903] 1916. In these sections the authors of *Geometrja* continued their practice of discussing logical principles in the context of geometry. For example, the use of necessary and sufficient conditions is discussed in detail in §6 in connection with equidistance from two points and incidence with their perpendicular bisector, the notions of converse and contrapositive are applied to organize §22, and conditions for existence and uniqueness of solutions to construction problems are a recurring theme.

Section §24 introduces the notion of ratio of commensurable segments, and begins to relate operations on segments and operations on ratios.

# PART THREE

# Measuring Segments and Areas

## § 25. Incommensurable Segments—Irrational Numbers

So far, we have been considering commensurable segments. We shall now prove that there are segments that are not commensurable.

We shall prove that [these] segments are not commensurable: the edge $AC = b$ of a square, and its diagonal $BC = a$ (figure 75).

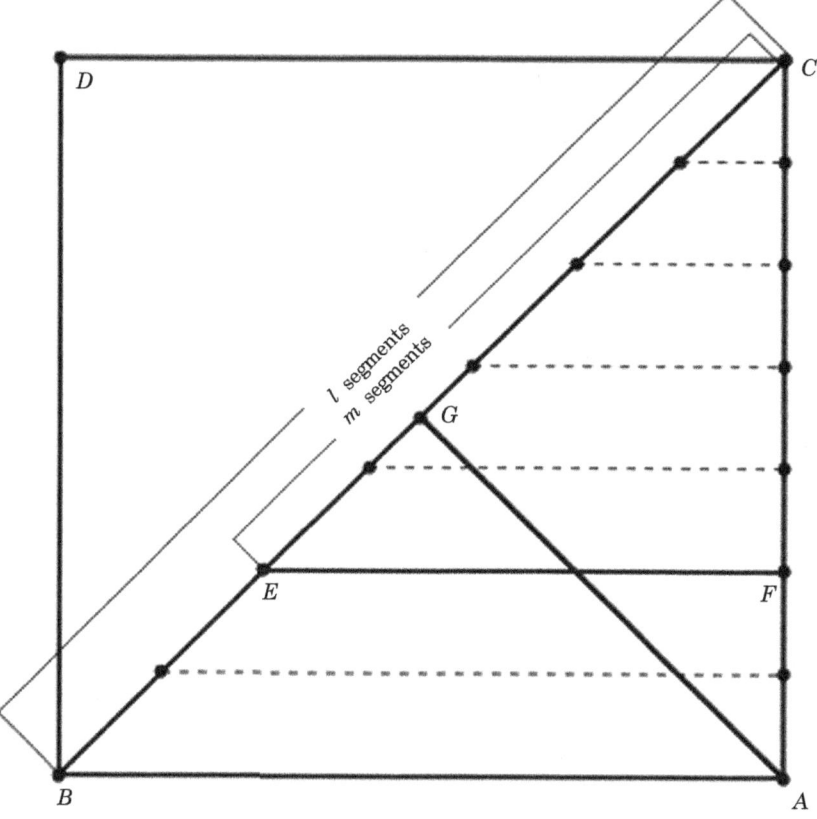

**Figure 75**

We suppose that $a$ and $b$ are commensurable, then let

$$a = \frac{l}{m}b.$$

We divide $BC$ into $l$ equal parts. Segment $CE$, containing $m$ such parts, is equal to $b$:

$$EC = b.$$

Through the subdivision points of $BC$ we draw parallels to $AB$. These lines divide $CA$ into $l$ equal parts (recall the method for dividing a segment into equal parts). [Let] $CF$ include $m$ of those parts: thus,

$$CF = \frac{m}{l} AC. \tag{1}$$

From point $A$ we drop $AG \perp BC$; therefore

$$BG = GC = \tfrac{1}{2}\, a.$$

Right triangles $CFE$ and $CGA$ are congruent, because $AC = EC = b$ [and] $\sphericalangle ECA$ is common [to them]; therefore, $CF = CG = \tfrac{1}{2}\,a$. By virtue of (1) we would have

$$\tfrac{1}{2}\, a = \frac{m}{l} b;$$

substituting $a = \dfrac{l}{m} b$, we would obtain

$$\frac{l}{2m} b = \frac{m}{l} b, \quad \text{or}$$

$$\left(\frac{l}{m}\right)^2 = 2. \tag{2}$$

Let us suppose that $l/m$ is a fraction in irreducible form. We shall prove that equation (2) is impossible.

From (2) it would follow that $l^2/m^2 = 2$, [and] $l^2 = 2m^2$.

Since the right-hand side of this equation is divisible by 2, the left side would also be divisible by 2, and hence $l$ would be divisible by 2. Then $l = 2l_1$ for some whole number $l_1$,

$$l^2 = 4 l_1^2, \quad \text{and thus}$$
$$4 l_1^2 = 2 m^2, \quad 2 l_1^2 = m^2.$$

The left side of the last equation is divisible by 2; therefore, the right [side] would also be divisible by 2, and thus

$$m = 2m_1, \text{ for some whole number } m_1.$$

We see that the fraction $l/m$ can be reduced by 2, which would contradict the assumption that $l/m$ should be an irreducible fraction.

Thus, the supposition that $BC$ and $AC$ are commensurable led us to a contradiction. Therefore, $BC$ and $AC$ are not commensurable.

We can provide many examples of pairs of segments that are not commensurable.

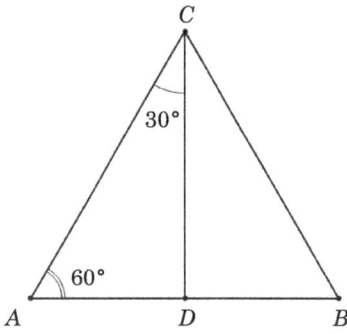

**Figure 76**

You can prove, for example, that in an equilateral triangle (figure 76) the edge $AB$ and the altitude $CD$ are not commensurable.[22]

Segments that are not commensurable are called **incommensurable**.

Let us take segment $AC$ for the unit length (figure 75 [page 292]). Because segments $BC$ and $AC$ are incommensurable, there can be no relationship between these segments [in the sense of §24].

We assume that the length $BC$ with respect to the unit $AC$ is a number. But it cannot be a whole or fractional number (that is, a ratio of whole numbers), because this would mean that $AC$ and $BC$ would be commensurable. Therefore, this is a new kind of number. Previously we did not know about such numbers, and only now do we become familiar with them.

Whole and fractional numbers, which are measures of segments commensurable with the unit, are called **rational** numbers; and those numbers that are measures of segments that are incommensurable with the unit, we call **irrational**.

Taking $AC = 1$ decimeter (figure 75) as unit of length, construct a square as in figure 75, as accurately as you can, and measure its diagonal with the aid of a millimeter scale.

The length of the diagonal of this square, as we know, is an irrational number.

You will convince yourself that the desired length of the diagonal is

|   |   |   |   |
|---|---|---|---|
| greater than | 1 decimeter [dm] | and less than | 2 dm |
| " " | 14 centimeters [cm] = 1.4 dm | " " " | 15 cm = 1.5 dm |
| " " | 141 millimeters [mm] = 1.41 dm | " " " | 142 mm = 1.42 dm. |

With [just] the aid of a millimeter scale you cannot determine the length of the diagonal any more accurately. With more accurate measuring devices, you could more accurately identify bounds between which is included the length of the diagonal.

If we want the longer segment to have greater length,[23] it is necessary to have accepted that the number which is the length of the diagonal (we denote this number by $a$) is larger than the numbers $1, 1.4, 1.41$, and less than the numbers $2, 1.5, 1.42$ —or

---

[22] [In this sentence in the original, $BC$ was misprinted for $CD$.]

[23] [The present editors are puzzled by this phrase—*Jeżeli chcemy, aby większy odcinek miał większą długość* in the original.]

$$1 < \alpha < 2$$
$$1.4 < \alpha < 1.5$$
$$1.41 < \alpha < 1.42.$$

Similarly, having constructed an equilateral triangle whose height is equal to 1 dm, we would find that the number $\beta$ (beta) that is the length of the edge of the triangle is contained within the following bounds:

$$1 < \beta < 2$$
$$1.1 < \beta < 1.2$$
$$1.15 < \beta < 1.16.$$

Carry out the indicated construction and convince yourself of it.

The numbers that are lengths of segments incommensurable with the unit are called the positive irrational numbers. Alongside them we consider the negative irrational numbers: namely, we assume that to any positive irrational number, such as $\gamma$ (gamma), corresponds the number $-\gamma$ (minus gamma). From now on, speaking of numbers, we shall have in mind both positive and negative numbers, rational or irrational.

We assume that irrational numbers can be compared with each other [according to size] and with rational numbers, and that the four arithmetic operations can be performed on irrational numbers; moreover, these operations are governed by the same laws that govern [corresponding] operations on the rational numbers.

Later we will learn how, in practice, to compare irrational numbers and perform operations on them.

**Exercises.** 1. We have shown that there is no rational number $l/m$ such that $(l/m)^2 = 2$. Prove in a similar way that there is no rational number $l/m$ such that $(l/m)^3 = 2$ or $(l/m)^3 = 4$.

2. Let $\beta$ be an arbitrary irrational number and $\gamma$, any rational number. Prove that $\beta + \gamma$ and $\beta - \gamma$ are irrational numbers. Similarly, if $\gamma \neq 0$, then $\beta \cdot \gamma$, $\beta/\gamma$, and $\gamma/\beta$ must be irrational. *Hint*: let $\beta + \gamma = \delta$, so that $\beta = \delta - \gamma$. What could be said about $\beta$, if not only $\gamma$ but also $\delta$ should be a rational number?

3. On the basis of the previous exercise, and knowing that the number $\alpha$ that was introduced previously is irrational, prove that $\alpha + 1$, $\alpha + 2$, ..., $2 + \alpha$, $3 + \alpha$, $2\alpha$, $2/3\,\alpha$, $1/\alpha$, $2/\alpha$, ... are irrational numbers. How many different irrational numbers are there? From among these indicated irrational numbers choose two whose sum, difference, product, and quotient, respectively, are rational numbers.[24]

---

[24] [The exercise was evidently garbled in the original: no two numbers in that list have a rational sum.]

## § 26. Measuring Incommensurable Segments with a Unit

We have established certain laws regarding segments whose lengths are commensurable with the unit. Now we suppose that these laws remain in force, but in this case the segment that we measure is incommensurable with the unit; thus we assume what follows.

Having selected an arbitrary segment as the unit of length, [we may assign] to each segment a number, called the length of the segment with respect to the given unit. Moreover, the following laws are satisfied.

**1.** To every segment corresponds exactly one positive number (rational or irrational), which is its length; we also call this number the ratio of that segment to the unit.

**2.** Every positive number corresponds to some segment, whose length is the given number.

**3.** The length of the segment chosen as unit is the number 1.

**4.** Equal segments have the same length.

**5.** The length of the sum of two segments is equal to the sum of the lengths of these segments.

**6.** The ratio of any two segments equals the ratio of their lengths.[25]

From the above assumptions follow [some] conclusions.

(1) The larger segment has the greater length. Rather than compare segments we can compare their lengths.

(2) The length of the difference of two segments is equal to the difference of their lengths.

Let us compare the two irrational numbers $\alpha$ and $\beta$ [on page 295] with each other: because $\beta < 1.2$ [and] $1.4 < \alpha$ it follows that $\beta < 1.2 < 1.4 < \alpha$ —that is, $\beta < \alpha$.

**Exercises.** 1. Prove that if we reduce the unit of length $n$-fold, the number expressing the length of the segment will increase $n$-fold.

Hint. Let $a$ be [the length of] an arbitrary segment and $b$, the length using the reduced unit; we apply condition 6.

---

[25] [The ratio of two commensurable segments was already defined in §24. Condition 6 connects length with that concept, and defines the ratios of incommensurable segments.]

2. We know that equal segments have equal length, and of two unequal segments the longer has greater length and the shorter, less. Formulate the converse theorems: are they true, and why?

Hint: a closed system of theorems. [See §3, page 283.]

## § 27. Rational Approximations of Irrational Numbers

### Operations on Irrational Numbers

In §26 we found decimal numbers between which lie irrational numbers $\alpha$ and $\beta$. We can find rational numbers arbitrarily close to any irrational number. These numbers are called **approximations** of the irrational numbers.

We establish some properties of irrational numbers. The following properties stem from theorems about incommensurable segments. Those results are presented at the end of the present section.

**Property 1. For each positive irrational number $\gamma$, given a whole number $m$ we can find a whole number $l$ such that $\gamma$ is contained between $l/m$ and $(l + 1)/m$.**

The numbers $l/m$ and $(l+1)/m$ are called the approximations of the number $\gamma$ with precision $1/m$ — $l/m$ **from below**, $(l+1)/m$ **from above**.

We have already learned about decimal numbers, which are **decimal approximations** of the numbers $\alpha$ and $\beta$.

**Property 2. If for each value of $m$ two irrational numbers $\gamma$ and $\delta$ have the same approximations**

$$\frac{l}{m} \quad \text{and} \quad \frac{l+1}{m}$$

**with precision $1/m$, then the numbers are equal to each other: $\gamma = \delta$.**

In practical calculations we can replace irrational numbers by their approximations and perform appropriate operations on the approximations. The results of the operations performed on approximations are approximate. If we add approximations of the numbers $\alpha$ and $\beta$ from below, we obtain an approximate value of the sum $\alpha + \beta$, also from below; and adding approximations of the numbers $\alpha$ and $\beta$ from above, we obtain an overestimate of the sum $\alpha + \beta$. For example,

$$1.41 + 1.15 < \alpha + \beta < 1.42 + 1.16$$
$$2.56 < \alpha + \beta < 2.58$$
$$2\,28/50 < \alpha + \beta < 2\,29/50.$$

In this way, we found approximations of the sum $\alpha + \beta$ with precision $1/50$.

Subtracting from an approximation of the number $\alpha$ from below an approximation to the number $\beta$ from above, we obtain an approximation of the difference $\alpha - \beta$ from below; subtracting from an approximation of the number $\alpha$ from above an approximation of the number $\beta$ from below, we obtain an approximation of the difference $\alpha - \beta$ from above:

$$1.41 - 1.16 < \alpha - \beta < 1.42 - 1.15$$
$$0.25 < \alpha - \beta < 0.27.$$

For the product $\alpha\beta$ and the quotient $\alpha:\beta$ we can write

$$1.41 \cdot 1.15 < \alpha\beta < 1.42 \cdot 1.16$$

$$\frac{1.41}{1.16} < \frac{\alpha}{\beta} < \frac{1.42}{1.15}.$$

**Exercises.** 1. As unit of length we take an edge of a square. Approximate the lengths of the following segments from above and below: (1) half of the diagonal of the square, (2) the sum of a diagonal and two edges of the square, (3) the difference of a diagonal and an edge of the square.

2. As unit of length we take half of an edge of the equilateral triangle in figure 76 [on page 294]. With precision one tenth, approximate the lengths of the radius of the inscribed circle of this triangle and the radius of the circumscribed circle.

---

We can deduce property 1 of irrational numbers from the following theorem.

**Theorem 1. For each segment $AB$ incommensurable with the unit $CD$ two segments commensurate with $CD$ can be found, one of which is longer and one shorter than $AB$, and whose difference is equal to $1/m\ CD$, where $m$ is any whole number.**

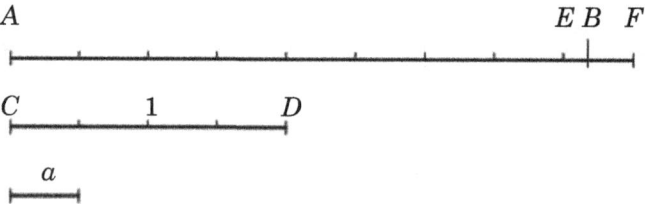

**Figure 77**

## 13.8 §27 Rational Approximations of Irrational Numbers

We divide segment $CD$ (figure 77) into $m$ equal parts [and] place one part [thus] obtained—segment $a$ —in segment $AB$, [starting] from point $A$, as many times as possible. We know from last year that by repeating this placement enough times, we will cross point $B$. Suppose that we place segment $a$ [there] $l$ times without reaching point $B$, but cross $B$ on the $(l+1)$st time, so that we can write

$$\frac{l}{m} \cdot CD < AB < \frac{l+1}{m} \cdot CD \quad \dots\dots\dots\dots\dots\dots\dots (1)$$

$$AE = \frac{l}{m} \cdot CD \qquad AF = \frac{l+1}{m} \cdot CD,$$

[and] therefore

$$AE < AB < AF.$$

Thus, $AE$ and $AF$ are the desired segments.[26]

Let the ratio $AB/CD =$ the irrational number $\gamma$; from (1) follows

$$\frac{l}{m} < \gamma < \frac{l+1}{m}.$$

From this, property 1 of irrational numbers follows directly.

The proof of property 2 is based on the following theorem.

**Theorem 2. Given two unequal segments incommensurable with the unit $CD$, we can find a segment, commensurable with the unit, that is smaller than one of them and larger than the other.**

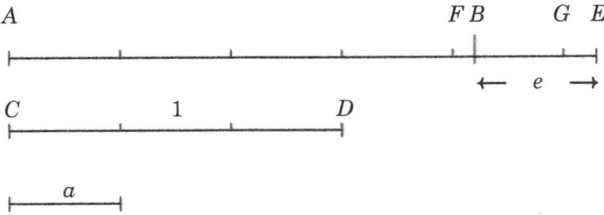

**Figure 78**

We place these segments, $AB$ and $AE$, as indicated in figure 78. We place segment $BE = e$ in $CD$, [starting] from point $C$, [just] as many times as needed to cross point $D$; suppose, to this end, it is necessary to place the segment $m$ times, so that

$$CD < me.$$

---

[26] [This and the preceding four lines were out of order in the original.]

Dividing $CD$ and $me$ by $m$, we get

$$(1/m)CD < e;$$

thus, dividing $CD$ into $m$ equal parts, we obtain one such part $a$, smaller than $e$.

We place segment $a$ in $AB$, [starting] from point $A$, [just] as many times as needed to cross point $B$. Let $F$ be the end of the last segment before $B$, [and let] $G$ be the end of the next segment, containing $B$:

$$FG = a = FB + BG < e,$$

and thus, moreover,

$$BG < e.$$

Therefore, point $G$ lies between $B$ and $E$, and so segment $AG$ is the required segment.

Letting the segments $AB$ and $AE$ have measures $\gamma$ and $\delta$ (irrational numbers), we can say this:

**Given two unequal irrational numbers, we can find some rational number that is larger than one of them and smaller than the other**—that is, which is contained between these numbers; we denote this rational number by $w$. Thus,

$$\gamma < w < \delta.$$

We say that between two unequal irrational numbers we can can locate at least one rational number.

Now property 2 can be easily proved. Indeed, let us suppose that $\gamma \neq \delta$ —for instance, $\gamma < \delta$ —then we can locate between $\gamma$ and $\delta$ a rational number $k/n$:

$$\gamma < k/n < \delta.$$

In this case the approximations of the numbers $\gamma$ and $\delta$ with precision $1/n$ would be different, [since $k/n$ would be larger than the approximation of the number $\gamma$ from below with precision $1/n$ and smaller than the approximation of the number $\delta$ from above with precision $1/n$.] This[27] is contrary to the assumption that the approximations of numbers $\gamma$ and $\delta$ are the same, so $\gamma < \delta$ cannot hold; similarly, $\gamma > \delta$ cannot hold, and therefore $\gamma = \delta$.

---

[27] [The present editors provided the text in brackets to replace a confusing passage in the original.]

### 13.9 Summary of §28–§32

This box describes the sections of the *Geometrja* text that lie between the previous and next translated sections.

The theorem of Thales presented in §28 is not the familiar proposition that an angle inscribed in a semicircle is right, but rather the following one: if two pairs of distinct parallel lines intersect two other lines at points $A,B; C,D$ and $E,F; G,H$ respectively, then
$$\frac{AB}{CD} = \frac{EF}{GH}.$$

For the case of congruent $AB, CD$ this result was discussed earlier, in §11. In §28 the authors extended it easily to the commensurable case. For the incommensurable case they showed that for any positive integers $m$ and $n$,
$$\frac{m}{n} < \frac{AB}{CD} < \frac{m+1}{n} \quad \text{implies} \quad \frac{m}{n} < \frac{EF}{GH} < \frac{m+1}{n}.$$

By property 2 in §27 this yields the equality of the ratios.

As consequences, §28 derives the converse of Thales's theorem and the following result:

> Given a triangle $ABC$ and points $B' \neq A$ and $C' \neq A$ on rays $AB$ and $AC$ respectively, then $BC \parallel B'C'$ if and only if $AB/A'B' = AC/A'C'$; and in that case these ratios are also equal to $BC/B'C'$.

The next section, §29, defines similarity of polygons in general and derives from this result the familiar AA, SAS, and SSS similarity theorems for triangles. Those two sections together contain fifty-one exercises, which are supplemented with ten more construction problems in §30. Two practical applications are discussed in §31: they use similarity to measure indirectly some objects, portions of which are inaccessible. The next section, §32, discusses one more application, the pantograph, continues with eighteen more excellent exercises, and contains one additional result that is required for the area theory later in the book:

> If through corresponding vertices in two similar convex polygons we pass all the diagonals, then they will divide the polygons into the same number of triangles, respectively similar and identically arranged.

## § 33. Equivalent Polygons

If two polygons I and II (figure 106) have no interior points in common and polygon III consists of all points of polygons I and II and contains no other points, we call polygon III the sum of these polygons (components).

In figure 107 polygon III' [is] the sum of polygons I' and II'. Adding polygon 3 to the sum of polygons 1 and 2 (figure 108), we obtain the sum of three polygons. In a similar way we can obtain the sum of any number of polygons.[28]

If polygon $W$ is the sum of two polygons $W_1$ and $W_2$ then we call $W_1$ a difference of polygon $W$ and polygon $W_2$.

For each of the polygons $W_1$ and $W_2$, we say it is part of polygon $W$.

If a given polygon is the sum of two congruent polygons (figure 109) then each of these polygons we call a half of that polygon; and if it is the sum of $n$ congruent polygons [then each of these polygons we call] an $n$th part of that polygon (figure 110).[29]

Polygons that are sums of polygons congruent in pairs (figures 106 and 107) we call **equivalent**: for example, polygons III and III'.

Equivalent polygons, for example III and III', might not be congruent.

**Exercises.** 1. Represent any triangle as the sum of (1) two triangles, (2) a triangle and a quadrilateral, (3) two quadrilaterals, (4) a quadrilateral and a pentagon, (5) two pentagons.

2. Represent any triangle as a difference of (1) two triangles, (2) a triangle and a quadrilateral, (3) a quadrilateral and a pentagon.

3. Divide any rectangle (1) into two equal trapezoids, (2) into five equal rectangles, (3) into six equal triangles.

4. Divide any isosceles triangle into two, into four, and into eight equal triangles.

5. A square with edge 6 cm and a rectangle with edges 4 cm and 9 cm are given. Prove that these two quadrilaterals are equivalent, dividing each of them into three respectively equal rectangles.

---

[28] [The authors evidently regarded the associative and commutative laws for this addition operation as inappropriate for discussion in the third gimnazjum class.]

[29] [This sentence was garbled in the original; the authors may have intended to write "$1/n$ częścią" for "$n$th part."]

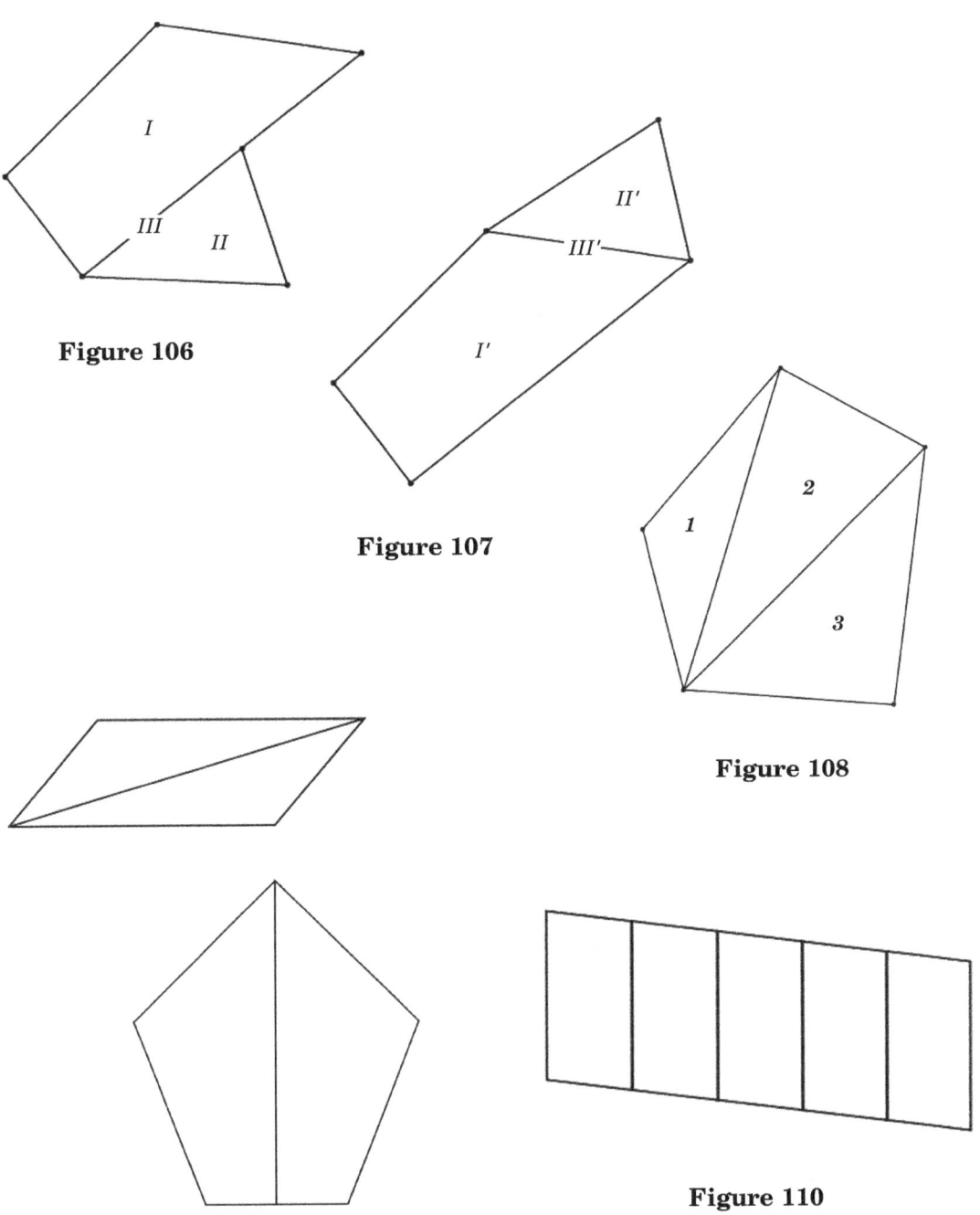

**Figure 106**

**Figure 107**

**Figure 108**

**Figure 109**

**Figure 110**

6. A square with edge 6 cm is given, as well as a rectangle with edges 4½ cm and 8 cm. Prove that these two quadrilaterals are equivalent, dividing each of them into four respectively equal rectangles.

7. A triangle $ABC$ is given. Construct a parallelogram equivalent to the given triangle, and having with it the common edge $BC$ and common angle $B$. Hint: let $D$ be the midpoint of $AB$, and $E$, the midpoint of $AC$; if we extend $DE$ by $EF = DE$, then $BCFD$ will be the desired parallelogram. Why?

8. A rectangle with edges 10 cm and 4 cm is given, as well as a parallelogram with edges of 10 cm and 8 cm and acute angle 30°. Show that these two quadrilaterals are equivalent. (Hint: place the quadrilaterals so that they have a common edge, and prove that the edges of the quadrilaterals parallel to the common edge must lie on one line.)

## § 34. On Measuring Areas of Polygons

We have already learned to measure segments: that is, to assign each segment a number called its measure or length. Not just segments can be measured, but also polygons: that means, assign each polygon a number called its measure or polygonal **area**.

With some polygon selected as the unit of area, each polygon can be assigned a number called the area of the given polygon with [respect to] the given unit.

Let us agree to adopt as unit the area of a square whose edge is equal to the unit length. And thus, if we take as unit the length 1 cm, 1 m, or the like, then we adopt as the unit area the area of the square with edge 1 cm, 1 m, or the like; we denote these units by 1 cm$^2$, 1 m$^2$, and the like.

It is necessary to remember this well while reading theorems that speak simultaneously of lengths of segments and areas of polygons.

### Area of a Rectangle

At the beginning, we shall deal with measuring the area of a rectangle. One of the edges of a rectangle we call the **base** and [one] perpendicular to this edge, the **altitude**.

The area of the unit square is equal to 1.

Consider a rectangle having edges 3 and 4 units in length (figure 111). This rectangle is the sum of 12 unit squares $(3 \cdot 4)$.

Let us suppose that

(1) congruent polygons have the same area;
(2) a polygon that is the sum of several polygons has area equal to the sum of areas of those polygons.

Then the area of that rectangle equals 12 —that is, $3 \cdot 4$.

The area (figure 112) that comprises $1/n$ of the unit square equals $1/n$, and the area of the rectangle that comprises $m/n$ parts of the unit square equals $m/n$.

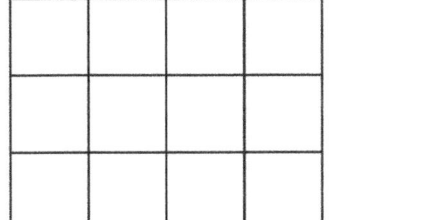
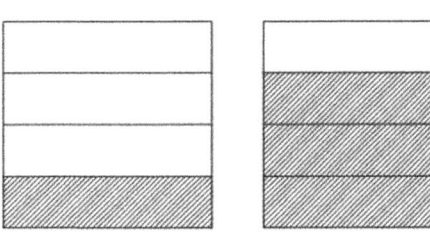

**Figure 111**            **Figure 112**

Now we prove the following theorem:

**Theorem. The area of a rectangle equals the product of the length of the base by the length of the altitude.**

Let us take a rectangle whose base has length 1 and altitude, length $a$. Consider two cases:

(1) $a$ is a rational number,
(2) $a$ is an irrational number.

[In case] (1) $a$ is a rational number—for example, $a = l/m$, where $l$ and $m$ are integers.

As shown in figure 113 [on the following page], we divided the altitude of the unit square into $m$ equal parts, the altitude of rectangle $ABCD$ contains $l$ such parts. We denote the area of rectangle $ABCD$ by $P$.

Through the points of division we passed parallels to the bases, which divided the square and rectangle into equal rectangles. The area of one such rectangle $= 1/m$; therefore, $P = l/m$ —that is, $P = a$.

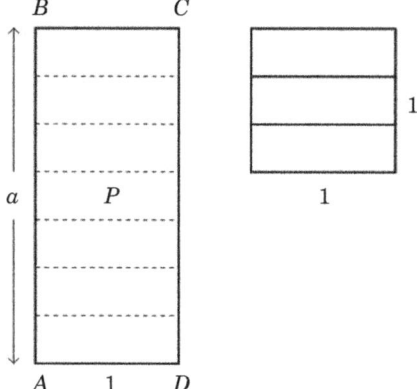

**Figure 113**

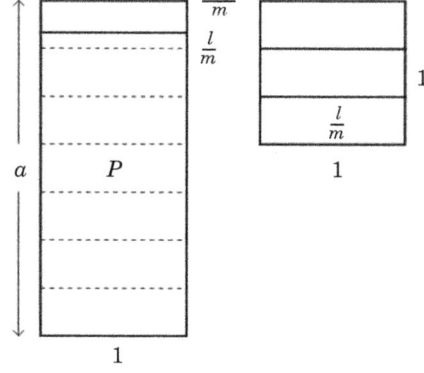

**Figure 114**

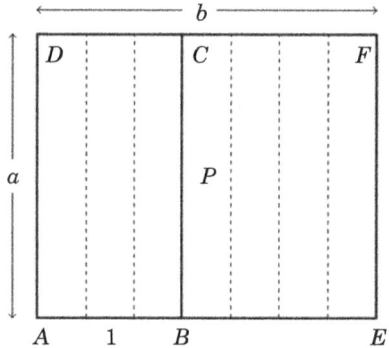

**Figure 115**

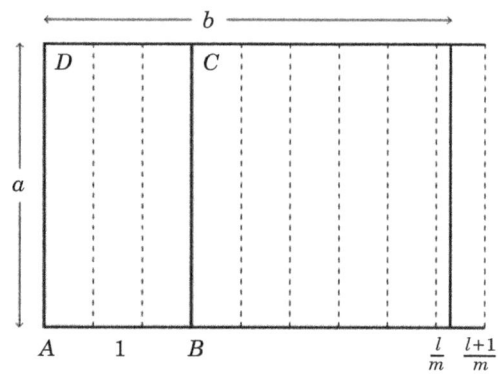

**Figure 116**

[In case] (2) $a$ is an irrational number (figure 114); let the approximations of this number with precision $1/m$ be $l/m$ and $(l+1)/m$, so that

$$\frac{l}{m} < a < \frac{l+1}{m}.$$

The segment with length $a$ is longer than the segment with length $l/m$ but shorter than [that of length] $(l+1)/m$. Then, as can be seen in the figure,

$$\frac{l}{m} < P < \frac{l+1}{m}.$$

We see that the numbers $a$ and $P$ have the same approximations [with precision] $1/m$, and specifically $l/m$ and $(l+1)/m$. Moreover, this will occur for every integer value $m$; therefore, according to property 2 of irrational numbers in §27 [page 297], numbers $a$ and $P$ are equal:

$$P = a = 1 \cdot a.$$

We see that if the base of the rectangle $= 1$, our theorem is true.

Now let us take a rectangle whose base has length $b$ and altitude, length $a$ —moreover, $a$ can be rational or irrational. Consider the two cases: (1) $b$ rational (figure 115), (2) $b$ irrational (figure 116).

[In case] (1) $b$ [is] rational: for example, $b = l/m$ (figure 115).

Denote by $P$ the area of the rectangle with edges $a$ and $b$, and by $P'$ [that of] rectangle $ABCD$ with base 1 and altitude $a$. On the basis of the previous [discussion], $P' = a$.

Lines parallel to $AD$ (figure 115) divide $AB$ into $m$ equal parts; in $b$ [the number of] such parts is $l$. These lines divide the area of rectangle $ABCD$ into $m$ equal parts, each of which is equal to $(1/m)a$; the area of rectangle $AEFD$ contains $l$ such parts.[30] Therefore,

$$P = \frac{l}{m} P',$$

and since $P' = a$,

$$P = \frac{l}{m} a = ba = ab.$$

[In case] (2) $b$ [is] an irrational number (figure 116); then, dividing $AB$ into $m$ equal parts we obtain

---

[30] [The authors neglected to specify points $E, F$ in figure 115.]

$$\frac{l}{m} < b < \frac{l+1}{m}.$$

Also, for areas, as seen in figure 116, we have

$$\frac{l}{m}a < P < \frac{l+1}{m}a.$$

Dividing by $a$, we obtain

$$\frac{l}{m} < \frac{P}{a} < \frac{l+1}{m}.$$

We see that the numbers $b$ and $P/a$ are contained between $l/m$ and $(l+1)/m$; moreover, this occurs for each value of $m$. In view of this, these numbers are equal: $P/a = b$, or

$$P = ab.$$

Thus, we have proved the theorem in all cases.

### Area of a parallelogram, triangle, trapezoid, and polygon

In a parallelogram, we call any two parallel edges the **bases**; then [we call] a segment that measures the distance between the bases an **altitude**. For example (figure 117), $AB$ is a base of the parallelogram $ABCD$ [and] $CE$ is an altitude.

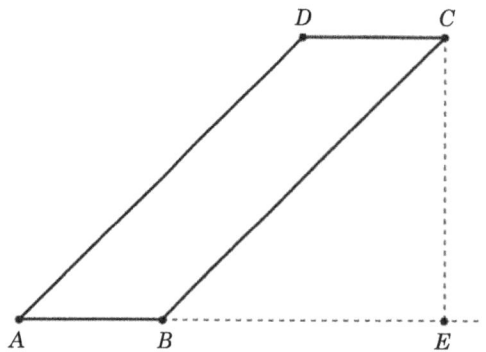

**Figure 117**

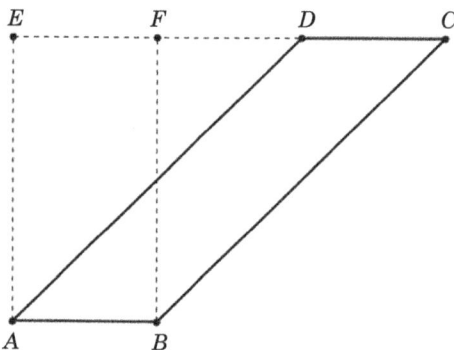

**Figure 118**

**Theorem.** The area of a parallelogram is equal to the product of the length of a base by the length of an altitude.

Parallelogram $ABCD$ (figure 118) has base $AB$ and altitude $AE$. Let us construct the rectangle $ABFE$ having base $AB$ and altitude $AE$. We shall represent the area of trapezoid $ABCE$ as the sum of areas of polygons in two ways:

area $ABCE$ = area $ABCD$ + area $ADE$ = area $ABFE$ + area $BCF$.

Since $\triangle ADE = \triangle BCF$, it follows that area $ABCD$ = area $ABFE$.

Therefore, the area of the parallelogram is equal to the area of the rectangle sharing a common base and altitude [length]. Thus, it is equal to the product of the length of a base by the length of an altitude.

**Corollary 1.** Parallelograms having equal bases and equal altitudes have equal areas.

**Corollary 2.** Dividing the area of a parallelogram by the length of a base, we obtain the length of an altitude. Therefore, if two parallelograms have equal areas and a common base and lie on the same side of the common base, then their other bases lie on one line (since the altitudes of the parallelograms are equal).

**Theorem. The area of a triangle is equal to half the product of the length of a base by the length of the altitude.**

Given triangle $ABC$ (figure 119) with base $AC$ and altitude $BD$, by constructing [lines] parallel to its edges we can create a parallelogram $ABEC$, half of which is triangle $ABC$ and which has with this triangle a common base and altitude.

Let $a$, $h$, $S$ denote the length of a base, length of the altitude, and the area of the triangle: then

$$S = \tfrac{1}{2}\,ah.$$

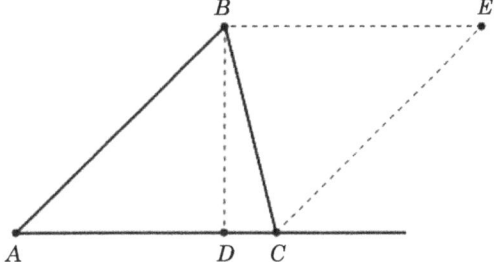

**Figure 119**

**Corollary.** Triangles having equal bases and altitudes have equal areas.

**Problem.** Replace a triangle $ABC$ by a rectangle with equal area (figure 120).

If we take a base $AC$ of the triangle as base of the desired rectangle, then half of the altitude $BF$ of the triangle—that is, $GF$—as an altitude of the rectangle, then $ADEC$ is the desired rectangle.

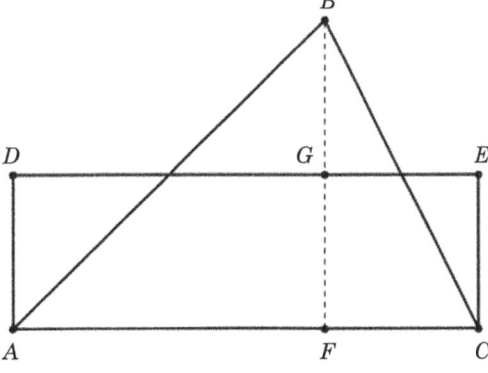

**Figure 120**

**Theorem. The area of a trapezoid is equal to half the product of the sum of the lengths of its bases by the length of an altitude.**

In trapezoid $ABCD$ (figure 121) parallel edges $AB$ and $CD$ are the bases; an altitude is $DE \perp AB$. We join $D$ with the midpoint $F$ of the edge $CB$, and extend $DF$ to [its] intersection with the line $AB$ at point $G$.

We have $\triangle FDC = \triangle FGB$ ($FC = FB$, $\sphericalangle 1 = \sphericalangle 2$, $\sphericalangle 3 = \sphericalangle 4$);[31] thus $BG = CD$ [and]

area $ADCB$ = area $ADFB$ + area $FDC$ = area $ADFB$ + area $FBG$ = area $ADG$.

We denote the length of $AB$ by $a$, of $DC$ or $BG$ by $b$, [and] of $DE$ by $h$.

The area of triangle $ADG$ equals half the product of the length of $AG$ (that is, $a + b$) by $h$, [and thus]

$$\text{area } ADCB = \tfrac{1}{2}(a+b)h.$$

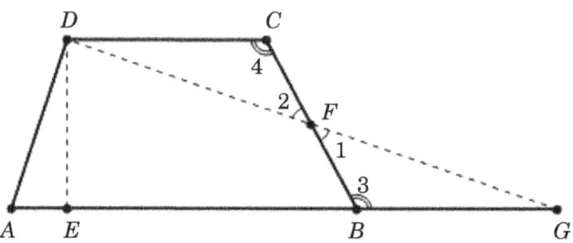

**Figure 121**

To calculate the area of a polygon, we decompose it as usual into triangles (figure 122) or trapezoids (figure 123) and calculate the sum of the areas of the triangles or trapezoids.

If the polygon is regular, the calculation of its area can be simplified. Let $AB$ (figure 124) be the edge of a regular polygon with $n$ edges; $O$, the center of the circumscribing circle of the polygon; [and] $OC$, an apothem.

We know that rays $OA$, $OB$, and so on, divide the polygon into $n$ congruent triangles. Denoting the length of an edge of the polygon by $a$, of an apothem by $r$, [and] the area by $S$, we have

$$S = n \cdot \tfrac{1}{2} a r = n a \cdot \tfrac{1}{2} r, \text{ and therefore}$$

**The area of a regular polygon equals the product of the length of its perimeter by half the length of an apothem.**

---

[31] [In the original, point $G$ was incorrectly labeled as $\sphericalangle 3$.]

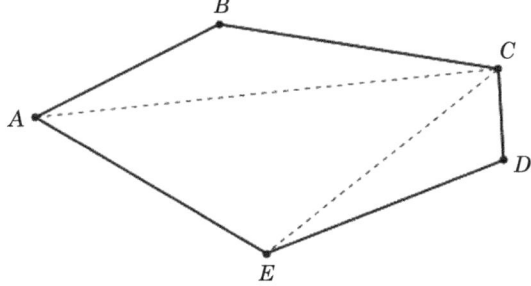

**Figure 122**

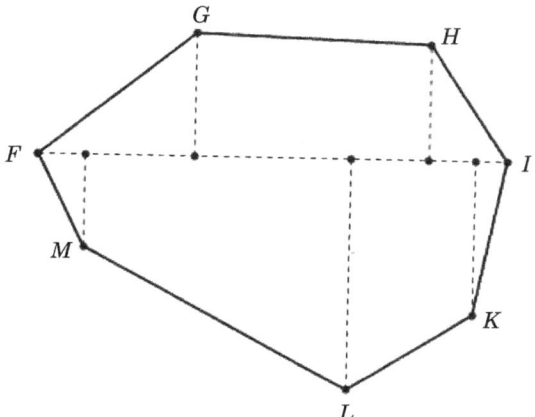

**Figure 123**

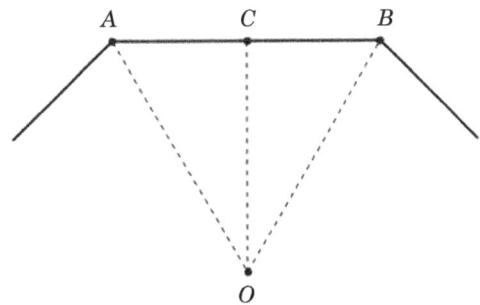

**Figure 124**

**Problem.** Replace a polygon by a rectangle having area equal to the area of the polygon.

In polygon *ABCDE* (figure 125) we draw a diagonal *CE* and a parallel *DF* to it through vertex *D*, to its intersection with the extension of *AE*. Triangles *CED* and *CEF* have the same base *CE* and the same altitude; therefore their areas are equal. If we remove triangle *CDE* from polygon *ABCDE* and replace it by triangle *CEF*, we will obtain the quadrilateral *ABCF*, which has area equal to the area of *ABCDE*. Replacing in a similar way triangle *ACB* by triangle *ACG*, we replace quadrilateral *ABCF* by the triangle *GCF*, which has area equal to the area of *ABCDE*. Using a familiar method we replace triangle *GCF* by a rectangle with the same area.

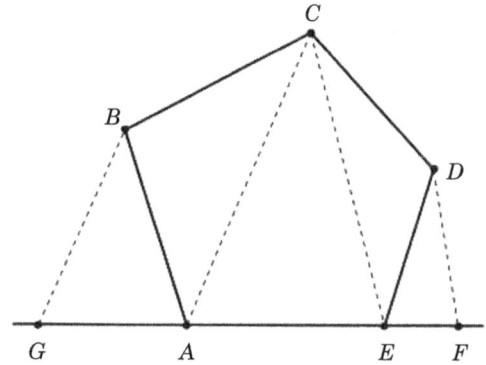

**Figure 125**

**Exercises.** 1. In square *ABCD*, chosen as the unit area, *O* is the point of intersection of the diagonals; *S*, the midpoint of edge *AB*; and *T*, the midpoint of edge *CD*. What areas should be assigned to (1) rectangle *ASTD*, (2) triangle *ABC*, (3) triangle *AOB*, (4) triangle *AOS*, (5) trapezoid *ASOD*?

2. A rectangle whose one edge is the unit length and the second, three times larger, is divided into four triangles by the diagonals. What are the areas of an undivided triangle and each of the four triangles?

3. Given a square, construct a triangle with an area equal to one eighth of the square.

4. One edge of a parallelogram is equal to an edge of a square, and a second, to the diagonal of the square; the acute angle of the parallelogram is equal to half of a right angle. Show that the square and the parallelogram have equal areas.

5. Given a parallelogram, construct a triangle that should have an edge in common with the parallelogram, as well as a common angle adjacent to this edge, and which should, in addition, have area equal to the area of the parallelogram.

6. Construct a square that should have an area two times larger than the area of a given square.

7. Through an arbitrary point *E* lying on the diagonal *AC* of a rectangle *ABCD* are drawn two lines parallel to the edges of the rectangle. One of them intersects the edge *AB* at point *F* and the edge *CD* at point *H*; tehe second [intersects] edge *BC* at point *G* and edge *AD* at point *I*. Prove that rectangles *FBGE* and *IEHD* have equal areas.

Hint: represent the considered rectangles as the differences of [respectively] equal triangles $FBGE = ABC - AFE - EGC$ and similarly for $IEHD$.

8. Indicate on a drawing made for the previous exercise two more pairs of rectangles with equal areas.

9. Can the theorem proved in exercise 7 be generalized, considering instead of the rectangle an arbitrary parallelogram?

10. Prove the theorem: if a rectangle with consecutive edges $a$ and $b$ and a rectangle with consecutive edges $c$ and $d$ have equal areas, then $c : a = b : d$. Formulate and prove the converse theorem.

11. Given a rectangle with edges $a$ and $b$; relying on the previous exercise, construct a rectangle of equal area, one edge of which would be the segment $c$. Hint: we apply one of the methods [in §28] for constructing the fourth proportional.

12. Explain geometrically the equation $(a + b)^2 = a^2 + 2ab + b^2$.

13. Prove that the area of a rhombus is equal to half of the product of the lengths of its diagonals. Generalize this theorem, considering instead of a rhombus a kite (quadrilateral of which two consecutive edges are equal and the other two consecutive edges are also equal).[32]

14. Prove that the area of a trapezoid with two nonparallel edges is equal to the length of an altitude, multiplied by the length of the segment joining the midpoints of the nonparallel edges.

15. Show that a median of a triangle divides this triangle into two triangles with equal areas.

16. Divide a triangle into five triangles of equal areas with the help of segments passing through one vertex.

17. In the convex quadrilateral $ABCD$, point $O$ is the midpoint of the diagonal $AC$. Show that quadrilaterals $ABOD$ and $BODC$ have equal areas.

18. Investigate whether the statement asserted in the previous exercise remains true in the case in which quadrilateral $ABCD$ is concave.

19. Two parallelograms $ABCD$ and $ABEF$ have equal areas and lie on the same side of the line $AB$. Show that edges $CD$ and $EF$ must lie on the same line.

20. A parallelogram $ABCD$ is given. Construct a parallelogram $ABEF$ whose area should be equal to the area of $ABCD$, and whose diagonal $AE$ should be equal to a given segment $k$. Investigate whether the problem is always possible and how many solutions it has.

21. Two triangles $ABC$ and $ABD$ have equal areas and lie on the same side of the line $AB$. Show that $AB \parallel CD$.

---

[32] [The authors' word for *kite* was *deltoid*.]

22. Triangle *ABC* is given, and an angle *D*. Construct a triangle *ABD* with area equal to the area of the triangle *ABC*. Give the condition of solvability of the problem in the case in which the angle *D* is right.

23. Triangle *ABC* is given. Construct a rectangle with equal area, whose base should be a given segment *k*. Hint: first we construct a rectangle of equal area and base *AB*, then apply exercise 11.

24. Solve the previous problem, taking instead of a triangle any polygon—for example, a pentagon. Hint: divide the pentagon into triangles and for each of those construct a corresponding rectangle; next, we add these rectangles to each other in order to obtain as sum a rectangle with the same base.

25. Two arbitrary polygons are given; we want to convince ourselves which of them has the greater area not with the help of measurements, but with geometric constructions. Based on exercise 24, describe how we should proceed.

26. A parallelogram has acute angle 30° and edges [with lengths] 12 cm and 6 cm. Calculate its area.

27. Two heights[33] *h* and *l* of a parallelogram are given, coming from a vertex *A* and edge $AB = c$. Calculate the remaining edges of the parallelogram.

28. The area of a rhombus is 42 cm$^2$ and one of the diagonals [measures] 7 cm. Calculate the [length of the] second diagonal (see exercise 13).

29. A diagonal of a parallelogram measures 50 cm and forms a 30° angle with an edge [having length] 36 cm. Calculate the area of a parallelogram.

30. In a triangle with base $AB = 40$ cm a rectangle with edges 10 cm and 8 cm has been inscribed so that the longer edge lies on *AB* and the ends of the opposite edge of the rectangle lie on the other edges of the triangle. Calculate the area of triangle *ABC*.

31. In an isosceles trapezoid the bases are equal to 56 cm and 42 cm. Calculate the area of the trapezoid, knowing that the arms form 45° angles with a base.[34]

32. Calculate the area of an isosceles trapezoid with height 56 cm, knowing that its diagonals are perpendicular to each other and their point of intersection divides them in the ratio 2:5.

33. In an isosceles triangle one arm is 8 cm and the angle between the arms is equal to 150°. Calculate the area of the triangle.

34. The edges of a parallelogram measure $a = 5$ cm and $b = 8$ cm. An altitude, dropped to the longer edge, measures $h = 3$ cm. How long is the altitude dropped to the shorter edge?

---

[33] [Here, *height* means *length of altitude*.]

[34] [Some text in this exercise was evidently scrambled in the original; the present editors have tried to restore it.]

35. The bases of a trapezoid have lengths $a = 7$ cm and $b = 11$ cm; the height $h = 4$ cm. Calculate the area of the trapezoid, and the area of the triangle that is formed if its edges are extended to their intersection.

36. Prove that the area of a trapezoid having two nonparallel edges may be expressed as the product of one of the nonparallel edges by the distance from this edge to the midpoint of the other edge.

37. An edge of a square measures $a$. Find the area of the square inscribed in it in such a way that the vertices of the inscribed square lie on the midpoints of the edges of the given square.

38. In a triangle $ABC$ in which $AB = 13$ cm, $BC = 14$ cm, $CA = 15$ cm, the height was calculated [as] $AD = 12$ cm. Find the areas of the parts into which the bisector of angle $A$ divides the triangle.

39. In parallelogram $ABCD$ the diagonals are given: $AC = 18$ cm and $BD = 12$ cm. The distance from point $A$ to diagonal $BD$ is 5 cm. What is the distance from point $B$ to diagonal $AC$?

40. In a trapezoid $ABCD$ in which the bases measure $AD = 9$ cm and $BC = 13$ cm and the height $h = 4$ cm, a line is drawn parallel to the bases, dividing the area of the trapezoid in half. How far from $AD$ was this line drawn?

41. In a quadrilateral $ABCD$ in which $AB = BC$ and $AD = DC$, the diagonals are given: $AC = 8$ cm, $BD = 15$ cm. Joining consecutive midpoints of the edges of this quadrilateral, we obtain a new quadrilateral. Determine its shape and find the area.

42. In an isosceles trapezoid the bases $AD = a = 24$ cm and $BC = b = 18$ cm and the height $h = 14$ cm are given. Find the area of the triangle between the edge $AB$ and the diagonals drawn through $A$ and $B$.

43. When designing navigation channels it is assumed that the cross section of the channel should be 5.5 times as large as the cross section of the submerged part of a passing freight ship. Assuming the ship's beam $h = 9$ m and draft $b = 1.75$ m, as well as the channel depth $l = 2.80$ m, calculate the width of the channel at the bottom and at the water level.[35]

The cross section of the channel is an isosceles trapezoid $ABCD$ (figure 126); in addition,
$$EB : AE = 3 \quad (EB \perp AE).$$

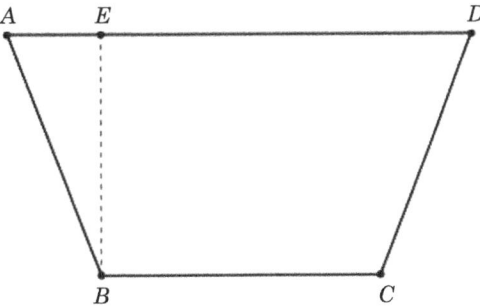

**Figure 126**[36]

---

[35] [A ship's *beam* and *draft* are its width at its widest point and the depth of water needed to float it. The authors' words for them were *szerokość* and *zanurzenie*.]

[36] [The trapezoid in the original was not isosceles, as required by the text.]

## § 35. Areas of Similar Polygons

**Theorem. The areas of similar triangles stand in the same ratio as the squares of the lengths of corresponding edges.**

We have two similar triangles $ABC$ and $A'B'C'$; moreover, corresponding vertices are denoted by $A$ and $A'$, $B$ and $B'$, $C$ and $C'$ (figure 127). Let the lengths of the bases be $a$ and $a'$, and their corresponding heights, $h$ and $h'$:

$$\frac{\text{area } ABC}{\text{area } A'B'C'} = \frac{\tfrac{1}{2}ah}{\tfrac{1}{2}a'h'} = \frac{a}{a'} \cdot \frac{h}{h'}.$$

Since $\triangle ADC \sim \triangle A'D'C'$ (why?),

$$\frac{AC}{A'C'} = \frac{h}{h'}.$$

From the assumptions, we have

$$\frac{AC}{A'C'} = \frac{AB}{A'B'} = \frac{a}{a'}.$$

Therefore

$$\frac{h}{h'} = \frac{a}{a'},$$

and thus

$$\frac{\text{area } ABC}{\text{area } A'B'C'} = \frac{a}{a'} \cdot \frac{a}{a'} = \frac{a^2}{a'^2}.$$

**Theorem. The areas of similar convex polygons stand in the same ratio as the squares of the lengths of corresponding edges.**

As we know, we can decompose similar polygons (figure 128) into similar triangles, with the help of the diagonals passing through corresponding vertices. The areas of these triangles stand in the same ratio as the squares of any two corresponding edges of these polygons. Denoting these areas by $P_1, P_2, \ldots, p_1, p_2, \ldots$, and[37] corresponding edges of these polygons by $a$ and $a_1$,

$$\frac{P_1}{p_1} = \frac{a^2}{a_1^2}, \quad \frac{P_2}{p_2} = \frac{a^2}{a_1^2}, \quad \ldots,$$

and therefore

$$\frac{P_1}{p_1} = \frac{P_2}{p_2} = \ldots.$$

---

[37] [In the original, the letters $P$ and $p$ were interchanged in this paragraph and the next.]

## 13.12 §35 Areas of Similar Polygons

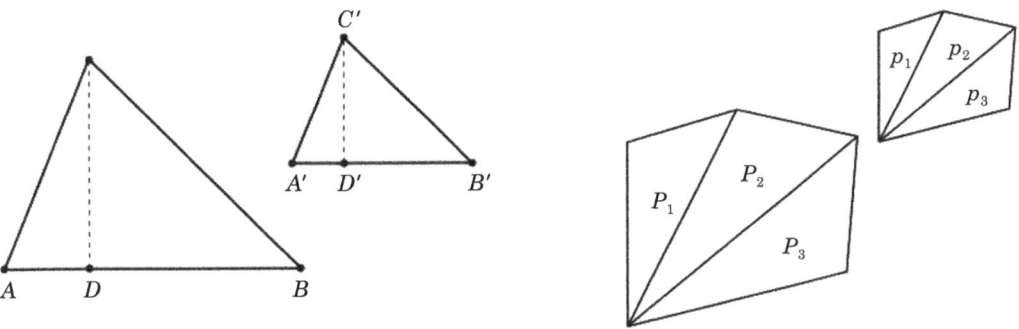

**Figure 127**    **Figure 128**

Thus, as we did similarly in §29 for perimeters of similar polygons, we can prove that

$$\frac{P_1 + P_2 + \cdots}{p_1 + p_2 + \cdots} = \frac{a^2}{a_1^2}.$$

Not proving in general the analogous theorem about similar nonconvex polygons (figure 129), we explain a method for such a proof for similar concave quadrilaterals.

We have two similar concave quadrilaterals $ABCD$ and $A'B'C'D'$ (figure 130).

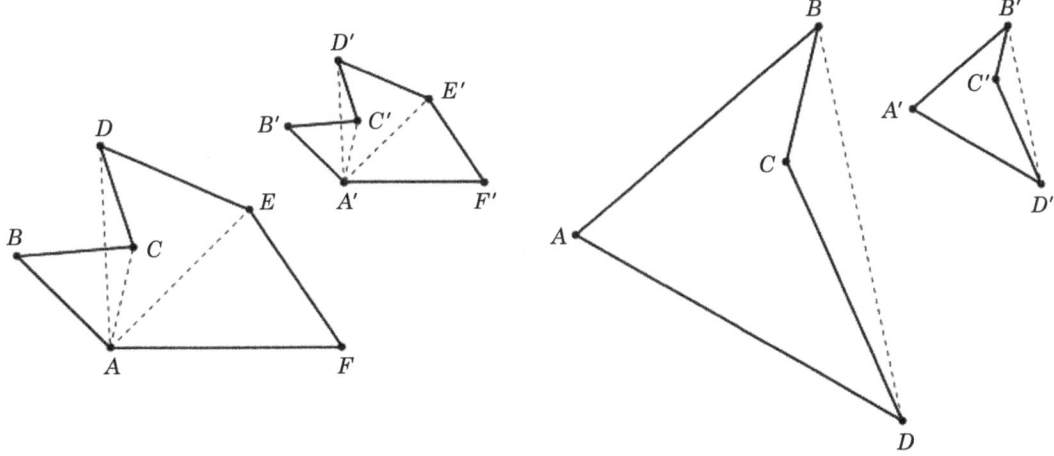

**Figure 129**    **Figure 130**

Denoting the lengths of any corresponding edges by $a$ and $a_1$,

$$\frac{\text{area } ABCD}{\text{area } A'B'C'D'} = \frac{\text{area } ABD - \text{area } BCD}{\text{area } A'B'D' - \text{area } B'C'D'}, \quad \dots\dots\dots\dots\dots \quad (1)$$

$$\frac{\text{area } ABD}{\text{area } A'B'D'} = \frac{a^2}{a_1^2}, \quad \frac{\text{area } BCD}{\text{area } B'C'D'} = \frac{a^2}{a_1^2},$$

$$\text{area } ABD = \frac{a^2}{a_1^2} \cdot \text{area } A'B'D', \quad \text{area } BCD = \frac{a^2}{a_1^2} \cdot \text{area } B'C'D'.$$

Substituting these values in (1), we obtain[38]

$$\frac{\text{area } ABCD}{\text{area } A'B'C'D'} = \frac{a^2}{a_1^2}.$$

**Exercises.** 1. Prove the theorem discussed here about areas of similar polygons, when the polygons considered are nonconvex pentagons. Consider three cases: (1) pentagons with one concave angle, (2) pentagons with two concave angles adjacent to the same edge, (3) pentagons with two concave angles not adjacent to the same edge.

2. In two circles are inscribed two regular polygons with the same number of edges. The radius of one circle is $k$ times as large as the radius of the second circle. What is the ratio of areas of these polygons?

3. Through the point of intersection of the medians of triangle $ABC$ is drawn a line parallel to $BC$, which intersects edge $AB$ at point $B'$, and edge $AC$ at $C'$. What part of the area of triangle $ABC$ is the area of triangle $A'B'C'$?

4. In trapezoid $ABCD$ points $E$ and $F$ are midpoints of the nonparallel edges $AD$ and $BC$, respectively; the base $AB$ is twice as long as the base $CD$. Show that the area of the trapezoid $ABFE$ is 1.4 times as large as area $EFCD$. Hint: we extend the edges $AD$ and $BC$ to intersect at the point $G$, and calculate the ratios of the areas of all triangles and trapezoids that we see in the figure[39] to the area of triangle $ABG$.

5. Calculate the ratio of the areas of the trapezoids $ABFE$ and $EFDC$ in the previous problem, assuming that the base $AB$ is four times as long as the base $CD$.

6. A plot of land has the shape of a pentagon $ABCDE$. The edges $AB = 75$ m, $BC = 80$ m, $CD = 95$ m, $DE = 112$ m, $EA = 67$ m, as well as the diagonals $AC = 126$ m and $AD = 149$ m were measured. Calculate the value of the plot if 1 m$^2$ costs 50 gr.[40] Hint: it is necessary to trace a pentagon with the given segments in a certain scale—for example, 1 : 1000 —and replace it by a rectangle with equal area.

---

[38] [In the original, equation (1) was typeset incorrectly, in an unfortunate way that seemed well-formed but appeared to beg the question.]

[39] [There was no figure in the original; moreover, point $G$ was called $E$ there.]

[40] [1 gr is one *grosz*: one hundredth of a zloty.]

# 14
# Teaching and Logic —To the New World

This book has pursued the following goals.

(1) Publish translations as necessary so that
- Alfred Tarski's works will all be accessible in English, French, or German; and
- his geometric works will all be accessible in English.

(2) Provide scientific and cultural background information about the works translated here: their origin, context, structure, and impact.

(3) Update the Givant 1986 bibliography of Tarski's publications, and include an annotated list of major works about Tarski's life and work.

The preceding portions of the book contain all of the translations and background information needed to accomplish the first two goals, except for a few short works, not concerned with geometry or teaching, that had been published only in Polish. Those are translated in chapter 15, with some background information for specific items. One of those short works has to do with politics; the rest, logic. The present chapter contains some general background for chapter 15 and concludes the biographical material in this book with an account of Tarski's 1939 voyage to the New World. The update for Tarski's bibliography is in chapters 16–18. For more information on Tarski's logical work in Poland and on his career in the United States, consult the books and papers listed in chapters 17 and 18—particularly the splendid biography, Feferman and Feferman 2004.

## 14.1 Teaching

During Alfred Tarski's career in Poland, the major component of his workload was his teaching at the Żeromski Gimnazjum. As noted in section 9.2, he probably spent as much as twenty-four hours per week in classes there, in addition to five hours in lectures and seminars at the university. After those commitments came class preparation and consultation with students, presumably some administrative chores, and activity with family and friends. Finally, there was his research and its presentation, evidently the intellectual high point of his life, which will be discussed in the next section. Little direct record remains of Tarski's teaching: most records were destroyed, and his students' impressions overshadowed, by the terrors of the coming cataclysm.

From the years 1937–1939 preceding his graduation from the Żeromski Gimnazjum, Tarski's student Jarosław Rudniański

> ...retained a memory of Tarski as a dignified and serene man (but somewhat nervous), with a "clouded" face.... As an instructor, he was patience itself as he set about explaining the meanders of mathematics, but one saw clearly that he was interested much less in the students than in the...mathematics. His lessons usually proceeded this way: Tarski would himself solve a problem at the board...while commenting clearly and abundantly about the methods.... Then a student (more often than not, an average one) would have to solve a problem of the same type, and Tarski would analyze the errors committed, according to the maxim, "Let us learn from others' mistakes." In this connection he was loved—but by the *gifted* students.[1]

Leon Błaszczyk, a Żeromski student during 1934–1939, reported that Tarski used textbooks by Władysław Nikliborc and by Stefan Banach, Wacław Sierpiński, and Włodzimierz Stożek. These two series entered publication after Tarski began teaching, in response to the lack of Polish texts described in section 9.2. Błaszczyk also commented that Tarski

> was known for always recognizing—even though he never monitored students during examinations—the pairs of works where one was copied to another; after correcting all of them, he...confined himself to asking who was the author of the "original."[2]

Witold Kozłowski, Tarski's Gimnazjum student during 1934–1938, has written,

> [Tarski] was characterized by the great clarity of his discourse, and he knew how to excite us in the discussions of various mathematical problems. He familiarized us with concepts such as complex numbers and the theory of sets. He told us about the history of mathematics and logic, going all the way back to Aristotle.[3]

According to Kozłowski, Tarski preferred not to use textbooks directly, because they were too distant from life. Students from his classes and seminars at the university would attend his lessons at the gimnazjum. Kozłowski confirmed Rudniański's observation about the gifted gimnazjum students: the most talented (about one in ten) would meet at Tarski's home. To Kozłowski, Tarski seemed cheerful and very witty; his language was beautiful, tended toward irony, and he made fine distinctions in conversation. Tarski's favorite mathematical subject at that level was the theory of measurement.[4] Tarski and his wife, Maria, both emphasized that athletics should be related to mathematics: ball games and gymnastics, for example, provide many example applications. Kozłowski observed that Tarski was an excellent dancer.

Kozłowski noted that Tarski was conscious of his own genius; but he told his students that he would treat them personally, and did so. Kozłowski believed that Tarski singled him out because he showed the ability to view the school and his peers as a social system.

---

[1] Jadacki 2003a, 145. **Jarosław Rudniański** was imprisoned by the Russians after their 1939 invasion of Poland, escaped from Siberia with General Władysław Anders's army, fought against the Germans in Italy, then returned to Warsaw after the war to study with Tadeusz Kotarbiński. He became professor of philosophy at the Polish Academy of Sciences (*Gazeta wyborcza* 2008).

[2] Jadacki 2003a, 145. **Leon Tadeusz Błaszczyk** became a philologist.

[3] Kozłowski 2003, 2010.

[4] Sections 13.6–13.8 and 13.10–13.13 of the present book are translations of the sections of Tarski's coauthored 1935 gimnazjum text *Geometrja* that cover the theory of measurement.

As a young lad, Kozłowski had shown great sensitivity and individuality, and had embraced atheism. He reported,

> Graduation was approaching, and I was worrying—even if I did pass the examinations, I wouldn't be given the Diploma of Maturation. The prosaic obstacle—I had not been baptized, and on my certificate my religious persuasion must be declared. I confessed my trouble to Professor Tarski and his wife.... I was baptized, in part thanks to Mrs. Tarski, who expressed the desire to be my godmother.[5]

Kozłowski seemed unaware that Tarski had confronted a similar situation at about the same age (see chapter 3).

## 14.2 Logic

Alfred Tarski's research in logic attracted worldwide attention during the 1930s. He published most of those results in French or German, and as his fame increased, many of them were translated into English. Nearly all of his works published after 1939 appeared originally in English; the others were in German or French. The translations from Polish in chapter 15 are all brief, and concerned with subjects not closely related to those covered earlier. The present section will describe their general context. Some minimal explanations of their specific features are included in chapter 15 with the translations.

Most of the works in chapter 15 originated in connection with colloquium presentations, sometimes by other scholars. In Warsaw, Tarski frequently reported his research results at the regular meetings of three different organizations. Listed below are the full names of the organizations and of the journals that published their proceedings:

- Warsaw Philosophical Institute, Logic Section
  *Przegląd filozoficzny, Ruch filozoficzny*[6]
- Polish Mathematical Society, Warsaw Section
  *Annales de la Société Polonaise de Mathématique*[7]
- Warsaw Society of Sciences and Letters
  *Comptes rendus de la Société des Sciences et des Lettres de Varsovie, classe III: sciences mathématiques et physiques*[8]

---

[5] Kozłowski 2003 and 2010; Zalasiewicz 2009. **Witold Kozłowski** was born in Oruro, Bolivia, in 1919. His father, Roman, was a noted paleontologist and geoscientist, and later a professor at the University of Warsaw. Witold started university studies in Polish literature in 1939; during the war, he was active in underground cultural life, particularly social services and journalism. He earned a master's degree after the war, in cultural education. He won many awards during his long career as journalist and author.

[6] Warszawski Instytut Filozoficzny, Sekcja Logiczna; *Philosophical Review, Philosophical Trends*, the latter published in Lwów.

[7] Polskie Towarzystwo Matematyczne, Oddział Warszawski; *Rocznik Polskiego Towarzystwa Matematycznego*.

[8] Towarzystwo Naukowe Warszawskie; *Sprawozdania z posiedzeń Towarzystwa Naukowego Warszawskiego, wydział III: nauk matematyczno-fizycznych*.

Tarski made his first scientific presentations at the Warsaw Philosophical Institute in 1921. Several of his publications mentioned in section 1.1 and chapter 3 stemmed from presentations at meetings of these three organizations. The discussions translated in sections 15.1, 15.4, and 15.6 took place at sessions of the Institute in 1924, 1928, and 1930, and section 15.7 contains a translation of the abstract of a presentation to the Warsaw Society of Sciences and Letters in 1932. Tarski continued reporting to Warsaw societies regularly throughout the 1930s. He served as vice-president of the Institute's logic section during the 1930s, while Jan Łukasiewicz was its president.[9]

During the year after the September 1929 Congress of Mathematicians of Slavic Countries Tarski continued to develop the fundamental concepts of the methodology of deductive sciences that he had described there. The subject was probably receiving much attention in his Warsaw research seminar. (For that background, see sections 8.1, 9.4, and 9.5.) Tarski presented his work in several venues, culminating in six lectures during the period from Wednesday through Tuesday, 10–16 December 1930, at meetings of the Lwów sections of the Polish Philosophical and Mathematical Societies:

| 1930–1931 | Venue | | Subject |
|---|---|---|---|
| Feb 20 | Vienna | University | Some basic concepts of metamathematics[10] |
| Mar 27 | Warsaw | Soc. of Sci. | Some basic concepts of metamathematics |
| Oct 8 | Warsaw | Phil. Inst. | Concept of a true sentence for formalized deductive systems[11] |
| Dec 10 | Lwów | Phil. Soc. | 1. Scope of research, basic concepts, and assumptions |
| .. 11 | .... | .... | 2. General properties of deductive systems |
| .. 12 | .... | .... | 3. Bases of deductive systems |
| .. 13 | .... | .... | 4. Consistency and completeness |
| .. 15 | .... | .... | Concept of truth in formalized deductive sciences |
| .. 16 | Lwów | Math. Soc. | Definable sets of real numbers |
| Mar 21 | Warsaw | Soc. of Sci. | Concept of truth in the languages of deductive disciplines |

The series of numbered lectures was called *Fundamental Concepts of the Methodology of Deductive Sciences*. They were published in Vienna as Tarski [1930] 1983a; according to the Philosophical Society's minutes, a second part of that paper would follow in the same journal, but that never occurred. Abstracts of the 15–16 December lectures were published with their respective societies' minutes. An abstract of the last lecture in this list was published in 1932 in Vienna.[12]

---

[9] Jadacki 2003a, 151, 156. For further information on the Warsaw logic milieu in this era, consult Woleński 1995a, section 4.

[10] This title is equivalent to that of Tarski's presentation to the 1929 congress. In Vienna, Tarski also gave presentations on other subjects. See Menger 1994, chapter 12; Menger 1998, 78, 131; Dawson 1998, 33–36, and Feferman and Feferman 2004, 76–84.

[11] Warszawski Instytut Filozoficzny 1931.

[12] Polskie Towarzystwo Filozoficzne w Lwowie 1930–1931, 209–210; Tarski [1930–1931] 2014 and 1930b. In April 1931 Tarski reported to the Warsaw Philosophical Institute about Kurt Gödel's closely related but independent [1931] 1967 work on the incompleteness of integer arithmetic: that was probably its first exposure outside Vienna (Tarski [1931] 2014h, Feferman and Feferman 2004, 84).

## 14.2 Logic

Within a decade, the material in Tarski's 1930 Lwów lectures became the framework in which many fields of logic are investigated and presented. The abstract of his 15 December lecture was the first appearance in print of his celebrated theory of truth, a particularly vital part of that framework: it supports much subsequent work in semantics, including that in philosophy, linguistics, and computer science. It is translated in section 15.6.

Chapter 15 also contains translations of Tarski's remarks about six presentations by other scholars at the Second and Third Polish Philosophical Congresses in 1927 and 1936. The third congress occurred in the middle of Tarski's major effort in the 1930s to present throughout Europe his results in mathematics and logic, and disseminate his point of view on logic. Together with reports of his Warsaw presentations, this portrays nearly frantic activity, a major presence in European mathematics and philosophy, all grafted onto his full-time occupation as a schoolteacher. Tarski's presentations and publications in logic and mathematics greatly overshadowed his work as a mathematics educator, and most of them quickly became available in Western European languages. His renown in logic explains why his work on geometry and teaching has been little known, even obscure.

The Second Polish Philosophical Congress took place during 23–28 September 1927 in Warsaw. Two hundred scholars attended. There were seventy lectures, apportioned among seven sections:

- history of philosophy
- metaphysics and theory of knowledge
- philosophy of natural sciences
- psychology[13]
- logic
- semantics
- aesthetics

According to the conference report by Warsaw logician Janina Hosiassonówna, the most notable sections were those on logic and psychology.[14] Jan Łukasiewicz gave the keynote address, on the traditional and the deductive approaches to philosophy.[15] He noted that the discipline of philosophy had given other sciences the scientific and deductive methods, but had never applied those to itself. As a consequence, traditional philosophy was dominated by meaningless answers to meaningless questions. He urged application of the deductive axiomatic method to all problems in philosophy. Although Łukasiewicz mentioned no other modern works, his prescription echoed the doctrines of the Peano school from a generation earlier, and reflected those then being formulated in Vienna by the logical positivist school.[16] Hosiassonówna reported that Łukasiewicz's address stirred up much controversy, and that he remained the central figure of the congress. He gave two talks in the logic section, comparing syllogistic to modern axiomatic logic, and outlining the theory of deduction that he and Tarski were formulating. Tarski presented yet another talk, continuing that discussion. (Two years later, Tarski would become Łukasiewicz's official assistant.) These three presentations were not published in the volume

---

[13] Until the 1930s, psychology was often regarded in academia as a branch of philosophy.
[14] Hosiassonówna 1928. For information about Hosiassonówna, see the box on page 335.
[15] Łukasiewicz [1927] 1928.
[16] See Marchisotto and Smith 2007, section 2.2, 126–128, and the introduction to Padoa [1900] 1901.

of congress proceedings. The latter two were probably subsumed by Tarski's later paper *On Some Fundamental Concepts of Metamathematics*.[17]

Sections 15.2 and 15.3 contain translations of Tarski's very brief remarks on two other presentations at the 1927 congress: by Henryk Greniewski on the theory of action, and by Tadeusz Czeżowski on causality. Both lay within the scope of Łukasiewicz's proposal. Specific background for them is included in those sections.

The table below lists some of the congresses and other meetings outside Warsaw in which Tarski participated, and indicates where some of those are discussed in the present book.[18] As noted, Tarski gave presentations at most of them. The list is surely not complete, if only because there is no clear borderline between "official" presentations at organized meetings and informal presentations in homes and offices of colleagues throughout Europe. Tarski often visited other researchers en route to meetings or on recreational excursions.

### Tarski's Colloquia Away from Warsaw

| Date | | Location | Meeting | Discussed in | |
|---|---|---|---|---|---|
| May | 1923 | Lwów | First Polish Philosophical Congress | chapter | 3 |
| Sep | 1927 | Lwów | First Polish Mathematical Congress* | section | 9.4 |
| Sep | 1928 | Bologna | International Congress of Mathematicians* | ..... | 8.8 |
| Feb | 1930 | Vienna | Mathematisches Kolloquium* | ..... | 9.5 |
| Dec | 1930 | Lwów | Polish Philosophical Society, Lwów* | ..... | 14.2 |
| Dec | 1930 | Lwów | Polish Mathematical Society, Lwów* | ..... | 14.2 |
| Sep | 1931 | Vilnius | Second Polish Mathematical Congress | | |
| Jun | 1932 | Lwów | Polish Philosophical Society, Lwów*,19 | | |
| Aug | 1934 | Prague | Vorkonferenz on Unity of Science* | ..... | 14.2 |
| Sep | 1934 | Prague | Eighth International Congress of Philosophy | ..... | 14.2 |
| Jan–Jun | 1935 | Vienna | University of Vienna* | ..... | 9.1, 14.2 |
| Sep | 1935 | Paris | First Congress for the Unity of Science* | ..... | 9.1, 14.2 |
| Jun | 1936 | Cracow | Third Polish Philosophical Congress | .. | 14.2, 15.8–11 |
| Jul | 1937 | Paris | Third Congress for the Unity of Science | ..... | 14.2 |
| Jul | 1937 | Paris | Ninth International Congress of Philosophy* | ..... | 14.2 |
| Sep | 1938 | Amersfoort | Modern Concepts of Reasoning | ..... | 14.2 |
| Sep | 1939 | Harvard | Fifth Congress for the Unity of Science* | ..... | 14.3 |

*Tarski presented a paper.

---

[17] Polski Zjazd Filosoficzny [1927] 1928. Tarski [1930] 1983. See also Jadacki 2003b, 119. Stanisław Leśniewski also presented at the congress an oral summary of three parts of the paper Leśniewski [1927–1930] 1992, introducing his theories of mereology, ontology, and logistic.

[18] The information in the table was gleaned largely from Feferman and Feferman 2004, Dawson 1998, and Jadacki 2003a, particularly the last. Jadacki claimed in error that Tarski had also presented a paper in Vienna in 1936.

[19] At the 1932 Lwów meeting Tarski spoke about categoricity of a theory and independence of its primitive notions.

During summer 1934, Tarski participated in two meetings in Prague. The first, held on 30 August and 1 September, was a *Vorkonferenz* devoted to planning the International Congress on Unity of Science to be held the following summer. There, Tarski presented his [1935–1936] 1983 paper "Some methodological investigations on the definability of concepts."[20] The Vorkonferenz was followed immediately by the Eighth International Congress of Philosophy in Prague, 2–7 September 1934. Many scholars associated with the Vienna Circle and logical positivism attended both meetings.

The International Congress of Philosophy attracted some four hundred participants from thirty-two countries, including all major countries in Europe except the Soviet Union. Continuing a series of conferences begun in 1900, it attracted much attention: nearly forty press representatives were listed as associates. The Czechoslovak foreign minister, Edvard Beneš, delivered the opening address.[21] Four of the twenty-eight congress sessions were devoted to *L'importance de l'analyse logique pour la connaissance*. Tarski participated in the discussions after the first and third of those, but presented no paper himself. The first "connaissance" session was chaired by Rudolf Carnap, professor at Prague and a leader in the Vienna Circle. The session featured papers on general subjects by Tarski's teacher and collaborator Łukasiewicz; by Moritz Schlick, a leading member of the Vienna Circle;[22] and by Jørgen Jørgensen, professor of philosophy at Copenhagen and the leading Danish exponent of logical positivism. The third "connaissance" session was chaired by Louis Rougier, from Besançon and Cairo, a conventionalist and associate of the Vienna Circle. It included general presentations by Ernest Nagel, of New York City, and by Eino Kaila, of Helsinki—both members of the logical positivist movement—and by Kazimierz Ajdukiewicz, from Lwów. The remaining papers, by Janina Hosiassonówna, Hans Reichenbach, and Zygmunt Zawirski, were devoted to the foundations of probability theory and multivalued logic.

Tarski commented on Jørgensen's paper and on the papers of Reichenbach and Zawirski. He published the first of these commentaries—[1934] 2003b—and two almost identical versions of the second—1935–1936 and [1934] 2003a—in the proceedings of the two Prague meetings. These remarks, originally in German, were translated into Polish and published in the collection Jadacki 2003b, 11–14. They are discussed in detail in section 16.3, which is devoted to posthumously published contributions of Tarski.

Tarski spent January through June 1935 in Vienna, supported by a grant from the Rockefeller Foundation.[23] Afterward, he participated in the conference that had been planned at the Prague *Vorkonferenz* the year before: the First Congress on the Unity of Science in Paris, 15–21 September 1935. There, Tarski presented two papers, which have

---

[20] See Frank 1935–1936 for an overview of the Vorkonferenz, and the individual papers in *Erkenntnis* 5(1) for reports of its proceedings.

[21] Before entering politics in 1918, Beneš had been a university professor. In 1936, he would become president of Czechoslovakia.

[22] Two years later, Schlick would become a famous victim of political assassination.

[23] Feferman and Feferman 2004, 91. Meanwhile, Tarski's wife, Maria, a new mother, had moved in with his parents in Warsaw (see section 9.1).

since become famous: *The Establishment of Scientific Semantics* and *On the Concept of Following Logically*.[24] A photograph on page 327 shows him in Paris with the Lwów philosopher Maria Kokoszyńska. Back in Warsaw the following November, Tarski presented a report about the congress, jointly with Janina Hosiassonówna and her husband, Adolph Lindenbaum, who had also presented papers there. Their text has not surfaced, but Kokoszyńska's analogous 1937 report to the Polish Philosophical Society in Lwów was published.[25]

On 24–26 September 1936, Tarski participated in the Third Polish Philosophy Congress in Cracow. It was somewhat larger than the 1927 Second Congress; all listed participants were Polish. Tarski presented no paper himself, but contributed to the discussions of the plenary-session papers of Ajdukiewicz, Witold Wilkosz, Zawirski, and Kokoszyńska. These presentations are described in detail in sections 15.8–15.11, and Tarski's remarks are translated there. The specialized logic section of that congress seems rather weaker. But there was a notable special section on *Catholic Thinking and Modern Logic* on the last day of the congress.[26]

The Unity of Science conferences were held every summer from 1935 until World War II. Tarski evidently missed the second one, in 1936 in Copenhagen. He attended the 1937 conference in Paris, which immediately preceded the Ninth International Congress of Philosophy. At the latter, he presented his 1937b expository paper, which was "a general exposition for the nonspecialist of the nature of completely formalized deductive systems in logic and mathematics."[27] At the 1938 Unity of Science conference, in Cambridge, England, Tarski was elected in absentia to the organizing committee for the next one, to be held in September 1939 in Cambridge, Massachusetts.[28]

One of Tarski's major goals throughout his career in Poland was to obtain a professorship in a university there: a position appropriate for conducting his research in mathematics, logic, and semantics; for guiding related research of other scholars; and for fostering worldwide acceptance of his framework for logical investigations. The first steps in that direction came easily. Tarski earned the *venia legendi* in 1925, which qualified him for teaching university courses, and in 1929, he was appointed assistant to Łukasiewicz.[29] Serving part-time in those capacities, Tarski offered some courses especially for teachers, some general courses on the methodology of the deductive sciences, and some courses on specific topics connected with his research, and he began supervising research of advanced students in Łukasiewicz's seminar. Positions of this sort paid very little. For his livelihood Tarski relied on his full-time employment as gimnazjum teacher, on his work training teachers during vacation periods, and, later, on his wife's employment as an elementary-school teacher.

---

[24] Tarski [1935] 1936a in the congress section on language, and [1935] 1936b in the section on logic.

[25] Warszawski Instytut Filozoficzny 1937; Kokoszyńska 1937.

[26] For further information about this meeting, see Polish Philosophy Congress 1936.

[27] Description from Langford 1938.

[28] Tarski [1956] 1983, viii; Jadacki 2003a, 158.

[29] See section 9.3.

*Maria Kokoszyńska,
Alfred Tarski
Paris, 1935*

**Maria Kokoszyńska** was born in 1905 in Bóbrka, a town near Lwów, in the Austrian Empire (now Bibrka, in Ukraine). At the University of Lwów, she studied with Kazimierz Ajdukiewicz and Kazimierz Twardówski. Under the latter's supervision, she earned the doctorate in philosophy in 1928. Kokoszyńska became an assistant at that university in 1930. In 1934, she began participating in meetings of the Vienna Circle. She met Alfred Tarski while there, and they became close personal friends. Around this time she married the historian and political activist Roman Lutman.* During the following years, Kokoszyńska contributed significantly to the literature of philosophical logic. In 1947, after World War II, she earned the *venia legendi* at the University of Poznań, and the next year she became a professor at the University of Wrocław. There she continued research in logic and served the university, the philosophical community, and the city schools in many leadership roles. Kokoszyńska retired in 1976 and died in 1981.[†]

---

*In the literature her surname is often listed as Lutman- or Lutmanowa-Kokoszyńska .
[†] Feferman and Feferman 2004, 88–92; Kokoszyńska c1961; Rojszczak 2002, 36; Wójcicki 1981.

Tarski entered the competition for a university professorship in Poland just after the establishment and expansion of academic institutions in Poland that followed its attainment of independence. But Poland faced hard economic times and difficult political questions during his entire career there. Early on, an opportunity appeared for a professorship in the philosophy faculty at the University of Lwów. Tarski was a strong contender, favored by Kazimierz Twardowski, the head of the logic program there and teacher of Tarski's own teachers and senior colleagues. But Tarski had not yet gained sufficient international recognition, and in fall 1930, after two years of deliberations, the position was awarded to an eminent and more senior candidate, Leon Chwistek.[30]

Tarski pursued a professorship throughout the 1930s, with no success. In such a quest, a scholar might expect to be supported by his own former research supervisor. But the previously cited sources clearly document Stanisław Leśniewski's antipathy toward his illustrious—but only—doctoral student, Alfred Tarski. In that connection and probably others, antisemitism played a role in frustrating Tarski's aspirations, in spite of his official conversion to Roman Catholicism and his change of surname. But the primary obstacle was simply the scarcity of professorial positions in Tarski's field in Poland, due to economic and political pressures.[31]

On the other hand, Alfred Tarski and Maria Tarska held reasonably congenial positions at the Żeromski gimnazjum, and his university research work and international reputation were soaring. In fall 1934 Tarski was named *adjunkt* of the Łukasiewicz seminar; this evidently provided more stature in the university hierarchy. Andrzej Mostowski entered the university in 1931, and completed the doctorate in 1938 with a dissertation on set theory supervised by Tarski. He was Tarski's first doctoral student, although his official sponsor was Kazimierz Kuratowski. Wanda Szmielew began studying logic in Warsaw in 1935; by 1938, she had published her first research paper.[32] Mostowski's dissertation and this research of Szmielew were very closely related to Tarski's early work in set theory. Moreover, Warsaw between the wars was a very exciting society, not just for

*[The narrative continues on page 330.]*

---

[30] The competition for the Lwów professorship is reported in Jadacki 2003a, 147–150, referring to Twardowski's correspondence. See also Feferman and Feferman 2004, 66–68, and Woleński 1995a, section 2. For more information on Chwistek, see page 329. (The self-portrait there is from Chwistek 1960.)

[31] See chapter 3. A professorship opened up in Poznań in 1937, for which Tarski was clearly the top candidate; it was left unfilled, probably to avoid offering it to him. For further information on Tarski's quest for a professorship, see Feferman and Feferman 2004, 100–108; Jadacki 2003a, 154–158; and Woleński 1995a, the first section 4, section 5, and notes 13–15.

[32] Mostowski 1938, reviewed in Tarski 1938c; Szmielew [1938] 1947—the initial date is from Kordos, Moszyńska, and Szczerba 1977, 241; Tarski 1924b, 1924e, and Lindenbaum and Tarski 1926—see section 8.3. For more information about Mostowski and Szmielew see section 9.4.

*Leon Chwistek*
*Self-portrait, 1929*

*Fencing*
*by Leon Chwistek, circa 1920*

**Leon Chwistek** was born in 1884 in Cracow, then part of the Austrian Empire. His father was a doctor who specialized in hydrotherapy; his mother, a pianist and artist. In 1891 the family built a house and clinic in Zakopane, a resort town in the Tatra mountains to the south. Chwistek was schooled in Cracow, then entered an art school and the university there in 1902, beginning studies in philosophy and mathematics. He earned the doctorate in 1906, with a dissertation *On Axioms* supervised by Stanisław Zaremba. Chwistek then taught in a gimnazjum. Leftist in politics, he was a long-time close friend of the noted artist and writer Stanisław Witkiewicz, also known as Witkacy. Chwistek spent the years 1908–1910 in Göttingen and Vienna, and 1913–1914 in Paris, studying mathematics, philosophy, and painting. There, he fought a famous duel, depicted above, with another Zakopane artist, Władysław Dunin-Borkowski. They had disagreed over a slight to Olga Steinhaus- ówna, the sister of the Lwów mathematician Hugo Steinhaus; Chwistek married her two years later. After service in Józef Piłsudski's Polish Legion during World War I, Chwistek returned to teaching. He established a reputation as artist, writer, and philosopher, publishing major works on art theory and the theory of types. Chwistek was a founder of the Formist movement in Polish art. In 1928 he was awarded the *venia legendi* in philosophy at the University of Cracow, and the next year won a competition with Alfred Tarski for a professorship at the University of Lwów. During the 1930s Chwistek continued developing his philosophy of logic and science, stressing that mathematical logic could not by itself form a basis for philosophy of the multiple realities of science. When the Germans and Russians conquered Poland in 1939, he emigrated to the Soviet Union. He taught for two years in Tiflis and a year in Moscow. Chwistek died there in 1944.*

---

*Estreicher 1971; Gromska 1948, 51–53; Skolimowski 1962. Witkacy created the portraits of Tarski and his wife, Maria, on the covers of Feferman and Feferman 2004.

*[continued from page 328]*

research, but in all cultural matters. In spite of outbreaks of antisemitism in the university, and the threats from Germany and the Soviet Union, which should have been all too plain, Alfred Tarski and Maria Tarska did not seek to emigrate.[33]

## 14.3 To the New World

Returning to Warsaw from a meeting at Amersfoort, in the Netherlands near Amsterdam, Alfred Tarski visited the philosopher Kurt Grelling in Berlin. There, on 1 October 1938, on the Wilhelmplatz, they witnessed a violent speech by Adolf Hitler, who had just returned to the Chancellery from his infamous Munich meeting with the leaders of Italy, Great Britain, and France. They had agreed to let Germany annex the Sudeten part of Czechoslovakia. The crowd in the adjacent Wilhelmplatz was ecstatic. Below are some headlines from the next day's Nazi party newspaper.[34]

**Die Reichshauptstadt huldigt dem Befreier des Sudetenlandes**

Berlin dankt Adolf Hitler im Namen des 80-Millionen-Volkes     Im Schutz der scharfen Waffe des Reiches

Wilhelmplatz: Ein einziger Orkan jubelnder Begeisterung

*The capital of the Reich pays homage to the liberator of the Sudetenland*

*Berlin thanks Adolf Hitler    Under protection
in the name of               of the sharp arms
the 80-Million-Volk          of the Reich*

*Wilhelmplatz: a unique hurricane of jubilant exaltation*

---

[33] For more information on Warsaw between the wars, see Wynot 1983, chapter 8; Nowicki 1992; and chapters 1, 3, and 9 of the present book. For information about the widespread movement during the 1930s to humiliate and restrict the education of Jewish students in Poland, see Natkowska 1999. Antisemitism penetrated deeply and peculiarly. For example, the following exercise (Rusiecki 1930b) was posed in *Parametr*, the journal that included many of the exercises translated in chapter 12:

> In a classroom the Jews constituted 22.58% of the total number of students. Determine how many students were in the class!

The problem really was to determine positive integers $m, n$ such that $0.22575 \leq m/n \leq 0.22585$ and $n$ could be the size of a school class. Most present-day readers would cringe at the gratuitous reference to Jews, but those in the intended audience could do so only in silence. The present editors found in that journal no other exercises posed in loaded ethnic terms.

[34] *Völkischer Beobachter* 1938. The Amersfoort congress took place during 19–25 September; the Munich agreement and acceptance by Czechoslovakia, on 30 September. According to Domarus 1992, 1213–1214, Hitler returned to Berlin on 1 October for two days only. See also Feferman and Feferman 2004, 103.

After the demonstration, Tarski spent a long evening in discussions with other scholars who had been there. Tarski's biographers Anita B. and Solomon Feferman expressed amazement that Tarski risked stopping in Berlin, but noted that he

> was not alone thinking that somehow Poland would be protected by England and France. This is not to say that he was unaware of...what was occurring in Europe and in his own country. Quite the opposite. He was well informed politically...But his actual involvement [in politics] was minimal; his own projects and those of his students consumed him, and he had those two...jobs.

Moreover, as late as 21 December 1938—after the terrible and widespread 10 November *Kristallnacht* pogroms in Germany—Tarski attended a reception celebrating the honorary doctorate that had been awarded to Jan Łukasiewicz by the University of Münster. Notable Germans were present, including the ambassador. Still, Tarski was evidently not considering emigration, beyond changing his publication language to English.[35]

What *were* Tarski's principal concerns during that academic year? The Tarskis' daughter Ewa Krystyna (Ina) was born in Warsaw on 3 October 1938.[36] Whether her mother still taught school after giving birth to her daughter is not known.[37] Alfred continued teaching at the gimnazjum and university. In fall 1938 he offered his course on teaching mathematics; in spring 1939, the one on methodology of deductive sciences.[38] His supervision of research in connection with Łukasiewicz's seminar continued, as mentioned in the preceding subsection. An outgrowth of that was his work with his former student Andrzej Mostowski on the metatheory of well-ordering, which would be continued a decade later (see section 8.2). Moreover, Tarski was continuing to collaborate with his colleague Adolf Lindenbaum, both on questions of equidecomposibility and on a large monograph about set theory, which they had begun several years before (see sections 8.5 and 15.12). Finally, Tarski was pursuing his own projects. A large paper in German was still being composed for publication in *Fundamenta Mathematicae*; it belongs to the filament of Tarski's research on the measure problem,[39] discussed in section 8.7. His [1940] 1967 monograph on the completeness of elementary arithmetic and geometry was under preparation for publication in English in Paris; he had also begun that work

---

[35] Jadacki 2003a, 158. Tarski's 1936a and 1936b papers and the original edition of his [1936] 1995 logic text were was his last major works in Polish; 1937b was his last in French. His first publication in English (1937a) was a technical appendix to Joseph H. Woodger's 1937 book *The Axiomatic Method in Biology*. Tarski had met Woodger at Vienna Circle functions.

[36] Feferman and Feferman 2004, 104; Ewa Krystyna's birthday is stated incorrectly there. The date given here was confirmed by Jan Tarski (2012).

[37] Witold Kozłowski (2010) recalled Maria Tarska as a secretary at the Żeromski Gimnazjum, not as a teacher. He may have confused Maria with her sister Józefa Zahorska, who was indeed a secretary there.

[38] The 1937 offering of the methodology course was described in section 9.3.

[39] Tarski 1939–1945 was Tarski's last paper in German. The editorial note *Fundamenta Mathematicae* 1945 explained its publication dates. The papers Tarski 1939a and 1939b and Mostowski and Tarski 1939, on logic, set theory, and Boolean algebras, respectively, were probably completed during the previous academic year.

*Voyage to New York*
*11–22 August 1939*

*Menu, Cover Art*

M/B „PIŁSUDSKI"

**OBIAD**

Consommé z diabłotkami                Bortsch with Potatoes
Barszcz zabielany z kartoflekami

Dorsz z wody po polsku                Cod-fish, Polish Style

Comber cielęcy, Renaissance           Saddle of Veal, Renaissance
Pularda z rożna z borówkami           Roast Chicken with Cranberries

Karotka    Fasolka zielo              Carrots         String Beans
      Sałata zielona                          Lettuce

Kartofle smażone                      French Fried Potatoes

Ciastko migdałowe                     Almond Cake

Owoce                                 Fruit
Kawa      Herbata                     Coffee      Tea

## 14.3 To the New World

a decade earlier: see section 9.4. Finally, Tarski was negotiating about publication of a paper on the significance of the concept of truth, which he had drafted the previous year.[40]

The Fefermans reported that Tarski had received an official invitation to speak at the upcoming September 1939 Fifth Congress on the Unity of Science in Cambridge, Massachusetts, but had delayed his response. They attributed that to Stanisław Leśniewski's death from thyroid cancer on 13 May 1939, and Tarski's desire to be present for any deliberations about Leśniewski's successor, a position for which Tarski would obviously be the leading candidate.[41] According to the biographers, the Harvard professor and logician Willard Van Orman Quine repeated the invitation and, in addition, urged Tarski to consider emigration—Quine would set up some appropriate contacts.

Tarski accepted only at the last minute. He obtained an exit visa on 7 August and probably boarded a train for the port city, Gdynia, on 10 August. He sailed on 11 August on the ocean liner M/S Piłsudski. Tarski packed only a single suitcase, with light clothing and some formal wear, intending to stay only about a month. On board, he was surprised to meet the Lwów mathematician Stanisław Ulam, en route to a position as lecturer at Harvard. Ulam was accompanied by his young brother Adam, ready to start university studies at Brown University, in Providence, Rhode Island.[42] Some news bulletins were aired on the ship, and passengers would have discussed the situation in Poland seriously at meals. A photograph in the Feferman biography shows those three dining, in formal dress. The ship itself and a menu are depicted on the facing page. The voyagers seemed to deny the reality of the danger at home. On their pictured 22 August arrival in New York City, they were met by very apprehensive colleagues, who did not: John von Neumann for the Ulams, and Carl Hempel for Tarski. Decades later, Adam Ulam, by then a distinguished Harvard political scientist, remarked,

> In hindsight we realized we must have been blind ... we really didn't think war was *immediately* imminent. Sure, everyone talked about the possibility of war ... but there was also a lot of talk and bravado about how France and England would defend Poland. In August ... when we left, all the Poles were going off on their traditional August vacations, as if everything was normal.

---

[40] According to Paolo Mancosu (2009, 132), Carl Hempel regarded the paper on truth as appropriate for the journal *Erkenntnis*, and Tarski had even in 1936 described the monograph with Lindenbaum as nearly ready for publication. No further trace of these drafts has surfaced. In the Tarski archive at the Bancroft Library in Berkeley is a complete draft of an unpublished set-theory monograph by Richard Montague, Dana S. Scott, and Tarski, circa 1970.

[41] Feferman and Feferman 2004, 106. Tarski was himself a member of the organizing committee for the congress.

[42] **Stanisław and Adam Ulam** were born in Lwów in 1909 and 1922, respectively, to a prosperous Jewish family. Stanisław earned the doctorate there in 1933 under supervision of Kazimierz Kuratowski. His early research interests were closely related to Tarski's work in set theory. Stanisław emigrated to the United States in 1935, spent a year at Princeton to collaborate with John von Neumann, and then three years at Harvard as a junior fellow. Stanisław became a leading figure in applied mathematics in the United States. Adam emigrated at the last possible moment, in August 1939, to start undergraduate studies at Brown University. He earned a doctorate in political science from Harvard in 1947 and spent a long career as professor there, an authority on Russia and the Soviet Union. Stanisław died in 1984 and Adam, in 2000. See Ulam 1976 and *Harvard University Gazette* 2000.

On the day after Tarski arrived in New York City, the Germans and Soviets concluded a secret agreement to destroy the Polish state.[43]

After sightseeing and socializing in New York City, Tarski traveled on 24 August by train to Boston. He spent a week in nearby Cambridge to prepare for the congress. On 1 September, the Germans invaded Poland. Stanisław Ulam reported that he and others spent hours listening to the radio news. The congress began on schedule on 3 September. The German philosopher Rudolph Carnap, who had emigrated to a professorship at the University of Chicago, commented,

> In spite of the exciting world events, we found it possible to devote ourselves to the theoretical discussions of the Congress.[44]

Tarski presented his paper *New Investigations on the Completeness of Deductive Theories* on 9 September, the last day of the congress, at a concurrent meeting of the Association for Symbolic Logic. Scheduled just after his talk were two by Kurt Grelling and Janina Hosiasson-Lindenbaumowa; another, by the Lwów logician Leon Chwistek, had been set on another day. Those speakers were ominously absent—read the boxes that follow.[45]

---

**Kurt Grelling** was born in Berlin in 1886. His father, Richard, was a lawyer, active in politics; his mother stemmed from a wealthy Jewish merchant family. Kurt finished his schooling in Gotha in 1904, then studied mathematics and philosophy at Göttingen. There, he discovered the antinomy about heterological concepts that is now named for him. It was published jointly with the philosopher Leonard Nelson. Under supervision of David Hilbert and Ernst Zermelo, Grelling earned the doctorate in 1910 for research on set theory and the axiomatics of arithmetic. He translated several philosophical treatises and wrote one on probability theory, but never habilitated, which was required for a professorship. Instead, he taught in secondary schools and immersed himself in the labor movement and politics. His argument against his father's indictment of the German instigation of World War I was publicized widely. Kurt Grelling became a leading exponent of the empiricist philosophy of the Berlin School, in association with Hans Reichenbach. During the 1930s he pioneered work on the logical foundations of gestalt psychology. He was fired from his teaching job in 1933, but by then was financially independent. But with no means of support outside Germany, and trouble obtaining travel documents, he and his wife left only in 1938, for Belgium. After the Germans conquered that country in 1940, they deported the Grellings to Vichy France, and thence to Auschwitz for extermination in 1942. For further information, see Luchins and Luchins 2000.

---

[43] Feferman and Feferman 2004, 124–127, which includes Adam Ulam's remark. The voyage is also recounted in Ulam 1976, 114–115, with the ship incorrectly identified. By 1939 Neumann and Hempel had emigrated from Hungary and Germany and had found positions at Princeton and City College of New York, respectively.

[44] Feferman and Feferman 2004, 128, which includes Carnap's remark; Ulam 1976, 119.

[45] Tarski 1939a is an abstract of Tarski's talk. For information about Chwistek see the box on page 335.

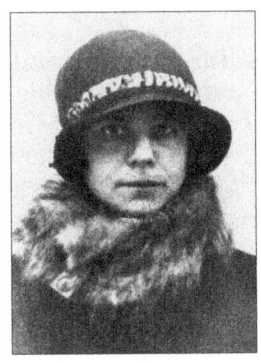 

*Janina Hosiassonówna*      *Adolf Lindenbaum*

**Janina Hosiassonówna** was born in 1899 in Warsaw. She studied mathematics and philosophy there, earned a doctorate in 1926 with a dissertation on inductive logic supervised by Tadeusz Kotarbiński, and became a secondary-school teacher. Hosiassonówna continued research on inductive logic in Warsaw, and spent some time around 1929–1930 in Cambridge, England, studying the work of several scholars there on probability and induction. She played a significant role in establishing such studies within the framework of formal logic. Hosiassonówna also translated into Polish several books by Bertrand Russell. After her marriage to Alfred Tarski's research collaborator Adolf Lindenbaum, she used the surname Hosiasson-Lindenbaumowa. She was scheduled to present her research at the Unity of Science Congress at Harvard in September 1939, but was unable to sail in time. After that month's invasions, she escaped the Nazi oppression temporarily by moving to Vilnius, which had been annexed by the Soviet Union. The Germans invaded that region in 1941; they arrested and murdered Janina Hosiasson-Lindenbaumowa in 1942.\*

**Adolf Lindenbaum** was born in Warsaw in 1904, to a Jewish family that became involved in the motion-picture business; he was independently wealthy. After graduating from gimnazjum in 1922, Lindenbaum entered the University of Warsaw to study mathematics. He became active in student organizations and left-wing politics, and earned the doctorate in 1927 under supervision of Wacław Sierpiński. Over the next decade, Lindenbaum contributed many results to the research seminar of Jan Łukasiewicz and Alfred Tarski, and published about twenty papers about general topology, set theory, and mathematical logic, including especially significant ones authored jointly with Tarski and Tarski's student Andrzej Mostowski. Some now-fundamental concepts in logic are named for Lindenbaum. He married another Warsaw logician, Janina Hosiassonówna. During 1940–1941, under the Soviet occupation, he taught at the pedagogical instutute in Białystok. Adolf Lindenbaum was murdered after the 1941 German invasion, near Vilnius.\*\*

---

\* Galavotti 2008; Gromska 1948, 57–59; Jedynak 2001.
\*\* Sękowska and Węglowska 2003; Woleński 1995a, 375–376.

## 14.4 Epilogue

After the Congress on the Unity of Science, Alfred Tarski was stranded in the United States. After some difficulties with immigration status, he was able to obtain barely adequate support while researching and teaching at various East Coast institutions. Eventually, in September 1942, he obtained a position at the University of California at Berkeley. The fighting in Europe ended on 30 April 1945; soon after that, Alfred became a citizen of the United States. After eight months' delay making arrangements, Tarski was reunited in Berkeley with his wife, Maria; son, Janusz; and daughter, Krystyna. After the German invasion in 1939, they had stayed in Warsaw with Maria's sister Józefa Zahorska. After the Warsaw Uprising, August–October 1944, they had found shelter away from that city. Maria had succeeded in concealing her children's Jewish ancestry.

Alfred Tarski's parents, brother, and his brother's family were murdered by the Germans. His colleagues Leon Chwistek, Janina Hosiasson-Lindenbaumowa, and Kurt Grelling had not been able to leave Poland and Germany for the congress in time. Chwistek escaped to Moscow, but sickened and died there. The others, and Tarski's colleague Adolf Lindenbaum, were murdered.

Tarski's students Andrzej Mostowski and Wanda Szmielew survived the war. She came to Berkeley in 1947, earned the doctorate there in 1950 as Tarski's fifth doctoral student, then returned to Warsaw. Both of those young scholars became leaders in logical research in postwar Poland.

In Berkeley, Alfred Tarski founded a research program in logic and methodology of science. Soon it became the world's leading program in that subject, and Tarski attained recognition as the world's preeminent logician during the next three decades.[46]

---

[46] Readers may obtain much further information about Tarski's work in logic and his career in the United States from the marvelous 2004 biography by Anita B. and Solomon Feferman. It will also be useful to consult the numerous other works listed in chapters 17 and 18 of the present book.

# Part Four
# Supplement

This book's concluding Part Four consists of four chapters with different purposes. The twelve sections of chapter 15 contain the remaining translations required to make all of Alfred Tarski's works accessible in at least one of the languages English, French, and German. All of these items are short. In content they are generally unrelated to each other and to the larger papers translated in previous parts of the book. Brief discussions are provided to set them in context. Chapter 16 updates Steven R. Givant's 1986 Tarski bibliography by identifying all known items authored by Tarski and first published after that date. Chapters 17 and 18 list and briefly annotate major published studies of Tarski's life and work. Their items were chosen because of unique relationships between the authors and Tarski or because of the breadth of their coverage. The lists in chapters 17 and 18 are a biproduct of the research that led to the background discussions in earlier parts of this book. The editors did not search exhaustively for material of the same sort not mentioned in items discovered in that research.

Part Four employs boxed text somewhat differently from the earlier parts of this book. As before, boxes with single borders contain informational essays that can be read independently. Readers are not expected to visit them in sequence. But doubly outlined boxes, as in chapter 12, enclose translations of Tarski's writing (and one abstract by Jan Łukasiewicz) and separate them from background discussions by the present editors. They are intended to be read in sequence. In several sections, doubly outlined translation boxes are continued from one page to the next by omitting the bottom border of the one and the top border of the next.

# 15
# Assorted Contributions

This chapter is devoted to translations from Polish and discussions of twelve minor publications of Alfred Tarski, in chronological order. They fulfill this book's goal of making all of Tarski's publications accessible in English, French, or German. They are included in this supplementary part of the book because they are not closely related to any of the works featured in earlier parts. The associated discussions of their context and content are sufficient to make them intelligible.

The items in sections 15.1–15.3 and 15.8–15.11 are Tarski's remarks included in published reports of other scholars' presentations at meetings in Warsaw in 1925 and 1927, and in Cracow in 1936. They are about various aspects of philosophical logic.

Section 15.4 features a translation and discussion of a 1928 Warsaw presentation by Jan Łukasiewicz: *On Definitions in the Theory of Deduction*. Łukasiewicz announced some results of Tarski, who contributed to the subsequent discussion. The whole paper is translated here because Tarski's work is not easily separated from it.

Tarski's 1929 paper on Polish pensions is translated in section 15.5. Published in a journal on economics, it revealed a mistake in a pension law. Its nature is that of a report on an involved applied exercise in elementary algebra.

The 1930–1931 paper *On the Concept of Truth in Reference to Formalized Deductive Sciences* is probably the most interesting item in this chapter. Translated in section 15.6, it is Tarski's report of a presentation in Lwów, the first appearance in print of his celebrated theory of truth. His full account would not reach publication for several years.

Tarski's 1932 Abstract *On Geometric Properties of Banach's Measure* is translated in section 15.7. The title of this record of his presentation to a Warsaw meeting was listed in French in earlier surveys, but its text was in Polish.

The chapter concludes with translations of two letters that Tarski wrote to Wacław Sierpiński from Berkeley in 1946 and 1947. They are included here because they show how a thread of Tarski's geometric research in Poland extended into his career in California.

## 15.1 Discussion of Łukasiewicz 1925, on Understanding Deduction

On 8 December 1924, Jan Łukasiewicz presented a paper to a meeting of the Logical Section of the Warsaw Philosophical Institute. This event belonged to the same series in which Alfred Tarski had presented two papers in 1921. The proceedings were edited by Bolesław Gawecki and published in volume 28 of the journal *Przegląd filozoficzny*. The published report of Łukasiewicz's presentation, entitled *On a Certain Way of Understanding the Theory of Deduction,* seems to be an edited version of a listener's notes.[1] It includes comments by Kazimierz Kuratowski, Stanisław Leśniewski, and Tarski, but no response from the speaker.

Łukasiewicz presented a graphical means of displaying formulas of Boolean logic with colored tiles representing two connectives and propositional variables, and of manipulating them according to the substitution and modus-ponens inference rules to provide proofs. He implemented this as a kind of single-player game. According to Łukasiewicz, the fact that the tiles convey no intuitive content shows that the process of deduction involves only the form, not the content, of the formulas.

Kuratowski remarked that it would be useful to include additional types of tiles to incorporate quantificational logic.

Leśniewski echoed and extended Kuratowski's comment. But he saw no advantage of this game over ordinary presentations. He noted that it shows clearly that grouping-symbols such as parentheses are unnecessary, and that Łukasiewicz's description of the game included a recursive definition of *formula* in this system of logic.

Tarski commented,

> The game suggests that for the future science of the theory of deduction, undeveloped at this point, it might be possible to find a geometric interpretation that would make it possible to carry out a proof of the consistency of this science.

That remark is listed as Tarski [1925] 2014 in the present book's bibliography.

Leśniewski's discussion reveals aspects of the state of development of logic in Poland in 1924. The ideas of parenthesis-free notation and recursive definition of linguistic constructs were new enough to warrant his attention. Łukasiewicz first published his

---

[1] Łukasiewicz 1925. The Polish title is *O pewnym sposobie pojmowania teorji dedukcji*.

now-famous parenthesis-free notation four years later.[2] On the other hand, in a 1936 letter to the Austrian philosopher Otto Neurath,[3] Tarski recalled that while the view of linguistic constructs as physical pictures had indeed been debated and regarded positively by the Vienna Circle before 1930, it had been common in Warsaw since at least 1918, as illustrated by Łukasiewicz's 1925 report.

Łukasiewicz's game was a precursor of a more elaborate and rather famous one published decades later by Layman E. Allen: *Wff 'N Proof*. Allen used and explicitly acknowledged Łukasiewicz's notation but was unaware of the earlier game.[4]

In the comment displayed on the preceding page, Tarski noted that geometric representation of formulas might lead to a method for proving consistency. That might be related to his later work on geometric dissections, in which he drew metamathematical conclusions from other geometric considerations: see section 8.5.

## 15.2 Discussion of Greniewski [1927] 1928, on Action

At the Second Polish Philosophical Congress in September 1927 in Warsaw, Henryk Greniewski presented the paper *An Attempt to Formulate Precisely Some Notions of the Theory of Action*. The congress is described in section 14.2. The present section summarizes Greniewski's talk, displays Alfred Tarski's [1927] 2014a comment about it (translated, in the box on the next page), and describes the speaker's response.[5]

Greniewski's goal was to make precise the notions *plan* and *method*, which were needed for a theory of action. He claimed that his work was inspired by Tadeusz Kotarbiński or Jan Łukasiewicz. According to Kotarbiński,

> ...if the plan is a certain description of a certain selection and sequence of actions, the method is nothing more nor less than that very planned selection and sequence of actions integrated by a common goal.[6]

Greniewski first noted that his system would consider not only individuals, but also generalized individuals denoted by pseudonyms such as *Zeus* that stand for certain classes of properties. Just as an event such as *Aristotle philosophized* should be a pair consisting of an individual and a class of individuals (for example, philosophers), a *generalized event*

---

[2] Łukasiewicz remarked ([1931] 1970, 180): "I came upon the idea of a parenthesis-free notation in 1924. I used that notation for the first time in [Łukasiewicz 1929, 610 ff]. See also [Łukasiewicz [1929] 1963, §§6 ff]...."

[3] Tarski [1930–1936] 1992, 26. These posthumously published letters are described in section 16.1.

[4] Allen 1965, 2010. The title *Wff 'N Proof* is a double pun: *Wff* is a common English abbreviation for the logical term *well-formed formula*; the Whiffenpoofs are a noted Yale University male a capella choral ensemble. The game originated at Yale.

[5] The Polish title of Greniewski [1927] 1928 was *Próba precyzacji pewnych pojęć teorji czynu*. For information about its author, see the box on page 343. An English translation of part of Tarski's remark was included in Jan Woleński's paper *Tarski as a Philosopher* (1993, 325–326).

[6] Kotarbiński [1955] 1965, 56.

should consist of a generalized individual and a class: *Zeus thundered*. Greniewski considered time-dependent generalized events, and sets of them that might ramify to represent possible future events. These would constitute *plans*, which are thus objects of higher-order types. Plans could be compared and judged equivalent by a suitable relation, and a *method* would be an equivalence class of plans.

The subsequent discussion included the following remark, which is listed as Tarski [1927] 2014a in the present book's bibliography.

> *Alfred Tarski* drew attention to the origin of the ideas contained in the paper, and pointed out that in interpreting the most varied terms in pure logic, a certain moderation is desirable, so that they retain only their formal properties. An interpretation is desired in those cases in which the terms satisfy two conditions: (1) these terms find an application in any field of knowledge; (2) in experience (in the world of concrete objects) no intuitive equivalents of these terms are found.

In response, Greniewski granted that too great a reliance on higher-order types would be harmful, but, as noted in a previous remark by Kazimierz Ajdukiewicz, abstraction with equivalence classes simplifies the theory, as it does for arithmetic. Moreover, Greniewski noted, there is no assurance that mental sensations should belong to the lowest type of logical objects: that is a very dark issue.

Tarski's remark certainly related to the subject at hand, particularly in view of Łukasiewicz [1927] 1928, the keynote address of the congress. But it does not seem directly related to the published version of Greniewski's presentation. Moreover, Greniewski's response does not seem to address Tarski's remark directly. Probably, other conversations were going on that were not recorded.

A recent survey[7] of the theory of action reveals a rich, broadly applicable subject that certainly contains traces of the ideas expressed in Kotarbiński's and Greniewski's early work but does not mention that as a forerunner. An investigation of this possibility would prove interesting. Both of those scholars continued working in this area and its applications for decades.

---

[7] Segerberg 2009.

> **Henryk Greniewski** was born in Warsaw, then a part of the Russian Empire, in 1903. He was schooled at the same Zamoyski Gimnazjum on Smolna Street that Tadeusz Kotarbiński had attended a generation earlier; somehow, they became lifelong friends. Greniewski studied briefly at the University of Warsaw, then completed doctoral studies in 1926 at Cracow with a dissertation on logical foundations of mechanics supervised (informally) by Leon Chwistek. A second concentration was the mathematics of finance and insurance, which gained him secure employment in the financial sector and at the Social Insurance Institution. During the 1930s, under the pseudonym Kozłowski, he was the author of many articles in the journal *Gospodarka narodowa* (*National Economy*).
>
> During World War II, Greniewski participated in underground socialist organizations. In 1948 he became director of the Group for Mathematical Apparatus of the Mathematical Institute of the University of Warsaw, supervising construction of electronic computers. During 1951–1968 he served the university as professor of education, then director of the Department of Econometrics. Greniewski maintained activity in logic, serving as an editor of *Studia Logica* from 1953 on, and lecturing on the foundations of computer science. He was known as an authority on the use of information technology in business planning. Greniewski died in 1972.*
>
> ---
> * Trzęsicki 2010; Chodorowski et al. 2005, 49.

## 15.3 Discussion of Czeżowski [1927] 1928, on Causality

In September 1927, Tadeusz Czeżowski presented the paper *On a Causal Relation* at a session on philosophy of natural sciences at the Second Polish Philosophical Congress in Warsaw. The congress is described in section 14.2. Czeżowski's paper is summarized here, followed by a translation of Alfred Tarski's [1927] 2014b comment about it, in the box on the next page. Two other comments and the speaker's response are paraphrased after that.[8]

Czeżowski discussed various properties of phenomena and relations between them. It is unclear what he meant by *phenomena*,[9] but they could be temporal, spatial, or repeating, and contiguous in time or space. He defined *direct formal causal relationship* to stand between two classes $A$ and $B$ of phenomena just in case for each $x \in A$ there exists $y \in B$ such that $x$ and $y$ are contiguous in time and space (or just in time if $x$ and $y$ are not both spatial). Further, two phenomena $a$ and $b$ stand in *direct material causal relationship* just in case there exist classes $A$ and $B$ of phenomena such that $a \in A$, $b \in B$, and $A$ and $B$ stand in direct formal causal relationship. *Indirect* formal and material causal relationship can be defined by generalizing the direct notions. Czeżowski did not specify the generalization in detail.

---

[8] The Polish title of Czeżowski [1927] 1928 was *O stosunku przyczynowym*. For information about its author, see the box on page 345. Tarski [1927] 2014b was listed as item 28ᶜa in the bibliography Givant 1986, but was omitted from the Tarski 1986a *Collected Papers* volumes.

[9] Czeżowski's word was *zjawiska*.

Czeżowski claimed that seeking a *formal* causal relationship between classes of phenomena is analogous to empirical inductive reasoning; no such relationship is logically necessary. However, mere contiguity does not suffice to establish a *material* causal relationship between two phenomena. One must verify that they constitute an example of a formal causal relationship; that step may involve reasoning from logical necessity. Czeżowski suggested that there is an analogy between the notions of material and formal causal relations and those of material and logical consequence.

The ensuing discussion included the following remark, which is listed as Tarski [1927] 2014b in the present book's bibliography.

---

*Alfred Tarski* pointed out the following consequences of the definitions given by the speaker:

(1) Every phenomenon is a direct material cause of every other phenomenon contiguous with it in space and time.

(2) Every phenomenon is an indirect material cause of every other phenomenon.

(3) Every class of phenomena is an indirect formal cause of every other class of phenomena.

---

In the discussion, Kazimierz Ajdukiewicz noted that Tarski's remark would apply even if in the definition of material causality, classes $A$ and $B$ were required to have more than one element.[10] Henryk Mehlberg[11] commented that the analogy between causality and implication is incomplete because the fact that a falsehood materially implies any proposition has no analogue for causality. Moreover, he doubted that there could be any direct causal relationship according to this definition if time varied continuously.

Czeżowski responded that the objection of Tarski and Ajdukiewicz applied only if *phenomenon* was understood as the state of the entire universe at a given time, in which case it said that the state at any moment is the cause of the state at the next. To Mehlberg he responded (roughly) that nevertheless there are similarities between the theory of causality and the theory of formal deduction and logical truth, and that the notion of contiguity is certainly related to that of continuity.

---

[10] For more information on Ajdukiewicz, see a box in chapter 3.

[11] Born in 1904 near Tarnopol, then in the Austrian Empire but now in Ukraine, **Henryk Mehlberg** studied with Kazimierz Twardowski and Kazimierz Ajdukiewicz in Lwów in the 1920s and remained there through World War II. In 1948 he emigrated to Toronto and soon moved to the University of Chicago, where he continued a distinguished career as a philosopher of science. He specialized in theories of time and causality. Mehlberg retired in 1970 and died in 1979. (Koterski 2001.)

The lack of precise language and references makes it difficult to interpret Czeżowski's paper, and must have contributed to the tone of Tarski's devastating remark. All three discussants' objections were valid. The paper is an early, but unsuccessful, step in lifting discussions in this area of traditional philosophy to the level of rigor that Jan Łukasiewicz demanded in his keynote address at the congress.[12]

*Tadeusz Czeżowski around 1925*

**Tadeusz Czeżowski** was born in Vienna in 1889; his father was a government official. Schooled in Lwów, Tadeusz attended the university there and earned a teaching credential in 1912. He taught in schools while he continued studies in logic with Kazimierz Twardowski. Czeżowski completed the doctorate in 1914 with a dissertation on the theory of classes. After military service, he worked with Twardowski as a university administrator. Then Czeżowski become director of the Department of Science and Secondary Eduation in the Ministry of Religious Denominations and Public Education. He was heavily involved in organizing the new Polish universities in Warsaw, Vilnius, and Poznań. Czeżowski habilitated in 1920, and in 1923 became professor at Vilnius. He was named full professor there in 1936 and held administrative posts during 1933–1939. Czeżowski contributed to many areas of philosophical logic and became a major figure in the Polish logic community. During World War II he taught philosophy clandestinely. Czeżowski was honored by the State of Israel with the designation "Righteous among the Nations" for undergoing great risk to save Vilnius Jews from the Holocaust. After the war Czeżowski transferred to Toruń, again helped establish a new university, and founded the Toruń Philosophical Society. For many years he edited the journal *Ruch filozoficzny*. Czeżowski retired in 1960 and died in 1981.*

* Gumański 1984

---

[12] Łukasiewicz [1927] 1928. See section 14.2.

## 15.4 Łukasiewicz 1928–1929, on Definitions

On 18 February 1928, Jan Łukasiewicz presented a paper to a meeting of the Logical Section of the Warsaw Philosophical Institute. This meeting belonged to the same series as the one reported by Łukasiewicz 1925, which is summarized in section 15.1. The proceedings of the meeting were published in the journal *Ruch filozoficzny*, volume 11, 177–178. The published report of Łukasiewicz's presentation, entitled *On Definitions in the Theory of Deduction*,[13] seems to be a loosely edited version of a listener's notes. Some of the research that it mentions is described only vaguely. It includes comments by Stanisław Leśniewski and Alfred Tarski, but no response from the speaker.[14]

In his presentation, Łukasiewicz compared two methods for handling definitions in propositional logic. One way, favored by Leśniewski during the discussion, is to build a notion of definitional equivalence into the system under study. Tarski provided some technical steps toward that end. Łukasiewicz favored regarding definitions as external to the system, keeping it simpler and more focused. But he presented the advantages and disadvantages of both methods. This paper provides an interesting preview of issues that would reappear a generation later, in the design of programming languages.

Łukasiewicz included discussion of some results of Tarski and of Mordchaj Wajsberg. For that reason, this paper is listed in the appendix of the Tarski bibliography, Givant 1986, but Tarski's discussion is not listed there. No part of the work is reproduced in Tarski's 1986a *Collected Papers*. Therefore, it is translated in its entirety in the following boxes, as Łukasiewicz [1928–1929] 2014. The translation adheres to the same standards as earlier ones in this book. Items in [square] brackets and all footnotes, which fall outside the boxes, are the work of the present editors.

---

### On Definitions in the Theory of Deduction

Prof. Dr. Jan Łukasiewicz

The speaker recognizes two types of definitions. [Definitions of type (1) are] found in *Principia Mathematica*: for example, $p \supset q .=: \sim p \vee q$ Df.[15] The authors of *Principia* say that the "=" and "Df" form a whole, and what is on the left side [of the "="] means the same as what is on the right. [Method (2)] defines primitive symbols in the system. It would seem that the only term suitable for use in a definition [in the original system] is equivalence. However, we do not

---

[13] The Polish title of this paper, Łukasiewicz [1928–1929] 2014, is *O definicyach w teoryi dedukcyi*. In that context, the term "theory of deduction" referred to propositional logic.

[14] Additional comments by Henryk Greniewski and Bronisław Knaster were mentioned but without detail.

[15] Whitehead and Russell 1910–1913, vol. 1, 11.

## 15.4 Łukasiewicz 1928–1929, on Definitions

know any theory of deduction with equivalence as a primitive term. In a system whose primitive term is implication, Alfred Tarski first used definitions in the form of two implications. That is the second method of definition.

The speaker gives an example that applies both of these methods of definition in the proof of a theorem—specifically, definitions[16]

(1)  *CNpq* || *Apq*         (2)  *CApqCNpq* and *CCNpqApq*.

[In this discussion he is] not treating the first definition as a hypothesis of the system, but as a rule that allows us to write everywhere, in place of what is on the left side of the definition, that which is on its right side, and vice versa. A proof turns out to be shorter with the first method of definition.

The mere fact that the first method of definition is more consistent with what we intuitively mean by a definition, and that the second method introduces definitions as hypotheses, similar to axioms, leads to [the speaker's] inclination to adopt the first method. It would seem, however, that definitions of the first type are logically stronger than [those of] the second. In fact, having the identity law *Cpp* in the system as well as a directive definition[17] such as *CNpq* || *Apq*, you can write the sentence *CCNpqCNpq* then replace the first occurrence of *CNpq* by *Apq* and [do this a second time with] the second one. Two hypotheses are obtained in this way, which constitute a suitable implicational definition.

Due to the following facts the speaker is definitely inclined to adopt the first type of definition.

(1) Mordchaj Wajsberg found a sentence, [using just the connectives] implication and negation, sufficient to build a theory of deduction, but only by adopting a definition of the second kind. He also showed that by using such a definition (namely, definitional hypotheses *CMqNq* and *CNqMq*) one of the speaker's axiom systems can be simplified. The speaker presents a way to simplify [this], which relies on [the possibility of] deducing from two axioms a third, independent of these,[18] expressed in terms of the primitive [connectives] only, using the above definition. According to the speaker, the purpose of these definitions lies in replacing the more complex expressions by simpler ones that have more intuitive value, rather than in introducing new theses into the system.[19]

---

[16] These are definitions of the disjunction connective *A* in terms of implication *C* and negation *N*: that is, of "*p* or *q*" as "not-*p* implies *q*."

[17] Łukasiewicz's term was *definicyę dyrektywalną*.

[18] Łukasiewicz's phrase was *niezależnego od nich*.

[19] The present editors have been unable to locate any other report of this work of Wajsberg.

Mr. Tarski stated that in writing a definition in the form of two implications he [would] adjoin *Cpp* to the system. What is adjoined, moreover, [should also be] adjoined with definitions of the first type, since, as has already been noted, from such definitions suitable implicational definitions can be obtained using *Cpp*. The speaker adds that directive definitions cannot imply any statements about the primitive terms independent of the axioms; indeed, having obtained a statement about primitive terms by such a definition, the same definition can be used to replace the introduced term by the primitive [one] everywhere in the proof. Therefore, considering the previously mentioned theorem of Mr. Tarski, a new hypothesis, the identity law, [should be] given with an implicational definition.[20]

(2) The speaker cites more facts confirming his conviction that we must use definitions of the first type. He assumes David Hilbert's axioms without [the law of] commutation.[21] Adjoining to these axioms a definition of the expresion *Apq* in the form of implicational hypotheses *CApqCNpq* and *CCNpqApq* does not allow a statement to be obtained that differs from one of the axioms only in that *Apq* stands in place of *CNpq*. (Of course, it can be obtained immediately by introducing the definition *CNpq* || *Apq*.) On the other hand, there are in this case as many as $2^{12}$ different interpretation tables[22] for *Apq*. Therefore, with the [implicational] definition above, the expression is ambiguous. But directive definition of the expression *Apq* allows here just one interpretation.

In the discussion, Stanisław Leśniewski, Henryk Greniewski, Mr. Tarski, and Bronisław Knaster spoke.

Prof. Leśniewski favors adopting a definition of the second type. He mentions four possible variations of the first type of definition, pointing to the difficulties associated with each of them. [The one] most similar to the type first presented by the speaker, according to Prof. Leśniewski, is to accept a directive in advance, according to which one is free to write as directed the replacements of some

---

[20] It is unclear whether the first two sentences of this paragraph are the speaker's or a comment from Tarski in the audience. The original wording of the paragraph's last clause was *nową tezą, daną przez definicyę implikacyjną, jest prawo tożsamości*.

[21] Łukasiewicz was probably referring to the axioms for the sentential calculus in Hilbert and Ackermann [1928] 1950, page 29. Łukasiewicz referred to that book in the introduction to his [1929] 1963 logic text, and on page 47 identified the law of commutation as *CCpCqrCqCpr*: that is, $(p \to (q \to r)) \to (q \to (p \to r))$.

[22] Łukasiewicz may have been alluding here to an investigation of the independence of the law of commutation, unknown to the present editors, which may have revealed very many ways to form multivalued interpretation tables for the connectives *C*, *N*, and *A*, under which the remaining axioms of Hilbert's system and the hypotheses for the implicational definition of *A* might all be valid.

expressions for others. There is a difficulty in formulating such a directive. Prof. Leśniewski further notes that it is possible to write a definition of the second type using [connectives] other than implication as primitive: for example, Sheffer's stroke. In that case the definition is only a single sentence. In Prof. Leśniewski's ontology, definitions have led to theses independent of the axioms. This is not a defect—on the contrary: the definitions being introduced should be as creative as possible. Finally, for a nonextensionalist, definitions of the first type are not acceptable.[23]

Mr. Tarski notes that with the help of implication the [second type of] definition can be written down in the form of only one hypothesis, specifically the hypothesis $CC\alpha C\beta rr$, where $\alpha$ is the hypothetical implication in the one direction and $\beta$, in the other.[24] He continues, giving a way to state all definitions in advance. Specifically, there is a countable number of possible definiens–definiendum pairs; arrange them in sequences $x_1, x_2, \ldots$ and $y_1, y_2, \ldots$, and accept that an appropriate $x$ can be replaced by [the corresponding] $y$: that is, that appropriate conditional [expressions] can be written down.

## 15.5 Tarski 1929, on Polish Pensions

This section is devoted to Alfred Tarski's 1929 article *Marginal Note on "Decree of the President of the Republic Concerning Insurance of Nonmanual Workers as of 24 November 1927,"* which appeared in the journal *Ekonomista, kwartalnik poświęcony nauce i potrzebom życia,* volume 29, pages 115–119.[25]

At that time *Ekonomista* was edited by Edward Lipiński, professor at the Warsaw School of Economics, and published by the Association of Polish Economists and Statisticians.[26] Still published, it has traditionally addressed a broad audience. The issue that contained Tarski's paper also included articles and book reviews on economic history, economics of municipal development, econometrics, aspects of the Polish economy and politics, and reports and announcements of the association.

---

[23] In the original, this sentence was *Dla nieekstensyonalisty wreszcie definicye pierwszego rodzaju są nie do przyjęcia.*

[24] This seems to require that $r$ be a contradictory sentence. Then, $Cpr$, or $p \to r$ in infix notation, is equivalent to $Np$ for any sentence $p$, and thus $CC\alpha C\beta rr$ is equivalent to $N(\alpha \to N\beta)$, which is in turn equivalent to $\alpha \& \beta$. Now let $\alpha$ and $\beta$ be the two implications that constitute the second type of definition.

[25] The Polish title of the article was *Na marginesie "Rozporządzenia Prezydenta Rzeczypospolitej o ubezpieczeniu pracowników umysłowych z dnia 24 listopada 1927 r."* The journal title means *The Economist: A Quarterly Devoted to Science and the Needs of Life.*

[26] Szkoła Główna Handlowa w Warszawie; Towarzystwo Ekonomistów i Statystyków Polskich.

Tarski's article pointed out a probably unintended unfair consequence of the wording of a government regulation that determined the amount of disability or old-age pension for a Polish white-collar worker. The amount depended on a base figure established elsewhere, on the number $m$ of months of the worker's contribution, and on the number of his or her children. Tarski noted that fairness, almost certainly intended by the legislature, would make the amount of a pension a nondecreasing function of $m$. He demonstrated that in fact, this was not the case, and expressed hope that the anomaly would be corrected.

Tarski was interested in other social issues too, possibly through his teacher colleagues at the gimnazjum. In 1928 and 1931, he was elected docents' representative to the faculty council of the University of Warsaw. In 1937 he joined a socialist political club.[27]

This article about pensions is similar in approach to that of some curriculum materials developed in the United States during the 1990s, which aimed at relating secondary and lower-level university mathematics instruction closely to "real-world" applications.[28] Tarski made his point through specific computational examples that used a precise algebraic description of the algorithm for the amount of a pension. Tarski hinted that elementary mathematics could yield a deeper and more general understanding of the anomaly in the regulation, which would be necessary to correct it. The present editors suggest this as an interesting project for a talented student at this level.

The present editors have not investigated the background of this regulation, determined the scope of its application, nor searched for any response to Tarski's article. He did not return to this subject in any later publication.

A full English translation of Tarski's article, Tarski [1929] 2014b, appears in the boxes that follow. This is its first translation. It is as close to the original as possible given the inexact correspondence between 1927 Polish government terminology and that of contemporary English. All text in [square] brackets or outside the boxes was provided by the present editors as explanation. Punctuation has been adapted to current English practice. White space was inserted liberally to enhance clarity. For some text, indentation has replaced emphasis by italics. Tarski's numbers for his footnotes (inside the boxes) have been replaced by distinctive symbols to avoid confusion with the present editors' footnotes (outside).

---

[27] Jadacki 2003a, 147, 152, 157. See sections 3.1–3.2.
[28] For example, Hughes-Hallett et al. 2003.

DR. ALFRED TARSKI

Docent, University of Warsaw

# Marginal Note on "Decree of the President of the Republic Concerning Insurance of Nonmanual Workers as of 24 November 1927"*

I would like to direct attention here to certain consequences that follow in an undeniable way, in my opinion, from the words included in the title Decree, and which remain, I presume, glaringly inconsistent with the intention of the legislators.

Let us take into consideration an insured who satisfies the following conditions:

(a)  [he] has the right to a disablility or old-age pension (by virtue of articles 22 and 24 of the Decree);

(b)  the "basis[29] of pension benefits" (article 33 paragraph 3) yields for him a specific sum, say $P$ zlotys;

(c)  his condition is of the sort that requires "constant care and assistance by other persons" (compare article 40 paragraph 1); and finally,

(d)  the insured has a specified number, for example $d$, of children younger than eighteen years of age, or even older than eighteen years of age, as long as they satisfy the conditions provided by article 28 paragraphs 3 and 4.

Under the above assumptions, according to Decree articles 38–40, the entire pension together with all allowances that the insured draws from the Institution for Nonmanual Workers' Insurance[30] is dependent solely on the number of "months of contribution" to the insurance. I will denote the number of "months of contribution" by the symbol $m$ and the pension together with allowances

---

* Poland President 1927.

---

[29] The Polish term was *podstawa wymiaru*.
[30] Zakład Ubezpieczeń Pracowników Umysłowych

(calculated in zlotys), as a function of the variable $m$, by the symbol $R(m)$. Then, according to article 16 paragraph 5 and article 24 paragraph 4 the function $R(m)$ is defined when $m$ assumes integer values in the interval $[60, 480]$: $60 \leq m \leq 480$.[†]

I shall now find the value $R(m)$ more precisely. In accordance with articles 38–40, this quantity is composed of three items:

(1) the proper disability or old-age pension;
(2) an allowance due to the fact that the insured "needs constant care and assistance by other persons";
(3) an allowance for children.

Since these three amounts, expressed in zlotys, depend more or less explicitly on the number $m$ of "months of contribution," I shall denote them by the symbols $R_1(m)$, $O(m)$, and $D(m)$ respectively. Thus, a formula is obtained:

$$R(m) = R_1(m) + O(m) + D(m). \tag{I}$$

Next (article 38 paragraph 1 and article 39) the proper pension $R_1(m)$ consists of a constant "principal amount" $Z$ and the "amount of increase in pension" $W(m)$, dependent on the number $m$:

$$R_1(m) = Z + W(m). \tag{II}$$

According to article 38 paragraph 2 we have

$$Z = 0.4\,P. \tag{III}$$

On the other hand, in accordance with paragraph 3 of that article—

> The increase in pension begins after completion of one hundred twenty months of contribution; the amount of increase in pension is $1/6\%$ of the basis for each additional month, and after four hundred eighty months of contribution reaches 60% of the basis.—

---

[†] To simplify considerations, I do not take into account here the exceptional cases provided for in article 16 paragraph 5 and article 33 paragraph 4.

$$W(m) = \begin{cases} 0, & \text{when } m < 120 \\ \dfrac{1}{600}P(m-120), & \text{when } 120 \leq m \leq 480.^{\ddagger} \end{cases} \qquad \text{(IV)}$$

Next, according to article 40 paragraph 1—

> A person receiving a disability or old-age pension, needing constant care and assistance of other persons, receives an allowance equal to the difference between the pension drawn and its basis (article 33 paragraphs 3 and 4).—

$$O(m) = P - R_1(m). \qquad \text{(V)}$$

Finally, in accordance with paragraph 2 [of article 40]—

> A person receiving a disability or old-age pension receives for each child (articles 28 and 29) younger than eighteen years of age, or older than eighteen years of age under the conditions provided in article 28 paragraphs 3 and 4, one tenth of the principal amount (article 38 paragraph 2) if the pension together with the allowance for children, but without the allowance provided for in the first paragraph, should not exceed the base pension (article 33 paragraphs 3 and 4).—

we conclude that

$$D(m) = \begin{cases} 0.1Zd, & \text{if } R_1(m) + 0.1Zd \leq P \\ P - R_1(m), & \text{if } R_1(m) + 0.1Zd > P. \end{cases} \qquad \text{(VI)}$$

The preceding formulas (I)–(VI) enable determination of each value $R(m)$ if $P$, $d$, and $m$ are given.[31]

It seems certain that according to the intent of the legislators, *the function $R(m)$ should be a nondecreasing function*. That is, expressed in ordinary language:

---

[‡] In order to satisfy article 38 paragraph 4, "Under no circumstance can the disability pension be less than fifty zlotys a month," some alterations should be introduced either in formula (II) or in formulas (III) and (IV). For this purpose, for example, it is possible to retain formulas (III) and (IV) but change (II) in the following way:

$$R_1(m) = \begin{cases} 50, & \text{if } Z + W(m) \leq 50 \\ Z + W(m), & \text{if } Z + W(m) > 50. \end{cases} \qquad \text{(II')}$$

Such an alteration would not make any impact on the further considerations.

---

[31] Tarski's variables are mnemonic:

| Variable | Polish | English | Variable | Polish | English |
|---|---|---|---|---|---|
| $d, D$ | dzieci | children | $R$ | renta | pension |
| $m$ | miesiący | months | $W$ | wzrost | increase |
| $O$ | opieka | care | $Z$ | kwota zasadnicza | principal |
| $P$ | podstawa | basis | | | |

Using the second alternative of (VI) in (I) would yield $R(m) = O(m) + P$.

(A) It should not happen that of two persons, having the right to a disability or old-age pension, and finding themselves in the very same situation with respect to all conditions that affect the size of pension and allowances, the only difference being that the first has been insured over a shorter time than the second,[§] the first should collect despite this a larger monthly sum from the Insurance Institution than the second.

The validity of the above principle is so obvious that arguing its elementary requirements of fairness or [discussing] examples of other laws and regulations of a similar nature seems unnecessary.

In view of the preceding remarks [it seems that] the author or authors of the Decree failed, no doubt, to notice the fact that *the function $R(m)$ determined by formulas* (I)–(VI) *in accordance with the wording of the Decree, being generally a nonincreasing function, actually decreases over some interval of variation of $m$; the size of the interval increases with the increase in the number $d$,*[∥] *and disappears only when $d = 0$.*

One can convince oneself of the truth of the preceding sentence with the help of completely elementary mathematical reasoning, which I will not include here. I will illustrate on the other hand the perceptible situation with a certain especially clear example. Specifically, I suppose that the number $d$ of children of the insured is 15; for definiteness I assume, moreover, that the "basis of pension benefits" [is] $P = 300$ zlotys. The course of values of the function $R(m)$ —that is, the total monthly pension together with allowances (expressed in zlotys), depending on the number $m$ of "months of contribution"—is then shown in the following table calculated on the basis of formulas (I)–(VI):

| $m$ | 60 | 120 | 180 | 240 | 300 | 360 | 420 | 480 |
|---|---|---|---|---|---|---|---|---|
| $R(m)$ | 480 | 480 | 450 | 420 | 390 | 360 | 330 | 300 |

The following [results], among others, [can be inferred] from the preceding table:

---

[§] More precisely, it should be expressed, *"the first person has been approved for insurance* [zaliczono do ubezpieczenia] *over a shorter time than the second."*

[∥] Here, $d$ assumes values in the interval $[0, 15]$: $0 \leq d \leq 15$; starting from $d = 15$, the size of the interval on which the function $R(m)$ decreases does not change.

(B) If two persons satisfy the following conditions—
- (a) both are entitled to a disability or old-age pension;
- (b) the first was insured over five years, and the second, over forty years—

moreover, all other circumstances having influence on the amount of the pension and allowances are in both cases completely identical, namely—
- (c) the "basis of pension benefits" for each person is 300 zlotys;[#]
- (d) both need "constant care and assistance of other persons"; [and] finally
- (e) both have the same number of children for whom they are entitled to an allowance, namely 15 each—

then the first person receives in total from the Insurance Institution 480 złotys per month, but the second, only 300 zlotys.

I shall present yet another example, less crude than the previous one, and likely to find significantly more applications in practical life. I shall suppose, specifically, that the number $d$ of children of the insured is 5, while the "basis of pension benefits" $P$ is 300 zlotys, as previously. The course of values of the function $R(m)$, depending on $m$, is then expressed in the following table:

| $m$ | 60 | 120 | 180 | 240 | 300 | 360 | 420 | 480 |
|---|---|---|---|---|---|---|---|---|
| $R(m)$ | 360 | 360 | 360 | 360 | 360 | 360 | 330 | 300 |

It follows that

(C) if two persons satisfy the following conditions—
- (a) both are entitled to a disability or old-age pension;
- (b) the first was insured over thirty years, and the second, thirty-five years—

moreover, whereas—
- (c) the "basis of pension benefits" is equal for both persons and is 300 zlotys;
- (d) both need "constant care and assistance of other persons"; [and] finally
- (e) each of them draws an allowance for 5 children—

then the first person receives in total from the Insurance Institution 360 zlotys per month, but the second, only 330 zlotys.

---

[#] Condition (c) is then satisfied, for example, when each of the persons considered earned 300 zlotys per month over the whole time and was approved for insurance (compare article 14 and article 33 paragraph 3).

> The contradiction between principle (A) and the concrete consequences (B) and (C) of the Decree is so obvious that it does not require comment.
>
> One ought to expect that the shortcoming of the "Decree Concerning Insurance of Nonmanual Workers" exposed here, having such great significance in general for nonmanual workers in Poland, will be removed in the near future through amendments to the law.

## 15.6 Tarski 1930–1931, on the Concept of Truth

At a December 1930 series of meetings of the Lwów section of the Polish Philosophical Society, Alfred Tarski presented the results of his investigations of fundamental concepts of the methodology of deductive sciences. He had begun reporting on this topic in 1928 and had probably been perfecting the work in his Warsaw research seminar. Polished versions of some of this material were being published already in 1930, in German.[32] But one of his lectures was aired only as an abstract in Polish, *On the Concept of Truth in Reference to Formalized Deductive Sciences*.[33] It appeared in the journal *Ruch filozoficzny*, volume 12 (1930–1931), pages 210–211. Tarski's celebrated theory of truth would not be fully published for several years. The abstract was its first appearance in print. At its conclusion it was labeled *auto-referat*: a report submitted by the speaker. A full translation, Tarski [1930–1931] 2014, appears in the boxes that follow. This is the first translation of the abstract.

Tarski presented another version of this talk to the Warsaw Society of Sciences and Letters in March 1931. His brief 1932 paper, published in Vienna in German, reported that talk; that paper is actually a description, not a repetition or a continuation, of the work translated here. These abstracts were greatly revised, amplified, and published in Polish as Tarski 1933. That monograph was edited and amplified further, translated, and published in German as Tarski [1933] 1935. The German version was translated into English and published in 1956; a revised English edition appeared as Tarski [1933] 1983.

The present translation, Tarski [1930–1931] 2014, is particularly faithful to the original abstract. Text in [square] brackets has been inserted sparingly for clarification. The only change in symbolic notation was italicizing variables. Tarski's styles of quotation marks have been maintained. Some other punctuation has been adapted to current English practice. White space is sometimes used in that way, and often inserted to display lists, a table, and the title. Even in those cases, Tarski's words have been translated strictly. Some particularly sensitive translation points have been explained in footnotes. Footnotes 38, 40, and 44 contain historical information, as well. All footnotes fall outside the boxes, and are the work of the present editors.

---

[32] Tarski [1930] 1983a,b; for further information about these presentations, see section 14.2 and Zygmunt 2010.

[33] The Polish title of the article was *O pojęciu prawdy w odniesieniu do sformalizowanych nauk dedukcyjnych*. The journal title means *Philosophical Trends*.

At the 307th plenary scientific meeting on 15 December 1930
Docent Dr. Alfred Tarski of the University of Warsaw gave a lecture

## On the Concept of Truth in Reference to Formalized Deductive Sciences

According to the intuitive understanding of the word „truth" we say that the sentence[34] „snow is falling" is true if and only if snow is falling. Generally, we could say that „$p$" is true if and only if $p$, or still more generally, $A$ is true if and only if for some $p$, $A =$ „$p$" and $p$. Such a definition of truth is incorrect, however, since the occurrence of the expression »„$p$"« in it must be considered a function[35] with a variable argument; and as Stanisław Leśniewski has demonstrated, this leads to antinomies—specifically to the antinomy of the liar and [to that] of heterological words.[36] Moreover, in both of these formulations, a non-extensional function occurs (that is, both »„$p$"« as well as »$A =$ „$p$"«, since both of these functions change their values even if for „$p$" we substitute only equivalent sentences).

The purpose of the lecture will not be to define the truth of any sentence whatever in everyday language; this we are not able to do. We shall define the truth of sentences occurring in the languages of certain deductive systems. We could try saying it thus: a sentence of a certain system is true if and only if it is a theorem[37] in this system. Such a definition, however, does not conform to intuitions relating to the word „truth". In fact, it does not support the law of excluded middle, which says that of two contradictory sentences one must be true: it is in fact not necessary that one of two contradictory sentences be a theorem of the system.

In order to give for the truth of the sentences of some deductive system a definition correct and consistent with our intuitions, we distinguish first of all the language of which we speak and the language in which we speak.[38] We define

---

[34] The word *zdanie*, in its various forms, is always translated here as *sentence*; this word occurs in the translation only in that way.

[35] In this paper Tarski used the word *function* (*funkcya*) only to denote certain linguistic expressions.

[36] Tarski's phrase was *do antynomii kłamcy i wyrazów heterosemantycznych*.

[37] Tarski's word was *teza*.

[38] Tarski's construction for this last clause was parallel and unemphasized: *odróżnimy przedewszystkiem język o którym mówimy i język w którym mówimy*. In the 1932 German abstract the language in which we speak is called the *Umgangssprache*. In the 1933 and [1933] 1935 Polish and German monographs it is the *metajęzyk* or *Metasprache,* respectively, meaning *metalanguage*. That term does not occur in the present paper.

the truth of a sentence belonging to the language of which we speak. In contrast, the definition that we give will now belong to the language in which we speak. We give this definition only for a certain specific, very elementary, deductive system. But, completely analogously, we could construct a definition of the truth of a sentence for many other deductive systems (namely, for all those systems in which the order of variables—in the sense of the so-called theory of types—does not exceed some natural number given in advance).

The language of which we speak (which could be called the language of the elementary algebra of classes) is presented as follows. It contains

    three variables   „$a$", „$b$", „$c$";<br>
    a primitive term   „$\subset$"   (we read this as „is part of");<br>
    logical terms[39]   „$\sim$"   (we read this as „it is not true that"),<br>
            „$\vee$"   (we read this as „or"),<br>
            „$\Pi$"   (we read this as „for all");

and, moreover, parentheses.

The axioms of this system express certain properties of the relation „being a proper part of". Here are some examples:

(1)   $\Pi a \sim (a \subset a)$,
(2)   $\Pi a \Pi b (\sim (a \subset b) \vee \sim (b \subset a))$,
(3)   $\sim \Pi a \sim \Pi b \sim (a \subset b)$.[40]

Among the expressions of the language we distinguish the so-called sentential functions. We call the inclusions „$a \subset b$", „$b \subset c$", and so on, functions of the first order, meaning elementary;[41] there are nine kinds of them, there being three variables. The functions of the second order appearing in this language are functions formed from functions of the first order with the help of one of the logical terms: for example, „$(a \subset b) \vee (b \subset a)$" is such a function.[42] Functions of the third order are functions formed from functions of the second order in the same way as functions of the second order from functions of the first order. Analogously, functions whose order is any natural [number] can be defined. Sentential functions not containing any so-called free variables are called sentences.

---

[39] Tarski's phrase was *terminy logiczne*.

[40] From (2), Tarski omitted the outermost parentheses, which are necessary. At the corresponding point in [1933] 1983, page 179, Tarski alluded to an axiom system of Edward V. Huntington (1904, 297) for the inclusion relation $\subseteq$. The present example axioms can probably be related to a counterpart of that system for the proper-inclusion relation $\subset$.

[41] Tarski's phrase was *funkcyami pierwszego rzędu czyli elementarnemi*. He used the same noun *rząd* for *order* of a variable in the previous paragraph.

[42] In error, Tarski wrote „$(a \subset b) \vee (b \supset a)$" here.

## 15.6 Tarski 1930–1931, on the Concept of Truth

The language in which we speak contains words and logical expressions to the extent in which they occur, for example, in *Principia Mathematica*,[43] thus in particular, expressions equivalent to the terms of the language of which we speak: for example, „is a part of" (that is, „⊂"), „or", and so on. Moreover, it contains

| | | |
|---|---|---|
| terms | „$z_1$", „$z_2$", „$z_3$" | which denote the variables „$a$", „$b$", „$c$", respectively; |
| a function of names[44] | „$I(x,y)$" | whose arguments are „$z_1$", „$z_2$", „$z_3$", while the values [are] names of the respective inclusions—that is, names of expressions „$(a \subset b)$", „$(b \subset a)$", and so on; |
| the expressions | „$\bar{x}$" | denoting the negation of $x$, |
| | „$x + y$" | which denotes the logical sum of $x$ and $y$; |
| and the term | „$\bigcap_x y$" | which denotes the generalization of the expression $y$ with respect to the variable $x$ —that is, an expression of type „$\Pi a\, p$". |

By $I_{n,p}$ we shall understand $I(z_n, z_p)$, where $n, p = 1, 2, 3$.

In constructing the definition of true sentence we shall use the notion of satisfaction of a sentential function by a certain sequence of objects.[45] Specifically, in an inductive manner, we shall define when some sequence of objects satisfies a sentential function of the $n$th order from the language of which we speak. Sequence $C$ satisfies a sentential function of first order—that is, an inclusion $I_{n,p}$ —if and only if

(1) the sequence $C$ is defined for numbers 1, 2, 3;

(2) $n, p = 1, 2, 3$;

(3) the $n$th term of the sequence is a part of the $p$th term of the sequence—that is, $C_n \subset C_p$.

---

[43] Whitehead and Russell 1910–1913.
[44] Tarski's term was *funkcya nazwowa*.
[45] Tarski's word for *objects* was *przedmioty*.

In order to say when some sequence satisfies a certain function of higher order, it suffices—in view of the way in which the functions were formed—to say when a negation, sum, or generalization of a function of a certain order is satisfied, assuming that we already know when a sequence satisfies a function of that order. Here, a sequence $C$ satisfies $\bar{x}$ if and only if

   (1) (as above);
   (2) the sequence $C$ does not satisfy $x$.

The sequence $C$ satisfies $x + y$ if and only if

   (1) (as above);
   (2) the sequence $C$ satisfies $x$ or satisfies $y$.

Finally, the sequence $C$ satisfies $\bigcap_{z_n} x$ (that is, the generalization of $x$ with respect to the $n$th variable) if and only if

   (1) (as above);
   (2) $n, p = 1, 2, 3$;
   (3) not only $C$, but also every sequence obtained from $C$ through a change just in the $n$th term, satisfies $x$.

We defined when a sequence of objects satisfies some sentential function in the language of which we speak. Since sentences in this language are also certain sentential functions, we have explained at the same time when a sequence of objects satisfies a sentence in the language of which we speak.

With this in mind, we define,

> A sentence in the language of which we speak is true if and only if every sequence satisfies this sentence.

Such a definition is consistent with intuitions connected with the word „truth". As an argument for this, one can use the fact (among others) that by such a definition of the truth of a sentence the law of excluded middle is satisfied. This holds because either every sequence satisfies sentence $x$, or none satisfies it. For if even one sequence does not satisfy $x$, then no sequence satisfies this sentence, and in that case every sequence of objects satisfies the sentence $\bar{x}$. Thus, if $x$ is not true, then $\bar{x}$ is true. Similarly, the principle of contradiction is satisfied. What is most important is that all the theorems constructed in the following way

> can be justified. We take under consideration the schema „$x$ is a true sentence if and only $p$" and any sentence $y$ in the language of which we speak, then in this schema we substitute for „$x$" the name of the sentence $y$ (expressed in the language in which we speak), and for „$p$" a sentence that is a translation of the sentence $y$ into the language in which we speak. Here is an example of such theorems: $\bigcap_{z_1} \bigcap_{z_2} I_{1,2}$ is a true sentence if and only if for any $a$ and $b$ whatever, $a$ is part of $b$. Therefore, on the basis of the [just] established definition of truth we are able to explain, in a way appropriate from the intuitive point of view, the meaning of the phrase „$x$ is a true sentence" as applied to every sentence in the language of which we speak.
>
> A further consequence of the above definition is a proposition according to which all theorems[46] of the system are true sentences; from this it is easy to conclude that the system of which we speak is consistent. In this way the construction of a definition of truth also yields a method for a proof of the consistency of the deductive system under consideration.

Tarski's presentation was discussed by Kazimierz Ajdukiewicz, Roman Ingarden, Maria Kokoszyńska, Kazimierz Kuratowski, Tadeusz Witwicki, and Tarski.[47] No further details were recorded.

Why did Tarski undertake to define *true sentence*? Solomon Feferman has concluded that Tarski and some other mathematicians had worked comfortably with informal versions of the notions explicated in this abstract at least since 1924. On the other hand, another student and colleague of Tarski, Robert L. Vaught, claimed that[48]

> Tarski appears to have been unhappy about various results obtained during the [1927–1929] seminar because he felt that he did not have a precise way of stating them... no one had made an analysis of truth [for example], not even of exactly what is involved in treating it [informally].... His major contribution was to show that the notion [of a true sentence] can simply be *defined* inside of ordinary mathematics.... Knowing that all of the semantical notions are just ordinary mathematical notions, logicians and mathematicians doubtless felt able to behave much more freely with them.... In historical fact, this was only a part of Tarski's motivation, for he was also very much concerned with the positions and attitudes on the notion of truth taken by various philosophers....

---

[46] Tarski's nouns *twierdzenie* and *tezy* have been translated here as *proposition* and *theorems*.

[47] Born in Cracow in 1893, **Roman Ingarden** earned the doctorate at Freiburg in 1918 from Edmund Husserl. He became professor of philosophy at Lwów and, after World War II, at Cracow. Ingarden was noted for his work in ontology and aesthetics (Thomasson 2012). He died in 1970. **Tadeusz Witwicki** was born in Lwów in 1902; he earned the doctorate in psychology in 1927 from the university there. Witwicki contributed to the literature of psychology and philosophy and taught at various institutions in Lwów, Warsaw, and Toruń until his death in 1970 (Hutnikiewicz 1965). For information about Ajdukiewicz, Kokoszyńska, and Kuratowski, see boxes in chapter 3 and sections 14.2 and 4.4, respectively.

[48] S. Feferman 2008, 79; Vaught [1971] 1974, 160–161. For more information about the 1927–1929 seminar, see section 9.4. The awkwardness of philosophical discussions involving the concept of truth, even by those aware of some contemporary discussions in logic, is illustrated in section 15.3.

Historical research reveals that considerations related to defining truth were much in the air in the philosophical circles of the 1920s and most were controversial. Tarski's definition made specific use of several of these. First, he defined *truth* in terms of *satisfaction*. Truth and satisfaction of what? For Tarski, truth-bearers are sentences, not thoughts or judgments, as some had held. His sentences could be expressed in any one of a rather broad class of languages. Tarski was not interested in colloquial languages, but in formal languages such as those employed in mathematics. Moreover, these languages were regarded as interpreted: their nonlogical symbols must all be assigned specific meanings not in question in his discussion. Since such languages were incapable of describing states of affairs dependent on factors beyond those meanings, Tarski's conception of truth was nearly absolute: it was relative only to the choice of language. This limited degree of relativity was sufficient for the scope of his study. Tarski emphasized that his discussion of truth would take place in a *metalanguage* distinct, in principle, from the *object language* under analysis. The metalanguage could be the informal language of ordinary mathematics or philosophy, or it could be formalized if that should be necessary for application of a truth definition to a specific problem. Tarski showed that the notion of satisfaction of a sentence in the formal object language could be defined recursively according to its complexity. That technique fit neatly into the apparatus then used in many elementary mathematical studies. Tarski noted in conclusion that his definition of truth is consistent with the intuitive Aristotelian concept[49] expressed in his opening paragraph; but unlike that concept, it can be used with many languages of interest without danger of contradiction.[50]

In the [1933] 1983 full monograph on truth, Tarski analyzed the concept of a truth definition further. There he formulated his now famous condition T,[51] which he later explained in simpler terms as follows:

> We stipulate that the use of the term "true" in its reference to sentences in [a given object language] then and only then conforms with the classical conception of truth if it enables us to ascertain every equivalence of the form
>
> (3)   "$p$" is true if and only if $p$
>
> in which "$p$" is replaced on both sides by an arbitrary ... sentence [in that language]. If this condition is satisfied, we shall say simply that the use of the term "true" is adequate.[52]

In that later discussion, Tarski first presented this condition for the English language, and used the antinomy of the liar to show that no adequate truth definition is possible.

---

[49] Tarski noted that this concept "can be found in Aristotle's *Metaphysics*: 'To say of what is that it is not, or of what is not that it is, is false, while to say of what is that it is, or of what is not that it is not, is true.' Here ... the word 'false' means ... 'not true' .... [This] is usually referred to as the *classical*, or *semantic conception of truth*." (Tarski 1969, 63; see also the reference to 1969 in [1933] 1983, 152, footnote †; and Aristotle 1971, book Γ, chapter 7, page 23.)

[50] The article Murawski and Woleński 2008 places Tarski's work in context, with many more details. It hardly mentions Stanisław Leśniewski's thoughts in this area, but Arianna Betti has considered Tarski's relationship to him in detail (2004, 2008).

[51] Tarski [1933] 1983, 187–188.

[52] Tarski 1969, 64–65. In [1933] 1983, Tarski used the phrase *materially adequate* in this context. Two years before, he had described a *materially adequate* definition as one "grasping the current meaning of the notion as it is known intuitively" ([1931] 1983, 128–129).

Then he rephrased the condition for object languages in general, and discussed how, for certain formalized object languages, an adequate truth definition could be achieved in the metalanguage. In [1933] 1983 he showed that the feature of the object- and metalanguages that is critical in this regard is their *order*: the maximum order of the logical types of the variables that they can handle. He concluded,

> A. For every formalized language a formally correct and materially adequate definition of true sentence can be constructed in the metalanguage with the help only of general logical expressions, of expressions of the language itself, and of terms from the morphology of language—but under the condition that the metalanguage possess a higher order than the language which is the object of investigation.
>
> B. If the order of the metalanguage is at most equal to that of the language itself, such a definition cannot be constructed.[53]

Writing in retrospect, Tarski clarified the intent of his definition:

> What will be offered can be treated in principle as a suggestion for a definite way of using the term "true", but the offering will be accompanied by the belief that it is in agreement with the prevailing usage of this term in everyday language.[54]

This analysis of truth is a vital part of Tarski's infrastructure for mathematical logic. It has also stimulated much subsequent semantic research in philosophy, linguistics, and computer science.

## 15.7 Tarski 1932, on Banach's Measure

On 23 February 1932, Alfred Tarski contributed the paper *On Geometric Properties of Banach's Measure* to a meeting of the Warsaw Society of Sciences and Letters. The paper was sponsored officially by Wacław Sierpiński; whether Tarski presented it in person is unclear. The published abstract,[55] in volume 25 of the *Comptes rendus des séances de la Société des Sciences et des Lettres de Varsovie, Classe III*, seems to be an edited version of a listener's notes. A full translation, its first, appears in the box on the next page.

In 1923 Stefan Banach had shown that there exist (finitely) additive extensions of Lebesgue measure, invariant under congruence, to the algebras of all bounded sets of real numbers and of all bounded sets of points in the Euclidean plane. That result played a major role in the 1924 works of Tarski and Banach about equidecomposability of point

---

[53] Tarski [1933] 1983, 273. For a nontechnical overview of that work, see Feferman and Feferman 2004, Interlude III, 109–123. For further information on Tarski's philosophical works, consult the sources listed in chapter 18 of the present book.

[54] Tarski 1969, 63.

[55] The Polish title was *O własnościach geometrycznych miary Banacha*. The journal also gave an equivalent French title.

sets.[56] In his 1932 presentation, Tarski outlined a plan for recasting Banach's work into a geometric framework that would unfold more naturally from the familiar theory of Peano–Jordan content. Tarski's abstract includes no details at all. The following translation, listed as Tarski [1932] 2014 in this book's bibliography, adheres to the same standards as earlier ones in this book. The word in [square] brackets was inserted by the present editors.

---

Alfred Tarski

**On Geometric Properties of Banach's Measure**
Presented by W. Sierpiński on 23 February 1932

**Sur les propriétés géométriques de la mesure de Banach**
Mémoire présenté par M. W. Sierpiński dans la séance de 23 février 1932

*Abstract*

In his 1923 work *Sur le problème de la mesure* Banach solved positively the broader problem of measure for linear and planar sets. Banach's construction method is not based on concepts having clear geometric content and does not resemble in any way the methods used in other constructions in this subject (for example, in the Peano–Jordan or Lebesgue theory of measure). In this paper the author outlines a plan for a certain reconstruction of Banach's results using concepts with clear geometric content. As a consequence of this reconstruction Banach's theory of measure becomes a natural generalization and development of the Peano–Jordan measure theory. To this end the author defines the class of absolutely measurable sets (that is, those that have the same measure for each [compatible] definition of measure) and he establishes some properties of this class.

---

Tarski returned to this project again in his 1938b paper on absolute measure; unfortunately, even there the exposition was meager.

---

[56] Tarski [1924] 2014b and Banach and Tarski [1924] 2014, translated in chapters 5 and 6. See chapter 4 for background information on these papers and Banach 1923.

## 15.8 Discussion of Ajdukiewicz 1936, on Idealism

On 25 September 1936 Kazimierz Ajdukiewicz presented the paper *The Problem of Idealism in a Semantic Formulation* at a plenary session of the Third Polish Philosophical Congress in Cracow. The congress is described in section 14.2. Ajdukiewicz's paper is described here, followed by a translation of Alfred Tarski's [1936] 2014a comment about it in the box on the next page. Ajdukiewicz's response is summarized after that.[57]

Ajdukiewicz's 1936 paper is vague, probably in part because he was discussing a subject very well known to his audience. In 1937 he published a longer, more informative, paper on the same subject. Ajdukiewicz was attempting to explicate, in the framework of contemporary logic, the thesis of transcendental idealism as formulated by the German philosopher Heinrich Rickert. Ajdukiewicz regarded Rickert's presentation of that philosophy as the easiest to understand but nevertheless complained about its lack of clarity. He cited no specific publication of Rickert.[58] The Polish philosopher Leszek Nowak has written,

> According to Rickert, there exists the objective spirit conceived of as a set of trancendental norms and the reality is to be a correlate of it. The latter claim is interpreted by Ajdukiewicz as saying that these norms show the set of all possible truths. These rules are transcendental in that sense that we, people, can have an access to the set of all possible truths only on the condition that we follow these rules. Therefore, they decide about the truthfulness of our judgements. Ajdukiewicz paraphrases Rickert's "transcendental norms" as the rules of logical consequence relative to the scientific language. Under this paraphrase, the thesis of transcendental idealism reads as the claim that in the language of natural sciences only those sentences are true that are shown in the rules of consequence characteristic of that language.[59]

Ajdukiewicz discerned certain features of Rickert's philosophy that correspond to the axioms (both logical and empirical) of some formal deductive system for natural science, and other features that correspond to its inference rules. Thus, Ajdukiewicz claimed that the judgments true in Rickert's sense correspond exactly to the sentences provable in that deductive system. Ajdukiewicz claimed that such a deductive system must include arithmetic, and by Kurt Gödel's famous incompleteness theorem,[60] it must therefore be incomplete: its language must include a sentence $S$ such that neither $S$ nor the negation of $S$ is provable by the inference rules. Therefore, Ajdukiewicz claimed, Rickert's philosophy of transcendental idealism is inadequate for natural science because it fails to satisfy the law of excluded middle: it cannot decide the truth or falsity of all sentences of natural science.

---

[57] The Polish title of Ajdukiewicz 1936 was *Zagadnienie idealizmu w sformułowaniu semantycznym*. It was published in volume 39 of the journal *Przegląd filozoficzny*. In the table of contents for the congress in the journal, the paper's title began instead with the word *Problemat*. For information about its author, see a box in chapter 3.

[58] Ajdukiewicz [1937] 1978, particularly 146 ff. Born in the Prussian city Danzig in 1863, **Heinrich Rickert** had been professor of philosophy at Heidelberg since 1916; he was regarded as a leader in the Baden school of neo-Kantian philosophy. He died in 1936.

[59] Nowak 1995, 213.

[60] Gödel [1931] 1967.

The ensuing discussion included the following remark, which is listed as Tarski [1936] 2014a in the present book's bibliography. The translation adheres to the same standards as earlier ones in this book. The word in [square] brackets was inserted by the present editors.

---

*Alfred Tarski* emphasized:

(1) On the characteristics of deductive systems:

Methodologists of the deductive sciences would prefer a characterization of deductive systems using a general concept of consequence over [one] that used a concept of direct consequence.

(2) On the relationship between the concepts of meaning and direct consequence:

(a) It would seem that the concept of meaning determines the concept of consequence, but not of direct consequence. (Let us assume that sentence $b$ follows from a class $A$ of sentences, but not directly; can we, accepting the sentences of class $A$, reject sentence $b$ without changing the meaning of the expressions occurring in these sentences?)

(b) Not only does the concept of meaning determine the concept of direct consequence, but vice versa (which the speaker did not mention in the talk, but which follows from the speaker's earlier work). Only because of this can metalogic be considered an adequate tool for carrying out epistemological research.

(3) The view that the language of the natural sciences is incomplete appears to be legitimate, but the given justification raises doubts. (For a certain definition of consequence, arithmetic is a complete system.)

---

Ajdukiewicz had alluded to some deductive rules that would distinguish certain consequences as indirect. Tarski's comment (1) indicated that he would not consider such rules in a characterization of deductive systems. His comment (2a) began with a verb in conditional mode: *It would seem*.⁶¹ Should Ajdukiewicz have really meant the clause that followed it, he would have been contradicting Tarski's comment (2b). Tarski's parenthetical remark in (2a) suggested a test that would judge that clause invalid. Together, (2a) and (2b) explain why Tarski favored removing indirect consequences from consideration.⁶²

Besides Tarski, ten others offered remarks about Ajdukiewicz's presentation:

| | | |
|---|---|---|
| Jerzy Braun | Henryk Mehlberg | Adam Wiegner |
| Piotr Chojnacki | Witold Steinberg | Helena de Willman-Grabowska |
| Roman Ingarden | Jarosław Stępniewski | Adam Żółtowski |
| Jan Leszczyński | | |

Ajdukiewicz replied to some of them at length. Responding to Tarski, he discussed alternative formulations using finitary and infinitary concepts of direct and general consequence. He agreed that invoking the incompleteness of arithmetic depended on the assumption of finitary inference rules in his proposed deductive system, and that incorporation of an infinitary rule would lead to reinterpreting the basic thesis of transcendental idealism. He concluded,

> The subtle conceptual nuances that Dr. Tarski raised in the discussion illustrate the increased level of philosophical investigations, when they are placed within the framework of modern logic.

## 15.9 Discussion of Wilkosz 1936, on the Significance of Logic

On 26 September 1936, Witold Wilkosz presented the paper *The Significance of Mathematical Logic in Mathematics and Other Exact Sciences* at a plenary session of the Third Polish Philosophical Congress in Cracow. The congress is described in section 14.2. His paper is summarized here, followed by a translation of Alfred Tarski's [1936] 2014b comment about it in the box on the next page. The speaker's response is paraphrased after that.⁶³

In his presentation, Wilkosz first summarized the development of logic as a tool for the justification of mathematics. He noted that mathematics always progressed in steps ahead of the logic it needed: for example, with Euclid, Newton, Cauchy, and Cantor. Researchers in logic had stressed consideration of axioms and primitive terms but neglected the rules of deduction. However, recent work had permitted study of the consistency of axiom systems and of logic itself. Logic should be, but was not yet, shaped to allow no doubt of its consistency.

---

⁶¹ In Polish, *wydawałoby się*.

⁶² In his longer 1937 paper, Ajdukiewicz used the term *consequence* (*wynikanie*) repeatedly, but only with the adjective *direct* (*bezpośredni*). He did not seem to consider indirect consequences.

⁶³ The Polish title of Wilkosz 1936 was *Znaczenie logiki matematycznej dla matematyki i innych nauk ścisłych*. For information about its author, see the box on page 369.

Wilkosz surveyed the state of the logical foundations of various disciplines. They were most highly developed for mathematics, but physics had not yet reached that level of precision and it was not clear whether that would even be beneficial. He mentioned the possibility of developing logical foundations for social and even humanistic sciences, and noted interesting work in foundations of theology, where the underlying assumptions were not in question! Wilkosz pointed out that the development of logic has ignored certain aspects of language—for example, shades of the grammar of conditional sentences—and suggested that the time will come for that study. He alluded to work in progress at Cracow on both broadening the scope of logic and simplifying its presentation.

After Wilkosz's presentation, three scholars offered comments: J. M. Bocheński, Benedykt Bornstein,[64] and Alfred Tarski. Bocheński noted, and the speaker agreed, that logic can and should also be developed as an independent science. Bornstein attempted to relate Wilkosz's paper to his own development of logic; the speaker pleaded unfamiliarity with that alternative approach. The following translation of Tarski's comment is listed as Tarski [1936] 2014b in this book's bibliography. It adheres to the same standards as earlier translations in this book.

---

*Alfred Tarski*:

(1) Metalogic and metamathematics can give much to mathematics: even already they have given it a thing or two. For example, theorems according to which all statements of a certain form can be proved in some theory can have an essential meaning for mathematics—and also theorems that allow deduction of the properties of mathematical notions from the content of a definition.

(2) Application of the axiomatic method and the framework of logic is being tried in the various sciences, even the nonmathematical ones. The research by Joseph H. Woodger in the field of foundations of biology deserves attention.

---

Responding, Wilkosz agreed with Tarski's first comment, but suggested that it stemmed from Tarski's mishearing him. Wilkosz mentioned that he could give other examples like those in Tarski's second comment, but he regarded them as still fragmentary.

---

[64] **Benedykt Bornstein** was born in 1880 in Warsaw, then part of the Russian Empire. He earned the doctorate in philosophy at Lwów in 1907, and from 1918 taught at the Free University in Warsaw. After World War II Bornstein taught at the University of Łódź. For information about Bocheński, see the box on the next page.

*Witold Wilkosz
Broadcasting, 1929*

**Witold Wilkosz** was born in 1891 in Cracow, in the Austrian Empire. His father taught Polish in a gimnazjum there. A school classmate of Stefan Banach, Witold had extremely broad interests and talents. While in gimnazjum, he won a study trip to Beirut with a paper on semitic linguistics. In 1910 he entered the university of Cracow to study philology. In 1912 he transferred to the University of Turin, in Italy, and studied analysis with Giuseppe Peano. Wilkosz served during 1914–1915 in Józef Piłsudski's Polish Legion, then returned to Cracow to study mathematics. After earning a credential in 1917, Wilkosz taught for three years in private secondary schools in nearby cities, and in Cracow he attended lectures on law. In 1918 he earned the doctorate there with a dissertation on measure theory, supervised by Stanisław Zaremba. He was awarded the *venia legendi* at Cracow in 1921 for further work in analysis, and in 1922 became a junior professor there. Wilkosz continued research in analysis and also logic, and wrote several texts on elementary and advanced mathematics and works popularizing mathematics. He taught courses for teachers for the Ministry of Religious Denominations and Public Education and was heavily involved in the development of radio, both in the technology and in production of educational broadcasts. In 1936 he became full professor. During World War II his health deteriorated. He continued to teach at the School of Economics but died of pneumonia in 1941.[*]

**Innocentius Maria Bocheński** was born in 1902 in Czuszów, in south-central Poland, became a Dominican monk, earned doctorates in philosophy and theology, then taught at the Pontifical College in Rome until 1940. He served with the Polish Army during World War II, then became professor and an academic leader at the University of Fribourg, in Switzerland. Fr. Bocheński wrote many major studies in the history of philosophy, often using his birth name Józef. He died in 1995.[†]

**Joseph Henry Woodger** was born in 1894 in Great Yarmouth, England. During 1911–1922, except for an interruption for World War I military service, he studied at University College, London. From then until his retirement he taught at the university's medical school. Woodger became a major scholar of the foundations of biology. In particular, he carried to that field the ideas of the Vienna Circle. Woodger met Tarski in 1935 at a function of that group. Tarski's first publication in English (1937a) was a technical appendix to Woodger's 1937 book *The Axiomatic Method in Biology*. Woodger edited the first published collection of Tarski's works, the [1956] 1983 book *Logic, Semantics, Metamathematics*. Woodger died in 1981.[‡]

---

[*] Pawlikowska-Brożek 2011, Albiński 1976, Rakoczy-Pindor 2003, Średniawa 1961.

[†] Morscher, Neumaier, and Simons 2011

[‡] Popper 1981.

## 15.10 Discussion of Zawirski 1936, on Synthesis

On 26 September 1936, Zygmunt Zawirski presented the paper *Concerning Scientific Synthesis* at a plenary session of the Third Polish Philosophical Congress in Cracow. The congress is described in section 14.2. Zawirski's paper is summarized here, followed by a translation of Alfred Tarski's [1936] 2014c remark about it. An explanation of that remark is included, and the speaker's response is paraphrased.[65]

Zawirski indicated that scientific synthesis and metaphysics cannot be strictly separated and spoke about providing a basis for metaphysics that could fit within the Vienna Circle guidelines. He suggested that although this subject is generally regarded as being concerned with hypotheses that are essentially untestable, some of their consequences may not be. He advocated a search for such statements, and underlined the delicacy and precision with which that goal should be pursued. He mentioned two recently published works that would help: Tarski's now famous monograph *The Concept of Truth in Formalized Languages* and Kazimierz Ajdukiewicz's paper *The Scientific World-View*. Zawirski also noted that care should be taken in selecting which of the logics now available should underlie a scientific metaphysics, and said, "we have a legitimate right to assume that only one of these systems corresponds best to reality."[66]

Besides Tarski, five others offered remarks about Zawirski's presentation:

| | | |
|---|---|---|
| Jerzy Braun | Bogumił Jasinowski | Jarosław Stępniewski |
| Piotr Chojnacki | Maria Kokoszyńska | |

The speaker responded to each of them. Maria Kokoszyńska noted that the speaker could be interpreted as suggesting that the Vienna Circle might have retreated from their position of empiricism. That would not be fair, she said, for the Circle demanded then, more than ever before, that scientific statements beyond those that are logically true be verifiable experimentally. In the box on the next page is a translation of Tarski's remark. Listed as Tarski [1936] 2014c in this book's bibliography, it adheres to the same standards as earlier translations in this book. Tarski addressed at once Zawirski's presentation, an earlier one by Jan Łukasiewicz, and a remark by Roman Ingarden about Łukasiewicz's paper.[67]

---

[65] The Polish title of Zawirski 1936 was *W sprawie syntezy naukowej*. For information about its author, see the box on page 373.

[66] Tarski [1933] 1983, Ajdukiewicz [1934] 1978; Zawirski 1936, 350.

[67] Łukasiewicz 1936. The journal identified Kokoszyńska as Maria Lutman-Kokoszyńska. For information about these two philosophers and Ingarden, see boxes in sections 9.4 and 14.2 and a footnote in 15.6, respectively.

## 15.10 Discussion of Zawirski 1936, on Synthesis

*Alfred Tarski.* Numerous talks and remarks in the current Congress include polemical moments with respect to the Vienna Circle, mainly due to the conflict this Circle has with metaphysics; such moments occurred for example in the talks of Professors Łukasiewicz and Zawirski as well as in Professor Ingarden's remarks following the talks of Professor Łukasiewicz. One of the reasons for this controversy seems to be a terminological misunderstanding. The Vienna Circle denies the scientific nature of any statements that cannot be checked empirically or[68] are not logical statements: in particular, metaphysical statements. Moreover, simply speaking, all statements that can be resolved—proved or refuted—by a pure deductive method based on just the axioms and inference rules accepted in the language are included as logical statements; therefore, in the understanding of the Vienna Circle, to the metaphysical statements belong only statements that cannot be proved either by a deductive or by an empirical method. On this point there is, as it seems, no essential difference between the Vienna Circle and Professor Łukasiewicz, who emphatically underlined that he knows of only two scientific methods to prove statements—the deductive and the empirical. Deductive metaphysics, whose possibility Professor Łukasiewicz raised, might be included by the representatives of the Vienna Circle in parts of the logical language; they would regard its propositions as logically true statements. There is also no essential difference between the Vienna Circle and the speaker, who recognizes the scientific nature of metaphysical statements provided that they admit empirical verification (or refutation); the difference is only that the Vienna Circle might include statements metaphysical in the speaker's sense in empirical, scientific knowledge. While between Professor Ingarden and the Vienna Circle there seems to be no significant difference in understanding of the term "metaphysics," on the other hand deeper essential differences are emphasized.

---

[68] Tarski may have meant *and* instead of *or*: in Polish, *i* instead of *lub*. The original text of this sentence is *Koło wiedeńskie odmawia charakteru naukowego wszelkim zdaniom, które nie dają się sprawdzić na drodze empirycznej lub nie są zdaniami logicznemi w szczególności zaś zdaniom metafizycznym.*

Contrary to Tarski's remark, the present editors do not regard the published versions of Zawirski's and Łukasiewicz's congress presentations as notably polemic. Ingarden's published remark, with sufficient emphasis indicated, would indeed seem vehement. These parts of the congress proceedings were probably redacted. Łukasiewicz noted that contemporary deductive logic, much more powerful than the old Aristotelian logic, can be used with any sentences that are not concerned with size and number: such sentences are logical truths.[69] Moreover, if we use the new logic more heavily, we must expect it to be applied to philosophical issues. Thus, many fundamental views should be reconsidered. Finally, the discovery that bivalence is not a necessary feature of logic should cause great upheaval. Ingarden's remark advocated resisting the influence of the Vienna Circle, which Łukasiewicz had not mentioned, and emphasized the search for appropriate methods of analyzing metaphysical questions, as had Łukasiewicz.

It is perhaps significant that Tarski was careful here to describe the philosophical position of the Vienna Circle without claiming it as his own.[70]

In response to Kokoszyńska and Tarski, the speaker acknowledged that defending metaphysics on the basis of experience does not affect the fundamental thesis of the Vienna Circle. But he stated two reservations about the Circle. The first, he said, coincided with that expressed in another recent article by Łukasiewicz:[71] the Circle favored the a priori sciences too much over the empirical and thus reduced too many claims of empirical science to purely linguistic statements. Zawirski's second reservation concerned the interdependence of mental and physical phenomena that underlie empirical psychology.

---

[69] Łukasiewicz's words were *Współczesna logika matematyczna okazała, że metodę demonstratywną, czyli matematyczną, można stosować i do zdań, które nie dotyczą wielkości i liczby. Takimi zdaniami są prawa logiczne,...*. They are remarkably evocative of the main point of Tarski's posthumous 1986b paper *What are logical notions?*

[70] According to Jan Woleński (1993, 322), Tarski "was very self-restrained in expressing his philosophical views." See also the annotation of Tarski [1965] 2007 in section 16.1.

[71] Łukasiewicz [1936] 1970.

## 15.10 Discussion of Zawirski 1936, on Synthesis

*Zygmunt Zawirski
around 1930*

**Zygmunt Zawirski** was born in 1882 near Tarnopol, then in the Austrian Empire (now Ternopil, in Ukraine). His father, a Polish patriot, had taken refuge there after the 1863 uprising against the Russian Empire. Zygmunt worked in schools during his 1901–1906 university studies in Lwów and for some years afterward; he earned the full teaching credential in 1907. By 1910 he had completed doctoral studies at Lwów under Kazimierz Twardowski, with a dissertation overlapping philosophy and psychology. For a year afterward, Zawirski studied in Berlin and Paris. During 1922–1928 he taught at the Lwów Polytechnic University. He earned the *venia legendi* in 1924 working with Cracow philosopher Władysław Heinrich, then was appointed professor at the University of Poznań in 1928. Zawirski continued research in fundamental areas of philosophy of science, emphasizing the application of formal logic to physics problems, and was particularly concerned with the interpretation of time. He continually and strongly emphasized an empirical approach, but stopped short of the extreme views of the logical positivists. Zawirski was promoted to full professor in 1934, and moved to the University of Cracow in 1937. He served as dean there both before and after World War II. During that conflict, he taught classes clandestinely. From 1934 on, Zawirski edited the journal *Kwartalnik filozoficzny*; from 1938, he was president of the Cracow section of the Polish Philosophical Society. Zawirski was an especially effective communicator and statesman, both within the nation and in its relations with foreign academics. He died in 1948.*

---

*Jadczak 1993a, Szumilewicz-Lachman 1994—these studies disagree on some details.

## 15.11 Discussion of Kokoszyńska 1936a, on Relative Truth

On 26 September 1936, Maria Kokoszyńska presented the paper *Concerning the Relativity and Nonrelativity of Truth* at a plenary session of the Third Polish Philosophical Congress in Cracow.[72] The congress is described in section 14.2. Kokoszyńska's paper is summarized here, and Alfred Tarski's [1936] 2014d remark about it is translated here, an explanation of that remark is included, and the speaker's response is paraphrased.[73]

A former student of Kazimierz Twardowski, Kokoszyńska had begun participating in meetings of the Vienna Circle in 1934, and soon earned a reputation as a representative of the Lwów–Warsaw School of Logic concerning such matters. In her presentation she did not refer explicitly to Tarski's new theory of truth.[74] She referred to an ongoing dispute about whether, in a description of a sentence $S$ as true, truth should be regarded as an absolute concept or as relative to other considerations. She indicated that the positivists tended to favor relativism. Truth, in her view, should be regarded as relative to the meanings of the terms in $S$. Investigation of meanings was prerequisite to further analysis of the language of $S$. Without that, the absolutist position would gain authority.

After Kokoszyńska's presentation, Walter Auerbach remarked that her concept of a relative term seemed to differ from the everyday one.[75] Tarski's much longer remark is translated in the box on the next page. Listed as Tarski [1936] 2014c in this book's bibliography, the translation adheres to the same standards as earlier ones in this book. Tarski first suggested that it would be simpler to construe truth as relative to the language in use. He did not say that this would suffice, though he may have implied tacitly that linguistic analysis would constitute a tool in studying meaning. Tarski's second comment was addressed not so much to Kokoszyńska's paper, but to her, personally, and to others doing similar studies. He did not want his claim of simplicity to appear belittling, particularly of researchers for whom he had great respect, and of one with whom he enjoyed very close acquaintance.[76]

---

[72] The congress proceedings list Kokoszyńska with this surname and under her married names Kokoszyńska-Lutmanowa and Lutman-Kokoszyńska (Polish Philosophy Congress 1936, 351, 424, 544).

[73] The Polish title of Kokoszyńska 1936a was *W sprawie względności i bezwzględności prawdy*. For information about its author, see a box in section 14.2. An English translation, by Jan Tarski, of part of Alfred Tarski's remark was included in Jan Woleński's paper *Tarski as a Philosopher* (1993, 325). The present translation is new; it seems consistent with that one.

[74] Rojszczak 2002, 36; Mancosu 2008, 195; Tarski [1933] 1983.

[75] **Walter Auerbach**, a former student of Kazimierz Twardowski and research collaborator of Tadeusz Witwicki, was then living in Warsaw and publishing on the philosophy and psychology of memory. He was murdered by the Nazis in Lwów in 1942. (Jadacki and Markiewicz 1993, 159–160.)

[76] Woleński has discussed this interchange, including Kokoszyńska's response. His comments (2001b, 73) on her paper and Tarski's first remark have a slightly different thrust. He also discussed Tarski's second remark, without regard to its context (1993, 325). For the personal relationship between Kokoszyńska and Tarski, see Feferman and Feferman 2004, 88–92.

## 15.11 Discussion of Kokoszyńska 1936a, on Relative Truth

> *Alfred Tarski.* (1) The speaker's words show, among other things, that the notion of truth—in one of its interpretations—should be made relative to the notion of meaning. Would it not be simpler to make the notion of truth relative to the notion of language, which seems to be the clearer notion and logically less complicated than the notion of meaning?
>
> (2) The second remark bears a more general character. In recent times one could observe more than once the following phenomenon: philosophers who are cultured in logic try to apply the apparatus and methods of contemporary logic to various classical problems of philosophy, and one finds then that these problems, which in some cases were the objects of philosophical investigations and speculations over long centuries, acquire a more or less trivial character after being subjected to logical analysis. By its nature, this phenomenon gives rise to various doubts. In the first place, the question arises whether by clarifying a problem and liberating its formulation from vagueness and from various imprecisions, one captures at the same time the "essential" intentions of those who posed this problem or thought about it in former times, even if they could not give to their investigation appropriate formal dress. It is possible, and even quite likely, that doubts of this kind are in many cases well founded. Nevertheless, even in these cases the work of those logicizing philosophers is surely not in vain: it forces opponents into making sufficiently precise the "essence" of the problems at issue, so as to correspond to the demands of logic and methodology, and so as to prevent future discussion of these problems from having the character of a continuous chain of never-ending misunderstandings.

Kokoszyńska's response has been described by the Catalan philosopher Manuel García-Carpintero as conceding that Tarski may be correct

> if languages are individuated not just by the well-formed combinations of sounds or inscriptions belonging to them, but also by their correct translations into the language in which the definition is given. If languages are only individuated "formally," i.e. in the first way, then any truth-definition for a language should be taken to be additionally relative to the way of translating its sentences into sentences of the language in which the definition is carried out.[77]

---

[77] García-Carpintero 1999, 143–144.

Kokoszyńska indicated that this second relativization is what she had called "relativization to the meanings." She said,

> I am inclined both to propose and to consider my "formal" formulation of the problem of the relativity of truth as an approximate approach to the traditional problem rather than a final clarification.[78]

Responding to Tarski's use of the word *trivial*, she claimed that granting the trivial nature of this issue would result in characterization of both the notion of truth and that of meaning as trivial, a position that would not garner much support.

## 15.12 Two Letters to Sierpiński, 1946–1947

Two letters from Alfred Tarski to Wacław Sierpiński are translated in the boxes in this section. This is their first publication. The letters connect Tarski's research in Poland to his career in Berkeley. These introductory paragraphs and some later footnotes continue the account of his life and career in Poland that was included in chapters 8 and 14. For further information about Tarski's relocation to the New World, see Feferman and Feferman 2004, particularly its chapters 6 and 7.

After the September 1939 conference at Harvard, Tarski spent three years in temporary research and teaching positions in the eastern United States. In September 1942 he took another temporary position as "utility" lecturer in mathematics at the University of California, Berkeley. To a large extent, the university's offerings were then a makeshift contribution to the war effort. Nevertheless, Tarski quickly began creating a research environment, to continue the work he had begun in Poland in earlier years. The university recognized Tarski's eminence and appointed him associate professor in 1945, then full professor in 1946. World War II ended in Europe in April 1945. Two months later Tarski earned United States citizenship. His wife and children finally arrived from Poland in January 1946. With his salary and loans he bought a house in the hilly neighborhood north of the Berkeley campus, at 1001 Cragmont Avenue. The Fefermans called it "charming, if slightly pretentious."[79]

Until the war's end, Tarski had achieved only sporadic contact with his former Polish teachers, colleagues, and students. When it became feasible, he resumed correspondence with them. His 1946 and 1947 letters to Sierpiński show through their tone the sort of relationships that Tarski would cultivate. They provide glimpses of

---

[78] In her much longer paper on this subject, published three months later, Kokoszyńska seemed to follow Tarski's suggestion (1936b, 155).

[79] Feferman and Feferman 2004, 175.

- Tarski's emerging research environment,
- his continuing research concerning cardinal arithmetic and the axiom of choice and concerning equidecomposability in set theory,
- Tarski's deep feeling for the murdered Adolf Lindenbaum and respect for his work,
- obstacles to communication and publication, and
- the start of Tarski's new research threads in universal algebra and the theory of relations.[80]

The original letters were handwritten in Polish on stationery provided by the University of California.[81] The translations, here designated Tarski [1946–1947] 2014, adhere to the same standards as earlier translations in this book. In this case, moreover, their layout is virtually identical with that of the originals. Personal names and bibliographic references have been altered to conform with the conventions of the present book. The present editors provided some bibliographical and historical information in footnotes that fall outside the boxes.

---

1001 Cragmont Avenue
Berkeley 8, California

**UNIVERSITY OF CALIFORNIA**

DEPARTMENT OF MATHEMATICS
BERKELEY 4, CALIFORNIA

30 October 1946

Dear Professor!

I received your letter of September 14 ten days ago. I sent the manuscript to the *Annals of Mathematics* while simultaneously expressing my willingness to relieve you of editing it. I have not yet received a response; moreover, I have asked to have the decision be announced to you and me at the same time.

I think your reasoning is really interesting for its simplicity and efficiency. I would just like to make one remark (which I told Paul Erdős in due time). As I recall, Adolf Lindenbaum had a more general result—a proof of the existence of two sets not equivalent by countable decomposition in relation to an arbitrary

---

[80] For biographical sketches of Sierpiński and Lindenbaum, see the boxes in section 4.1 and chapter 14, respectively. Concerning Tarski's continuing other research threads to Berkeley, particularly completeness and decidability, see the long introduction to Tarski [1946] 2000 by its editor, Hourya Sinaceur.

[81] Documents APAN sygn. III-194 in the Archives of the Polish Academy of Sciences.

family of bijective transformations—not necessarily a family of isometric transformations (and perhaps even a more general result for arbitrary cardinal numbers). I do not remember the proof at all, and also do not remember whether the family of bijective transformations was subjected to some additional assumptions. I am under the impression (but can be completely mistaken) that Lindenbaum announced his result without proof either in an article in *Fundamenta Mathematicae* or in the reports of talks in the *Annales de la Société Polonaise de Mathématique*. In any event it would be worthwhile to reconstruct and announce the result. Overall it seems to me that there is an obligation to mathematics and to the memory of Lindenbaum to encourage people to become acquainted with what Lindenbaum left behind in print and to publish proofs of results [that he] supplied without proof. In this way, several appropriate topics for master's theses and perhaps even dissertations can be found. I spread some propaganda to this end here. (Erdős's article is to a certain extent the result of this.) I will publish proofs—as a matter of fact I am doing this now—of Lindenbaum's results that are closely associated with my own research or were announced previously in our joint work.[82]

I really must congratulate you on the invitation to lecture at the Sorbonne. I sincerely thank you for the card from Switzerland, the volume of *Fundamenta* (it made a big impression here) and for sending copies of articles. I am promising myself to reconsider some of your works in the near future—particularly those that are related to Stefan Banach's and my paradox. Please note that I have always had the impression that the number 9 given by John von Neumann and reduced by you can be easily obtained by directly analyzing the proof by Banach and me.[83]

I sent a few of my articles printed in the United States along with general shipments of books and articles for Polish mathematicians and logicians; I hope that they have already arrived in Warsaw, or will arrive in the near future. On the articles, I wrote in pencil the names of people for whom they are intended; I do not need to mention that your name is on all the works which I believe may interest you.

---

[82] The first paragraph of this letter referred to Sierpiński's 1947 paper *Sur un théorème de A. Lindenbaum*. There, Sierpiński referred to Erdős's 1943 paper *Some Remarks on Set Theory*, in the same journal. With the aid of the axiom of choice, Erdős had proved (page 144) that for every cardinal $m$ there is a family of $2^{2^{\aleph_0}}$ sets of real numbers, no two of which can be decomposed into $m$ disjoint mutually congruent subsets. Erdős had credited Lindenbaum for announcing a less general version of that theorem, without proof. Sierpiński proved the same result without using that axiom. Tarski thought that Lindenbaum had achieved an even more general result.

[83] *Sorbonne* refers to the University of Paris. Neumann's number 9 is explained later, in footnote 89.

> I am unsure whether I mentioned in my last letter to you that my book on "cardinal algebras" should appear in print around 1 January 1947. It is a book on the border between set theory and algebra, and contains reconstructions of all of my and some of Lindenbaum's results in the theory of bijective transformations, but in completely new algebraic clothing. In a few weeks my student Bjarni Jónsson's and my seventy-page monograph on decompositions of simple finite algebraic systems will appear in print. I am working on a book on the theory of relations.[84]
>
> I am curious whether Karol Borsuk has already left. In December, a mathematics conference will be held at Princeton University in honor of its two-hundredth anniversary. I was invited and hope that I can meet him there.[85]
>
> I add sincere regards, and I ask [you] to pass along [my] regards to all [our] colleagues.
>
> [signed] Alfred Tarski
>
> PS. For the information about the circumstances surrounding my book about the concept of truth,[86] thank you very much.

Tarski's March 1947 letter to Sierpiński begins on the next page.

---

[84] Tarski 1949a, *Cardinal Algebras*. Jónsson and Tarski 1947, *Direct Decompositions of Finite Algebraic Systems*. Tarski had reported about his work on the theory of relations already in 1940, in an introductory philosophy course at City College of New York: see Rosen 1985.

[85] **Karol Borsuk**, born in 1905 in Warsaw, then part of the Russian Empire, earned the doctorate in 1930 from the University of Warsaw with a dissertation on topology supervised by Stefan Mazurkiewicz. Borsuk soon joined the faculty there and in 1946 became full professor. He contributed many major research results and was instrumental in rebuilding Polish mathematics after World War II. He spent 1946–1947 at the Institute for Advanced Study in Princeton. Borsuk died in 1982. For further information, see Keesling 1990. Tarski attended the Princeton Bicentennial Conference in summer 1946, and presented the paper Tarski [1946] 2000.

[86] Tarski 1933 or [1933] 1935.

**UNIVERSITY OF CALIFORNIA**

1001 Cragmont Avenue
Berkeley 8, California

DEPARTMENT OF MATHEMATICS
BERKELEY 4, CALIFORNIA

18 March 1947

Dear Professor!

Thank you sincerely for your letter, papers, and books, as well as the invitation to publish an article in *Fundamenta Mathematicae* about the equivalence of the axiom of choice and the theorem quoted in your letter. I have already outlined an article. I included certain results related to the older result, but obtained only recently. Consequently, the article has grown slightly: I suppose it will take about twenty pages to print. It will take some time, however, before someone here corrects the article in terms of its style, and before I can rewrite it cleanly and send it to you. I will of course be very grateful if anyone in Warsaw can read the manuscript over carefully, check the accuracy of the reasoning and later relieve me of doing the corrections; I would be extremely glad if Andrzej Mostowski should take this up. Minor cosmetic changes may be made without consulting me. In the article I use the symbolism of *Fundamenta*, varying slightly from that which is used here (thus, symbols such as $\subset$, $+$, $\cdot$, $\Sigma$, $\Pi$, and not $\subseteq$, $\cap$, $\cup$, $\bigcup$, $\bigcap$). Just one remark about cosmetics: I've heard here several times that for British and American readers, use of the so-called expanded spacing is incomprehensible—the spaces are usually considered as errors made by the printing machine. Instead of expanding the spacing, they use so-called "small caps"—that is, the font is the size of the lowercase letters of the alphabet but has the capitalized shape. Because my article will be written in English, I will be glad if the expanded spacing is replaced either by "small caps" or by the usual italic.[87]

I am truly happy that you are announcing other results by Lindenbaum and me. You may be interested in knowing that some of those results (that is, those that do not require the axiom of choice, and among others, a proof of the theorem on the relationship between $k \cdot m \leq k \cdot n$ and $m \leq n$) will appear simultaneously elsewhere—in my book *Cardinal Algebras* (Oxford University Press, New York).* They will appear there as an application of a certain general theory of an abstract and algebraic character. It will not even be immediately apparent that the proof does not require the axiom of choice—because in putting these results in abstract

---

* Unfortunately, the printing of this book has not yet begun. (It was to be published 1 January 1947). As you see, even here these things are not easy.

---

[87] The article was Tarski 1948, *Axiomatic and Algebraic Aspects of Two Theorems on Sums of Cardinals*. Tarski's request that SMALL CAPS be used instead of e x p a n d e d  l e t t e r s p a c i n g was not heeded.

algebraic terms, some limited use of the axiom of choice is necessary. But, following the comments that I included in the book, the intelligent reader will likely come up easily with purely set-theoretic proofs, not relying on the axiom of choice.[88] The material that I included in articles in *Fundamenta* will be helpful.

You probably heard from Borsuk [about] a result by a local colleague of mine, Raphael M. Robinson (who is lecturing this year at Princeton). The result is really beautiful. It concerns the equivalence between a ball and the union of two balls with the same radius. Not being familiar with your earlier work in this field, he obtained a stronger result—the number 5 instead of 9 or 8. Moreover, one part consists of just a single point—the center of the ball—and this part can be eliminated completely if, instead of balls, one considers their surfaces. It is easy to see that these results are already the best possible. What is more important, Robinson shows that the Cantor–Bernstein method, which Banach and I have used in our work, is not needed in the treatment of this problem (it is probably important when considering the general question of the equivalence of two "solids"); [Robinson's] method has a direct character, and depends on modifying Felix Hausdorff's methods.[89] Another colleague of mine, Anthony P. Morse (at Berkeley, already published in *Fundamenta*), has also obtained a certain result in this field which, in my opinion, deserves to be published. As you know, the assertion about the nonexistence of a "paradoxical decomposition" of a segment can be proved without aid of the axiom of choice. With my advice, Morse took up a similar problem for the square and obtained an identical result (although with quite different methods). I urged the two— Robinson and Morse—to send their results to Warsaw. In principle they are inclined to the idea, but are quite overwhelmed with writing. Perhaps if you could write to the both of them (if you consider it appropriate to do so), we may obtain a better outcome.[90]

---

[88] To analyze the relationships between cardinal-arithmetic theorems and the axiom of choice, the then-standard definition of cardinal number must be avoided because it depends on the axiom. Alternatives are to introduce cardinals axiomatically, as in Tarski 1924e, or through the notion of rank, as in Lévy 1979, chapters II–III.

[89] In his 1947 paper *On the Decomposition of Spheres*, Robinson showed that a ball can be decomposed into five pieces, one consisting of a single point, that can be reassembled to form two balls of the same radius. Neumann noted in his 1929 paper *Zur allgemeinen Theorie des Masses*, without proof, that the task could be done with nine pieces. In his 1945 paper *Sur le paradoxe de MM. Banach et Tarski,* Sierpiński proved that eight pieces were sufficient. Robinson also showed that four pieces were not enough. His methods stemmed from those of Hausdorff 1914a.

[90] The paper Morse and Randolph 1940, on measure theory, had been scheduled to appear in *Fundamenta Mathematicae* in 1939, but was delayed until 1945; it was published elsewhere in 1940. In his 1949 paper *Squares are Normal*, Morse showed, without the aid of the axiom of choice, that no decomposition of a square is paradoxical in the sense of Banach and Tarski [1924] 2014. Morse and Randolph had derived that result as a corollary of a more general one that required the axiom: see sections 5.2 and 6.2.

From time to time I receive inquiries about current and past volumes of *Fundamenta*. Among others, the University of Nevada, Reno, took my advice and wrote directly to Warsaw a few months ago, expressing their desire to acquire all of the volumes of *Fundamenta*. I do not know whether their letter arrived, because there has as yet been no response.

I am unsure whether I communicated to you the following result related to a problem of Mikhail Suslin.[91] The problem can be summarized as follows: let $Z$ be an ordered set satisfying condition

($W_1$) every transfinite sequence of disjoint intervals is at most countable;

is this set ordered like the set of real numbers? Instead of ($W_1$) we may consider an equivalent condition ($W_2$) by replacing the word "disjoint" with the word "decreasing." Of course, both conditions ($W_1$) and ($W_2$) are necessary for $Z$ to be like a certain set of real numbers. Now consider condition ($W_3$) that is stronger than ($W_1$) and ($W_2$): namely,

($W_3$) if $P_0, P_1, \ldots, P_\xi, \ldots$ ($\xi < \tau$) is a sequence of intervals in $Z$ such that for each pair of intervals $P_\xi$ and $P_\eta$ where $\xi < \eta$ we have $P_\xi \cdot P_\eta = \phi$ or $P_\xi \supsetneq P_\eta$, then the sequence is at most countable.

It can be shown that this condition ($W_3$) is necessary and sufficient for $Z$ to be similar to a certain set of real numbers. The proof is not too difficult, but it seems to me it is not trivial (in either direction). From here it is easy to obtain a set of conditions characterizing the order of the set of all real numbers.

My wife and I send warm greetings to both of you. Please pass on my regards to all your colleagues; I am writing to Kazimierz Kuratowski at the same time.

[signed] A. Tarski

---

[91] The famous *Suslin's Problem* has been the subject of many investigations in set theory.

*Alfred Tarski in 1968*

# 16
# Posthumous Publications

Chapter 16 is an annotated bibliography of works by Alfred Tarski that were not listed or were incompletely described in the definitive 1986 bibliography by Steven R. Givant.[1] None were reprinted in Tarski's 1986a *Collected Papers*. Each listed item begins with the author–date designation used in this book, continues with a description adapted from this book's bibliography,[2] and is followed by a brief annotation. In some cases, the annotation includes Tarski's own words, enclosed in a box with a double border.

Many new editions and translations are included.[3] However, the numerous translations in this book are not listed here, though they logically belong with the items that are. Readers may use this book's bibliography to construct items for them analogous to those in this chapter. For example, the following item would represent the first translation in this book:

> Tarski [1921] 2014. A contribution to the axiomatics of well-ordered sets. In *Alfred Tarski: Early Work in Poland—Geometry and Teaching*, translated and edited by Andrew McFarland, Joanna McFarland, and James T. Smith, chapter 2. New York: Birkhäuser. Translation of "Przyczynek do aksjomatyki zbioru dobrze uporządkowanego," *Przegląd filozoficzny* 24, 85–94, Givant 1986 item 21. This is Tarski's first published paper.

## 16.1 Papers

For additional translations of papers by Tarski into Polish, see the annotation for Tarski 1995–2001 in section 16.2.

---

[1] Four works listed here were published during Tarski's lifetime, but not listed in Givant 1986: Tarski [1930] 2014h, 1965a, Tarski 1972, and Mates, Henkin, and Tarski 1959. Tarski 1995–2001 contains a less extensively updated bibliography.

[2] This book's bibliography includes more details of publication.

[3] Some translations into Polish are mentioned only in the annotations. Reprints are mentioned in this chapter only when particularly germane to the discussions in this book. Chapters 16–18 are a by-product of the research that underlies earlier chapters. The editors did not search exhaustively for other works of Tarski not mentioned in items discovered during that work.

Tarski [1923] 1998. On the primitive term of logistic. In *Leśniewski's Systems: Protothetic*, edited by Jan T. J. Srzednicki and Zbigniew Stachniak, 43–68. Dordrecht: Kluwer Academic Publishers.

Stachniak translated this from Tarski 1923a, "O wyrazie pierwotnym logistyki," *Przegląd filozoficzny* 26: 68–89, Givant 1986 item 23. That was the original published version of Tarski's Warsaw doctoral dissertation, which was supervised by Stanisław Leśniewski. The original was not reprinted in Tarski's 1986a *Collected Papers*. The French version Tarski 1923b and its English adaptation in [1956] 1983, 1–23, do not include the entire dissertation.

Tarski, Alfred. [1928] 2010. Remarks on fundamental concepts of the methodology of mathematics. Translated by Robert Purdy and Jan Zygmunt. In *Universal Logic: An Anthology, from Paul Hertz to Dov Gabbay*, edited by Jean-Yves Béziau, 67–68. Basel: Birkhäuser.

Purdy and Zygmunt translated this from Tarski's 1928 abstract: his first published contribution to general metamathematics. For information on the sequence of lectures that this presentation inaugurated, see section 9.4 and Zygmunt 2010.

Tarski, Alfred. [1931] 2014h. O pewnym systemie logiki matematycznej i wynikających z niego zagadnieniach metodologicznych i semantycznych. *Ruch filozoficzny* 12 (1930–1931): 232. Abstract of a 15 April 1931 presentation to the Logic Section of the Warsaw Philosophical Institute. Omitted from the Givant 1986 bibliography and Tarski's 1986a *Collected Papers*, this abstract is translated in the box below. This is probably the first exposure of Gödel's famous incompleteness theorem outside Vienna (Feferman and Feferman 2004, 84).

---

Dr. A. Tarski: On a certain system of mathematical logic and the resulting methodological and semantic issues. The Speaker had previously shown how to build a very simple fragment of logic that is sufficient for building all known mathematics. Such a system, although perhaps inconvenient in practice, is very convenient for examining mathematics. After outlining such a system, the Speaker acquainted the audience with the latest result of Gödel in Vienna, based on such a system. Namely, Gödel has proved that the system of integer arithmetic is undecidable,[4] even the sentences just about addition and multiplication. After sketching a proof in the framework of the system, the Speaker discussed further momentous consequences that follow from Gödel's result.

---

Tarski, Alfred. [1934] 2003a. Discussion of two papers:

Reichenbach [1934] 1936, "Die Bedeutung des Wahrscheinlichkeitsbegriffs für die Erkenntnis";

Zawirski [1934] 1936, "Bedeutung der mehrwehrtigen Logik für die Erkenntnis und ihr Zusammenhang mit der Wahrscheinlichkeitsrechnung."

---

[4] Tarski's word was *nierozstrzygalny*.

With a Polish translation by Mariusz Grygianiec. In *Alfred Tarski: dedukcja i semantyka (déduction et sémantique)*, edited by Juliusz Jacek Jadacki, 12–14. Warsaw: Wydawnictwo Naukowe Semper. Reprint and translation of Tarski [1934] 1936a.[5]

This is Tarski's contribution to a discussion that took place after presentations by the German and Polish philosophers Hans Reichenbach[6] and Zygmunt Zawirski at the Eighth International Congress of Philosophy in Prague, 2–7 September 1934.[7] As a way of surmounting certain difficulties with the foundations of probability theory, Reichenbach and Zawirski each had suggested formulating the theory within some version of multivalued logic. Tarski claimed that those measures would not suffice. Moreover, the difficulties would disappear if the theory were based instead on a function that assigns to a sentence $s$ and a set $Y$ of sentences the probability of $s$ *under the conditions* $Y$; ordinary logic would then suffice. Tarski referred to Keynes 1921, which followed that approach, and the more recent studies Ajdukiewicz 1928 and Mazurkiewicz 1932, which extended it. Thus, Tarski claimed, reformulating logic to meet the needs of probability theory was neither necessary nor desirable.

Tarski [1934] 2003b. Discussion of Jørgensen [1934] 1936, "Die logischen Grundlagen der Wissenschaft." With a Polish translation by Mariusz Grygianiec. Ibid., 11. Reprint and translation of Tarski [1934] 1936b.

This is Tarski's contribution to a discussion that took place after a presentation by the Danish philosopher Jørgen Jørgensen[8] at the congress mentioned in the previous item. Jørgensen had considered in turn the meanings of the terms that made up the title of his presentation: *science*, *foundations* of science, *logical* foundations of science. He regarded a science as a set of sentences in a given language $L$ that are held to be true and can be ordered as a kind of deductive system. Tarski objected that the description was too general: *any* set of sentences of $L$ could be so ordered. Jørgensen went to great length to describe the role logic should play in that ordering, and in the ensuing discussion emphasized that *held to be true* is different from *true*. Tarski noted that most of Jørgensen's claims depended on the notion of "tautologous" sentence, which for most such languages had not yet been adequately defined. Tarski had already published papers defining the notion of *true sentence*, and would soon publish another that would define the notion of *tautologous (logically true) sentence* and distinguish it from the former notion: see section 15.6 and the following two items.

---

[5] The text of Tarski [1934] 1936a is nearly identical to that of Tarski 1935–1936. Tarski [1934] 1936a was described incorrectly in Givant 1986 as a discussion of a paper by Janina Hosiasson-Lindenbaumowa. It was reprinted in Tarski 1986a, but misidentified in the same way.

[6] **Hans Reichenbach** was born in Hamburg in 1891 and schooled there. His father was a merchant. After studying at several universities, Hans earned the doctorate at Erlangen in 1916 for studies in probability theory. By 1926 he was wielding major influence in the philosophy of science and was appointed professor at the University of Berlin. Dismissed in 1933 because of his Jewish ancestry, he emigrated to Istanbul, and five years later to Los Angeles, where he helped establish a major program of philosophical studies. Reichenbach died in 1953. (Schuster 1975.)

[7] For biographical information about Zawirski, see a box in section 14.10. The congress is described in section 14.2.

[8] **Jørgen Jørgensen** was born in Haderup, in northeastern Denmark, in 1894. His father was a vicar. Jørgen was graduated from the Soro Academy in 1912, earned a master's degree in philosophy in 1918, worked as a secretary for eight years, and in 1926 was named professor of philosophy at the University of Copenhagen. A leading exponent of logical positivism, he died in 1969. (See Witt-Hansen 1969.)

Tarski [1935] 1984. Sobre el concepto de consecuencia logica. In *Lecturas de lógica II: Selección, introducción y traducción*, edited by Pilar Castrillo and Luis Vega Reñon, section A.2, 178–192. Madrid: Universidad Nacional de Educación a Distancia.

Vega translated this from Tarski [1935] 1936b, "Über den Begriff der logischen Folgerung," *Actes du Congrès International de Philosophie Scientifique, Sorbonne, Paris, 1935*, volume 7, 1–11, Paris: Hermann & Cie., Givant 1986 item 36g. Vega's translation is preceded by a significant introduction, pages 155–177.

Tarski [1935] 2002. On the concept of following logically. Translated from the German and Polish, with commentaries, by Magda Stroińska and David Hitchcock. *History and Philosophy of Logic* 23: 155–196.

Stroińska and Hitchcock translated this from Tarski's German version, cited in the previous item, and from Tarski 1936b, "O pojęciu wynikania logicznego," *Przegląd filozoficzny* 39: 58–68, Givant 1986 item 36a. In addition to analyzing the content of this paper, they systematically noted how the two originals differ from each other and from the English version in Tarski [1956] 1983, 409–430. The translators' account of their methodology is particularly useful: it guided the translations from Polish in the present book.

Tarski [1939–1940] 1995. Some current problems in metamathematics. Edited by Jan Tarski and Jan Woleński. *History and Philosophy of Logic* 16: 159–168. The editors constructed this article from Tarski's almost-complete notes for a presentation, probably given at Harvard University. Here is their summary (page 159):

"In this article the author first described the developments which brought to focus the importance of consistency proofs for mathematics, and which led Hilbert to promote the science of metamathematics. Further comments and remarks concern the (partly analogous) beginnings of the work on the decision problem, Gödel's theorems and related matters, and general metamathematics. An appendix summarizes a text by the author on completeness and categoricity."

Tarski [1944] 1985. Definition. *American Mathematical Monthly* 92: 358. This consists merely of the following quotation from Tarski 1944, 359.

> In fact, I am rather inclined to agree with those who maintain that the moments of greatest creative advancement in science frequently coincide with the introduction of new notions by means of definition.

Tarski [1946] 2000. Address at the Princeton University Bicentennial Conference on Problems of Mathematics (December 17–19, 1946). Edited and with an introduction by Hourya Sinaceur. *Bulletin of Symbolic Logic* 6: 1–44.

According to the editor (page 1), "This article presents Tarski's address...together with a separate summary. Two accounts of the discussion which followed are also included. The central topic...is decision problems. The introductory note gives information about the Conference, about the ... subjects discussed ... and about subsequent developments ... ." The editor's commentary details the continuity of Tarski's research activity as he moved from Poland to California.

Tarski [1955] 1993. Sur la théorie des modèles. Translated and edited by Anne Preller. In *Philosophie de la logique et philosophie du langage*, edited by Jacques Bouveresse, volume 2, 137–158. Paris: Éditions Odile Jacob.

This survey of the basic concepts of model theory emphasizes algebraic more than logical aspects. According to the editor's note, it "is the translation of a communication presented by Alfred Tarski in Paris between 26 September and 1 October 1955 at the Seventieth International Colloquium, *Le raisonnement en mathématiques et en sciences expérimentales*, of the Centre National de la Recherche Scientifique. The typescript with some annotations and corrections in the hand of Alfred Tarski was sent to me by Jan Tarski." The typescript is in the Bancroft Library of the University of California at Berkeley, Tarski Archive 84/69, Container 12.

Mates, Henkin, and Tarski 1959. Jan Kalicki, philosophy, mathematics: Berkeley and Davis. Berkeley: University of California Digital Archives.

**Jan Kalicki** was born in Warsaw in 1922. He was educated there and at the University of London, where he earned a doctorate in mathematics in 1948. A prolific scholar in logic, he soon emigrated to the United States. In 1953 Kalicki became assistant professor of philosophy at the University of California, Davis. Three months later, he died in an automobile accident. Tarski and his colleagues Benson Mates and Leon Henkin constructed this obituary for the Berkeley community.

Tarski [1965] 2007. Two unpublished contributions by Alfred Tarski. Edited by Francisco Rodríguez-Consuegra. *History and Philosophy of Logic* 28: 257–264.

These are edited transcriptions of remarks delivered in April 1965 in Chicago, and in July 1965 in London, at symposia devoted to implications of Kurt Gödel's incompleteness theorems and to recent results in set theory, respectively. Tarski contributed to the discussions after the invited talks. In the first discussion, according to the editor (page 257), Tarski "takes advantage of the opportunity to announce his general philosophical position, very likely for the first time in public: Tarski claims to be an extreme anti-Platonist." The bulk of Tarski's comments responded to previous talks by Hilary Putnam, Andrzej Mostowski, and Georg Kreisel.

Tarski 1986b. What are logical notions? Edited by John Corcoran. *History and Philosophy of Logic* 7: 143–154. Givant 1986 item 87?. According to the editor's introduction (page 143),

"In this article Tarski proposes an explication of the concept of logical notion. His earlier well-known explication of the concept of logical consequence presupposes the distinction between logical and extra-logical constants (which he regarded as problematic at the time). Thus, the article may be regarded as a continuation of previous work.... [Felix] Klein's Erlanger Programm for classifying geometrical notions is sketched.... Then, generalizing beyond geometry, a notion (individual, set, function, etc.) based on a fundamental universe of discourse is said to be *logical* if and only if it is carried onto itself by each one-one function whose domain and range both coincide with the entire universe of discourse."

Henkin, Monk, and Tarski 1986. Representable cylindric algebras. *Annals of Pure and Applied Logic* 31: 23–60.

With Leon Henkin and J. Donald Monk, Tarski had written the 1971–1985 monograph *Cylindric Algebras*. This article investigates which cylindric algebras are isomorphic to algebras of relations.

Tarski and Givant 1999. Tarski's system of geometry. *Bulletin of Symbolic Logic* 5: 175–214.

From the abstract (page 175, written by Steven R. Givant): "This paper is an edited form of a letter written by [Tarski and Steven R. Givant] to Wolfram Schwabhäuser around 1978. It contains extended remarks about Tarski's system of foundations for Euclidean geometry, in particular its distinctive features, its historical evolution, the history of specific axioms, the questions of independence of axioms and primitive notions, and versions of the system suitable for the development of 1-dimensional geometry." The definitive presentation of Tarski's system is the 1983 monograph *Metamathematische Methoden in der Geometrie* by Schwabhäuser, Wanda Szmielew, and Tarski.

## 16.2 Monographs

Tarski 1965a. *Bibliography of Alfred Tarski*. Berkeley: Department of Mathematics, University of California, Berkeley. Superseded by the Givant 1986 bibliography.

Tarski 1986a. *Collected Papers*. Four volumes. Edited by Steven R. Givant and Ralph N. McKenzie. Basel: Birkhäuser.

Givant 1986 item $81^m$a was a preliminary edition. Maddux 1993 is a thorough review, with corrections, of Tarski 1986a and the 1986 bibliography by Steven R. Givant.

Tarski and Givant 1987. *A Formalization of Set Theory without Variables*. American Mathematical Society colloquium publications, 41. Providence: American Mathematical Society. Item $8\_^m$ in Givant 1986, with incomplete publication data.

According to the 1989 review by J. Donald Monk, the main ideas and results in this book stem from Tarski's work during 1940–1945. Givant made many independent contributions while writing the book with Tarski. Monk explained,

> This last book written (in part) by Tarski ... contributes both to our understanding of what mathematics is, as well as to technical foundational research. ... A simple equational language $\mathcal{L}^\times$ is introduced, and it is shown that set theory (for example, ZFC) can be translated into equations of $\mathcal{L}^\times$ which have no variables, so that sentences derivable in set theory are translated into equations derivable by equational rules of inference from the translates of the set-theoretical axioms. ... It is shown that, in principle, mathematics can be developed in the very simple framework of equations and substitution of equals for equals, rather than the customary basis in set theory formalized in first-order logic.

The 1990 review by István Németi continues,

> All quantifiers, individual variables, and so forth disappear from the language of mathematics, and all mathematical statements appear in the form of equations (and the only proofs are ... derivations

of equations from equations following...age-old rules). Surprisingly many mathematical statements become shorter, more intuitive, and [more] suggestive.

Further, according to Németi,

> The book provides a synthesis of the seemingly diverse schools of thought and research directions [that] Tarski...created or pursued seriously....The importance of finitistic methods in computer science is well known, and...the book provides a wealth of tools and results in this area. Another connection is the theories and calculi of relations, the modern form of which Tarski originated. And indeed, the book contains such a wealth of profound results and methods of this area not available before (or anywhere else yet) that after reading the book one's perspective of the whole area is fundamentally transformed.

Tarski 1994. *Introduction to logic and to the methodology of the deductive sciences*. Fourth edition. Edited and with a preface and biographical sketch by Jan Tarski. Oxford Logic Guides, 24. New York: Oxford University Press. Item $41^m$ in Givant 1986 lists previous editions.

The third edition, Givant 1986 item $41^m(10)$, was reprinted as an inexpensive paperback by Dover Publications in 1995; it is still in print in 2014. According to the 1995 review by Elliott Mendelson, this new edition, Tarski 1994,

> ...differs considerably from the third....The editor, Tarski's son, has added a four-page "Short biographical sketch" and a large number of new footnotes. Stylistic changes have been made and the terminology and notation have been brought up to date....Some of the original exercises...have been replaced by somewhat more elementary problems.

Tarski 1996 (see a following item) is an augmented Polish translation of this work.

Tarski 1995–2001. *Pisma logiczno-filozoficzne*. Volume 1: *Prawda*. Volume 2: *Metalogika*. Translated and annotated, with introduction, by Jan Zygmunt. With a bibliography by Steven R. Givant, supplemented by Zygmunt. Warsaw: Wydawnictwo Naukowe PWN. The title means *Logical-Philosophical Writings*; *Prawda* means *Truth*.

These volumes (1, 2) contain reprints ($r$) or new Polish translations ($t$) of the following works of Tarski—two are excerpts ($x$):

| | | |
|---|---|---|
| [1930] 1983a ........ $2t$ | [1938] 1939 ......... $2t$ | 1953 .............. $2t$ |
| [1930–1931] 2014 .... $1r$ | [1938] 1983 ......... $2t$ | [1955] 1993 ........ $2t$ |
| [1930–1936] 1992 ... $1tx$ | 1939b ............. $1t$ | 1965b ............. $2t$ |
| 1932 ............... $1t$ | [1939–1940] 1995 .... $2t$ | 1969 ............... $1t$ |
| 1933 ............... $1r$ | [1940] 1967 ......... $2t$ | 1986b ............. $2t$ |
| 1934 ............... $2r$ | [1942–1947] 1999 ... $2tx$ | Lindenbaum/Tarski |
| [1934] 1983 ......... $2t$ | 1944 ............... $1t$ |   [1935] 1983 ....... $2t$ |
| 1936a .............. $1r$ | [1944] 1987 ......... $1t$ | Łukasiewicz/Tarski |
| 1936b .............. $1r$ | [1946] 2000 ......... $2t$ |   [1930] 1983 ....... $2t$ |
| [1936] 2014d ........ $1r$ | [1948] 1957 ......... $2t$ | |

Reviewed in Cesarski1995, this work also incorporates the biographical sketch Zygmunt 1995 and informative notes on all included papers. It contains Givant 1995, a version of the Givant 1986 Tarski bibliography, updated and tailored to fit a more limited scope.

Tarski 1996. *Wprowadzenie do logiki i do metodologii nauk dedukcyjnych*. Second edition. Translated by Monika Sujczyńska from the fourth English edition, Tarski 1994 (see an earlier item). Edited and with a preface by Jan Tarski, with a supplement on elements of predicate logic by Witold Marciszewski. Warsaw: Filia Uniwersytetu Warszawskiego, Philomath, Aleph.

> This was reviewed extensively in A. Wójtowicz 1996. The first edition was issued in 1995 by the same publisher with a different supplement by Marciszewski: "Logika a dzielność umysłu"—"Logic and the boldness of mind." In his letter 1996a, Jan Tarski complained about the inappropriateness of the supplement. That may have occasioned the second edition.

## 16.3 Letters

Tarski [1918] 2014. Curriculum Vitae. Translation of handwritten document "Życiorys" in the Archive of the University of Warsaw, sygnatura Teitelbaum-Tarski Alfred: RP 2909. Displayed in the box below and discussed in section 1.1. This is its first publication. The original has no other marks. Tarski's signature is reproduced in chapter 3, page 38.

---

Capital City of Warsaw, 30 September 1918

*Curriculum Vitae*

I was born on 14 January 1901 in Warsaw, to my father Ignacy (Izaak) and mother Róża (Rachel) Prussak.

I received my first instruction at home. In September 1910, I was sent to first grade at State Gimnazjum IV in Warsaw. There I passed five classes. In September 1915, having been promoted, I was accepted into grade VI of the eight-grade School of the Mazovian Land.[9] I attended this school for three years. In the spring of the current year I passed the final exams successfully and obtained a certificate of maturation.

Now I plan to study with the Philosophical Faculty of the University of Warsaw, wishing to devote myself to scientific work in the field of biology.

[signed] Alfred Teitelbaum

---

Tarski [1924] 2004. Letter to a Warsaw city government office, 18 March. In Feferman and Feferman 2004, 40. The original handwritten document "Komisariat Rządu na m. st. Warszawa" is in the Archive of the University of Warsaw, sygnatura Teitelbaum-Tarski Alfred: RP 2909. This is Alfred Tajtelbaum's application for changing his surname to Tarski.

---

[9] In Polish, *Szkoła Ziemi Mazowieckiej*.

Tarski [1924] 2014a. Letter to the dean of the Philosophical Faculty, 2 June 1924. In section 16.1 of the present book, Tarski 2014. Translation of handwritten document "Do Pana Dziekana Wydziału Filozoficznego Uniwersytetu Warszawskiego" in the Archive of the University of Warsaw, Signatura Teitelbaum-Tarski Alfred: RP 2909. This is displayed in the box below and discussed in chapter 3. Alfred's signature is reproduced there. This is its first publication. The dean was Wiktor Lampe, a chemist.

---

To
Dean of the Philosophical Faculty of the University of Warsaw

Application

After obtaining the doctoral degree, wishing to complete my education in certain branches of mathematics, especially in the field of theoretical physics, I respectfully ask the Dean for permission to continue my studies in the mathematics and natural (physical) sciences division of the Philosophical Faculty.

[signed] Dr. Alfred Tarski

Warsaw, 2 June 1924

I agree to the continuation of studies            [Stamp of approval]
in the mathematics division.

3 June 1924                                       [signed] W. Lampe

---

Tarski [1930–1936] 1992. Drei Briefe an Otto Neurath. Edited, with a foreword, by Rudolf Haller. Translated by Jan Tarski. *Grazer philosophische Studien* 43:1–32. The letters are transcribed, then translated into English. According to the editor (pages 1–2),

"There are in the main two themes treated in Tarski's letters to Neurath.[10] The first presents information about scientific philosophy in Poland and its possible links to the work of the Vienna Circle.... The second theme touches on syntax and semantics and Tarski's influence on research and discussion in the Vienna Circle."

Tarski [1933] 2003. Letter to Kazimierz Twardowski, 12 July. In *Alfred Tarski: dedukcja i semantyka (déduction et sémantique)*, edited by Juliusz Jacek Jadacki, 19. Warsaw: Wydawnictwo Naukowe Semper. This is about preparing Tarski 1933 for publication. (For information on Twardowski, see a box in section 1.2.)

---

[10] **Otto Neurath**, born in Vienna in 1882, was the son of a noted political economist. He earned the doctorate in that same subject from the University of Berlin in 1906. Throughout his life, he held a variety of positions in the intellectual and socialist-political communities. He habilitated in 1917 but never became a professor. Neurath was particularly influential in the study of symbolic languages and their use in education, and in the philosophy of logical positivism. During the 1920s, with Hans Hahn and others, he helped found the Vienna Circle. After the Nazi takeover in 1934, he fled to the Netherlands with his wife, Hahn's sister. She died there, and he then married another philosopher, the sister of the German mathematician Kurt Reidemeister. In exile, Neurath founded the *Encyclopedia of Unified Science*, a major organ of logical positivism. He fled again to England in 1940, and died there suddenly in 1945. (See Cat 2010.)

Tarski [1934] 2014. Letter to the Ministry of Religious Denominations and Public Education, 29 November. Translation of handwritten document N921/R15345, page 8, preserved in the archival fund No. 14, Ministerstwo Wyznań Religijnych i Oświecenia Publicznego w Warszawie 1917–1939, sygnatura 6216, of the Archiwum Akt Nowych (Central Archives of Modern Records) in Warsaw. Displayed in the box below, and discussed in section 9.1. At the bottom is the signature of the rector, the noted physicist Stefan Pieńkowski, and a red stamp indicating that he read the letter on the same day. The original document and its attached letter by Jan Łukasiewicz, page 11 of the same archive file, was transcribed in Jadacki , op. cit., 25–26.

---

To the Ministry of Religious Denominations and Public Education
(through the Dean of the Division of Mathematical and Natural Sciences
of the University of Warsaw)

I wish to go abroad, to Austria, France and Czechoslovakia, for scientific studies. Please be so kind as to grant me permission to obtain a discounted passport for the period between 15 December 1934 until 15 December 1935.

I have a serious chance of receiving a Rockefeller grant for a period of eleven to twelve months, starting 1 January 1935. The formal decision of the committee is to take place no earlier than December, and news of the decision on that date would be impossible to use to obtain a passport on 1 January. Regardless of the grant, I want to leave in mid-December for Prague, Czechoslovakia, and Vienna, to conduct with professors Rudolf Carnap and Karl Menger a series of scientific conversations that are important to me.

[signed] Dr. Alfred Tarski
29 November 1934     docent, University of Warsaw
Two enclosures     Warsaw, Koszykowa 51, apartment 14

I strongly endorse this application, referring to the attached
letter from Prof. Jan Łukasiewicz.
[signed] Stefan Mazurkiewicz

---

Tarski [1935] 2014. Two letters to the Ministry of Religious Denominations and Public Education, 25 November and 7 December. Translations of handwritten documents, N840/R12478, pages 21–22, and N873/R13421, page 24, preserved in the archival fund No. 14, Ministerstwo Wyznań Religijnych i Oświecenia Publicznego w Warszawie 1917–1939, sygnatura 6216, Archiwum Akt Nowych (Central Archives of Modern Records) in Warsaw. Displayed in the box on the facing page and discussed in section 9.1.

Transcriptions of the original letters were published in Jadacki 2003b, 26–27. Jadacki mistook the date of the first letter. Below the body of second letter, in a position analogous to Mazurkiewicz's endorsement of the first, someone added the word *Czytałem* (*I have read*) and an illegible signature. Both letters carry a red stamp with that word, a date, and the signature of the Rector, Stefan Pieńkowski. Both letters carry various ministry stamps and endorsements.

## 16.3 Letters

To the Ministry of Religious Denominations and Public Education
(through the administration)

In January of this year, I went abroad for several months to study, using the leave granted me by the Ministry. As a consequence of this trip, I was forced to give up the apartment which I previously occupied. Incidentally, the apartment was too tight for me as my family increased in size with the birth of a son. After returning to the country, I encountered great difficulties in attempts to rent a new apartment. For housing that remains under the protection of the Tenant Protection Act, a large sum was required of me as a so-called consideration[11] (the amount of which in recent years has significantly increased). For housing in new homes, exorbitant rent was required. Therefore I was forced to live temporarily with my wife and child with a larger family in one sublet room.[12] Of course, under these housing conditions, I can neither prepare properly to fulfill my professional duties—university lectures—nor continue my research.

Recently, an opportunity for obtaining housing opened up to me: the Social Insurance Institution granted me the right to purchase a property, an apartment of two and one-half rooms in one of its newly built homes. An essential requirement for purchasing the apartment is that I pay the sum of 3500 zlotys. (The rest I can pay off in long-term installments.) Not having any savings, I cannot make this payment without financial aid from the Ministry.

In this situation, I most politely appeal to the Ministry, asking for the advance of six months salary, which I wish to pay off in thirty-six monthly installments. I dare to say that this matter is extremely urgent for me: if I do not make the payment within the shortest possible time, the apartment will be given to someone else.

With this application I enclose a certificate[13] issued by the Social Insurance Institution.

Warsaw, November 25, 1935

Dr. Alfred Tarski,
Docent, Józef Piłsudski University,
Adjunct, Philosophical Seminar,
the same University.

I support this application
[signed] Stefan Mazurkiewicz

..................................................................

To the Ministry of Religious Denominations and Public Education
(through the administration)

In connection with the application submitted by me on November 26 of this year and containing a request for a salary advance for the purchase of an apartment, I am pleased to announce that I managed to obtain the necessary amount through a private loan, and therefore I withdraw my request.

Warsaw, December 7, 1935

Docent Dr. Alfred Tarski
Adjunct, Philosophical Seminar
Józef Piłsudski University

---

[11] Tarski's word was *odstępne*.

[12] Tarski's words were *podnajętym przy większej rodzinie*. It is not clear whether he was emphasizing or downplaying his reliance on his parents.

[13] See Zakład Ubezpieczeń Społecznych 1935.

Tarski [1937] 2003. Two letters to Kazimierz Twardowski, 1 and 24 May. In Jadacki, op. cit., 19–20. These notes are about some publications of the Polish Logic Society.

Tarski [1939] 2003. Letter to the dean of the Division of Mathematics and Natural Sciences of the University of Warsaw, 16 January. Ibid., 27. Tarski requested that a record of his university studies be sent to the state pension system.

Tarski [1940] 1996–1999. Letter to Józef M. Bocheński, 26 April. *Filozofia nauki* 4: 123–125, 7: 197–199. Polish transcription, with an English translation by Jan Tarski.

> This letter, written from New York, was described and quoted in the Fefermans' biography of Tarski:
>
>> In understandably obsessive detail he told of his position in the United States ... and his so-far unsuccessful attempts to get his family out of Poland ... and he concluded by saying he will be grateful for anything Bocheński can do even if it is only to convey the information in the letter to his wife.[14]

Tarski [1942–1947] 1999. Letters to Kurt Gödel, 1942–47. Edited by Jan Tarski. In *Alfred Tarski and the Vienna Circle: Austro–Polish Connections in Logical Empiricism*, edited by Jan Woleński and Eckehart Köhler, 261–273. Dordrecht: Kluwer Academic Publishers. According to the editor (page 261), these

> "... sixteen letters, which include notes on postcards, form (apparently) the only surviving series of Tarski's correspondence from the difficult transitional period, when he was in the course of arranging his life in his new country. ... The letters split naturally into two groups. Those of the first group, consisting of ten letters from June 1942 to April 1944, were written in German; this language was Tarski's principal international language in his former years. ... The six letters of the second group, March 1945–January 1947, were written in English. All of the letters were handwritten, and they deal with a mixture of professional and personal matters. ... However, the reader will note a definite shift of point of view, from the primarily personal in the first group of letters to the primarily professional in the second."

Tarski [1944] 1987. A philosophical letter of Alfred Tarski. Edited by Morton White. *Journal of Philosophy* 84: 28–32. Item 87$^l$ in Givant 1986, with incomplete publication data. According to the 1988 review by William S. Hatcher, this letter to White[15]

> "... discusses certain philosophical questions, mainly related to semantical and logical issues. Adopting a generally common-sensible and practical-minded approach throughout, Tarski opts unequivocally for an empirical-linguistic rather than a purely linguistic origin for logical truth.

---

[14] Feferman and Feferman 2004, 136–137. For information about Bocheński, see a box in section 15.9.

[15] **Morton White** was born in New York City in 1917. He was educated there and earned a doctorate in philosophy at Columbia University in 1942. During 1940 he was assistant to Tarski at City College of New York. (See Rosen 1985, annotated in chapter 17, for a student's view of Tarski's course.) White achieved note in several areas of philosophy and its history, particularly its relation to cultural and social matters. He served at Harvard during 1953–1970, and after that at the Institute of Advanced Study in Princeton. (White 1999.)

In particular, Tarski sees no 'difference in principle' between the status of logical and nonlogical premises of science or of a given scientific discipline." [16]

Tarski [1946–1947] 2014. Two letters to Wacław Sierpiński, 30 October 1946 and 18 March 1947. In section 15.12 of the present book, Tarski 2014. Translations of handwritten documents, APAN signature III-194, in the Archiwum Polskiej Akademii Nauk (Archive of the Polish Academy of Sciences) in Warsaw. This is their first publication. They are discussed at length in section 15.12.

Tarski [1947–1955] 2003. Four letters to Tadeusz Kotarbiński, 6 December 1947 and 16 October 1953 from Berkeley, and 7 September 1954 and 10 December 1955 from Amsterdam. In Jadacki, op. cit., 20–24.

These are about daily life, philosophical issues, conferences, colleagues, and publication of Tarski [1956] 1983. For information on Kotarbiński, see a box in section 1.1.

Tarski [1963] 2003. Letter to Tadeusz Kotarbiński, 7 June. Ibid., 24.

Tarski signed this letter for the Group in Logic and Methodology at the University of California, Berkeley. In English, it asked Kotarbiński to send any available reprints of his work and to add the group to his mailing list.

Tarski et al. [1971] 2003. Letter to Tadeusz Kotarbiński, 24 April. Ibid., 25. This postcard conveys warm wishes and greetings from eleven participants in a congress at a Polish research institute in New York. One of the authors was Feliks Gross.

Tarski 1972. O szkicu Feliksa Grossa w *Wiadomościach*. *Wiadomości* 27(1347, 23 January): 6. In English, with Polish commentary by the journal editor.

Tarski had written this letter to Feliks Gross urging republication of Gross's article "Trip to the moon: Why?" That article had appeared in the same journal in 1969 just before men first stepped on the moon. Tarski wrote,

> You concern yourself there with the impact of fantasy and creative myth on the cultural development of mankind.... This was the time when much shallow and demagogical criticism of the present efforts to conquer space could be heard..... This superficial polemic...has caused a great deal of harm by weakening the...support of the human community for the monumental efforts needed to realize the dream.... Your article...can serve as a potent antidote for narrow-minded criticism.

Gross was a noted sociologist who had escaped Poland during World War II. The journal, published in London, addressed the Polish emigré community. Gross was one of the authors of Tarski et al. [1963] 2003.

---

[16] This view is often associated with W. V. O. Quine. About his relationship to Tarski, see Mancosu 2005.

Tarski [1972–1973] 1995. Two letters to Svätoslav Mathé, 21 July 1972 and 14 February 1973. *Organon F* 2: 56–58. The first is in Slovak, translated from Polish by Pavel Cmorej. The second is handwritten in Polish. The journal is published in Bratislava.

Mathé is a Slovak political scientist and journalist. At the time he was an editor for Slovak Television in Bratislava. Preparing a television film about Tarski, he had sent a script to Tarski for comment. In the first letter Tarski expressed delight, no objections, and related some pleasant experiences in Slovakia before World War II and in Prague the previous summer. Months later, Mathé sent Tarski the film and an article about him. In the second letter, Tarski conveyed his appreciation and assured Mathé that he should not be troubled by some words of a certain Dr. Hajek in the film.[17]

---

[17] Jan Zygmunt has suggested that this was the Czech logician Petr Hajek.

# 17
# Biographical Studies

This chapter lists some personal biographies and sketches of Alfred Tarski. It is a by-product of the research that underlies earlier chapters. The editors did not search exhaustively for other sketches of Tarski. The list includes all longer works that they encountered, but shorter ones only if their authors appear to have possessed unique information or distinction. Each listed item begins with the author-date designation used in this book and continues with a description adapted from this book's bibliography;[1] it is followed by a brief annotation.

- Addison, John W. 1983. Eloge: Alfred Tarski, 1901–1983. *California Monthly* 94 ( 2, December): 28. Reprinted in *Annals of the History of Computing* 6 (1984): 335–336. Translated by Irena Czubińska as "Jeden z największych logików," *Przegląd techniczny* 20 (1985): 39. That title means "One of the greatest logicians."

  Addison was Tarski's colleague for twenty years at the University of California at Berkeley. The journal was the university's news magazine for alumni. Addison suggested that Tarski's success as a teacher was due to Tarski's

  > dogged insistence on precision and clarity in the way his students expressed their ideas. In a seminar he would never be satisfied with an "almost" clear account accompanied by a wave of the hand to indicate "you see what I mean." "No," he would rejoin, "you must *say* what you mean."

  Addison discussed Tarski's truth definition:

  > The seeming simplicity of his famous example that the sentence "Snow is white" is true just in case snow is white belies the depth and complexity of the consequences....[2]

  Finally, Addison suggested an epitaph for Tarski: *He Sought Truth and Found It*.

- Chuaqui Kettlun, Rolando. 1984. Alfred Tarski, matemático de la verdad. *Revista universitaria* 11: 30–34. Translated into Portuguese by S. P. Monoide as "Alfred Tarski, matemático da verdade," *Boletim da Sociedade Paranaense de Matematica* 6 (1985): 1–10.

---

[1] This book's bibliography includes more details of publication.

[2] See section 15.6. Addison's 55-word sentence was soon quoted in wonder by San Francisco's celebrated humor columnist Herb Caen, who quipped, "A snow job if I ever read one" (*San Francisco Chronicle*, 15 January 1984).

After training as a surgeon in Chile, Chuaqui earned a doctorate from Berkeley in 1965 in foundations of probability theory. He became a professor in Chile, and exerted major influence on the growth of interest in logic in South America. The *Revista* is published by the Universidad Católica de Chile. This elegant resumé of Tarski's research career emphasized those aspects readily graspable by nonspecialists. Chuaqui particularly mentioned Tarski's 1974–1975 visit to Chile and Brazil and Tarski's continued interest in their scientific communities. Chuaqui concluded (page 34): "that is what I, along with all who knew him, feel for him: admiration, gratitude, and friendship."

- Feferman, Anita B. [1997] 1999. How the unity of science saved Alfred Tarski. In *Alfred Tarski and the Vienna Circle: Austro–Polish Connections in Logical Empiricism*, edited by Jan Woleński and Eckehart Köhler, 43–52. Dordrecht: Kluwer Academic Publishers.
- ———. 1999. Alfred Tarski. *American National Biography* 21: 330–332.

This lively article and professional biographical sketch are by-products of the research that supported the full biography described in the next item. Feferman used her article title for chapter 5 of that book.

- Feferman, Anita B., and Solomon Feferman. 2004. *Alfred Tarski: Life and Logic*. Cambridge, England: Cambridge University Press. Feferman and Feferman 2009, described in the next item, is a Polish translation. The following annotation is constructed largely from five reviews.

This is the only full-length biography of Alfred Tarski. Anita B. Feferman is a historical researcher and writer, with a previous biography of the logician Jean Van Heijenoort. Her husband Solomon was Tarski's tenth doctoral student, from 1948 to 1957 at Berkeley. Since then Solomon has been a professor of mathematics and philosophy at nearby Stanford University, and a noted researcher in foundations of mathematics and computer science. The American mathematician Anil Nerode wrote,

> No one could be better qualified than Anita to write his personal biography. No one could be better qualified than Sol to write his scientific biography.[3]

The book's dust cover displays remarkable psychedelic portraits of Alfred Tarski and his wife, Maria, which were created in 1934 and 1938, respectively, by the Polish artist and writer Stanisław Ignacy Witkiewicz. Famous under the name Witkacy, he was their close friend.[4]

According to Roger D. Maddux, Tarski's twenty-second doctoral student,

> The story of Tarski's life and loves is told in 15 chapters, while his work in philosophy, logic, and mathematics is described in 6 interludes. All mathematical and logical symbolism is confined to the interludes. This structure makes this book accessible to the widest possible audience.[5]

The Moroccan mathematician and historian Hourya Sinaceur described the book as

> an enthralling success story of a self-confident, enterprising, untiring, and entrepreneurial scientist, and a rich and scrupulous account of [his] numerous achievements.... It is particularly remarkable that the Fefermans were able to reconstruct so vividly Tarski's childhood

---

[3] Nerode 2010, 286–288, published in the *American Mathematical Monthly*.

[4] The original portraits are in the possession of Tarski's son, Jan, and daughter, Ewa Krystyna.

[5] Maddux 2005, 536, 540, published in the *Journal of Symbolic Logic*. Maddux earned the PhD in 1978. The Fefermans included a list of doctorates completed under Tarski's supervision (pages 385–386).

in Warsaw; the brilliant *gymnasium* and university years ... in short, the scientific, political, cultural, and artistic atmosphere ... The reader follows month by month or year by year the irresistible, though treacherous, ascension of a strong and passionate character.[6]

Nerode, a long-standing acquaintance of Tarski, reported,

> I am amazed that Anita was able to uncover his personal life in such detail. People were more discreet in previous generations.[3]

Sinaceur described the Fefermans' interludes as

> an introduction to the main problems and results that Tarski pushed to the forefront of logical research. Written with a minimum of technicalities, [they] meet the demanding standards of the exceptional teacher that Tarski was and serve perfectly the purpose of pedagogical presentation. Simple but self-contained, explaining step-by-step everything that is needed and leaving nothing fuzzy nor obscure, they are easily understandable even by those not previously acquainted with the subject matters. They furnish a beautiful, though partial, survey of a whole century of logic in Europe and America.[6]

The present editors note that the Fefermans did not uncover much about the critical years 1918–1920 in the lives of Alfred and his wife, Maria, nor about Alfred's work as a full-time gimnazjum teacher and teacher trainer. The present book partially fills those gaps.[7] Reviewers in general have found few faults with the Fefermans' biography; none mentioned those. The American historian Irving Anellis would have appreciated more information about certain publication and priority disputes and about influence on Polish thinkers by the American philosopher Charles Sanders Peirce.[8] According to the Brazilian philosopher Walter Carnielli,

> The book probably overstresses the role of anti-semitism as a universal explanation for everything that happened to Tarski in Poland.[9]

Moreover, he regretted the book's "somewhat excessive insistence on Tarski's love affairs, on what he smoked and drank."[10]

Sinaceur concluded her review,

> The complex and varied portrait the Fefermans have painted is a rigorous attempt at capturing with greatest objectivity the complex socio-psychological facts that can help us understand Tarski's personal leanings, his high professional conscientiousness, his "unending concern for clarity, precision, and rigor," and, last but not least, his strong will to put his mark on his time.[11]

Tarski's former student Maddux finished,

> Many times while reading this book I thought, "Yes! That's what he was like!"[5]

• ———. 2009. *Alfred Tarski: Życie i logika*. Translated by Joanna Golińska-Pilarek and Marian Srebrny. With a preface by Jan Woleński. Warsaw: Wydawnictwa Akademickie i Profesjonalne. Polish translation of Feferman and Feferman 2004.

---

[6] Sinaceur 2007, 986–988, published in the *Notices of the American Mathematical Society*.

[7] Sections 1.1 and 9.1 of the present book describe the 1919–1920 Polish-Soviet War from the viewpoints of Alfred and Maria, respectively; Part Three is mostly devoted to Alfred's teaching and teacher training.

[8] Anellis 2005, 121–122, published in the *Review of Modern Logic*.

[9] Carnielli 2006, 95, published in the journal *Logic and Logical Philosophy*. Carnielli was echoing the opinion expressed in Woleński 1995a, notes 13–15. See also Jan Woleński's preface to the Polish translation described in the next item.

[10] Ibid., 92.

[11] Sinaceur 2007, 989, quoting Feferman and Feferman 2004, 41.

Golińska-Pilarek is a logician at the University of Warsaw. Srebrny is a theoretical computer scientist at the Polish Academy of Sciences. The expansive preface by the noted Polish philosopher Jan Woleński is remarkable for these features:
- an account of some of the early steps in the United States and Poland in the creation of the Fefermans' 2004 biography of Tarski;
- a discussion from an alternative viewpoint of the effects of antisemitism on Tarski's life in Poland;
- a discussion of nationalism in academic culture.

- Feferman, Solomon. [1997] 1999. Tarski and Gödel: between the lines. In Woleński and Köhler, op. cit., 53–63.

This conference report could be regarded as a by-product of the author's research for the Tarski biography described in the previous two items. But it has another source, as well: Solomon Feferman was the editor of the 1986–2003 *Collected Works* of Kurt Gödel. Feferman's paper presents bird's-eye views of several personal encounters of Tarski and Gödel. It contains a transcription of Gödel's January 1931 letter to Tarski, in which Gödel first mentioned his incompleteness result. For each thread of Tarski's research that intertwined with one of Gödel's, Feferman explained the interaction.[12] He concluded by comparing and contrasting the public and private philosophies that underlay Tarski's and Gödel's mathematical works. Each mathematician's public practice was apparently opposite to his private philosophy, and the private philosophies of the two were opposite, as were their public practices.

- ———. 2003. Alfred Tarski and a watershed meeting in logic: Cornell, 1957. In *Philosophy and Logic in Search of the Polish Tradition: Essays in Honour of Jan Woleński on the Occasion of his 60th Birthday*, edited by Jaakko Hintikka et al., 151–162. Dordrecht: Kluwer Academic Publishers.

This report, too, was a by-product of Solomon Feferman's work on the Tarski biography described in previous items. It is an account of the five-week 1957 Summer Institute for Symbolic Logic held at Cornell University in New York. Feferman emphasized the institutional background for the session, the political process of organizing it, and the vibrant social interactions among the eighty-five participants. About ten of those, including him, were Tarski's research collaborators or students. Twenty others came from the computer industry: this was the first organized interaction between pure and applied logical researchers. Feferman stressed Tarski's role in organizing the session and ensuring the participation of his associates. Tarski stayed for three weeks, gave three presentations and participated significantly in the discussions. However, Feferman noted, Tarski seemed to take little interest in the papers about practical applications. It is remarkable how many concepts and results, soon to become core elements of mathematical logic or theoretical computer science, received early exposure in the eighty papers presented at Cornell. Feferman concluded by listing several similar conferences that occurred during 1957–1960.

- Givant, Steven R. 1991. A portrait of Alfred Tarski. *Mathematical Intelligencer* 13 (3): 16–32. Translated into Czech by Helena Nešetřilová as "Portrét Alfréda Tarského,"

---

[12] For a corresponding discussion from Gödel's viewpoint, see Krajewski, S. 2004.

*Pokroky matematiky, fysiky a astronomie* 37 (1992): 185–205. Givant and Huber-Dyson 1996, described in the next item, is an augmented Polish translation.

Givant's doctoral research at Berkeley during 1967–1975 was significantly influenced though not formally supervised by Tarski. After that, Givant became a professor of mathematics at nearby Mills College. He maintained an intense collaboration and friendship with Tarski until the latter's death. According to the American philosopher John Corcoran,

> Givant knew Tarski and Tarski's later mathematical work perhaps better than anyone else; he is well qualified to present to the mathematical community this brief portrait of Tarski.[13]

The article contains many good photographs of Tarski, his family, and his associates. It is an emotional and powerful presentation of major aspects of Tarski's life, from a markedly personal viewpoint. While many of its details have been subsumed by the Fefermans' full biography of Tarski, described in previous items, the tone of Givant's study remains striking. For example, here is Givant's concluding sentence:

> I don't know if anyone who interacted with [Tarski] on a deeper level can completely sort out the mixture of admiration, exasperation, loyalty, affection, frustration, anger, and gratitude that he evoked. I know I can't.

Corcoran, who also knew Tarski well, suggested that limitations of Givant's relationship with Tarski prevented Givant from presenting Tarski's philosophical positions effectively; nor did Givant sufficiently emphasize Tarski's

> enthusiasm for elementary algebra, elementary geometry, elementary analysis and elementary set theory...[That] informed many of his most sophisticated mathematical achievements and it was important, sometimes decisive, in his philosophy.

Corcoran concluded,

> Appreciative and intimate but not uncritical, the portrait reveals various occasionally unflattering features of Tarski's habits and character. The author resisted any temptation to be hagiographic; his well-written, richly informative, and tasteful treatment of one of the world's greatest mathematicians is a pleasure to read.

• Givant, Steven R., and Verena Huber-Dyson. 1996. Alfred Tarski w kalejdoskopie impresji osobistych. Translated from the English by Adelina Morawiec. *Wiadomości matematyczne* 32 (1996), 95–127. The title means "Alfred Tarski in a kaleidoscope of personal impressions."

The Swiss mathematician Huber-Dyson met Tarski at the 1957 Cornell meeting described by Solomon Feferman in a previous item. Later she enjoyed a long career in Canada as professor of philosophy at the University of Calgary. For more about her relationship with Tarski, see Feferman and Feferman 2004, chapters 9 and 11. She wrote a new introduction, coda, and twenty-two long footnotes for the article Givant 1991 described in the previous item. The whole was translated by the Polish logician Morawiec and published in the annual journal of the Polish Mathematical Society. Unfortunately, the translated article does not include any of the illustrations from the original.

• Golińska-Pilarek, Joanna, Joanna Porębska-Srebrna, and Marian Srebrny. 2009a. Król logiki. *Rzeczpospolita* (4–5 April): A20–A21. The title means "King of logic." The journal is a daily newspaper with nationwide circulation; its name means *The Republic*.

---

[13] Corcoran 1993, published in *Mathematical Reviews*.

Two of these authors are the translators of the Tarski biography Feferman and Feferman 2009 described in an earlier item. The third is an architect and city planner. This article was published on the same day as another newspaper article, Woleński 2009, described in a later item. (Woleński had written the preface for that translated biography.) Readers can see how Tarski is regarded by the educated public in Poland, and glimpse from a Polish viewpoint the prewar society of Warsaw described by the Fefermans.

This article also details how its authors corrected an error in the Tarski biography. Tarski's home for nearly thirty years—his parents' apartment building, Koszykowa 51—was not destroyed in World War II, as the Fefermans had claimed, but stands today as Koszykowa 51a in an attractive little residential court in downtown Warsaw. The authors provide a history of that neighborhood. Many notable figures of an astonishingly vibrant society trod there in Tarski's time.[14]

•Hiż, Henryk. 1971. Jubileusz Alfreda Tarskiego. *Kultura* 288: 134–140. The title means "Jubilee of Alfred Tarski." *Kultura* was a literary and political magazine published monthly in Paris for the worldwide Polish emigré community.

As a Warsaw undergraduate in 1937, Hiż attended Tarski's course on the methodology of the deductive sciences.[15] After the war he emigrated to the United States, earned a doctorate in logic, and became a noted researcher in linguistics. This article was prompted by the international symposium held in Berkeley during 23–30 June 1971 to honor Tarski on the occasion of his seventieth birthday. Its proceedings, edited by Leon Henkin and others, were published in 1974.[16] Describing the conference for his Polish emigré readers, Hiż quipped that viewing the crowd of participants gave him the impression that he was actually visiting a Berkeley campus of the University of Warsaw! Hiż's account overlaps those proceedings only to a small extent, mainly in the philosophical aspects of Tarski's work. Hiż began with a brief reminiscence about Tarski's 1937 lectures, then continued with a discussion of the notion of truth and a general survey of Tarski's publications of interest to philosophers. Hiż concentrated on relationships among the views of Tarski and Polish and Viennese philosophers contemporary with him. Finally, Hiż indicated those works of Tarski best suited for introducing a wide audience to philosophical logic.

•Hodges, Wilfrid. 1986. Alfred Tarski. *Journal of Symbolic Logic* 51: 866–868.

This is the first of a series of essays by distinguished logicians about Tarski and his work, published in volumes 51 (1986) and 53 (1988) of the journal. The others are described in section 18.1 of the present book. Hodges also coauthored the one on decidable theories. The content of this introduction is subsumed by the Fefermans' 2004 biography of Tarski, except for a list of about a dozen researchers who collaborated with Tarski on projects at Berkeley, and for an invitation to contribute to a fund in honor of Tarski that would be used to support further such projects. This article is also the source of the Fefermans' list (2004, 385–386) of Tarski's doctoral students.

---

[14] This article includes a contemporary photograph of Koszykowa 51a. In section 9.1 of the present book is a 1946 photograph of the building, only slightly altered from its form in Tarski's time. The same three authors constructed in 2009 a presentation, *Szlakiem Alfreda Tarskiego po Warszawie* (*Along the Route of Alfred Tarski in Warsaw*) with striking illustrations of Tarski's residences and neighborhoods. Unfortunately, it has not been published.

[15] He is quoted in section 9.3, page 191.

[16] This symposium was also reported by the article Melnick 1971, described in a later item.

- Jadacki, Jacek Juliusz. 2003a. Alfred Tarski à Varsovie. In *Alfred Tarski: dedukcja i semantyka (déduction et sémantique)*, edited by Jacek Juliusz Jadacki, 139–180. Warsaw: Wydawnictwo Naukowe Semper. Translated by Wanda Jadacka from the article "Alfred Tarski w Warszawie," ibid., 112–137.

    The author is a professor of philosophy at the University of Warsaw, specializing in logical semiotics and the history of Polish philosophy. The article is in the proceedings of a 15 January 2001 symposium in Warsaw in honor of the centenary of Tarski's birth. It is a wonderfully detailed and richly documented chronology of Alfred Tarski's life in Warsaw, plus snippets from his American years. Without its guidance, assembling the background material for the present book would have been impossible. The present editors owe a great debt to Jadacki.

    The symposium proceedings also include a large number of fascinating illustrations and transciptions or translations of letters and small works of Tarski. Many of those are cited elsewhere in the present book.

    Jadacki's article is a lightly edited transcription of historical research notes, in day-by-day sequence, with a multitude of references. He graciously acknowledged the assistance of Alfred Tarski's son, Jan, and daughter, Ewa Krystyna, a number of former students of Tarski's university and gimnazjum classes, and a number of relatives of Tarski's wife, Maria. With their help he also dug deeply into university and government archives and genealogical records. Jadacki is to be congratulated and thanked for publishing his data in this form, for others to use to add to the stories of Alfred and Maria. Jadacki indicated where his account differs from others, and where he filled gaps. Particularly notable are a report of Maria's involvement in the 1919–1920 Polish–Soviet War, an alternative interpretation of Alfred's conversion to Catholicism, records of Alfred's university teaching, and information that led the present editors to a greater understanding of Alfred's gimnazjum teaching.

- James, Ioan Mackenzie. 2009b. Napoleon of logic: Alfred Tarski (1901–1983). In *Driven to Innovate: A Century of Jewish Mathematicians and Physicists*, 284–288. Witney, Great Britain: Peter Lang.

    Apparently derived from Feferman and Feferman 2004, this essay concentrates on Tarski's experiences as a Jew. It is virtually undocumented, and contains numerous small errors. The author is a noted Oxford topologist. His historical introduction to the containing publication is informative. But readers should also consider the highly critical review Siegmund-Schultze 2010.

- Kozłowski, Witold. 2003. Wspomnienie (14 1 1901 – 27 X 1983) Alfred Tarski. *Gazeta wyborcza* 264 (November 13): 11. The first word in the title means "A memory of." The journal is a daily newspaper with nationwide circulation; its name means *Electoral Gazette*.

    An award-winning journalist, Kozłowski was Tarski's gimnazjum student during 1934–1938. Section 14.1 contains a discussion of Tarski as a teacher based on this article and a 2010 personal interview with Kozłowski. That section includes a brief biographical sketch of him.

- Melnick, Norman. 1971. The 'Einstein of Mathematics' at UC. *San Francisco Chronicle* (27 June).

    This half-page summary of Tarski's career publicized the symposium then underway in Berkeley to honor Tarski's seventieth birthday. Melnick was a science writer for the *Chronicle*, the leading daily newspaper in the Bay Area, where Berkeley is located. He had evidently interviewed the Hungarian mathematician Paul Erdős, who described a dinner he had enjoyed in Princeton in 1942 with Tarski and Albert Einstein. An Einstein headline was certain to attract attention.[17]

- Pawlikowska-Brożek, Zofia. 2003. Tarski (Tajtelbaum) Alfred (1901–1983): logik, filozof, matematyk. In *Słownik biograficzny matematyków polskich*, edited by Stanisław Domoradzki, Zofia Pawlikowska-Brożek, and Danuta Węglowska, 244–245. Tarnobrzeg, Poland: Państwowa Wyższa Szkola Zawodowa.

  This is a brief account of Tarski's research career. The author is a historian of mathematics, then on the staff of the State Advanced School of Professional Education, which published this *Biographical Dictionary of Polish Mathematicians*.

- Polska Akademia Umiejętności. 1948. Alfred Tarski. In "Skład Polskiej Akademii Umiejętności (stan w marcu 1948)," *Rocznik Polskiej Akademii Umiejętności* 1946–1947: xli–xlii.

  In 1946, Tarski was selected for the Polish Academy of Arts and Sciences in Cracow. This biographical sketch is included in its March 1948 membership roster. It appears to be the first biographical sketch of Tarski published in Poland; its details seem correct except for Tarski's year of birth. In 1949 Tarski's membership was terminated because of inactivity.[18]

- Rosen, Saul. 1985. Alfred Tarski in 1940. *Annals of the History of Computing* 7: 364–365.

  Rosen described one of the first classes that Tarski taught in the United States, at City College of New York. Led to expect an introduction to the philosophy of mathematics, students instead experienced Tarski's carefully prepared English exposition of his current research in the theory of relations. Conversation with most students in the class was effective only through intermediaries who could translate between English and German. Rosen earned a bachelor's degree there in 1941. After service in the Army Signal Corps, he completed a doctorate and became a pioneer in computer science, working particularly at Purdue University in Indiana.

- *San Francisco Chronicle*. 1983 (28 October). Alfred Tarski: Expert logician, UC Professor.
- *San Francisco Examiner*. 1983 (28 October). Alfred Tarski.

  These two sober and factual obituaries appeared in the Bay Area's leading daily newspapers.[19]

- Sinaceur, Hourya. 2000. Introduction. In "Address at the Princeton University Bicentennial Conference on Problems of Mathematics (December 17–19, 1946)," by Alfred Tarski, edited by Hourya Sinaceur, *Bulletin of Symbolic Logic* 6: 1–20.

  This introduction by a noted Moroccan mathematician and historian provides much information about the effect on Tarski's research career of his transition to the New World.

---

[17] This symposium was also reported by the article Hiż 1971, described in an earlier item.

[18] Jadacki 2003a, 160.

[19] In contrast, see footnote 2.

- ———. 2009. Tarski's practice and philosophy: between formalism and pragmatism. In *Logicism, Intuitionism, and Formalism: What Has Become of Them?* edited by Sten Lindström et al., 357–396. New York: Springer.

    According to the American philosopher John Corcoran,[20] who knew Tarski well, this paper is
    > a rather rambling... report on, among other things, the author's efforts to glean insights into Tarski's philosophy by combing his mathematical and logical works, his correspondence, his spontaneous remarks at public conferences and meetings, and even things Tarski read.

    Corcoran explained that in this paper, *pragmatism*
    > refers vaguely to practice-oriented opportunism (see page 390) and not to the American philosophical movement initiated by [Charles Sanders] Peirce and [William] James... the conclusion is that Tarski's basically mathematical agenda required strong Platonist premises temporarily and pragmatically assumed for the sake of argument as long as they contributed to the development of his program. This conclusion rings true.

    Corcoran reported that Tarski once told him directly that "the closest thing to a statement of his philosophy" might be the article Kotarbiński [1935] 1956 on pansomatism, which Tarski helped translate. Corcoran criticized Sinaceur for not considering it. He also complained about the inadequate editing of her paper.

- Suppes, Patrick, Jon Barwise, and Solomon Feferman. 1989. Commemorative meeting for Alfred Tarski: Stanford University, November 7, 1983. In *A Century of Mathematics in America*, edited by Peter Duren et al., volume 3, 393–403. Providence: American Mathematical Society.

    This article is the proceedings of a meeting that consisted of talks by the three authors, all Stanford professors of philosophy. Suppes described his experience as a professor sitting in on Tarski's Berkeley research seminars in the 1950s; only a summary of his talk is included. Feferman described his own experience as a student in those seminars. Both of those accounts of Tarski emphasize his personal characteristics and scientific and teaching style, from the authors' specific viewpoints. Feferman also discussed with some speculation Tarski's private philosophy. Barwise, who earned the doctorate in 1967 under Feferman's supervision, spoke about Tarski's contributions to semantics and the theory of models.

- Tarski, Jan. 1994a. A short biographical sketch of Alfred Tarski. In *Introduction to Logic and to the Methodology of the Deductive Sciences*, fourth edition, edited by Jan Tarski, xviii-xxii. New York: Oxford University Press.

    This edition of Alfred Tarski's classic text is described in section 16.2. Alfred's son, Jan, edited it and wrote this very personal account. A Polish translation appears in the monograph Tarski 1996, which is also described in 16.2.

- ———. 1996b. Philosophy in the creativity of Alfred Tarski. In *Truth after Tarski*, edited by Michał Hempoliński, *Dialogue and Universalism* 6 (1–2), 157–159.

    Published by the Polish Academy of Sciences, this is the journal of the International Society for Universal Dialogue. The issue surveyed developments in the theory of truth since Tarski's 1933 work (see section 15.6). The author, Alfred Tarski's son, Jan, speculated about the

---

[20] Corcoran 2011.

reasons underlying Alfred's transition away from conceptual and philosophical investigations, which he had emphasized during the 1930s, toward his more strictly mathematical and technical research of the 1940s. This coincided with Alfred's moving to the New World but, Jan claimed, may have been only partly caused by that displacement.

- Wells, Benjamin. 2007. Alfred Tarski, friend and daemon. *Notices of the American Mathematical Society* 54: 982–984.

  Tarski's last doctoral student, from 1963 to 1982, Wells became professor of mathematics and computer science at the University of San Francisco. He explained his use of *daemon*: "a leonine externalized conscience." His intensely personal reminiscence describes the daemon as mathematical, political, cultural, philosophical, moral, and spiritual.[21]

- Wojciechowska, Agnieszka. 2001, 2011. Alfred Tarski (1901–1983). *Matematyka: Czasopismo dla Nauczycieli* (2001) 54 (3): 132–135; (2011) 64 (9): 3-9. In Polish.

  These two articles are different. The journal's title means *Mathematics: Journal for Teachers*. The journal *Parametr*, to which Tarski had contributed in 1931–1932, ceased publication with the World War II: see sections 9.7–9.8. The exciting journal *Matematyka* was founded in 1948 to fill that role. Dr. Wojciechowska is its current editor. The cover of its 2011 issue 9 carries a fine color portrait of Tarski.

- Wójcik, Andrzej. 1985. Alfred Tarski: Życie i dzieło. *Pismo literacko-artystyczne* 4 (1): 174–178. The title phrase means "Life and work." *Pismo* means *Writings*.

  This account of Tarski's scientific career was written when the author was a doctoral student at the University of Silesia in Katowice. Later, he became a professor of philosophy there.

- Woleński, Jan. 1984. Alfred Tarski (1901–1983). *Studia filozoficzne* (2): 3–8. In Polish.

  The author of this professional biographical sketch is a noted philosopher at the University of Cracow. The journal was published monthly by the Polish Academy of Sciences.

- ———. 1995b. On Tarski's background. In *From Dedekind to Gödel: Essays on the Development of the Foundations of Mathematics*, Synthese library 251, edited by Jaakko Hintikka, 331–341. Dordrecht: Kluwer Academic Publishers Group.

  This paper is partly about Tarski's life and partly about his philosophical research. Woleński aimed

  > to throw light on [the] cognitive conflict or dissonance of Tarski between his nominalistic and empiricistic sympathies and his "Platonic" mathematical practice as well as why he was so parsimonious in expressing his philosophical views.

- ———. 2001a. Alfred Tarski–książę logików: wzór na prawdę. *Polityka* 2281 (20 January): 76–77. The title phrases mean "Prince of logicians: pattern for truth." The journal is a weekly newsmagazine.

---

[21] Readers may want to explore the use of *daemon* in mythology, religion, and software engineering.

- ———. 2009. Alfred Tarski: Logik z fantazją. *Gazeta wyborcza* (4–5 April): 24–25. The title phrase means "Logician with flair."

This essay appeared in one Polish national daily newspaper on the same day that Golińska-Pilarek et al. 2009a appeared in another. The occasion was publication of the 2006 Tarski biography by the Fefermans. Golińska-Pilarek and her colleague Marian Srebrny had translated the biography into Polish; their article and that book were discussed in earlier items in this section. Woleński had written the book's new preface. Under the title of this article, alongside a famous dark portrait of Tarski as a thoughtful sage, Woleński quoted in bold letters,

> Once someone asked him, "How do I become a great logician like you?" Tarski replied, "That's simple. You must be either a Jew or a Pole, and preferably both."

From his section headings, readers may see what Woleński aimed to convey to the public:

| *Logik i sybaryta* | (Logician and Sybarite) |
| *Żyd i Polak* | (Jew and Pole) |
| *Geniusz bez profesury* | (Genius without Professorship) |
| *Pedagog i bohater* | (Teacher and Hero) |

- Woytak, Lidia. 1987. Alfred Tarski—matematyk o języku. *Przegląd polonijny* 1987 (4): 67–75. The title phrase means "mathematician of language." The journal title means *Polonia Review*.

The author, a writer and editor at the Defense Language Institute in Monterey, California, has a doctorate from the University of Poznań. The journal is published quarterly by the Polish Academy of Sciences; its name refers to the Polish emigré community. Tarski's thorough critique of an early draft of this article is in the Tarski archive in the Bancroft Library in Berkeley: carton 13, folder 6.

- Zygmunt, Jan. 1995. Szkic biograficzny. In *Pisma logiczno-filozoficzne*, by Alfred Tarski, edited by Jan Zygmunt, volume 1, vii–xx. Warsaw: Wydawnictwo Naukowe PWN. The first words of the titles of the article and the containing publication mean "sketch" and *Writings*.
- ———. 2000. Alfred Tarski. *Edukacja filozoficzna* 29: 274–307. In Polish. Reprinted, slightly updated, in 2001 in *Polska filozofia powojenna*, edited by Witold Mackiewicz, volume 1, 342– 375. Warsaw: Agencja Wydawnicza Widmark.

The author is a noted Polish logician and historian of logic at the University of Wrocław. The first book, *Pisma logiczno-filozoficzny*, is described in section 16.2. The second, *Polska filozofia powojenna* (*Postwar Polish Philosophy*), is a two-volume set of similar articles about other Polish logicians and philosophers. According to the author's summary of the five parts of the latter article,

- I     lists the main body of Tarski's published works…
- II     gives a selection of papers on Tarski's life and work…
- III     is a short [scientific] biography of Alfred Tarski…
- IV     surveys five areas of Tarski's work…
- V     traces Tarski's influence….

Part III is a dense stream of facts. The author's 1995 essay is smoother, more polished.

# 18
# Research Surveys

This final chapter lists some general surveys of Alfred Tarski's research and legacy. It is a by-product of the present editors' investigations that underlie earlier chapters. They did not search exhaustively for other such surveys. Those that they encountered are listed here if they cover substantial areas of Tarski's work, not just single results. Each listed item begins with the author-date designation used in this book and continues with a description adapted from this book's bibliography;[1] it is followed by a brief annotation.

Of Tarski's twenty-four doctoral students, eleven are authors of works about Tarski mentioned in this or the previous chapter. Seven more of these authors were Tarski's "grand-students." Those numbers attest to the persistence, respect, and loyalty that Tarski inspired among his students and colleagues.

## 18.1  1986,1988 JSL Surveys

Major surveys of Alfred Tarski's work in several areas of mathematics and logic were published in volumes 51 (1986) and 53 (1988) of the *Journal of Symbolic Logic* (JSL).

- Blok, Willem J., and Don Pigozzi. 1988. Alfred Tarski's work on general metamathematics. 53: 36–50.

  Blok earned the doctorate in 1976 at the University of Amsterdam for research in general algebra. Pigozzi was Tarski's nineteenth doctoral student, with a 1970 dissertation on cylindric algebras.

- Doner, John, and Wilfrid A. Hodges. 1988. Alfred Tarski and decidable theories. 53: 20–35.

  Tarski's eighteenth doctoral student, Doner earned his degree in 1968 and collaborated with Tarski on the theory of ordinal algebras. Hodges earned the doctorate in 1970 from Oxford University, in model theory.

---

[1] The entries in the present book's bibliography include more publication details.

- Etchemendy, John. 1988. Tarski on truth and logical consequence. 53: 51–79.

    Etchemendy earned the doctorate in1982 from Stanford University, working in model theory and the theory of truth.

- Jónsson, Bjarni. 1986. The contributions of Alfred Tarski to general algebra. 51: 883–889.

    Tarski's second doctoral student, Jónsson earned his degree in 1946. He collaborated with Tarski on the theory of cardinal algebras and on general algebra, and spent much of his career at Vanderbilt University in Nashville,Tennessee.

- Lévy, Azriel. 1988. Alfred Tarski's work in set theory. 53: 2–6.

    Lévy earned the doctorate in 1958 from the Hebrew University of Jerusalem, in metamathematics of set theory, and spent his career there.

- McNulty, George F. 1986. Alfred Tarski and undecidable theories. 51: 890–898.

    McNulty was Tarski's twentieth doctoral student, with a 1972 dissertation on general algebra.

- Monk, J. Donald. 1986. The contributions of Alfred Tarski to algebraic logic. 51: 899–906.

    Monk was Tarski's thirteenth doctoral student, with a 1961 dissertation on cylindrical algebras. He continued collaborating with Tarski in that area. Monk is a professor at the University of Colorado at Boulder.

- Suppes, Patrick. 1988. Philosophical implications of Tarski's work. 53: 80–91.

    Suppes earned the doctorate in 1950 from Columbia University in New York City, in the philosophy of physics. He became noted in several areas of philosophy and mathematics, and was the director of the Institute for Mathematical Studies in the Social Sciences at Stanford University during 1959–1992.

- Szczerba, Lesław W. 1986. Tarski and geometry, 51: 907–912.

    Szczerba collaborated with Tarski on foundations of geometry. He had studied with Wanda Szmielew, who had been Tarski's fifth doctoral student. Szczerba had a distinguished career in Poland as a researcher and teacher in logic and as a university administrator.

- Van den Dries, Lou. 1988. Alfred Tarski's elimination theory for real closed fields. 53: 7–19.

    Van den Dries earned the doctorate in 1978 at the University of Utrecht, with a dissertation on model theory and field theory.

- Vaught, Robert L. 1986–1987. Alfred Tarski's work in model theory. 51: 869–882; 52(4): vii.

    Vaught was Tarski's eighth doctoral student, with a 1954 dissertation on model theory. From 1958 on he was Tarski's colleague and collaborator at Berkeley.

## 18.2 Other Surveys

This chapter concludes with a list of other surveys whose scope is comparable to those listed in the previous section.

- Adamowicz, Zofia, et al., editors. 2004. *Provinces of Logic Determined: Essays in the memory of Alfred Tarski. Annals of Pure and Applied Logic*, volumes 126 and 127.

    Adamowicz earned the doctorate in 1975 under supervision of Andrzej Mostowski, Tarski's first doctoral student. She is now a researcher at the Polish Academy of Sciences. These volumes consist of papers presented at the Alfred Tarski Centennial Conference held in Warsaw, 27 May–1 June 2001. Eighteen papers are on topics relevant to this section; most of them are too specialized to be featured individually. However, taken as a group they constitute a major survey of issues surrounding some aspects of Tarski's research and its impact. They are listed below.

| | | Vol. 126 |
|---|---|---|
| Solomon Feferman | Tarski's conception of logic | 5–13 |
| Jens Erik Fenstad | Tarski, truth, and natural languages | 15–26 |
| Mario Gómez-Torrente | The indefinability of truth in the *Wahrheitsbegriff* | 27–37 |
| Henryk Hiż | Reexamination of Tarski's semantics | 39–48 |
| Ilkka Niiniluoto | Tarski's definition and truth-makers | 57–76 |
| John W. Addison | Tarski's theory of definability: common themes in descriptive set theory, recursive function theory, classical pure logic, and finite-universe logic | 77–92 |
| Wilfrid Hodges | What languages have Tarski truth definitions? | 93–113 |
| Roman Kossak | Undefinability of truth and nonstandard models | 115–123 |
| Jan Mycielski | On the tension between Tarski's nominalism and his model theory (definitions for a mathematical theory of knowledge) | 215–224 |
| Benjamin Wells | Applying, extending, and specializing pseudo-recursiveness | 225–254 |
| Pietro Benvenuti and Radko Mesiar | On Tarski's contribution to the additive measure theory and its consequences | 281–286 |
| Andrzej Grzegorczyk | Decidability without mathematics | 309–312 |
| | | Vol. 127 |
| Ivo Düntsch and Ewa Orlowska | Boolean algebras arising from information systems | 77–98 |
| Roger D. Maddux | Finite, integral, and finite-dimensional relation algebras: a brief history | 117–130 |
| George F. McNulty | Minimum bases for equational theories of groups and rings: the work of Alfred Tarski and Thomas Green. | 131–153 |
| Leo Esakia | Intuitionistic logic and modality via topology | 155–170 |
| Urszula Wybraniec-Skardowska | Foundations for the formalization of metamathematics and axiomatizations of consequence theories | 243–266 |
| Arianna Betti | Leśniewski's earlier liar, Tarski, and natural language | 267–287 |
| Stanisław Krajewski | Gödel on Tarski | 303–323 |

- Chang, Chen-Chung. [1971] 1974. Model theory 1945–1971. In *Proceedings of the Tarski Symposium: An International Symposium Held to Honor Alfred Tarski on the Occasion of His Seventieth Birthday*, edited by Leon Henkin et al., 173–186. Providence: American Mathematical Society.

  Chang was Tarski's ninth doctoral student, with a 1955 dissertation on model theory. They continued to collaborate. Chang became professor at the University of California, Los Angeles.

  > The major features of this survey are an extensive diagrammatic view of the influences among different aspects of model theory in the last 20 years (grouped under the four broad headings of first-order logic, algebra and syntax, compactness theorems, and Löwenheim–Skolem–Tarski theorems) and some comments on the possible directions of future research. There is an extensive bibliography of 119 items.[2]

  See Vaught [1971] 1974, described in a later item, for an account of Tarski's earlier work on model theory.

- Czelakowski, Janusz, and Grzegorz Malinowski. 1985. Key notions of Tarski's methodology of deductive systems. *Studia Logica* 44: 321–351.

  The authors are professors of mathematics and logic in Poland. According to the American mathematician John W. Dawson, Jr.,

  > The aim of this article, as the authors state in their abstract, is "to outline the historical background and the present state of the methodology of deductive systems invented by Alfred Tarski." As such, it provides a survey that is at once broad but (necessarily, in view of its brevity) shallow. Basic notions are defined, major directions of research indicated, and subsequent developments traced through citations of important results.

  Dawson criticized the authors' unsubstantiated claims that various later developments were "implicit" in Tarski's work.[3]

- Feferman, Solomon. 2006. Tarski's influence on computer science. *Logical Methods in Computer Science* 2(3): 1-1-1-13.

  Feferman was Tarski's tenth doctoral student, with a 1957 dissertation on model theory and proof theory. Since then he has been a professor at Stanford University and a noted researcher in foundations of mathematics and computer science. From his first page:

  > Here we survey Tarski's work on the decision procedure for algebra and geometry, the method of elimination of quantifiers, the semantics of formal languages, model-theoretic preservation theorems, and algebraic logic; various connections of each with computer science are taken up.

- Givant, Steven R. [1988] 1991. Tarski's development of logic and mathematics based on the calculus of relations. In *Algebraic Logic: Colloquia Mathematica Societatis János Bolyai, 54 (Budapest, 8–14 August 1988)*, edited by Hajnal Andréka et al., 189–215. Amsterdam: North–Holland Publishing Company.

  By including the operation of relation composition, the calculus of binary relations transcends the power of Boolean algebra and incorporates some of that of first-order logic. The author

---

[2] *Mathematical Reviews* 1979.

[3] Dawson 1987.

presents a sketch of the history of works dedicated, by Tarski or at his instigation, to the delineation of this part, as well as the results obtained (the strongest is that in this fragment of first-order logic it is still possible to formalize set theory and arithmetic, and, ultimately, all that can be formalized in mathematics by means of the latter)....[4]

- ———. 1999. Unifying threads in Alfred Tarski's work. *Mathematical Intelligencer* 21 (1): 47–58.

The chapter 17 annotation of Givant 1991, a personal biographical sketch of Tarski, describes the close relationship of this author and his subject. The present article is a deeper study of Tarski's vast research corpus, an attempt "to find underlying unity in Tarski's work, to trace steps that may have led to some of his discoveries." Givant tackled the following questions:

> How did a logician end up working in so many different areas? Were there interconnections in his work that led him from one field to another? What drew him to set theory in the first place, and what drew him away from it a few years later? What sparked his interest in algebra and geometry? How did he become involved in the problem of defining truth? Why did he work so intensively in algebraic logic... and what did this work have to do with his other research? Just why did he go into logic, anyway?

- Gómez-Torrente, Mario. 2006. Alfred Tarski. *Stanford Encyclopedia of Philosophy*. On the Internet at `http://plato.stanford.edu/entries/tarski/`.

The author earned the doctorate at Princeton University in 1996; he now serves at the Universidad Nacional Autónoma de México. This article surveys only Tarski's research in logic.

- Jadacki, Jacek Juliusz, editor. 2003b. *Alfred Tarski: dedukcja i semantyka (déduction et sémantique)*. Warsaw: Wydawictwo Naukowe Semper.

This book consists of materials for a January 2001 symposium sponsored by the University of Warsaw, the Polish Philosophical Society, and the Warsaw Society of Sciences and Letters to commemorate the centennial of Tarski's birth. The author is professor of philosophy at the university. The book contains many historical materials on Tarski: photographs, transcriptions, Polish translations of short Tarski papers, and the indispensible chronological record "Alfred Tarski in Warsaw." Many of these are cited or described in preceding chapters. The materials also include nine papers presented at the symposium, all related to Tarski's research:

| Subject | Authors |
|---|---|
| Tarski's foundation of the theory of deduction | Urszula Wybraniec-Skardowska[5] |
| Tarski's definition of true sentence | Adam Nowaczyk |
| | Joanna Odrowąż-Sypniewska |
| | Jan Woleński |
| Tarski's definition of logical constant | Janusz Maciaszek |
| Other aspects of Tarski's semantics | Adam Grobler |
| | Artur Rojszczak |
| | Marcin Selinger |
| Tarski and the definition of computability | Andrzej Grzegorczyk |

All these materials are in Polish, except for "Alfred Tarski in Warsaw," which is provided in both Polish and French.

---

[4] Guillaume 1993.

[5] This paper is similar to her English paper mentioned in the annotation of Adamowicz 2004 earlier in this section.

- Łos, Jerzy. 1986. O Alfredzie Tarskim. *Ruch filozoficzny* 43: 3–10.

    The author, a noted Polish logician, was not a direct academic descendent of Tarski; nevertheless, he stressed (page 4): "I feel like his disciple." The journal is published quarterly in Toruń by the Polish Philosophical Society. In this biographical sketch, Łoś emphasized Tarski's research on the borderline of mathematics and philosophy.

- Mancosu, Paolo. 2009. Tarski's engagement with philosophy. In *The Golden Age of Polish Philosophy: Kazimierz Twardowski's Philosophical Legacy*, edited by Sandra Lapointe et al., chapter 9, 131–153. Dordrecht: Springer.

    The author earned the doctorate in 1989 at Stanford University, supervised by Tarski's former student Solomon Feferman. Mancosu is now professor of philosophy at Berkeley. In this study he started from Jan Woleński's significant analysis of Tarski's work in philosophy (Woleński 1993, discussed later in this section, and 1995b, discussed in chapter 17). Mancosu noted that Woleński's discussion was based almost entirely on published works of Tarski. In this article, Mancosu reported his own further pursuit of these matters using various archival materials. He indicated that further investigations of that nature might prove productive.

- Moore, Gregory H. 1990. Alfred Tarski. In *Dictionary of Scientific Biography* 1990, volume 18: supplement II, 893–896.

    The author is a noted historian of logic and mathematics at McMaster University in Hamilton, Ontario. The article's synopses of the impact of Tarski's research are very effective: for example,

    > He brought clarity and precision to the semantics of mathematical logic, and in so doing he legitimized semantic concepts, such as truth and definability, that had been stigmatized by the logical paradoxes.

    Readers should be aware that there are a number of errors in historical detail.

- Mostowski, Andrzej. 1967. Alfred Tarski. In *The Encyclopedia of Philosophy*, edited by Paul Edwards, volume 8, 77–81. New York: Macmillan Publishing Company.

    Mostowski was Tarski's first doctoral student, during 1931–1938, in Warsaw. After World War II he continued collaboration with Tarski and led the redevelopment of logic research in Poland. This article surveys only Tarski's research in logic.

- Murawski, Roman. 2011. Alfred Tarski. In *Filozofia matematyki i logiki w Polsce międzywojennej*, section 3.6, 134–155. Toruń: Wydawnictwo Naukowe Uniwersytetu Mikołaja Kopernika. The book's title means *Philosophy of Mathematics and Logic in Interwar Poland*.

    The author is a noted logician at the University of Poznań. The article is one of many biographical sketches in his comprehensive book.

- Patterson, Douglas, editor. 2008. *New Essays on Tarski and Philosophy*. Oxford, U.K.: Oxford University Press.

    Most of the essays in this high-quality collection are too specialized to merit individual inclusion in this section. However, taken as a group they consistitute a major survey of issues surrounding Tarski's philosophical work and its impact. Several of them have served as resources for earlier sections of the present book. Here is the table of contents:

| | | |
|---|---|---|
| Roman Murawski and Jan Woleński | Tarski and His Polish Predecessors on Truth | 21–43 |
| Arianna Betti | Polish Axiomatics and Its Truth: On Tarski's Leśniewskian Background and the Ajdukiewicz Connection. | 44–71 |
| Solomon Feferman | Tarski's Conceptual Analysis of Semantical Notions | 72–93 |
| Wilfrid Hodges | Tarski's Theory of Definition | 94–132 |
| Marian David | Tarski's Convention T and the Concept of Truth | 133–156 |
| Douglas Patterson | Tarski's Conception of Meaning | 157–191 |
| Paolo Mancosu | Tarski, Neurath, and Kokoszyńska on the Semantic Conception of Truth | 192–224 |
| Greg Frost-Arnold | Tarski's Nominalism | 225–246 |
| Panu Raatikainen | Truth, Meaning, and Translation | 247–262 |
| John Etchemendy | Reflections on Consequence | 263–299 |
| Gila Sher | Tarski's Thesis | 300–339 |
| Mario Gómez-Torrente | Are There Model-Theoretic Logical Truths That Are Not Logically True? | 340–368 |
| Peter Simons | Truth on a Tight Budget: Tarski and Nominalism | 369–389 |
| Jody Azzouni | Alternative Logics and the Role of Truth in the Interpretation of Languages | 390–429 |

• Patterson, Douglas. 2012. *Alfred Tarski: Philosophy of Language and Logic*. New York: Palgrave Macmillan.

According to the book's cover, Tarski conceived of his logical work in the late 1920s as

a contribution to a view that he called "Intuitionistic Formalism." The book explains this view in terms of the views of Tarski's teachers, in particular as found in the work of Stanisław Leśniewski and Tadeusz Kotarbiński. These figures conceived of meaning not in terms of the semantic relation of words to the world, but in terms of the expression of thoughts. Questions about Tarski's work are then addressed by applying this reading of Tarski. Throughout the book, close attention is paid both to the development of Tarski's work through [the mid-1930s] and to the details of the expression of his ideas....

• Pla i Carrera, Josep. 1984. Alfred Tarski i la lògica contemporània. Primera part. *Butlletí de la Societat Catalana de Matemàtiques* 17: 26–46. In Catalan.
• ———. 1989. Alfred Tarski i la teoria de conjunts. *Theoria* (Spain) (series 2) 4: 343-417. The title means "Alfred Tarski and set theory." In Catalan.

The author, professor of logic, history, and philosophy of science at the University of Barcelona, published several articles about Tarski in Catalonian and Spanish journals. The first article discusses Tarski's work on propositional and first-order logic, and on consequence operations and cylindric algebras. There appears to be no second part of that article. The second article emphasizes Tarski's work on the axiom of choice and related topics.

- Rodríguez-Consuegra, Francisco. 2005. Tarski's intuitive notion of set. In *Essays on the Foundations of Mathematics and Logic*, edited by Giandomenico Sica, 227–266. Monza, Italy: Polimeric International Scientific Publisher.

  The author, professor of philosophy at the University of Valencia, has published major studies of Bertrand Russell's mathematical philosophy. He noted (pages 227–228) that
  > Tarski made important contributions to set theory.... Also...set theory was the main instrument used by Tarski in his most significant contributions which had philosophical implications and presuppositions.... [For] Tarski set theory was reliable as a working instrument, then presumably as a conceptual ground.... Like Russell and [Kurt] Gödel, Tarski was much more sincere about his philosophical tendencies when no publication was involved.... Thus, in this case the archival work has been once again the only effective way to try to understand Tarski's actual ideas. The resulting picture is a fascinating struggle between his nominalistic tendencies and his professional need to behave *as if* the mathematical entities with which he was working actually existed.

- Scanlan, Michael. 2003. American postulate theorists and Alfred Tarski. *History and Philosophy of Logic* 24: 307–325.

  The author spent a long career at Oregon State University. He had earned the doctorate in 1982 under supervision of John Corcoran at the State University of New York at Buffalo, with a dissertation on the American postulate theorists. This article outlines their results and
  > their influence on Tarski's work in the 1930s that was to be foundational for model theory. The American Postulate Theorists were influenced by the European foundational work of the period around 1900.... Their work served as paradigm examples of the theories and concepts investigated in model theory. The article also examines the possibility of a more specific impetus to Tarski's model theoretic investigation, arising from his having studied in 1927–1929 a paper by C. H. Langford proving completeness for various axiom sets for linear orders. This used the method of elimination of quantifiers.[6]

- Simmons, Keith. 2009. Tarski's logic. In *Handbook of the History of Logic*, edited by Dov M. Gabbay and John Woods, volume 5: *Logic from Russell to Church*, 511–616. Amsterdam: Elsevier.

  The author is professor of philosophy at the University of North Carolina. This detailed outline covers of all of Tarski's research except some works of strictly technical mathematics. Moreover, the author assesses its impact and describes its relationship to some later works of others.

- Sinaceur, Hourya. 1996. Mathématiques et métamathématique du congrès de Paris (1900) au congrès de Nice (1970): Nombres réels et théorie des modèles dans les travaux de Tarski. *Rendiconti del Circolo Matematico di Palermo* (serie 2 supplemento) 44: 113–132.

  This study by a noted Moroccan mathematician and historian is devoted to Tarski's contributions to the metamathematics of the real-number system and to model theory: in particular, the decidability and completeness of the elementary theory of real numbers.

---

[6] From the author's abstract. Langford 1926, 1927. For information on Tarski's 1927–1929 seminar see section 9.4.

• ———. 2001. Alfred Tarski: semantic shift, heuristic shift in metamathematics. *Synthese* 126: 49–65.

In his review of this paper, the American–French logician Yehuda Rav wrote,

> The author traces, with impeccable scholarship, the remarkable shift in the place of metamathematics within the mathematical sciences due to the work of Tarski. The emphasis by [David] Hilbert and his school was on syntactic methods, and the early [Rudolf] Carnap followed this trend.... Tarski... brought about a major shift...through the introduction of semantic methods, culminating in the development of model theory, with its interplay of semantic and syntactic relations. Thus, metamathematics... became part and parcel of mathematics, on a par with any other branch of mathematics.[7]

• Suppes, Patrick, Jon Barwise, and Solomon Feferman. 1989. Commemorative meeting for Alfred Tarski: Stanford University, November 7, 1983. In *A Century of Mathematics in America*, edited by Peter Duren et al., volume 3, 393–403. Providence: American Mathematical Society.

This article is the proceedings of a meeting that consisted of talks by the three authors, all Stanford professors of philosophy. The biographical talks by Suppes and Feferman were described in chapter 17. Barwise, who earned the doctorate in 1967 under Feferman's supervision, spoke in more technical terms about Tarski's contributions to semantics and the theory of models.

• Vaught, Robert L. [1971] 1974. Model theory before 1945. In Henkin et al., op. cit., 153–172.

The section 18.1 annotation of the later study Vaught 1986–1987 describes the author's relationship to Tarski. The present article emphasizes the Löwenheim–Skolem theorem, its elaboration in Tarski's 1926–1928 Warsaw seminar (see section 9.4), Tarski's theory of truth, and related developments. See Chang [1971] 1974 for an account of later work on model theory by Tarski and his associates.

• Woleński, Jan. 1987. Alfred Tarski jako filozof. *Wiadomości matematyczne* 27: 247–259.
• ———. 1993. Tarski as a philosopher. In *Polish Scientific Philosophy: The Lvov–Warsaw School*, edited by Francesco Coniglione et al., 319–338. Amsterdam: Rodopi.

The second paper is a translated and greatly expanded version of the first. The author, a leading Polish philosopher,

> describes the intellectual climate in which Tarski worked (the Vienna Circle, the Lwów–Warsaw logic school), studies his philosophical opinions concerning mathematics, and discusses the philosophical components of Tarski's works in logic (primarily of his semantics).[8]

---

[7] Rav 2001. The issue *Synthese* 126 (1/2) is devoted to recent considerations of the concept of truth.

[8] Murawski 1988. Tarski's son, Jan, also published comments about this paper (Tarski, J. 1994b).

•Zygmunt, Jan. 2009. Alfred Tarski—logik i metamatematyk. In *O przyrodzie i kulturze*, edited by Ewa Dobierzewska-Mozrzymas and Adam Jezierski, 305–327. Wrocław: Wydawnictwo Uniwersitytetu Wrocławskiego. The article's title phrase and book's title mean "logician and metamathematician" and *On Nature and Culture*.

The author is a noted Polish logician and historian of logic at the University of Wrocław. Contents:
1. Scientific biography
2. Set theory
3. Geometry and measure theory
4. Decidable and undecidable theories
5. Formalizing foundations of mathematics based on the relational calculus

# Bibliography

This bibliography includes all, and *only*, works referred to in the present volume. Since most of the book is about the first third of Alfred Tarski's career, in Poland, very many works that Tarski completed later, in the United States, are not mentioned here. In 1986, Steven R. Givant prepared an authoritative, complete bibliography of Tarski's works. That is updated in this book by lists of additional Tarski publications in chapter 16 and supplemented with lists of major works about Tarski in chapters 17 and 18. The entries of those lists are related to corresponding entries in the following bibliography, but were tailored to suit their special purposes.

In this book, bibliographic references are cited by giving a short version of the author's name and a date, nearby and usually, but not always, in that order. The book mentions more than one person named Tarski; citations that include this surname only are references to Alfred Tarski. Joint authors, up to three, are listed in title-page order. Alphabetization of the bibliography ignored diacritical marks, punctuation, and spaces within names. Warning: this resulted in alphabetization different from the Polish standard!

Some cited works have appeared in several versions. Information about the first often has historical interest, even when reference to a later one is more appropriate. In such cases, both dates are given, as in

Ajdukiewicz, Kazimierz. [1921] 1966....

The remaining data in such entries refer to the later version unless otherwise specified. Some journals cited here identify their volumes by year ranges rather than single years. An article in such a journal is identified by its year of publication if possible; the range is then given in parentheses following the volume number. Often, when a cited item is contained in a larger publication, both are cited, with cross-references. The entries for five large collections of Tarski's works are displayed in boxes.

After author, date, and title, each entry of the bibliography includes information about the item's publication, or its location in case it has not been published. This is followed by an annotation that explains its origin, import, and relationship to other items. When a title does not clearly suggest the language of a non-English item, the annotation does so. For an item listed in the Givant 1986 Tarski bibliography, the first element in the annotation is the identification code that Givant assigned. An annotation may include a code such as "JFM: Name" or "MR: Name" to indicate that the cited work was reviewed in the *Jahrbuch über die Fortschritte der Mathematik* or in *Mathematical Reviews*, by the named reviewer. Many annotations include English translations of Slavic titles or institutional names. Each annotation identifies sections of this book or other bibliography items that refer to it.[1]

---

[1] The terms *przegląd* and *ruch*—review and trends—recur in Polish journal titles. A section number of the form $n.0$ refers to text in chapter $n$ before the first numbered section.

As much as possible, titles appear here as they did on their published title pages. This practice, advocated by current documentation authorities,[2] is consistent with major catalogs now searchable electronically. But readers may encounter difficulty matching titles, particularly of journals, with nomenclature in past literature, or in catalogs based on it, which often differs considerably from that of title pages. For example, the *Giornale di matematiche* was often referred to as the *Giornale di Battaglini*, after the original editor; and the journal published by the organization known as the Dorpater Naturforschergesellschaft was sometimes listed by an institutional name—Universität Tartu—or in the language—Estonian—more politically correct at the time the catalog was constructed.[3]

Adamowicz, Zofia, et al., editors. 2004. *Provinces of Logic Determined: Essays in the Memory of Alfred Tarski.* Same as volumes 126–127 of *Annals of Pure and Applied Logic.* Cited in 18.2.

Addison, John W. 1983. Eloge: Alfred Tarski, 1901–1983. *California Monthly* 94 (2, December): 28. Reprinted in *Annals of the History of Computing* 6 (1984): 335–336. Translated by Irena Czubińska as "Jeden z największych logików," *Przegląd techniczny* 20 (1985): 39. That title means "One of the greatest logicians." Cited in chapter 17.

Ajdukiewicz, Kazimierz. [1921] 1966. From the methodology of the deductive sciences. Translated by Jerzy Giedymin. *Studia Logica* 19: 9–46. Originally published as *Z metodologii nauk dedukcyjnych*, publication 10, by the Polish Philosophical Society in Lwów (Polskie Towarzystwo Filozoficzne we Lwowie). Cited in chapter 3.

——. 1928. *Główne zasady metodologji nauk i logiki formalnej*. Lectures delivered during academic year 1927–1928. Authorized transcript by Mojżesz Presburger. Publication 16. Warsaw: Faculty of Mathematics and Physics of the University of Warsaw. The title means *Main Principles of the Methodology of Science and Formal Logic*. Cited in 9.4 and 16.1.

——. [1934] 1978. The scientific world-perspective. Translated by Wilfrid Sellars. In Ajdukiewicz 1978, 111–117. Originally published as "Naukowa perspektywa świata," *Przegląd filozoficzny* 37 (1934): 409–416. Cited in 15.10.

——. 1936. Zagadnienie idealizmu w sformułowaniu semantycznym. In Polski Zjazd Filozoficzny 1936, 334–340. Contains the original version of the comment Tarski [1936] 2014a. The title means "The problem of idealism in a semantic formulation." This was published in *Przegląd filozoficzny* 39. In the table of contents for the congress in the journal, the paper's title began instead with the word "Problemat." Cited in 15.8.

——. [1937] 1978. A semantical version of the problem of transcendental idealism. Translated by Jerzy Giedymin. In Ajdukiewicz 1978, 140–154. Originally published as "Problemat transcendentalnego idealizmu w sformułowaniu semantycznym," *Przegląd filozoficzny* 40: 271–287. Cited in 15.8.

——. 1978. *The Scientific World-Perspective and Other Essays, 1931–1963*. Synthese library 108. Edited by Jerzy Giedymin. Dordrecht: D. Reidel. Contains Ajdukiewicz [1934] 1978, [1937] 1978, and Giedymin 1978.

Albiński, Marian. 1976. Wspomnienia o Banachu i Wilkosu. *Wiadomości matematyczne* 19: 133–135. The title means "Memories of Banach and Wilkosz." Cited in 15.9.

Allen, Layman E. 1962. *Wff 'N Proof: The Game of Modern Logic*. New Haven, Connecticut: Wff'n Proof. Cited in 15.1.

——. 2010. Personal communication to James T. Smith, 25 June 2010.

---

[2] Chicago Manual 1993, 15.208–209; Mann 1998.

[3] The serials list in May 1973, Appendix 2, may help with this problem.

Amaldi, Ugo. [1900] 1912. Sulla teoria della equivalenza. In Enriques [1900] 1912–1914, volume 1, 145–198. Revision of the original version in Enriques 1900. The 1911 first volume of Enriques [1900] 1907–1911 contains a German translation. Amaldi [1900] 1914 is a Polish translation.

———. [1900] 1914. O teorji równoważności. Translated by Stefan Kwietniewski and Władysław Wojtowicz. In Enriques [1900] 1914–1917, article 6, 132–181. Translation of Amaldi [1900] 1912. Cited in 4.1, 4.3, and chapter 5.

Amaldi, Ugo, and Tullio Levi-Cività. 1920. *Lezioni di meccanica razionale*. Padova: "La litotipo" Editrice Universitaria. Cited in 9.2.

Andréka, Hajnal, J. Donald Monk, and István Németi, editors. 1991. *Algebraic Logic: Colloquia Mathematica Societatis János Bolyai, 54 (Budapest, 8–14 August 1988)*. Amsterdam: North-Holland Publishing Company. Contains Givant [1988] 1991. Cited in 18.2.

Anellis, Irving H. 2005. Review of Feferman and Feferman 2004. *Review of Modern Logic* 10: 117–129. Cited in chapter 17.

Apostol, Tom M., and Mamikon A. Mnatsakanian. 2011. Complete dissections: converting regions and their boundaries. *American Mathematical Monthly* 118: 789–798. MR: Richter. Cited in 8.5.

Apt, Krzystof R. 2007. Review of Feferman and Feferman 2004. *Mathematical Intelligencer* 29(2): 79–80. Cited in chapter 17.

Arboleda, Luis Carlos. 1990. Kuratowski, Kazimierz. In *Dictionary of Scientific Biography* 1970–, volume 17, supplement 2: 519–521. Cited in 4.4.

Archimedes. [1897] 2002a. Measurement of a circle. In Archimedes [1897] 2002b, 91–98. Cited in 9.8.

———. [1897] 2002b. *The Works of Archimedes*, edited by Thomas L. Heath. Mineola, New York: Dover Publications. JFM: Gibson. MR: Schneider. Originally published in 1897 in Cambridge, England, by Cambridge University Press, with a supplement in 1912. Contains Archimedes [1897] 2002a.

Aristotle. 1971. *Metaphysics, Books Γ, Δ, and E*. Translated with notes by Christopher Kirwan. Oxford: Clarendon Press. Cited in 15.6.

Bafia, Stanisław, and Andrzej Maryniarczyk. 2001. *Powszechna encyklopedia filozofii*. Nine volumes. Lublin: Polskie Towarzystwo Tomasza z Akwinu. The first word of the title means *Universal*. The publisher is the Polish Society of Thomas Aquinas. Contains Majdański and Lekka-Kowalik 2001.

Banach, Stefan. 1923. Sur le problème de la mesure. *Fundamenta Mathematicae* 4: 7–33. MR: Rosenthal. Reprinted in Banach 1967, 66–89. Cited in 4.2, ..., 4.4, 5.0, 6.0, 6.2, 8.7, and 15.7.

———. 1924. Un théorème sur les transformations biunivoques. *Fundamenta Mathematicae* 6: 236–239. Reprinted in Banach 1967, 114–117. Cited in 4.4, 6.1, and 8.8.

———. [1931] 1987. *Theory of Linear Operations*. Translated by Francis Jellett. Contains "Some aspects of the present theory of Banach spaces," by Aleksander Pelczyński and Czesław Bessaga, 161–237. Amsterdam: North-Holland Publishing Company. Translated from the 1932 French edition. Originally published in Warsaw by the Mianowski Fund as *Teorja operacyj*, volume 1: *Operacje linjowe*. JFM: Aumann. Cited in 4.2.

Banach, Stefan. 1967. *Oeuvres*. Volume 1: *Travaux sur les fonctions réeles et sur les séries orthogonales*. Edited by Stanisław Hartman and Edward Marczewski. Warsaw: PWN—Éditions scientifiques de Pologne. MR: Zygmund. Contains reprints of Banach 1923 and 1924 and of the original version of Banach and Tarski [1924] 2014.

Banach, Stefan, Wacław Sierpiński, and Włodzimierz Stożek. 1933. *Arytmetyka dla I klasy gimnazjalnej*. Lwów: Książnica-Atlas. The title means *Arithmetic for the First Class in Gimnazjum*. Cited in 4.2.

Banach, Stefan, and Alfred Tarski. [1924] 2014. On decomposition of point sets into respectively congruent parts. Chapter 6 of the present book, Tarski 2014. Translation of "Sur la décomposition des ensembles de points en parties respectivement congruents," *Fundamenta Mathematicae* 6: 244–277, Givant 1986 item 24d, reprinted in Banach 1967, 118–148, and in Tarski 1986a, volume 1, 119–154. JFM: Rosenthal. Cited in 4.0, ..., 4.4, 5.3, 8.4, 8.6, 15.7, and 15.12.

Bartocci, Claudio. 2012. *Una piramide di problemi: Storie di geometrie da Gauss a Hilbert*. Milan: Raffaello Cortina Editore. Cited in 4.1.

Barzin, Marcel. [1938] 1939. Langage et réalité; discussion. In Institut International de Collaboration Philosophique 1939, 20–27, 35–65. Contains Tarski [1938] 1939.

Baumslag, Naomi. 2005. *Murderous Medicine: Nazi Doctors, Human Experimentation, and Typhus*. Westport, Connecticut: Praeger Publishers. Cited in 4.2.

Benko, David. 2007. A new approach to Hilbert's third problem. *American Mathematical Monthly* 114: 665–676, 770. MR: Linhart. Cited in 4.1.

Bettazzi, Rodolfo. 1890. *Teoria delle grandezze. Opera premiata dalla Reale Accademia dei Lincei*. Pisa: Enrico Spoerri. JFM: Vivanti. Cited in 4.2.

Betti, Arianna. 2004. Leśniewski's early liar, Tarski, and natural language. *Annals of Pure and Applied Logic* 127: 267–287. MR: Murawski. Cited in 1.1, chapter 3, 15.6, and 18.2.

———. 2008. Polish axiomatics and its truth: On Tarski's Leśniewskian background and the Ajdukiewicz connection. In Patterson 2008, chapter 3, 44–71. Cited in 15.6 and 18.2.

———. 2011. Kazimierz Twardowski. In Zalta 2011 at http://plato.stanford.edu/archives/sum2011/entries/twardowski/.

Béziau, Jean-Yves, editor. 2010. *Universal Logic: An Anthology, from Paul Hertz to Dov Gabbay*. Basel: Birkhäuser. Contains Tarski [1928] 2010.

Birkhoff, Garrett. 1948. *Lattice Theory*. Revised edition. Colloquium publication 25. Providence, Rhode Island: American Mathematical Society. MR: Frink. Cited in 4.4.

Błażejewski, Stanisław, Janusz Kutta, and Marek Romaniuk. 1994–2006. *Bydgoski słownik biograficzny*. Seven volumes. Bydgoszcz: Kujawsko-Pomorskie Towarzystwo Kulturalne Wojewódzki Ośrodek Kultury. The title and publisher's name mean *Bydgoszcz Biographical Dictionary* and Cultural Center of the Kujawsko-Pomorskie Regional Cultural Association. Contains Kutta 1997.

Blok, Willem J., and Don Pigozzi. 1988. Alfred Tarski's work on general metamathematics. *Journal of Symbolic Logic* 53: 36–50. MR: Mendelson. Cited in 18.1.

Bocheński, J. M. 1994. Morals of thought and speech—reminiscences. In Woleński 1994, 1–8. MR: Veldman. Cited in 1.1.

Bojarski, Bogdan, Julian Ławrynowicz, and Yaroslav G. Prytula, editors. 2009. *Lvov Mathematical School in the Period 1915–1945 as Seen Today*. Warsaw: Polish Academy of Sciences, Institute of Mathematics. Contains Schinzel 2009.

Boltyanskiĭ, Vladimir Grigorevich. 1978. *Hilbert's Third Problem*. Translated by Richard A. Silverman. Introduced by Albert B. J. Novikoff. Washington: V. H. Winston and Sons. MR: Firey. Cited in 4.4.

Bolyai Farkas. [1832] 1904. *Tentamen Iuventutem Studiosam in Elementa Matheseos Purae Elementaris ac Sublimioris Methodo Intuitiva Evidentiaque Huic Propria Introducendi, cum Appendice Triplici*. Volume two: *Elementa Geometriae et Appendices*. Second edition. Edited by József Kürschák, Móricz Réthy, and Béla Tőtössy de Zepethnek. Budapest: Hungarian Academy of Sciences. The main title means *Essay for Studious Youth on the Elements of Elementary and Higher Pure Mathematics, Properly Introduced Here by an Intuitive and Clear Method, with Three Appendices*. Cited in 4.1.

Bolyai János. [1832] 1891. The science absolute of space independent of the truth or falsity of Euclid's axiom XI (which can never be decided a priori). Translated by George Bruce Halsted. *Scientiae Baccalaureus* 1: 203–260. Reprinted in Bonola [1906] 1955. Originally published as "Scientiam spatii absolute veram exhibens: a veritate aut falsitate axiomatis XI Euclidiei (a priori haud unquam decidenda) independentem," appended to the [1832] 1904 book by Bolyai's father Farkas.

Bonola, Roberto. [1906] 1955. *Non-Euclidean Geometry: A Critical and Historical Study of Its Developments.* Translated, with additional appendices by H. S. Carslaw. With an introduction by Federigo Enriques. With a supplement containing the translations J. Bolyai [1832] 1891 and Lobachevsky [1840] 1914 by George Bruce Halsted. New York: Dover Publications. Bonola's book was first published by Nicola Zanichelli in Bologna as *La geometria non-euclidea: Esposizione storico-critico de suo sviluppo*, JFM: Lampe. Cited in 12.11.

Borel, Émile. 1914. *Leçons sur la théorie des fonctions.* Second edition. Paris: Gauthier-Villars. The first edition was published in 1898. JFM: Lampe. Cited in 4.4.

Borowski, Marjan. 1922. Letter to Kazimierz Twardowski, 24 February 1922. In the "Documentation on Leśniewski" section of Coniglione and Betti, editors 2001–. Translated by Jennifer Smith. The original is in the Twardowski Archives in Warsaw. Cited in 1.1.

Bouveresse, Jacques, editor. 1991–1993. *Philosophie de la logique et philosophie du langage.* Two volumes. L'âge de la science 4–5. Paris: Éditions Odile Jacob. Contains Tarski [1955] 1993.

Brandes, Hans. 1907. *Über die axiomatische Einfachheit mit besonderer Berücksichtigung der auf Addition beruhenden Zerlegungsbeweise des Pythagoräischen Lehrsatzes.* Inaugural dissertation presented to the Philosophical Faculty of the University of Halle–Wittenberg. Brunswick: Friedrich Vieweg und Sohn. JFM: Zacharias. Cited in 7.1.

Braun, Stephania, Edward Szpilrajn, and Kazimierz Kuratowski. 1935. Annexe. *Fundamenta Mathematicae* 1 (second edition of 1920 volume): 225–254. Cited in 4.4.

Brieskorn, Egbert, editor. 1996. *Felix Hausdorff zum Gedächtnis*, volume 1: *Aspekte seines Werkes*. Braunschweig: Friedrich Vieweg & Sohn. MR: Scriba. Contains Schreiber 1996.

Brzęk, Gabriel. 1983. *Henryk Raabe: 1882–1951.* Lublin: Uniwersitet Marii Curie-Skłodowskiej. Cited in 9.2.

Bury, Jan. 2004. Polish codebreaking during the Russo-Polish war of 1919–1920. *Cryptologia* 28: 193–203. Cited in 1.1.

Cantor, Georg. 1883. Über unendliche lineare Punktmannigfaltigkeiten, 5: Grundlagen einer allgemeinen Mannigfaltigkeitslehre. *Mathematische Annalen* 21: 51–58, 545–586. JFM: Schlegel. Reprinted in Cantor 1962, 165–209. Cantor [1883] 1996 is an augmented translation. Cited in 1.2.

———. [1883] 1996. Foundations of a general theory of manifolds: A mathematico-philosophical investigation into the theory of the infinite. Translated by William B. Ewald. In Ewald 1996, volume 2, section 19.C, 878–920. The original, published by B. G. Teubner, is a reprint, with additional preface and footnotes, of Cantor 1883.

———. 1895–1897. Beiträge zur Begründung der transfiniten Mengenlehre, *Mathematische Annalen* 46 (1895): 481–512; 49 (1897): 207–246. JFM: Vivanti. Reprinted in Cantor 1962, 282–356. Cantor [1895–1897] 1952 is a translation.

———. [1895–1897] 1952. *Contributions to the Founding of the Theory of Transfinite Numbers.* Translated and with a preface, introduction, and notes by Philip E. B. Jourdain. New York: Dover Publications. JFM: Vivanti. Translation of Cantor 1895–1897, originally published in 1915 by the Open Court Publishing Company. Cited in 1.2.

———. 1962. *Gesammelte Abhandlungen mathematischen und philosophischen Inhalts, mit erläuternden Anmerkungen sowie mit Ergänzungen aus dem Briefwechsel Cantor–Dedekind.* Edited by Ernst Zermelo. With a biography of Cantor by Abraham A. Fraenkel. Hildesheim: Georg Olms Verlagsbuchhandlung. JFM: Pannwitz. Reprint of the original 1932 edition. Contains reprints of Cantor 1883 and 1895–1897.

Carnielli, Walter. 2006. Review of Feferman and Feferman 2004. *Logic and Logical Philosophy* 15: 91–96. Cited in chapter 17.
Castelnuovo, Guido. 1947. Federigo Enriques. *Accademia Nazionale dei Lincei, classe di scienze fisiche, matematiche e naturali: Rendiconti* (series 8) 2: ix–xix. In Italian. Cited in 9.2.
Castelnuovo, Guido, and Federigo Enriques. 1908. Grundeigenschaften der algebraischen Flächen. In Meyer and Mohrmann 1907–1934, part 2, half 1, 635–673. Article III C, 6a. JFM: Ostrowski. Cited in 9.2.
———. 1914. Die algebraischen Flächen vom Gesichtspunkte der birationalen Transformationen aus. In Meyer and Mohrmann 1907–1934, part 2, half 1, 674–767. Article III C, 6b. JFM: Ostrowski. Cited in 9.2.
Castrillo, Pilar, and Luis Vega Reñon, editors. 1984. *Lecturas de lógica II: Selección, introducción y traducción.* Madrid: Universidad Nacional de Educación a Distancia. Contains Tarski [1935] 1984. Cited in 16.1.
Cat, Jordi. 2010. Otto Neurath. In Zalta 2011– at http://plato.stanford.edu/entries/neurath/. Cited in 16.3.
Cesarski, Wojciech. 1995. Review of Tarski 1995–2001, volume 1. *Gazeta wyborcza* 267 (17 November): 2. In Polish. The journal is a daily newspaper with national circulation. Cited in 16.2.
Chang, Chen-Chung. [1971] 1974. Model theory 1945–1971. In Henkin et al. 1974, 173–186. MR: Anonymous. Cited in 18.1 and 18.2.
Chatterji, Srishti D. 2001. Commentary on Hausdorff 1914b. In Hausdorff 2001–2002, volume 4, 11–18. In English. Cited in 4.4.
———. 2002. Hausdorff als Masstheoretiker. *Mathematische Semesterberichte* 49: 129–143. Cited in 4.2.
*Chicago Manual of Style, The.* 1993. Fourteenth edition. Chicago: University of Chicago Press. Cited in the preface and this bibliography's introduction.
Chicago, University of, News Office. 2008. Izaak Wirszup, 1915–2008. Press release, 31 January. Chicago: University of Chicago News Office. Cited in 7.1.
Chodorowski, Barbara, et al. 2005. *100 lat na Smolnej czyli dzieje gimnazjum i liceum im. Jana Zamoyskiego w Warszawie.* Warsaw: XVIII Liceum im. Jana Zamoyskiego. The title means *100 Years of History of the Jan Zamoyski Gimnazjum and Liceum on Smolna Street.* Cited in 9.9 and 15.2.
Choquet, Gustave, et al., editors. 2004. *Autour du centenaire Lebesgue.* Panoramas et synthèses 18. Paris: Société Mathématique de France. MR: Anonymous. Contains item De la Harpe 2004.
Chuaqui Kettlun, Rolando. 1984. Alfred Tarski, matemático de la verdad. *Revista universitaria* 11: 30–34. Translated into Portuguese by S. P. Monoide as "Alfred Tarski, matemático da verdade," *Boletim da Sociedade Paranaense de Matematica* 6 (1985): 1–10. Cited in chapter 17.
Church, Alonzo. 1941–1947. Reviews of editions of Tarski [1936] 1995. *Journal of Symbolic Logic* 6(1941): 30–32; 12(1947): 61.
———. 1956. *Introduction to Mathematical Logic.* Princeton: Princeton University Press. MR: Gál. There is no volume two. Cited in chapter 3.
Chwiałkowski, Zygmunt. 1914. *О функціональныхъ уравненіяхъ.* Warsaw: Типографія Варшавскаго Учебнаго Округа. These are the original spellings of the title and publisher's name; the author's name was spelled С. А. Хвялковскій. The title and publisher's name mean *On Functional Equations* and Typographer of the Warsaw Schools. Cited in 3.1.
Chwiałkowski, Zygmunt, Wacław Schayer, and Alfred Tarski. [1935] 1946. *Geometrja dla trzeciej klasy gimnazjalnej.* Second edition, reprinted. Hanover: Polski Związek Wychodźctwa Przymusowego. Givant 1986 item $35^m$. The title and publisher's name mean *Geometry for the Third Gimnazjum Class* and Polish Association of Forced Emigration. The 1935 edition was

published in Lwów by the Państwowe Wydawnictwo Książek Szkolnych (National School Books Publisher). The 1946 edition was originally published in Jerusalem in 1944 by the Ministerstwo Wyznań Religijnych i Oświaty Publicznej, Sekcji Wydawniczej Armii Polskiej na Wschodzie (Ministry of Religious Denominations and Public Education, Publishing Section for the Polish Army in the East). For an advertisement, see Państwowe Wydawnictwo Książek Szkolnych 1937. Cited in 9.7, 9.9, 13.0, and 14.1.

Chwistek, Leon. 1960. *Wielość rzeczywistości w sztuce i inne szkice literacke*. With preface by Karol Estreicher. The title means *The Multiplicity of Reality in Art, and Other Literary Essays*. Warsaw: Czytelnik. Cited in 14.2.

Congrès Polonais de Mathématiques. [1937] 1938. Comptes-rendus et analyses. *Annales de la Société Polonaise de Mathématique* 16: 182–212. Contains Lindenbaum [1937] 1938.

Coniglione, Francesco, and Adrianna Betti. 2001. Kazimierz Ajdukiewicz. In the "Main Polish Philosophers of the 20th Century" section of Coniglione and Betti, editors 2001–. Cited in chapter 3.

Coniglione, Francesco, and Arianna Betti, editors. 2001–. *Polish Philosophy Page*. Catania, Italy: University of Catania. Internet: http://segr-did2.fmag.unict.it/~polphil/PolHome.html. Contains Borowski 1922 and Coniglione and Betti 2001.

Coniglione, Francesco, Roberto Poli, and Jan Woleński, editors. 1993. *Polish Scientific Philosophy: The Lvov–Warsaw School*. Poznań studies in the philosophy of the sciences and the humanities 28. Amsterdam: Rodopi. Contains Woleński 1993. Cited in 18.2.

Corcoran, John. 1991. Review of Tarski 1986a. *Mathematical Reviews* 91: MR1015501 (91h: 01101). Cited in the preface.

———. 1993. Review of Givant 1991. *Mathematical Reviews* 93: MR1117874 (93a: 01046). Cited in chapter 17.

———. 2011a. Personal communication to James T. Smith, 20 February 2011. Cited in chapter 3.

———. 2011b. Review of Sinaceur 2009. *Mathematical Reviews* 2011: MR2509665 (2011b: 03006). Cited in chapters 3 and 17.

Couturat, Louis. [1905] 1918. *Algebra logiki*. Translated by Bronisław Knaster and Jan Łukasiewicz. Warsaw: Wydawnictwo Koła Matematyczno-Fizycznego Słuchaczów Universytetu Warszawskiego. The publisher's name means "Publisher of the Circle of Mathematics and Physics Students of the University of Warsaw." Translation supported by the Mianowski Fund. Originally published in Paris by Gauthier-Villars as *L'algèbre de la logique*. Cited in chapter 3.

Cwojdziński, Kazimierz. 1930. Kilka słów o ciągach. *Parametr* 1: 300–305. The title means "Some words about sequences." Cited in 12.5.

Czelakowski, Janusz, and Grzegorz Malinowski. 1985. Key notions of Tarski's methodology of deductive systems. *Studia Logica* 44: 321–351. MR: Dawson. Cited in 18.2.

Czeżowski, Tadeusz. [1927] 1928. O stosunku przyczynowym. In Polski Zjazd Filozoficzny 1928, 165–168. The title means "On a causal relation." Contains the original version of the comment Tarski [1927] 2014b. Cited in 15.3.

Czyż, Janusz. 1990. "Enigmatyczna" wojna 1920 r. *Wiedza i życie* 44: 44–48. The paper and journal titles mean "The 'enigmatic' war of 1920" and *Knowledge and Life*. Cited in 1.1.

———. 1994. *Paradoxes of Measures and Dimensions Originating in Felix Hausdorff's Ideas*. Singapore: World Scientific Publishing Company. MR: Grattan-Guinness. Cited in 4.4 and 8.6.

Davies, Charles. 1890. *Elements of Geometry and Trigonometry from the Works of A. M. Legendre, Adapted to the Course of Mathematical Instruction in the United States*. Edited by J. Howard van Amringe. New York: American Book Company. Cited in 9.8.

Davies, Norman. 1972. *White Eagle, Red Star: The Polish–Soviet War, 1919–20*. With forward by A. J. P. Taylor. New York: St. Martin's Press. Cited in 1.1.

———. 1982. *God's Playground: A History of Poland*. Volume 1: *The Origins to 1795*. Volume 2: *1795 to the Present*. New York: Columbia University Press. Cited in 1.1, chapter 3, 9.1, and 9.9.

Davis, Anne C. 1955. A characterization of complete lattices. *Pacific Journal of Mathematics* 5: 312–319. MR: Jónsson. Cited in 8.8.

Davis, Martin. 2005. The man who defined *truth*. *American Scientist* 93 (2, Scientist's Bookshelf) 1. Review of Feferman and Feferman 2004. Cited in chapter 17.

Davis, Philip J. 2005. A life of logic and the illogic of life. Review of Feferman and Feferman 2004. *SIAM News* 38 (2, March). Cited in chapter 17.

Dawson, John W., Jr. 1987. Review of Czelakowski and Malinowski 1985. *Mathematial Reviews* 87: MR0832393 (87f:03083). Cited in 18.2.

———. 1998. Logical contributions to the Menger colloquium. In Menger et al. 1998, 33–42. Cited in 14.2.

Dehn, Max. 1901–1902. Ueber den Rauminhalt. *Mathematische Annalen* 55: 465–478. Cited in 4.1, 4.3, and 5.3.

De la Harpe, Pierre. 2004. Mesures finiment additives et paradoxes. In Choquet 2004, 39–61. MR: Rosenblatt. Cited in 8.6.

De Zolt, Antonio. 1881. *Principii della eguaglianza di poligoni preceduti da alcuni critici sulla teoria della equivalenza geometrica*. Milan: Briola. This booklet is rare. Cited in 4.1.

———. 1883. *Principii della eguaglianza di poliedri e di poligoni sferici*. Milan: Briola. This booklet is rare. Cited in 4.1.

Derkowska, Alicja. 2001. Zenon Waraszkiewicz. *Wiadomości matematyczne* 37: 135–137. In Polish. The published title included an incorrect date for Waraszkiewicz's death: see Tatarkiewicz 2003. Cited in 7.1.

*Dictionary of Scientific Biography*. 1970–. Eighteen volumes. Charles C. Gillispie et al., editors. New York: Charles Scribner's Sons. Contains Eisele 1971, Keesling 1960, Knaster 1973, and Moore 1990. Cited in chapter 17 and 18.2.

Dierkesmann, Magda, et al. 1967. Felix Hausdorff zum Gedächtnis. *Jahresbericht der Deutschen Mathematiker-Vereinigung* 69: 51–76. Cited in 4.4.

Dobierzewska-Mozrzymas, Ewa, and Adam Jezierski. 2009. *O przyrodzie i kulturze*. Studium Generale 13. Wrocław: Wydawnictwo Uniwersytetu Wrocławskiego. The title means *On Nature and Culture*. Contains Zygmunt 2009.

Dobrzycki, Stanisław. 1971. Kwietniewski Władysław (1840–1902). *Polski słownik biograficzny* 16: 379–380. In Polish. Cited in 9.3.

Dobrzyńska, Bożena, and Olszewska, Irena. 1998. Chwiałkowski Zygmunt. *Bibliografia polska 1901–1939*. Volume 4: *Ch–Cy*. In Polish. Warsaw: Biblioteka Narodowa. Cited in 9.9.

Domarus, Max. 1992. *Hitler: Speeches and Proclamations, 1932–1945*. Volume 2: *The Years 1935 to 1938*. Translated by Chris Wilcox and Mary Fran Gilbert. Wauconda, Illinois: Bolchazy–Carducci Publishers. Cited in 14.3.

Domoradzki, Stanisław. 2011. Personal communication to Andrew McFarland, May 2011. Cited in 9.4.

Domoradzki, Stanisław, Zofia Pawlikowska-Brożek, and Danuta Węglowska. 2003. *Słownik biograficzny matematyków polskich*. Tarnobrzeg, Poland: Państwowa Wyższa Szkola Zawodowa. The title and publisher's name mean *Biographical Dictionary of Polish Mathematicians* and State Higher Vocational School. Contains Pawlikowska-Brożek 2003a and 2003b, W. Piotrowski 2003, Sękowska and Węglowska 2003, and Pawlikowska-Brożek 2011 (in a future edition). Cited in chapter 17.

Doner, John E., and Wilfrid A. Hodges. 1988. Alfred Tarski and decidable theories. *Journal of Symbolic Logic* 53: 20–35. MR: Plotkin. Cited in 18.1.

Doner, John E., Andrzej Mostowski, and Alfred Tarski. 1978. The elementary theory of well-ordering—a metamathematical study. In Macintyre, Pacholski, and Paris 1978, 1–54. Givant 1986 item 78. MR: Slomson. Cited in 8.2.

Drozdowski, Marian, editor. 1968–1973. *Warszawa II Rzeczypospolitej 1918–1939*. Five volumes. Warsaw: Państwowe Widawnictwo Naukowe. The title means *Warsaw under the Second Republic*. Contains Konarski 1971 and 1973.

Dubiel, Władysław. 1975. *Młody matematyk*—czasopismo dla młodzieży szkolnej. *Delta* 4: 2–6. The title phrase means "magazine for young scholars." Contains a reprint of the original version of Tarski [1931] 2014d. Cited in 9.7 and 12.0.

———. 1990. *Polskie czasopisma poświęcone matematyce i jej nauczaniu (1911–1939)*. Lublin: Wydawnictwo Uniwersytetu Marii Curie-Skłodowskiej. The title means *Polish Periodicals Dedicated to Mathematics and Teaching (1911–1939)*. Cited in 9.7.

Duda, Roman. 1996. Fundamenta Mathematicae and the Warsaw school of mathematics. In Goldstein, Gray, and Ritter 1996, chapter 21, 479–498. Cited in 4.4.

———. 2009. Facts and myths about Stefan Banach. Translated by Joseph Pomianowski. *Newsletter of the European Mathematical Society* 71 (March): 29–34. Cited in 2.1.

Duren, Peter, Richard A. Askey, Harold M. Edwards, and Uta C. Merzbach, editors. 1989. *A Century of Mathematics in America*. Three volumes. Providence: American Mathematical Society. Contains Suppes, Barwise, and Feferman 1989.

Edwards, Paul. 1967. *The Encyclopedia of Philosophy*. Eight volumes. New York: Macmillan Publishing Company. Contains Mostowski 1967. Cited in 18.2.

Ehrenfeucht, Andrzej, Wiktor W. Marek, and Marian Srebrny, editors. 2008. *Andrzej Mostowski and Foundational Studies*. Amsterdam: IOS press. Contains Krajewski and Srebrny 2008 and Woleński 2008.

Eisele, Carolyn. 1971. Federigo Enriques. In *Dictionary of Scientific Biography* 4: 373–375. Cited in 9.2.

Enriques, Federigo. 1907. Prinzipien der Geometrie. In Meyer and Mohrmann 1907–1934, part 1, half 1, 1–129. Article III A, B 1. JFM: Lampe. Cited in 9.2.

Enriques, Federigo, editor. 1900. *Questioni riguardanti la geometria elementare*. Bologna: Nicola Zanichelli. JFM: Loria. Also reviewed in *Revue de métaphysique* 1904. Enriques [1900] 1907–1911 is a German translation. Enriques [1900] 1912–1914 is a revised and augmented edition. The [1900] 1914–1917 Polish translation incorporates those revisions. Contains the original versions of Amaldi [1900] 1912 and [1900] 1914.

———. [1900] 1907–1911. *Fragen der Elementargeometrie*. Volume 1 (1911): *Die Grundlagen der Geometrie*, translated and edited by Hermann Thieme, incorporating revisions drafted for Enriques [1900] 1912–1914. Volume 2 (1907): *Die geometrischen Aufgaben, ihre Lösung und Lösbarkeit*, translated and edited by Hermann Fleischer. Leipzig: B. G. Teubner. Translation of Enriques 1900. Volume 1 contains a translation of Amaldi [1900] 1912. JFM: Lampe.

———. [1900] 1912–1914. *Questioni riguardanti le matematiche elementari*. Volume 1: *Critica dei principii*. Volume 2: *Problemi classici della geometria, numeri primi e analisi indeterminata, massimi e minimi*. Bologna: Nicola Zanichelli. JFM: Vivanti. Revised and augmented edition of Enriques 1900. Partially translated in Enriques [1900] 1914–1917. Volume 1 contains Amaldi [1900] 1912.

———. [1900] 1914–1917. *Zagadnienia dotyczące gieometrji elementarnej*. Volume 1: *Krytyka podstaw*. Volume 2: *Konstrukcje gieometryczne i teorja izoperymetrów*. Translated by Stefan Kwietniewski and Władysław Wojtowicz. Warsaw: Skład główny w Księgarni Gebethnera i Wolfa. Translation of those articles in Enriques [1900] 1912–1914 whose original versions were in Enriques 1900. Volume 1 contains Amaldi [1900] 1914 and Kwietniewski and Wojtowicz 1914. Supported by the Mianowski Fund. Cited in 9.3.

Enriques, Federigo, and Ugo Amaldi. [1903] 1913. *Elementi di geometria ad uso delle scuole secondarie superiori.* Sixth edition. Bologna: Nicola Zanichelli. This book appeared in many editions, tailored to various audiences. Enriques and Amaldi [1903] 1916 is a Polish translation of one of them. The first edition was reviewed in *Revue de métaphysique* 1904. Cited in 4.2, 9.2, and 9.3.

———. [1903] 1916. *Zasady gieometrji elementarnej do użytku szkół średnich.* Translated by Władysław Wojtowicz. Warsaw: E. Wende i Spółka. Translation of the fifth edition of Enriques and Amaldi [1903] 1913. Polish libraries list this as published in 1914 or 1916. The book the present editors have studied carries no such information, but indicates that it was cleared by the German censors in 1916. Cited in 4.1, 4.2, chapter 5, 9.2, 9.8, 11, 12.11, and 12.13.

Erdős, Paul. 1943. Some remarks on set theory. *Annals of Mathematics* (2) 44: 643–646. MR: Todd. Cited in 15.12.

Estreicher, Karol. 1971. *Leon Chwistek: Biografia artysty (1884–1944).* Cracow: Państwowe Wydawnictwo Naukowe. Cited in 14.2.

Etchemendy, John. 1988. Tarski on truth and logical consequence. *Journal of Symbolic Logic* 53: 51–79. MR: Luce. Cited in 18.1.

Euclid. [1908] 1956. *The Thirteen Books of Euclid's Elements.* Translated from the text of Heiberg, with introduction and commentary by Thomas L. Heath. Second edition. Three volumes. New York: Dover Publications. Cited in 2.1, 12.11, 12.13, and 13.1.

Everett, Hazel, Christian Gillot, et al. 2009. The Voronoi diagram of three arbitrary lines in $\mathbb{R}^3$. To appear in *Proceedings of the 25th European Workshop on Computational Geometry (EuroCG'09).* Brussels: Université Libre de Bruxelles. Cited in 12.12 and 12.13.

Everett, Hazel, Daniel Lazard, et al. 2009. The Voronoi diagram of three lines. *Discrete and Computational Geometry* 42: 94–130. MR: Anton. Cited in 12.12 and 12.13.

Eves, Howard W. 1963–1965. *A Survey of Geometry.* Two volumes. Boston: Allyn and Bacon. MR: Kelly. Cited in 8.5.

Ewald, William B., editor. 1996. *From Kant to Hilbert: A Source Book in the Foundations of Mathematics.* Two volumes. Oxford: Clarendon Press. Contains Cantor [1883] 1996.

Feferman, Anita B. [1997] 1999. How the unity of science saved Alfred Tarski. In Woleński and Köhler 1999, 43–52. Cited in chapter 17.

———. 1999. Alfred Tarski. *American National Biography* 21: 330–332. Cited in chapter 17.

Feferman, Anita B., and Solomon Feferman. 2004. *Alfred Tarski: Life and Logic.* Cambridge, England: Cambridge University Press. MR: Dauben. Also reviewed in Anellis 2005, Apt 2007, Carnielli 2006, M. Davis 2005, P. Davis 2005, Maddux 2005, Nerode 2010, and Sinaceur 2007. Feferman and Feferman 2009 is a Polish translation. Contains Tarski [1924] 2004. Cited in the preface, 1.1, chapter 3, 4.1, 4.3, 9.1, 14.0, 14.2, ..., 14.4, 15.6, 15.11, 15.12, 16.1, 16.3, and chapter 17.

———. 2009. *Alfred Tarski: Życie i logika.* Translated by Joanna Golińska-Pilarek and Marian Srebrny. With preface by Jan Woleński. Warsaw: Wydawnictwa Akademickie i Profesjonalne. Polish translation of Feferman and Feferman 2004. Cited in the preface and chapter 17.

Feferman, Solomon. [1997] 1999. Tarski and Gödel: between the lines. In Woleński and Köhler 1999, 53–63. Cited in chapter 17.

———. 2003. Alfred Tarski and a watershed meeting in logic: Cornell, 1957. In Hintikka et al. 2003, 151–162. Cited in chapter 17.

———. 2006. Tarski's influence on computer science. *Logical Methods in Computer Science* 2(3): 1-1-1-13. Originally published in *Proceedings of the 20th Annual IEEE Symposium on Logic in Computer Science, 2005,* 342 ff. MR: S. Feferman. Cited in 8.8 and 18.2.

———. 2008. Tarski's conceptual analysis of semantical notions. In Patterson 2008, chapter 4, 72–93. Cited in 15.6 and 18.2.

Ford, Lester R., and Carver, Walter B. 1957. Laying a rug. *American Mathematical Monthly* 64: 114–116. Cited in 12.9.

Forsythe, Alexandra. 1969. What points are equidistant from two skew lines? *Mathematics Teacher* 62: 97–101. Cited in 12.13.

Frąckowiak, Danuta. 1998. Aleksander Jabłoński in the eyes of his daughter. In Szudy 1998, 19–40. Cited in chapter 3.

Frank, Phillip. 1935–1936. Prager Vorkonferenz der Internationalen Kongresse für Einheit der Wissenschaft—1934: Vorbemerkungen. *Erkenntnis* 5: 1–2. Cited in 14.2.

Fryde, Matthew M. 1964. Wacław Sierpiński—mathematician. *Scripta Mathematica* 27: 105–111. Cited in chapter 3.

Fuchs, Eduard. 1921. *Die Juden in der Karikatur: ein Beitrag zur Kulturgeschichte.* Munich: Albert Langen Verlag. Cited in 1.1.

*Fundamenta Mathematicae*. 1934 (volume 23, 161). Bemerkung der Redaktion. Cited in 9.4.

———. 1945 (volume 33, ix). Editorial note. This explains the gap in publication between the two parts of Tarski 1939–1945. Cited in 14.3.

Gabbay, Dov M., and John Woods. 2009. *Handbook of the History of Logic.* Volume 5: *Logic from Russell to Church.* Amsterdam: Elsevier. MR: Kahle. Contains Simmons 2009. Cited in 18.2.

Galavotti, Maria Carla. 2008. A tribute to Janina Hosiasson Lindenbaum: A philosopher victim of the Holocaust. In Scazzieri and Simili 2008, 179–194. Cited in 14.3.

García-Carpintero, Manuel. 1999. The explanatory value of truth theories embodying the semantic conception. In Peregrin 1999, 129–148. Cited in 15.11.

Gardner, Richard J., and Stan Wagon. 1989. At long last, the circle has been squared. *Notices of the American Mathematical Society* 36: 1338–1343. MR: Sah. Cited in 4.3.

Garlicki, Andrzej, et al. 1982. *Dzieje Uniwersytetu Warszawskiego 1915–1939.* Warsaw: Państwowe Wydawnictwo Naukowe. The title means *History of Warsaw University 1915–1939.* Cited in 1.1.

Garnett, F. M., and Walter B. Carver. 1925. Problem 3036 [1923, 337]. *American Mathematical Monthly* 32: 47–49. Cited in 12.9.

*Gazeta Warszawska.* 1920 (20 July: 3). Profesorowie na front. The title means "Professors on the front." The journal is a Warsaw daily newspaper. Cited in 1.1.

*Gazeta wyborcza.* 2008 (25 April: 3). Profesor Jarosław Rudniański. In Polish. The journal is a daily newspaper with national circulation. Cited in 14.1.

Gerwien, P. 1833a. Zerschneidung jeder beliebigen Anzahl von gleichen geradlinigen Figuren in dieselben Stücke. *Journal für die reine und angewandte Mathematik* 10: 228–234, 301. Cited in 4.1.

Gerwien, P. 1833b. Zerschneidung jeder beliebigen Menge von verschieden gestalteter Figuren von gleichem Inhalt auf der Kugelfläche in dieselben Stücke. *Journal für die reine und angewandte Mathematik* 10: 235–241, 303. Cited in 4.1 and 6.2.

Giedymin, Jerzy. 1978. Ajdukiewicz's life and personality. In Ajdukiewicz 1978, xiii–xvi. Reprinted in Sinisi and Woleński 1995, 10–12. Cited in chapter 3.

Givant, Steven R. 1986. Bibliography of Alfred Tarski. *Journal of Symbolic Logic* 51: 913–941. MR: anonymous. Also reviewed, with corrections, in Maddux 1993. Supersedes Tarski 1965a. Reprinted in Tarski 1986a, volume 4, 729–757. Updated and partially translated into Polish in Givant 1995. Cited in the preface, chapter 11, 12.0, the introduction to part four, 15.3, 15.4, 16.0, …, 16.2, and this bibliography's introduction.

———. [1988] 1991. Tarski's development of logic and mathematics based on the calculus of relations. In Andréka et al. 1991, 189–215. MR: Guillaume. Cited in 18.2.

———. 1991. A portrait of Alfred Tarski. *Mathematical Intelligencer* 13(3): 16–32. MR: Corcoran. Translated into Czech by Helena Nešetřilová as "Portrét Alfréda Tarského," *Pokroky matematiky, fysiky a astronomie* 37 (1992): 185–205. Givant and Huber-Dyson 1996 is an augmented Polish translation. Cited in 12.0, chapter 17, and the permissions section.

———. 1995. Bibliografia prac Alfreda Tarskiego. In Tarski 1995–2001, volume 1, 333–372. Translated and updated from Givant 1986, and trimmed to the narrower scope of Tarski 1995–2001. Cited in 16.2.

———. 1999. Unifying threads in Alfred Tarski's work. *Mathematical Intelligencer* 21(1): 47–58. MR: Murawski. Cited in 1.1, chapter 3, 8.0, 9.1, 9.3, 18.2, and the permissions section.

Givant, Steven R., and Verena Huber-Dyson. 1996. Alfred Tarski w kalejdoskopie impresji osobistych. Translated from English by Adelina Morawiec. *Wiadomości matematyczne* 32 (1996), 95–127. The title means "Alfred Tarski, a kaleidoscope of personal impressions." Givant 1991 stemmed from the English draft of this work. Cited in chapter 17.

Gödel, Kurt. [1931] 1967. On formally undecidable propositions of *Principia Mathematica* and related systems I. Translated by Jean van Heijenoort. In Van Heijenoort [1967] 1970, 592–617. There is no part II. Cited in 14.2 and 15.8.

Goldstein, Catherine, Jeremy Gray, and Jim Ritter, editors. 1996. *L'Europe mathématique: Histoires, mythes, identités*. Paris: Éditions de la Maison des Sciences de l'Homme. MR: Lewis. Contains Duda 1996.

Golińska-Pilarek, Joanna, Joanna Porębska-Srebrna, and Marian Srebrny. 2009a. Król logiki. *Rzeczpospolita* (4–5 April): A20–A21. The titles mean "King of logic" and *The Republic*. The journal is a daily newspaper with national circulation. Cited in 1.1 and chapter 17.

———. 2009b. *Szlakiem Alfreda Tarskiego po Warszawie*. Illustrated presentation, 1 April. The title means *Along the Route of Alfred Tarski in Warsaw*. Cited in 1.1 and chapter 17.

Gómez-Torrente, Mario. 2006. Alfred Tarski. In Zalta 2011– at http://plato.stanford.edu/entries/tarski/. Cited in 18.2.

Goodall, E. W. 1920. Typhus fever in Poland, 1916 to 1919. *Proceedings of the Royal Society of Medicine, Section of Epidemiology and State Medicine* 13: 277–289. Cited in 1.1.

Grattan-Guinness, Ivor. 1989. Review of Tarski [1956] 1983. *Journal of Symbolic Logic* 54: 281–282.

———. 2000. *The Search for Mathematical Roots 1870–1940: Logics, Set Theories and the Foundations of Mathematics from Cantor through Russell to Gödel*. Princeton: Princeton University Press. MR: Murawski. Cited in the foreword.

Greniewski, Henryk. [1927] 1928. Próba precyzacji pewnych pojęć teorji czynu. In Polski Zjazd Filozoficzny 1928, 157–160. The title means "An attempt to formulate precisely some notions of the theory of action." Contains the original version of the comment Tarski [1927] 2014a. Cited in 15.2.

Gromska, Daniela. 1948. Philosophes polonais morts entre 1938 et 1945. *Studia Logica* 3: 31–97. Cited in 14.2 and 14.3.

Gross, Feliks. 1969. Podróz na księżic—dlaczego? *Wiadomości* 24 (22 June, issue 1212): 3. The title means "Trip to the moon—why?" Cited in 16.3.

Grosse Brockhaus. 1928–1935. *Der Grosse Brockhaus: Handbuch des Wissens*. Fifteenth, fully revised edition. Twenty volumes. Leipzig: F. A. Brockhaus. Cited in item Grosse Brockhaus 1931.

———. 1931. Felix Hausdorff. In Grosse Brockhaus 1928–1935, volume 8, 239. In German. Cited in 4.4.

Gruszczyński, Rafał, and Andrzej Pietruszczak. 2008. Full development of Tarski's geometry of solids. *Bulletin of Symbolic Logic* 14: 481–540. MR: Pambuccian. Cited in 9.4.

Guillaume, Marcel. 1993. Review of Givant [1988] 1991. *Mathematical Reviews* 93: MR1153425 (93d: 03066). Cited in 18.2.

Gumański, Leon. 1984. Tadeusz Czeżowski: an advocate of scientific philosophy. Translated by Sylwia Twardo. *Dialectics and Humanism* 11: 657–664. Cited in 15.3.

Gupta, Haragauri N. 1965a. *Contributions to the Axiomatic Foundations of Geometry*. PhD dissertation. Berkeley: University of California. Cited in 12.11.

Haaparanta, Leila, editor. *The Development of Modern Logic*. Oxford: Oxford University Press. Contains Mancosu, Zach, and Badesa 2009.

Hadwiger, Hugo. 1957. *Vorlesungen über Inhalt, Oberfläche und Isoperimetrie*. Grundlehren der Mathematischen Wissenschaften 93. Berlin: Springer-Verlag. MR: Hewitt. Cited in 8.7.

*Harvard University Gazette*. 2000 (6 April). Adam Ulam, authority on Russia, dies at 77. Cited in 14.3.

Hatcher, William S. 1988. Review of Tarski [1944] 1987. *Mathematical Reviews* 88: MR876004 (88d: 03008). Cited in 16.3.

Hauber, Karl Friedrich. 1829. *Scholae Logico-Mathematicae, in Quibus Ars Cogitandi et Eloquendi, Inveniendi et Demonstrandi circa Unam Propositionem Quae Est Euclidis Elmentorum Theorema Primum, Multis Modis et Magna Exemplorum Varietate Exercetur, Proponuntur et Varia Generalia de Methodo, et Nova Quaedam Tum ad Logicam Theoreticam Pertinentia, Tum de Porismatibus in Analysi Geometrica Antiquorum*. Reutlingen, Württemberg: Officina Literaria. Cited in 9.9.

Hausdorff, Felix. 1914a. *Grundzüge der Mengenlehre*. Leipzig: Veit & Comp. JFM: Lampe. Reprinted in Hausdorff 2001–2002, volume 2, 90–476, with extensive notes and commentaries. Cited in 4.1, ..., 4.4, 6.0, 6.2, 6.3, and 15.12.

———. 1914b. Bemerkung über den Inhalt von Punktmengen. *Mathematische Annalen* 75: 428–433. Reprinted, with a 2001 commentary by Srishti D. Chatterji, in Hausdorff 2002, volume 3, 3–18. JFM: Lampe. Cited in 4.4.

———. 2001–2002. *Gesammelte Werke*, volume 2: *Grundzüge der Mengenlehre*, volume 4: *Analysis, Algebra, und Zahlentheorie*. Edited by Egbert Brieskorn et al. Berlin: Springer-Verlag. Contains Chatterji 2001, Purkert 2002, and reprints of Hausdorff 1914a and 1914b.

Heath, Thomas L. 1921. *A History of Greek Mathematics*. Two volumes. Oxford: Clarendon Press. Cited in chapter 11.

Heflich, Aleksander, and Stanisław Michalski, editors. 1915–1932. *Poradnik dla samouków*. Ten volumes. Warsaw: Gebethner i Wolff, Mianowski Fund. The title means *Handbook for Self-Education*. Supported by the Mianowski Fund, which took over publication after volume 2. Contains Kwietniewski 1923. Cited in 9.3.

Heller, Celia S. 1994. *On the Edge of Destruction: Jews of Poland between the Two World Wars*. Detroit: Wayne State University Press. Originally published in 1977. Cited in 1.1 and chapter 3.

Hempoliński, Michał, editor. 1996. *Truth after Tarski*. Dialogue and Universalism 6 (1–2): 160 pages. Contains J. Tarski 1996a. Cited in chapter 17.

Henkin, Leon, et al., editors. 1974. *Proceedings of the Tarski Symposium: An International Symposium Held to Honor Alfred Tarski on the Occasion of His Seventieth Birthday*. Proceedings of symposia in pure mathematics 25. Providence: American Mathematical Society. MR: Anonymous. Held at the University of California, Berkeley, 23–30 June 1971. Contains Chang [1971] 1974 and Vaught [1971] 1974. Cited in chapter 17.

Henkin, Leon, J. Donald Monk, and Alfred Tarski. 1971–1985. *Cylindric Algebras*. Two parts: *Part I with an Introductory Chapter, General Theory of Algebras*. Studies in logic and the foundations of mathematics 64 and 115. Amsterdam: North–Holland Publishing Company. Givant 1986 items 71$^m$ and 85$^m$. MR: Preller, Comer. Cited in 16.1.

———. 1986. Representable cylindric algebras. *Annals of Pure and Applied Logic* 31: 23–60. MR: Georgescu. Cited in 16.1.

Henkin, Leon, Patrick Suppes, and Alfred Tarski, editors. 1959. *The Axiomatic Method with Special Reference to Geometry and Physics: Proceedings of an International Symposium Held at the University of California, Berkeley, December 26, 1957– January 4, 1958.* Amsterdam: North–Holland Publishing Company, 1959. Contains Tarski [1957] 1959.

Hessenberg, Gerhard. 1906. *Grundbegriffe der Mengenlehre: zweite Bericht über das Unendliche in der Mathematik.* Göttingen: Vandenhoeck & Ruprecht. JFM: Meyer. Cited in chapter 3.

Hilbert, David. [1899] 1900. *Les principes fondamentaux de la géométrie.* Translated by Léonce Laugel. Paris: Gauthier-Villars. JFM: Engel. Translation of the first edition, revised, of Hilbert [1899] 1922. Cited in 9.3.

———. [1899] 1922. *Grundlagen der Geometrie.* Fifth, enlarged, edition. Wissenschaft und Hypothese 7. Leipzig: B. G. Teubner. The original 1899 edition is reprinted in Hilbert 2003, chapter 5. JFM: Engel, Dehn. Hilbert [1899] 1900 is a French translation of the original edition. Hilbert [1899] 1971 is an English translation of the tenth edition. Cited in 4.1, chapter 5, 9.3, and 12.11.

———. [1899] 1971. *Foundations of Geometry (Grundlagen der Geometrie).* Second edition. LaSalle, Illinois: Open Court. Translated by Leo Unger from the 1968 tenth German edition of Hilbert [1899] 1922, which was revised and enlarged by Paul Bernays and published in Stuttgart by B. G. Teubner. Cited in 9.3, 9.4, and 12.11.

———. [1900] 2000. Mathematical problems. Translated by Mary Winston Newson. *Bulletin of the American Mathematical Society* (new series) 37: 407–436. Reprinted from *Bulletin of the American Mathematical Society,* 8 (1902): 437–479. Originally published as "Mathematische Probleme," *Nachrichten von der Königlichen Gesellschaft der Wissenschaften zu Göttingen, Mathematisch-physikalische Klasse* 1900: 253–297, JFM: Wallenberg; reprinted in Hilbert [1932–1935], volume 3, 290–329. A shorter French version is included in International Congress of Mathematicians 1902, 58–114. Cited in 4.1.

———. [1932–1935] 1970. *Gesammelte Abhandlungen.* Three volumes. Second edition. Berlin: J. Springer. Contains a reprint of the original version of Hilbert [1900] 2000.

———. 2003. *David Hilbert's Lectures on the Foundations of Geometry, 1891–1902.* Edited by Michael Hallett and Ullrich Majer. Berlin: Springer-Verlag. Contains a reprint of the original edition of Hilbert [1899] 1922. Cited in 12.11.

Hilbert, David, and Wilhelm Ackermann. [1928] 1950. *Principles of Mathematical Logic.* Translated and edited by Lewis M. Hammond, George G. Leckie, F. Steinhardt, and Robert E. Luce. New York: Chelsea Publishing Company. JFM: Dürr. Originally published by Springer-Verlag as *Grundzüge der theoretischen Logic.* Cited in 15.4.

Hinkis, Arie. 2013. *Proofs of the Cantor–Bernstein Theorem: A Mathematical Excursion.* Science networks historical studies, 45. Basel: Birkhäuser. MR: Plotkin. Cited in 8.8.

Hintikka, Jaakko, editor. 1995. *From Dedekind to Gödel: Essays on the Development of the Foundations of Mathematics.* Synthese library 251. Dordrecht: Kluwer Academic Publishers Group. Contains Woleński 1995b. Cited in chapter 17.

Hintikka, Jaakko et al., editors. 2003. *Philosophy and Logic in Search of the Polish Tradition: Essays in Honour of Jan Woleński on the Occasion of his 60th Birthday.* Synthese library 323. Contains S. Feferman 2003. Dordrecht: Kluwer Academic Publishers. Cited in chapter 17.

Hiż, Henryk. 1971. Jubileusz Alfreda Tarskiego. *Kultura* 288: 134–140. The title means "Jubilee of Alfred Tarski." Cited in 9.3 and chapter 17.

Hodges, Wilfrid A. 1986. Alfred Tarski. *Journal of Symbolic Logic* 51: 866–868. Cited in chapter 17.

Holdener, Judy. 2009. Review of Sally and Sally 2007. *American Mathematical Monthly* 116: 754–758.

Hoormann, Cyril F. A., Jr. 1971. On Hauber's statement of his theorem. *Notre Dame Journal of Formal Logic* 12: 86–88. Cited in 9.9.

Horn, Alfred, and Alfred Tarski. 1948a. Measures in Boolean algebras. *Bulletin of the American Mathematical Society* 54: 79. Givant 1986 item 48[a]. Abstract of a presentation at an American

Mathematical Society meeting in Pasadena, California, 29 November 1947. Reprinted in Tarski 1986a, volume 4, 574. This does not cover the material in Horn and Tarski 1948b that is relevant to the present book. Cited in 8.7.

———. 1948b. Measures in Boolean algebras. *Transactions of the American Mathematical Society* 64: 467–497. MR: Loomis. Givant 1986 item 48c. Reprinted in Tarski 1986a, 199–232. Horn and Tarski 1948a is an abstract for part of this. Cited in 8.7.

Hosiassonówna, Janina. 1928. Der II. polnische Philosophenkongress (Warschau, 23.–28. September 1927). *Archiv für systematische Philosophie und Soziologie* (new series) 31: 98–103. Polski Zjazd Filozoficzny 1928 is the congress proceedings. Cited in 14.2.

Hübner, Piotr, Jan Piskurewicz and Leszek Zasztowt. 1992. Józef Mianowski Fund: a foundation for the promotion of science, established 1881. History. Translated by Jacek Soszyński. On the Internet at http://www.mianowski.waw.pl/. Originally published as "Zarys historii Kasy im. Józefa Mianowskiego," in *Kasa imienia Józefa Mianowskiego Fundacja Popierania Nauki 1881–1991*, Warsaw: Kasy im. J. Mianowskiego, 7–54. Cited in 9.3.

Hughes-Hallett, Deborah, et al. 2003. *Applied Calculus*. Second edition. Hoboken, New Jersey: John Wiley & Sons. Cited in 15.5.

Huntington, Edward V. 1904. Sets of independent postulates for the algebra of logic. *Transactions of the American Mathematical Society* 5: 288–309. JFM: Jahnke. Cited in 15.6.

Hutnikiewicz, Artur, editor. 1965. Witwicki Tadeusz. *Sprawozdania Towarzystwa Naukowego w Toruniu* 17 (1963): 53–55. In Polish. The journal title means *Reports of the Toruń Scientific Society*. The article was anonymous. Cited in 15.6.

Infeld, Leopold. 1924. O dowodzie tw. Talesa w szkole średniej. *Przegląd matematyczno-fizyczny* 2: 87–89. The title means "On the proof of Thales's theorem in secondary school." Cited in 4.3.

———. 1980. *Quest: An Autobiography*. Second edition. New York: Chelsea Publishing Company. The first edition was published in 1941. Cited in chapter 3 and 4.3.

Institut International de Collaboration Philosophique. 1939. *Les conceptions modernes de la raison: Entretiens d'été, Amersfoort 1938*. Volume 1: *Raison et monde sensible*. Edited by Raymond Bayer. Actualités scientifiques et industrielles 849. Paris: Hermann & Cie. Contains Barzin [1938] 1939 and Tarski [1938] 1939.

International Congress for the Unity of Science. 1936. *Actes du Congrès International de Philosophie Scientifique, Sorbonne, Paris, 1935*. Eight volumes. Actualités scientifiques et industrielles 388–395. Paris: Hermann & Cie. Contains Tarski [1935] 1936a and [1935] 1936b.

International Congress of Mathematicians. 1902. *Compte rendu du Deuxième Congrès International des Mathématiciens tenu à Paris du 6 au 12 août 1900. Procès-verbaux et communications*. Edited by Ernest Duporcq. Paris: Gauthier-Villars. Contains a French translation of part of the original version of Hilbert [1900] 2000.

———. 1930. *Atti del Congresso Internazionale dei Matematici, Bologna, 3–10 settembre 1928*, six volumes. Bologna: Nicola Zanichelli Editore. Contains Tarski [1928] 1930.

International Congress of Philosophy. 1900–1903. *Bibliothèque du Congrès International de Philosophie*. Four volumes. Paris: Librairie Armand Colin. Contains Padoa [1900] 1901.

———. 1936. *Actes du Huitième Congrès International de Philosophie à Prague, 2–7 septembre 1934*. Prague: Comité d'Organisation du Congrès. Reprinted in 1968 in Nendeln, Liechtenstein, by Kraus Reprint Limited. Contains Jørgensen [1934] 1936, Reichenbach [1934] 1936, Tarski [1934] 1936a and [1934] 1936b, and Zawirski [1934] 1936. Cited in 16.1.

———. 1937. *Travaux du IX$^e$ Congrès International de Philosophie: Congrès Descartes*. Twelve fascicles. Edited by Raymond Bayer. Actualités scientifiques et industrielles 530–541. Paris: Hermann & Cie. Contains Tarski 1937b.

Iwaszkiewicz, Bolesław. 1935. *Geometrja dla III klasy gimnazjalnej.* Lwów: Wydawnictwo Zakładu Narodowego Imienia Ossolińskich. The title means *Geometry for the Third Year of Gimnazjum.* Cited in 9.9.

———. 1949a. Od Redakcji. *Matematyka: Czasopismo dla nauczycieli* 2: 6–7. The article and journal titles mean "From the Editor" and *Mathematics: A Journal for Teachers.* The article was unsigned; Iwaszkiewicz was the journal editor. Marxist commentary on Jaśkowski 1949. Cited in 9.7.

———. 1949b. Rozwiązania zadań z zeszytu nr. I. Zadania z *Parametra. Matematyka: Czasopismo dla nauczycieli* 2(4): 39–42. The article and journal titles mean "Solutions of exercises in volume 1. Exercises from *Parametr*" and *Mathematics: A Journal for Teachers.* Contains a solution of Tarski [1930] 2014c. This material was unsigned. Cited in 9.7, 12.0, and 12.3.

Jadacki, Jacek Juliusz. 2003a. Alfred Tarski à Varsovie. In Jadacki 2003b, 139–180. Translation of "Alfred Tarski w Warszawie" by Wanda Jadacka, ibid., 112–137. Cited in 1.1, chapter 3, 9.1, 9.3, 9.4, 14.1, ..., 14.3, 15.5, and chapter 17.

---

Jadacki, Jacek Juliusz, editor. 2003b. *Alfred Tarski: dedukcja i semantyka (déduction et sémantique). Materiały Sympozjum Instytutu Filozofii Uniwersytetu Warszawskiego, Polskiego Towarzystwa Filozoficznego i Towarzystwa Naukowego Warszawskiego odbytego 15 stycznia 2001 roku w Sali Lustrzaney Pałacu Staszica w Warszawie, Nowy Świat 72 (I piętro) z okazji senej rocznicy urodzin* ALFREDA TARSKIEGO *(14 I 1901, Warszawa—27 X 1983, Berkeley).* Studia semiotica 25–26. Warsaw: Wydawnictwo Naukowe Semper. This contains

- Jadacki 2003a; Tarski et al. [1971] 2003;
  Tarski [1933] 2003, [1937] 2003, [1939] 2003, [1947–1955] 2003, [1963] 2003;

- Polish translations Tarski [1934] 2003a, [1934] 2003b, with reprints of their original versions;

- and reprints of the original versions of
  Łukasiewicz [1928–1929] 2014 (in part);
  Tarski [1925] 2014, [1927] 2014a, [1927] 2014a, [1934] 2014,
      [1935] 2014, [1936] 2014a, ..., [1936] 2014c.

Cited in 3.1, 14.2, 14.3, 16.1, 16.3, chapter 17, and 18.2.

---

Jadacki, Jacek Juliusz, and Barbara Markiewicz, editors. 1993. *A mądrości zło nie przemoże.* Warsaw: Polskie Towarzystwo Filozoficzne. The title means *Evil Never Overpowers Wisdom.* Cited in 15.11.

Jadczak, Ryszard. 1993a. Lwowski i Poznański okres w działalności naukowej Zygmunta Zawirskiego. *Edukacja filozoficzna* 16: 197–205. The title means "The Lwów and Poznań period in the research of Zygmunt Zawirski." Cited in 15.10.

———. 1993b. Pozycja Stanisława Leśniewskiego w Szkole Lwowsko-Warszawskiej (Polskie Towarzystwo Filozoficzne, oddział w Toruniu, 9 grudnia 1992 r.). *Ruch filozoficzny* 50: 311–316. The title means "The position of Stanisław Leśniewski in the Lwów-Warsaw school (Polish Philosophical Society, Toruń section, 9 December 1992). Cited in 1.1.

Jahnke, Hans Niels, editor. 2003. *A History of Analysis.* History of Mathematics 24. Providence and London: American Mathematical Society and London Mathematical Society. MR: Barbeau. Cited in the foreword.

Jakimowicz, Emilia, and Adam Miranowicz, editors. 2007. *Stefan Banach: Remarkable Life, Brilliant Mathematics*. Gdańsk: Gdańsk University Press. Cited in 4.2.

Jakubowski, J. L. 1964. Docent Doktor Roman Hampel. In Śmigielski 1964, 335–339. In Polish. Cited in 9.1.

James, Ioan Mackenzie. 2009a. *Driven to Innovate: A Century of Jewish Mathematicians and Physicists*. Witney, Great Britain: Peter Lang. MR: Lewis. Also reviewed in Siegmund-Schultze 2010. Contains James 2009b. Cited in chapter 17.

———. 2009b. Napoleon of logic: Alfred Tarski (1901–1983). In James 2009a, 284–288. Cited in chapter 17.

Janiszewski, Zygmunt. [1918] 1968. The needs of mathematics in Poland. Translated by Mary Grace Kuzawa. In Kuzawa 1968, appendix, 112–118. The original paper, "O potrzebach matematyki w Polsce," was published in *Nauka polska* 1: 11–18. Cited in 1.1 and 4.4.

Jaroszyńska-Kirchmann, Anna D. 2002. Patriotism, responsibility, and the Cold War: Polish schools in DP camps in Germany, 1945–1951. *Polish Review* 47: 35–66. Cited in 9.9.

———. 2004. *The Exile Mission: The Polish Political Diaspora and Polish Americans, 1939–1956*. Cited in 9.9.

Jaśkowski, Stanisław. 1948. Une modification des définitions fondamentales de la géométrie des corps de M. A. Tarski. *Annales de la Société Polonaise de Mathématique* 21: 298–301. MR: Blumenthal. Cited in 9.4.

———. 1949. Geometria brył. *Matematyka: Czasopismo dla nauczycieli* 2: 1–6. The titles mean "Geometry of solids" and *Mathematics: A Journal for Teachers*. Iwaszkiewicz 1949a is a Marxist commentary on this. Cited in 9.4 and 9.7.

Jastrzębowski, Wojciech. 1929. Powszechna Wystawa Krajowa. Poster for the General Polish Exposition, Poznań, May–September. Bydgoszcz, Poland: Zakłady Graficzne "Bibljoteka Polska". Cited in 9.5.

Jedynak, Anna. 2001. Janina Hosiasson-Lindenbaum: The Logic of Induction. In Krajewski 2001, 97–101. Cited in 14.3.

Jerzmanowski, Andrzej. 2013. Personal communication to James T. Smith, 2 June. Cited in 9.9.

Jónsson, Bjarni. 1986. The contributions of Alfred Tarski to general algebra. *Journal of Symbolic Logic* 51: 883–889. MR: Hawkins. Cited in 18.1.

Jónsson, Bjarni, and Alfred Tarski. 1947. *Direct Decompositions of Finite Algebraic Systems*. Notre Dame mathematical lectures 5. Notre Dame: University of Notre Dame Press. Givant 1986 item $47^m$. MR: Birkhoff. Cited in 15.12.

Jørgensen, Jørgen. [1934] 1936. Die logischen Grundlagen der Wissenschaft. In International Congress of Philosophy 1936, 100–116. Discussed in Tarski [1934] 1936b and [1934] 2003b. Cited in 16.1.

Kac, Mark. 1931. O nowym sposobie rozwiązywania równań stopnia trzeciego. *Młody matematyk* 1 (1931–1932): 69–71. Kac was then a gimnazjum student; the journal listed his name as Marek Katz. The title means "A new method for solving third-degree equations." Cited in 9.7.

———. 1985. *Enigmas of Chance: An Autobiography*. New York: Harper & Row, Publishers. MR: Hobbs. Cited in chapter 3 and 9.7.

Kałuża, Roman. 1996. *Through a Reporter's Eyes: The Life of Stefan Banach*. Translated and edited by Ann Kostant and Wojbor Woyczyński. Boston: Birkhäuser. MR: Cooke. Cited in 2.1.

Kalicun-Chodowicki, Bazyli. 1925. *Geometrja wykreślna: dla wyższych klas szkół średnich (kurs systematyczny, według programu MWRiOP*. Lwów: Księgarnia Naukow. The title means *Descriptive Geometry for the Higher Classes of Secondary Schools (A Systematic Course Following the Program of the Ministry of Religious Denominations and Public Education)*. Cited in 9.2.

Kaptur, Leon. 1930. IV zadanie konkursowe: Dobór parametra. *Parametr* 1: 397. The title means, "Fourth contest problem: selection of a parameter." Cited in 12.1.

Karski, Jan. 1944. *Story of a Secret State*. Boston: Houghton Mifflin Company 1944. Cited in the preface and 9.9.

Kasperowiczowa, Helena. 1969. Warszawskie szkoły średnie ogólnokształcące w latach 1915–1944. *Przegląd historyczno-oświatowy* 12(2): 187–202. The titles mean "Warsaw general secondary schools in the years 1915–1944" and *Historical-educational review*. Cited in 9.1.

Keesling, James. 1990. Karol Borsuk. In *Dictionary of Scientific Biography* 1970–, volume 17, supplement 2: 93–95. Cited in 15.12.

Kępno. Państwowe Gimnazjum Koedukacyjne. 1931–1933. *Sprawozdanie dyrekcji Państwowego Gimnazjum Koedukacyjnego w Kępnie*. Volumes for academic years 1929/1930–1930/1931, 1931/1932, and 1932/1933. Edited by Henryk Moese. Kępno: Drukarnia Dziennika Poznańskiego Spółka Akcyjna, Drukarnia Spółkowa w Kępnie. The title means *Report of the directorate of the National Coeducational Gimnazjum in Kępno*. Cited in 7.1.

Keynes, John Maynard. 1921. *A treatise on probability*. London: Macmillan and Company. Cited in 16.1.

Kiselev, Andrej Petrovič. 1917. *Gieometrja elementarna: do użytku szkół średnich*. Two volumes. Translated by Bolesław Herman. Łódź: Księgarnia Ludwika Fiszera. Originally published in 1892 as Элементарная геометрия, для среднихъ учебныхъ заведений, Moscow: Dumnov's Bookstore (В. В. Думнова). The title means *Elementary Geometry for Use in Secondary Schools*. Kiselev 2006–2008 is an English translation. Cited in 9.2.

———. 2006–2008. *Kiselev's Geometry*. Book 1: *Planimetry*. Book 2: *Stereometry*. Adapted from a later edition of Kiselev 1917 by Alexander Givental. El Cerrito, California: Sumizdat. Cited in 9.2 and 12.11.

Knaster, Bronisław. 1927. Un théorème sur les fonctions d'ensembles. *Annales de la Société Polonaise des Mathématique* 6: 133–134. Givant 1986 item 27. JFM: Hurewicz. Reprinted in Tarski 1986a, 546–547. Abstract of a 9 December talk to the Warsaw section of the Polish Mathematical Society presenting joint results of Knaster and Tarski. Cited in 8.8.

———. 1973. Janiszewski, Zygmunt. *Dictionary of Scientific Biography* 7: 71–73. Cited in 1.1.

Kokoszyńska, Maria. 1936a. W sprawie względności i bezwzględności prawdy. In Polski Zjazd Filozoficzny 1936, 424–426. The title means "Concerning the relativity and nonrelativity of truth." Contains the original version of the comment Tarski [1936] 2014d. Cited in 8.1 and 15.11.

———. 1936b. Über den absoluten Wahrheitsbegriff und einige andere semantische Begriffe. *Erkenntnis* 6 (December): 143–165. Cited in 15.11.

———. 1937. Wrażenia z Pierwszego Międzynarodowego Kongresu Filozofii Naukowej (Paryż, 15.–23. września 1935). *Ruch filozoficzny* 13: 77–76. The title means "Impressions of the First International Congress of Scientific Philosophy (Paris, 15–23 September 1935). Cited in 14.2.

———. c1961. Résumé. Item RC 088-57-07 in the Carnap Collection of the library of the University of Pittsburgh. Cited in 14.2.

Komjáth, Péter. 2006. Review of Wapner 2005. *Notices of the American Mathematical Society* 53: 1035–1036.

Konarski, Stanisław. 1971. Warszawskie szkolnictwo powszechne w latach 1918–1939. In Drozdowski 1968–1973, volume 3, 215–246. The title means "Warsaw elementary schools in the years 1918–1939." There is a preprint of this dated 1966. Cited in 9.1.

———. 1973. Warszawskie średnie szkolnictwo ogólnokształcące w latach 1918–1939. In Drozdowski 1968–1973, volume 5, 179–252. The title means "Warsaw general secondary schools in the years 1918–1939." Cited in 9.1.

——— 1994. Schayer Stanisław. *Polski słownik biograficzny* 35: 411–412. In Polish. Cited in 9.9.

Kordos, Marek, Maria Moszyńska, and Lesław W. Szczerba. 1977. Wanda Szmielew 1918–1976. Translated by J. Smólska. *Studia Logica* 36: 241–244. Originally published in Polish in *Wiado-*

*mości Matematyczne* (2) 21 (1978): 25–28. Establishes the date of the work in Szmielew [1938] 1947. Cited in 9.4 and 14.2.

Kotarbiński, Tadeusz. [1935] 1955–1956. The fundamental ideas of pansomatism. *Mind* 64: 488–500, 65: 288. Translated by Alfred Tarski and David Rynin. Givant 1986 item 56. Originally published as "Zasadnicze myśli pansomatyzmu," *Przegląd filozoficzny* 38, 283–284. Reprinted in Tarski 1986a, volume 3, 577–592. Cited in chapter 17.

Kotarbiński, Tadeusz. [1955] 1965. *Praxiology: An Introduction to the Sciences of Efficient Action*. Translated by Olgierd Wojtasiewicz. Oxford: Pergamon Press. The original edition was published in Wrocław as *Traktat o dobrej robocie* by the Zakład Narodowy im. Ossolińskich Wydawnictwo Polskiej Akademii Nauk. Cited in 15.2.

———. 1966. Garstka wspomnień o Stanisławie Leśniewskim. *Ruch filozoficzny* 24: 155–163. The title means "A handful of memories about Stanisław Leśniewski." A typewritten English translation has been circulated but evidently not published. Cited in 1.1.

Koterski, Artur. 2001. Henryk Mehlberg—*The Reach of Science*. In W. Krajewski 2001, 121–127. Cited in 15.3.

———. 1977. Philosophical self-portrait. *Dialectics and Humanism* 4: 22–26. Cited in 1.1.

Kötter, Fritz. 1891. Review of Réthy 1891. *Jahrbuch über die Fortschritte der Mathematik* 23: review 23.0532.01. Cited in 4.3.

Kozłowski, Witold. 2003. Wspomnienie. Alfred Tarski (14 1 1901–27 X 1983). *Gazeta wyborcza* 264 (November 13): 11. The first word in the title means "A memory of." The journal is a daily newspaper with national circulation. Cited in 14.1 and chapter 17.

———. 2010. Personal interview with Andrew and Joanna McFarland, 6 February, in Warsaw. Cited in 14.1 and 14.3.

Krajewski, Stanisław. 2004. Gödel on Tarski. *Annals of Pure and Applied Logic* 127: 303–323. Cited in chapter 17.

Krajewski, Stanisław, and Marian Srebrny. 2008. On the life and work of Andrzej Mostowski (1913–1975). In Ehrenfeucht, Marek, and Srebrny 2008, 3–14. MR: Surma. Cited in 9.4.

Krajewski, Władysław, editor. 2001. *Polish Philosophers of Science and Nature in the 20th Century*. Amsterdam: Rodopi. Contains Jedynak 2001 and Koterski 2001.

Królikowski, Lech. 1991. Rusiecki Antoni Marian (1892–1956). *Polski słownik biograficzny* 33(1): 125–126. In Polish. Cited in 9.7.

Krysicki, Włodzimierz. 1930. List do Redakcji dotyczący zadania Nr. 1. *Parametr* 1: 196–197. The title means "Letter to the Editor about exercise 1." That exercise is Rusiecki 1930a. Cited in 12.9.

Księgarnia Św. Wojciecha. 1930. Słowo wstępne. *Parametr* 1(1930): 1. The author name and title mean "St. Wojciech's Bookstore" and "Foreword." St. Wojciech is often referred to as St. Adalbert. Cited in 9.7.

Kuratowski, Kazimierz. 1924. Une propriété des correspondances biunivoques. *Fundamenta Mathematicae* 6: 240–243. JFM: Rosenthal. Cited in 4.4, 6.1, and 8.8.

———. 1958. *Topologie*. Volume 1. Fourth edition, corrected and augmented, with an appendix and two notes by Andrzej Mostowski and Roman Sikorski. Monografie matematyczne 20. Warsaw: Państwowe Wydawnictwo Naukowe. MR: Roberts. Cited in 4.4.

———. 1974. Wacław Sierpiński. In Sierpiński 1974–1976, volume 1, 9–14. In French. Cited in chapter 3.

———. 1980. *A Half Century of Polish Mathematics: Remembrances and Reflections*. Translated by Andrzej Kirkor. Warsaw: Polish Scientific Publishers. MR: Mendelson. The original Polish edition was published in 1973. Cited in 1.1, chapter 3, and 4.4.

———. 1981. *Notatki do autobiografii*. Warsaw: Czytelnik. Cited in 1.1.

Kurzawa, Jan, and Stanisław Nawrocki. 1978. *Dzieje Kępna*. Warsaw: Państwowe Wydawnictwo Naukowe, oddział w Poznaniu. The title means "History of Kępno." Cited in 7.1.

Kutta, Janusz. 1997. Pasenkiewicz Kazimierz (1897–1995). In Błażejewski et al. 1997, volume 4, 81–83. In Polish. Cited in chapter 3.

Kuzawa, Mary Grace. 1968. *Modern Mathematics: The Genesis of a School in Poland*. New Haven, Connecticut: College and University Press. Contains Janiszewski [1918] 1968. Cited in 9.3.

Kwietniewski, Stefan. 1923. Review of W. Wojtowicz [1919] 1926. In Heflich and Michalski 1915–1932, volume 3, 19. Cited in 9.2.

Kwietniewski, Stefan, and Władysław Wojtowicz. 1914. Od tłumaczów. In Enriques [1900] 1914–1917, volume 1, i. The title means "From the translators." Cited in 9.3.

Laczkovich, Miklós. 1990. Equidecomposability and discrepancy: a solution of Tarski's circle-squaring problem. *Journal für die reine und angewandte Mathematik* 404: 77–117. MR: Wagon. Cited in 4.3 and 8.6.

Langford, C. H. 1926. Some theorems on deducibility. *Annals of Mathematics* 28: 16–40. Cited in 18.2.

———. 1927. Theorems on deducibility. *Annals of Mathematics* 28: 459–71. Cited in 18.2

———. 1938. Review of Tarski 1937b. *Journal of Symbolic Logic* 3: 56. Cited in 14.2.

Lapointe, Sandra, et al., editors. 2009. *The Golden Age of Polish Philosophy: Kazimierz Twardowski's Philosophical Legacy*. Dordrecht: Springer. Contains Mancosu 2009. Cited in 18.2.

Lebesgue, Henri. 1904. *Leçons sur l'intégration et la recherche des fonctions primitives, professées au Collège de France*. Paris: Gauthier-Villars. JFM: Stäckel. Cited in 4.2.

———. 1922. Apropos d'une nouvelle revue mathématique: *Fundamenta Mathematicae*. *Bulletin des sciences mathématiques* (2) 46: 35–48. JFM: Lichtenstein. Cited in the foreword.

Leja, Franciszek, editor. 1930. *Comptes-rendus du I Congrès des Mathématiciens des Pays Slaves, Warszawa, 1929*. Warsaw: Książnica Atlas. Contains Sierpiński [1929] 1930 and the original version of Presburger [1929] 1991. Cited in 9.5 and chapter 10.

Leśniewski, Stanisław. [1911] 1992. A Contribution to the Analysis of Existential Propositions. Translated by Stanisław J. Surma and Jan W. Wójcik. In Leśniewski 1992, volume 1, 1–19. Originally published as "Przyczynek do analizy zdań egzystencjalnych," *Przegląd filozoficzny* 14 (1911), 329–345. Edited version of Leśniewski's doctoral dissertation. Cited in 1.1.

———. [1927–1930] 1992. On the foundations of mathematics. Translated by D. I. Barnett. In Leśniewski 1992, volume 1, 174–382. Originally published as "O podstawach matematyki," *Przegląd filozoficzny* 30 (1927): 164–206; 31 (1928): 261–322; 32 (1929): 60–101; 33 (1930): 77–105. Cited in chapter 3 and 14.2.

———. [1929] 1992. Fundamentals of a new system of the foundations of mathematics. Translated by Michael P. O'Neil. In Leśniewski 1992, volume 2, 410–605. Originally published as "Grundzüge eines neuen Systems der Grundlagen der Mathematik," *Fundamenta Mathematicae* 14: 1–81, JFM: Skolem. Cited in chapter 3.

———. 1992. *Collected Works*. Two volumes. Edited by Stanisław J. Surma, Jan T. J. Srzednicki, D. I. Barnett, and V. Frederick Rickey. Nijhoff international philosophy series 44. Dordrecht: PWN-Polish Scientific/Kluwer Academic Publishers. Reviewed in Woleński 2000–2001. Contains Leśniewski [1911] 1992, [1927–1930] 1992, and [1929] 1992. Cited in 1.1.

Levin, Aleksey E. 1988. Expedient catastrophe: a reconsideration of the 1929 crisis at the Soviet Academy of Science. *Slavic Review* 47: 261–279. Cited in 9.5.

Lévy, Azriel. 1979. *Basic Set Theory*. Berlin: Springer-Verlag. MR: Kanamori. Cited in 15.12.

———. 1988. Alfred Tarski's work in set theory. *Journal of Symbolic Logic* 53: 2–6. MR: Surma. Cited in 18.1.

Lindenbaum, Adolf. [1937] 1938. Sur l'équivalence de deux figures par décomposition en nombre fini de parties respectivement congruents. In Congrès Polonais de Mathématiques [1937] 1938, 197. Abstract. Cited in 7.1.

Lindenbaum, Adolf, and Alfred Tarski. 1926. Communication sur les recherches de la théorie des ensembles. *Comptes rendus de la Société des Sciences et des Lettres de Varsovie, classe III* 19: 299–330. JFM: Ruziewicz. Givant 1986 item 26. Reprinted in Tarski 1986a, volume 1, 171–204. Cited in 8.3, 8.6, and 14.2.

———. [1935] 1983. On the limitations of the means of expression of deductive theories. In Tarski [1956] 1983, 384–392. The original paper, "Über die Beschränktheit der Ausdrucksmittel deduktiver Theorien," *Ergebnisse eines mathematischen Kolloquiums* 7: 15–22, Givant 1986 item 36b, JFM: Bachmann, was reprinted in Menger et al. 1998, 333–340, and Tarski 1986a, volume 2, 203–212, and translated into Polish in Tarski 1995–2001, volume 2, pages 147–157. Cited in 16.2.

Lindström, Sten, et al., editors. 2009. *Logicism, Intuitionism, and Formalism: What Has Become of Them?* Synthese library 341. New York: Springer. Contains Sinaceur 2009. Cited in chapter 17.

Lobachevsky, N. I. [1840] 1914. *Geometrical Researches on the Theory of Parallels.* Translated by George Bruce Halsted. Chicago: Open Court Publishing Company. Originally published by G. Fincke in Berlin as *Geometrische Untersuchung zur Theorie der Parallellinien.* Reprinted in Bonola [1906] 1955.

Łódź. Rząd Miejski. 1946. Świadczenie o czasowym pobycie na terenie m. Łodzi: Szmielew Wanda ur. Montlak. Document K. 7175, Zespół Archiwalny Akt Pracowniczych, Archiwum Uniwersytetu Warszawskiego. The author name, title phrase, and source names mean "Municipal administration of Łódź," "Certification for temporary residence in Łódź," and "Archival Unit for Personnel Records of the University of Warsaw." Cited in 9.4.

Łomnicki, Antoni. 1919. O układach zasad koniecznych i dostatecznych, służących do definicyi pojęcia wielkości. (Sur les systèmes de principes, nécessaires et suffisants servant à la définition de la notion de grandeur.) *Wiadomości matematyczne* 23: 37–70. Cited in 4.2.

———. 1922. O zasadzie dysjunkcyi w logistyce i matematyce. *Ruch filozoficzny* 6 (1921–1922): 144–146. Presented before the Lwów section of the Polish Philosophical Society on 21 January 1922. The title means, "On the principle of disjunction in logic and mathematics." Cited in 4.2.

———. 1923. *Geometrja: Podręcznik dla szkół średnich.* Volume 1: *Planimetrja-stereometrja.* Lwów: Książnica Polska Towarzystwa Nauczycieli Szkół Wyższych. The title phrase and publisher's names mean *Textbook for Secondary Schools* and "Library of the Polish Society of Teachers in Higher Schools." Revision of an edition published in Lwów in 1911. Cited in 4.2, 9.2, and 12.11.

———. 1930. *Tablice matematyczno-fizyczne czterocyfrowe, dla użytku szkół średnich.* Second edition. Lwów: Książnica-Atlas. The title means *Four-digit matematical and physical tables for use in secondary schools.* The first edition was published in 1926. Cited in chapter 11.

———. 1934. *Geometrja dla II klasy gimnazjalnej.* Lwów: Książica-Atlas. The title means *Geometry for the Second Year of Gimnazjum.* Cited in 9.9, 13.1, and 13.4.

———. 1935. *Geometrja dla III klasy gimnazjalnej.* Lwów: Książica-Atlas. Cited in 9.9.

Lorenz, Johann Friedrich. 1798–1807. *Grundriss der reinen und angewandten Mathematik, oder der erste Cursus der gesamten Mathematik.* Two volumes. Volume I: *Der erste cursus der Arithmetik und Geometrie.* Volume II: *Der erste cursus der angewandten Mathematik.* Helmstädt: Carl Gottfried Gleckeisen. Cited in 12.11.

Łoś, Jerzy. 1986. O Alfredzie Tarskim. *Ruch filozoficzny* 43: 3–10. Cited in 18.2.

Lowry, John, and William Wallace. 1814. Question 269 and solution. *New Series of the Mathematical Repository* 3: 44–46. Cited in 4.1.

Luchins, Abraham S., and Edith H. Luchins. 2000. Kurt Grelling: steadfast scholar in a time of madness. *Gestalt Theory* 22: 228–281. Cited in 14.3.

Luciano, Erika, and Clara Silvia Roero. 2010. La scuola di Peano. In Roero 2010, xi–xviii, 1–212. Cited in the foreword.

Łukasiewicz, Jan. 1910. *O zasadzie sprzeczności u Arystotelesa: Studyum krytyczne.* Cracow: Akademia Umiejętności. The title means *The Principle of Contradiction in Aristotle: A Critical Study.* Cited in 9.4.

———. 1916. O pojęciu wielkości. *Przegląd filozoficzny* 19: 1–70. The title means "On the concept of magnitude." Cited in 4.2.

———. 1925. O pewnym sposobie pojmowania teorji dedukcji. *Przegląd filozoficzny* 28: 134–136. Contains the original version of the comment Tarski [1925] 2014. The title means "On a certain way of understanding the theory of deduction." Cited in 8.1 and 15.1.

———. [1927] 1928. O metodę w filozofji. In Polski Zjazd Filozoficzny 1928, 3–5. The title means "On method in philosophy." Cited in 14.2, 15.2, and 15.3.

———. [1928–1929] 2014. On definitions in the theory of deduction. In section 15.4 of the present book, Tarski 2014. Translation of "O definicyach w teoryi dedukcyi," *Ruch filozoficzny* 11: 177–178, Givant 1986 item 28/29. This abstract announced some results of Tarski, which are reprinted in Jadacki 2003b, 10. Cited in 8.1, 15.0, and 15.4.

———. 1929. O znaczeniu i potrzebach logiki matematycznej. *Nauka polska* 10: 604–620. The title means "On the significance and needs of mathematical logic." Cited in 15.1.

———. [1929] 1963. *Elements of Mathematical Logic*, translated by Olgierd Wojtasiewicz, International series of monographs on pure and applied mathematics 31. New York: Macmillan Company. Originally published as *Elementy logiki matematycznej*, edited by Mojżesz Presburger, publication 18, by the Publishing Commission of the Mathematics and Physics Students of the University of Warsaw. Cited in 15.1, 15.4, and 9.4.

———. [1931] 1970. Comments on Nicod's axiom and on "generalizing deduction." Translated by Olgierd A. Wojtasiewicz. In Łukasiewicz 1970, 179–196. Originally published as "Uwagi o aksjomacie Nicoda i 'dedukcji uogólniającej,'" in Polskie Towarzystwo Filozoficzne we Lwowie 1931, 366–382. The translator is not identified. Cited in 15.1.

———. 1936. Co dała filozofii logika matematyczna? In Polski Zjazd Filozoficzny 1936, 325–329. The title means "What has mathematical logic given to philosophy?" Cited in 15.10.

———. 1936 [1970]. Logistic and philosophy. Translated by Olgierd A. Wojtasiewicz. In Łukasiewicz 1970, 218–235. Originally published as "Logistyka a filozofja," *Przegląd filozoficzny* 39: 115–131. Cited in 15.10.

———. [1951] 1957. *Aristotle's Syllogistic from the Standpoint of Modern Formal Logic.* Second, enlarged edition. MR: Martin. Cited in 9.4.

———. [1953] 1994. Curriculum vitae of Jan Łukasiewicz. *Metalogicon* 7: 133–137. MR: Anonymous. Incorrectly reprinted in B. Sobociński 1956. Cited in 1.1 and 9.4.

———. 1970. *Selected Works.* Edited by Ludwik Borkowski. Amsterdam: North–Holland Publishing Company. MR: anonymous. Contains Łukasiewicz [1931] 1970, [1936] 1970, and a reprint of Łukasiewicz and Tarski [1930] 1983.

Łukasiewicz, Jan, and Alfred Tarski. [1930] 1983. Investigations into the sentential calculus. In Tarski [1956] 1983, chapter 4, 38–59. Givant 1986 item 30d. Translation of "Untersuchungen über den Aussagenkalkül," *Comptes rendus de la Société des Sciences et des Lettres de Varsovie, classe III*, 23: 30–50, reprinted in Tarski 1986a, volume 1, 321–344. This translation was reprinted in Łukasiewicz 1970, 131–152. The original paper was translated into Polish in Tarski 1995–2001, volume 2, pages 3–30. Cited in 9.4.

Macintyre, Angus, Leszek Pacholski, and Jeff Paris, editors. 1978. *Logic Colloquium '77: Proceedings of the Colloquium Held in Wrocław, August 1977.* Amsterdam: North–Holland Publishing Company. Contains Doner, Mostowski, and Tarski 1978.

Mackiewicz, Witold, editor. 2001. *Polska filozofia powojenna.* Two volumes. Warsaw: Agencja Wydawnicza Widmark. The title means *Postwar Polish Philosophy.* Cited in chapter 17.

Mac Lane, Saunders. 1938. Review of Tarski [1936] 1995. *Journal of Symbolic Logic* 3: 51–52.

Maddux, Roger D. 1993. Review of Alfred Tarski, *Collected Papers*. *Modern Logic* 3: 174–188. Review, with corrections, of Givant 1986 and Tarski 1986a. Cited in 16.2.

———. 2005. Review of Feferman and Feferman 2004. *Bulletin of Symbolic Logic* 11: 535–540. Cited in chapter 17.

Mahlo, Paul. 1908. *Topologische Untersuchungen über Zerlegung in ebene und sphaerische Polygone*. Inaugural dissertation presented to the Philosophical Faculty of the University of Halle–Wittenberg. Halle: Hofbuchdruckerei von C. A. Kaemmerer & Co. JFM: Dehn. Cited in 7.1.

Majdański, Stanisław, and Agnieszka Lekka-Kowalik. 2001. Drewnowski Jan Franciszek. In Bafia and Maryniarczyk 2001, volume 2, 717–721. In Polish. Cited in chapter 3.

Maligranda, Lech. 2008. Antoni Łomnicki (1991–1941). *Wiadomości matematyczne* 44: 61–112. In Polish. Cited in 4.2.

Mancosu, Paolo. 2005. Harvard 1940–1941: Tarski, Carnap and Quine on a finitistic language of mathematics for science. *History and Philosophy of Logic* 26: 327–357. Cited in 16.3.

———. 2008. Tarski, Neurath, and Kokoszyńska on the semantic conception of truth. In Patterson 2008, 192–224. Cited in 15.11 and 18.2.

———. 2009. Tarski's engagement with philosophy. In Lapointe et al. 2009, chapter 9, 131–153. Cited in 14.3 and 18.2.

Mancosu, Paolo, editor. 1998. *From Brouwer to Hilbert*. With the collaboration of Walter P. van Stigt. New York: Oxford University Press. MR: Pambuccian. Papers translated from Dutch, French, and German. Cited in the foreword.

Mancosu, Paolo, Richard Zach, and Calixto Badesa. 2009. The development of mathematical logic from Russell to Tarski: 1900-1935. In Haaparanta 2009, 318–470. Cited in 9.4.

Mann, Thomas. 1998. *The Oxford Guide to Library Research*. New York: Oxford University Press. Cited in this bibliography's introduction.

Manteuffel, Tadeusz. 1936. *Uniwersytet Warszawski w latach 1915/16—1934/35: Kronika*. Warsaw: Nakładem Uniwersytetu Józefa Piłsudskiego. *Latach* means *years*. Cited in 1.1 and chapter 3.

Maramorosch, Karl, and Farida Mahmood, editors. 1999. *Maintenance of Human, Animal, and Plant Pathogen Vectors*. Enfield, New Hampshire: Science Publishers. Contains Szybalski 1999.

Marchisotto, Elena A., and James T. Smith. 2007. *The Legacy of Mario Pieri in Geometry and Arithmetic*. Boston: Birkhäuser. MR: Lewis. Contains Pieri [1908] 2007. Cited in 4.3, 9.3, 9.4, and 14.2.

Marcus, Joseph. 1983. *Social and Political History of the Jews in Poland, 1919–1939*. Berlin: Mouton Publishers. Cited in 9.1.

Marczewski, Edward. 1971. Kwietniewski Stefan (1874–1940). *Polski słownik biograficzny* 16: 378–379. In Polish. Cited in 9.3.

Maryniarczyk, Andrzej, et al. 2011. *Encyklopedia filozofii polskiej*. Two volumes. Lublin: Polskie Towarzystwo Tomasza z Akwinu. The publisher is the Polish Society of St. Thomas Aquinas. Contains an abridgment of Trzęsicki 2010.

Mates, Benson, Leon Henkin, and Alfred Tarski. 1959. Jan Kalicki, philosophy, mathematics: Berkeley and Davis. Berkeley: University of California Digital Archives. Cited in 16.1.

*Mathematical Reviews*. 1979. Review of Chang [1971] 1974. MR0472502 (57 #12200). Cited in 18.2.

May, Kenneth O. 1973. *Bibliography and Research Manual of the History of Mathematics*. Toronto: University of Toronto Press. MR: anonymous. Cited in this bibliography's introduction.

Mazurkiewicz, Stefan. 1932. Zur Axiomatik der Wahrscheinlichkeitsrechnung. *Comptes rendus de la Société des Sciences et des Lettres de Varsovie, classe III* 25: 1–4. JFM: Dörge. Cited in 16.1.

———. 1939. Stanisław Leśniewski (1886–1939). *Przegląd filozoficzny* 42: 115. In Polish. Cited in 1.1.

———. 1956. *Podstawy rachunku prawdopodobieństwa.* Edited by Jerzy Łoś. Monografie matematyczne 32. Warsaw: Państwowe Wydawnictwo Naukowe. MR: Birnbaum. The title means *Foundations of the Calculus of Probability.* Cited in 1.1.

Mazurkiewicz, Stefan, and Wacław Sierpiński. 1914. Sur un ensemble superposable avec chacune de ses deux parties. *Comptes rendus hebdomadaires des séances de l'Académie des Sciences* 158: 618–619. JFM: Korn. Reprinted in Sierpiński 1974–1976, volume 1, 87–88. Cited in 4.1.

McCall, Storrs, editor. 1967. *Polish Logic 1920–1939: Papers by Ajdukiewicz, Chwistek, Jaśkowski, Jordan, Leśniewski, Łukasiewicz, Słupecki, Sobociński, and Wajsberg.* With an introduction by Tadeusz Kotarbiński. Translated by Bohdan Gruchman, Henryk Hiż, Zbigniew Jordan, Eugene C. Luschei, Storrs McCall, Wendy Teichmann, Horst Weber, and Peter Woodruff. London: Oxford University Press. Cited in the foreword.

McNulty, George F. 1986. Alfred Tarski and undecidable theories. *Journal of Symbolic Logic* 51: 890–898. MR: Anshel. Cited in 18.1.

Mehrtens, Herbert. 1980. *Felix Hausdorff: Ein Mathematiker seiner Zeit.* With forewords by Egbert Brieskorn and Eberhard Voigt. Bonn: Fachschaftsrat Mathematik und Mathematisches Institut der Universität Bonn. MR: Kline. Cited in 4.4.

Melnick, Norman. 1971. The 'Einstein of mathematics' at UC. *San Francisco Chronicle* (27 June). Cited in chapter 17.

Mendelson, Elliott. 1995. Review of Tarski 1994. *Mathematical Reviews* 95: MR1269114 (95e: 03005). Cited in 16.1.

Menger, Karl. 1934. Ist die Quadratur des Kreises lösbar? In *Alte Probleme—neue Lösungen in den exacten Wissenschaften: fünf Wiener Vorträge, zweiter Zyklus*, edited by Karl Menger, 1–28. Leipzig: Franz Deuticke. JFM: Pinl. Cited in 4.3 and 8.6.

———. 1994. *Reminiscences of the Vienna Circle and the Mathematical Colloquium.* Edited by Louise Golland, Brian McGuinness, and Abe Sklar. Dordrecht: Kluwer Academic Publishers. MR: Harkleroad. Cited in 14.2.

Menger, Karl, Egbert Dierker, and Karl Sigmund, editors. 1998. *Ergebnisse eines mathematischen Kolloquiums.* With contributions, foreword, and afterword by John W. Dawson Jr., Richard Engelking, W. Hildenbrand, Gérard Debreu, and Franz Alt. Vienna: Springer-Verlag. MR: Duda. Contains Dawson 1998, and reprints of Tarski 1931 and the original version of Lindenbaum and Tarski [1935] 1983. Cited in 14.2.

Meschkowski, Herbert. 1966. *Unsolved and Unsolvable Problems in Geometry.* Translated by Jane A. C. Burlak. New York: Frederick Ungar Publishing Co. Originally published in 1960 as *Ungelöste und unlösbare Probleme der Geometrie*, in Brunswick, Germany, by Friedrich Vieweg & Sohn. MR: Dyer-Bennet. Cited in 4.3.

Meyer, Franz, and Hans Mohrmann, editors. 1907–1934. *Encyklopädie der mathematischen Wissenschaften mit Einschluss ihrer Anwendungen.* Volume 3: *Geometrie.* Leipzig: B. G. Teubner. Three parts; parts 1 and 2 have two halves; part 2, half 2, has subdivisions A and B. Other volumes and editions have different editors. Contains Castelnuovo and Enriques 1908 and 1914 and Enriques 1907.

Mihułowicz, Jerzy. 1930. O określeniu obwodu koła. *Parametr* 1 (1930–1931): 219–223. The title means, "On defining the circumference of a circle." Cited in chapter 11.

Ministerstwo Wyznań Religijnych i Oświecenia Publicznego. 1930. Czasopismo *Parametr. Oświata i wychowanie* 2: 983. The author name means Ministry of Religious Denominations and Public Education; *Czasopismo* means *journal.* The article is unsigned. *Oświata i wychowanie* was a journal published by the Ministry. Cited in 9.7.

Močnik, Frančišek. 1896. *Geometrya poglądowa dla klas niższych szkół średnich*. Adapted by Grzegórz Maryniak from a German edition. Lwów: Księgarnia Seyfartha i Czajkowskiego. JFM: Maynz. The title means *Intuitive Geometry for the Lower Classes of Secondary Schools*. The German version of the author's name is Franz, Ritter von Močnik. The original publication was probably the 24th edition of *Geometrische Anschauungslehre für Untergymnasien*, published in Vienna by F. Tempsky. Cited in 9.2.

Moese, Henryk. 1929. Problem z Lîlâvatî: Problem siedmiu siódemek. In Śrem 1929, 9–24. The title means "Problem from Lîlâvatî: The problem of seven sevens." Cited in 7.1.

———. [1932] 2014. Contribution to A. Tarski's problem "On the degree of equivalence of polygons." Section 7.3 of the present book, Tarski 2014. Translation of "Przyczynek do problemu A. Tarskiego: 'O stopniu równoważności wielokątów,'" *Parametr* 2 (1931–1932): 305–309. A draft translation by Izaak Wirszup, with the same title, appeared in Tarski and Moese 1952, 9–14. Cited in 4.3, 7.0,..., 7.4, 8.5, 9.7, 9.8, 12.0, and 12.7.

Moise, Edwin E. [1963] 1990. *Elementary Geometry from an Advanced Standpoint*. Third edition. Reading, Massachusetts: Addison–Wesley Publishing Company. Cited in 13.1.

Monk, J. Donald. 1986. The contributions of Alfred Tarski to algebraic logic. *Journal of Symbolic Logic* 51: 899–906. MR: Surma. Cited in 18.1.

———. 1989. Review of Tarski and Givant 1987. *Bulletin of the American Mathematical Society* 20: 236–239. Cited in 16.2.

Montague, Richard, Dana S. Scott, and Alfred Tarski. Circa 1970. *An Axiomatic Approach to Set Theory*. Berkeley: Bancroft Library, Tarski Archive 84/69, Container 1. Cited in 14.3.

Moore, Gregory H. 1982. *Zermelo's Axiom of Choice: Its Origins, Development, and Influence*. New York: Springer-Verlag. MR: Smoryński. Cited in 4.4.

———. 1990. Alfred Tarski. *Dictionary of Scientific Biography* 18 (supplement 2): 893–896. Cited in 18.2.

Morscher, Edgar, Otto Neumaier, and Peter M. Simons. 2011. *Ein Philosoph mit "Bodenhaftung": zu Leben und Werk von Joseph M. Bocheński*. Projekte zur Philosophie 9. Sankt Augustin, Germany: Academia. Cited in 15.9.

Morse, Anthony P. 1949. Squares are normal. *Fundamenta Mathematicae* 36: 35–39. MR: Randolph. Cited in 15.12.

Morse, Anthony P., and Randolph, John F. 1940. Gillespie measure. *Duke Mathematical Journal* 6: 408–419. JFM: Haupt. MR: Jeffrey. Also published in *Fundamenta Mathematicae* 33 (1945): 12–26. Cited in 15.12.

Mostowski, Andrzej. 1938. *O niezależności definicji skończoności w systemie logiki*. Dodatek do *Rocznika Polskiego Towarzystwa Matematycznego* 11. Cracow: Drukarnia Uniwersytetu Jagiellońskiego. The title means "On the Independence of the Definitions of Finite in a System of Logic: Supplement to *Annales de la Société Polonaise de Mathématique*." Doctoral dissertation, University of Warsaw. Reviewed in Tarski 1938c. Cited in 9.4, 14.2, and 14.3.

———. 1967. Alfred Tarski. In Edwards 1967, volume 8, 77–81. Cited in 18.2.

Mostowski, Andrzej, and Tarski, Alfred. 1939. Boolesche Ringe mit geordneter Basis. *Fundamenta Mathematicae* 32: 69–86. Givant 1986 item 39a. JFM: Skolem. Reprinted in Tarski 1986a, volume 2, 529–548. Cited in 14.3.

———. 1949. Arithmetical Classes and Types of Well-Ordered Systems. *Bulletin of the American Mathematical Society* 55: 65, 1192. Cited in 8.2.

Murawski, Roman. 1988. Review of Woleński 1987. *Mathematical Reviews* 88: MR0908888 (88k:01047). Cited in 18.2.

———. 2011. Alfred Tarski. In *Filozofia matematyki i logiki w Polsce międzywojennej*, section 3.6, 134–155. Toruń: Wydawnictwo Naukowe Uniwersytetu Mikołaja Kopernika. In Polish. MR: Anonymous. The book's title means *Philosophy of Mathematics and Logic in Interwar Poland*. Cited in 18.2.

Murawski, Roman, and Jan Woleński. 2008. Tarski and his Polish predecessors on truth. In Patterson 2008, chapter 2, 21–43. Cited in 15.6 and 18.2.
Mycielski, Jan. 1979. Finitely additive measures, I. *Colloquium Mathematicum* 42: 309–318. MR: Ramsey. Cited in 4.2.

Nastasi, Pietro. 2005–. *Biografie di matematici Italiani.* On the Internet at http://matematica-old.unibocconi.it/presentazione.htm. This website augments Tricomi 1962.
Natkowska, Monika. 1999. *Numerus clausus, getto ławkowe, numerus nullus, "paragraf aryjski": Antysemityzm na Uniwersytecie Warszawskim 1931–1939.* Warsaw: Żydowski Instytut Historyczny. The Polish phrases in the first clause of the title mean "ghetto benches" and "Aryan paragraph"; the publisher's name means "Institute of Jewish History." Cited in 14.2.
Németi, István. 1990. Review of Tarski and Givant 1987. *Journal of Symbolic Logic* 55: 350–352. Cited in 16.2.
Nerode, Anil. 2010. Review of Feferman and Feferman 2004. *American Mathematical Monthly* 117: 286–288. Cited in chapter 17.
Neumann, Johann von. 1929. Zur allgemeinen Theorie des Masses. *Fundamenta Mathematicae* 13: 73–116, 133. JFM: Pannwitz. Cited in 4.2, 8.6, and 15.12.
Nikodym, Otton. 1930–1937. *Dydaktyka matematyki czystej.* Volume 1: *W zakresie gimnazjum wyższego, liczby naturalne.* Volume 2: *W zakresie wyższych klas szkoły średniej, ułamki oraz ich algebra.* Lwów, Warsaw: Książnica–Atlas, Nasza Księgarnia. The titles mean *Teaching pure mathematics*; *Natural Numbers in the Higher Gimnazjum Classes*; and *Fractions and Their Algebra In the Higher Classes of Secondary School.* Cited in 9.3.
Nowak, Leszek. 1995. Ajdukiewicz and the status of the logical theory of natural language. In Sinisi and Woleński 1995, 205–219. Cited in 15.8.
Nowicki, Ron. 1992. *Warsaw: The Cabaret Years.* San Francisco: Mercury House. Cited in 9.1 and 14.2.
Nowik, Grzegorz. 2004–2010. *Zanim złamano* ENIGMĘ... *Polski radiowywiad podczas wojny z bolszewicką Rosją 1918–1920.* Two volumes. The title of volume 2 includes after the ellipsis the clause, *rozszyfrowano* REWOLUCJĘ. Warsaw: Oficyna Wydawnicza RYTM. That title means *Before* ENIGMA *Was Broken*... REVOLUTION *Was Decoded: Polish Radio-Intelligence during the War Against Bolshevik Russia 1918–1920*; the uppercase names refer to a German cypher machine and a Russian code. Cited in 1.1.
Nye, Mary Jo, editor. 2003. *The Cambridge History of Science.* Volume 5: *The Modern Physical and Mathematical Sciences.* Cambridge, United Kingdom: Cambridge University Press. MR: Gray. Contains Rowe 2003.

*Ogniwo: Organ Związku Zawodowego Nauczycielstwa Polskich Szkół Średnich i biuletyn rządu głównego Z. Z. N. P. S. Śr.* 1931 (volume 11, issue 3, 112–113). Drugie zawody strzeleckie nauczycieli szkół średnich w Warszawie. The titles mean *Organ of the Trade Union of Polish Secondary School Teachers and Bulletin of the General Administration* (of the same organization) and "Second shooting competition for Warsaw secondary-school teachers." Cited in 9.1.

Padoa, Alessandro. [1900] 1901. Essai d'une théorie algébrique des nombres entiers, précédé d'une introduction logique à une théorie déductive quelconque. In International Congress of Philosophy 1900–1903, volume 3 (1901), 309–365. The introduction, "Logical introduction to any deductive theory," translated by Jean van Heijenoort, is in Van Heijenoort [1967] 1970, 118–123. Cited in 14.2.
Pambuccian, Victor. 2004. Early examples of resource-consciousness. *Studia Logica* 77: 81–86. MR: Anonymous. Cited in 7.1.

Państwowe Wydawnictwo Książek Szkolnych we Lwowie. 1937. Advertisement. *Przegląd pedagogiczny* 56: 221. The author name means "National Schoolbook Publisher in Lwów." This lists Chwiałkowski, Schayer, and Tarski [1935] 1946. Cited in 9.9.

Parafia Wszystkich Świętych w Warszawie, Proboszcz. [1929] 1940. Akt Małżeństwa. Nr. 367. Berkeley: Bancroft Library, Tarski Archive 84/69, Carton 13. The author and title mean *Parish of All Saints in Warsaw, Rector; Marriage Ceremony*. Handwritten official copy of parish records, accompanied by a typewritten German translation, both evidently made in 1940 for the German occupation authorities. Cited in 9.1.

Pasenkiewicz, Kazimierz. 1935. *Wykaz wykładów i ćwiczeń, pod L. 7615*. In the Archive of the University of Warsaw, Sygnatura Pasenkiewicz Kazimierz: RP 7615. The title phrase means *List of Lectures and Recitations*. Pasenkiewicz's handwritten official record of studies, 1921–1935. Cited in chapter 3.

———. 1984. Recollections about Alfred Tarski from the years 1921–1925. Handwritten translation, eight pages, by Jan Tarski. In the Alfred Tarski Archive, Bancroft Library, University of California at Berkeley, Carton 15. Cited in 1.1 and chapter 3.

Patterson, Douglas. 2012. *Alfred Tarski: Philosophy of Language and Logic*. New York: Palgrave Macmillan. Cited in 18.2.

———, editor. 2008. *New Essays on Tarski and Philosophy*. Oxford: Oxford University Press. Contains Betti 2008, S. Feferman 2008, Mancosu 2008, and Murawski and Woleński 2008. Cited in 4.6 and 18.2.

Pawlikowska-Brożek, Zofia. 1975. Mazurkiewicz Stefan (1888–1945). *Polski słownik biograficzny* 20: 313–314. In Polish. Cited in 1.1.

———. 2003. Krysicki Włodzimierz (1905–2001); Tarski (Tajtelbaum) Alfred (1901–1983): logik, filozof, matematyk. In Domoradzki, Pawlikowska-Brożek, and Węglowska 2003, 119, 244–245. In Polish. Cited in 12.9 and chapter 17.

———. 2011. Wilkosz Witold. To appear in the next edition of Domoradzki, Pawlikowska-Brożek, and Węglowska 2003. In Polish. Cited in 15.9.

Peano, Giuseppe. 1903. De latino sine flexione: lingua auxiliare internationale. *Revista de mathematica* 8: 74–83. JFM: Vivanti. Included in Peano 2002 as file 1903d.pdf. Reissued in Roero 2003. This specification starts in Scholastic Latin; as it proposes syntax simplifications, it implements them, so that it ends in Latino sine Flexione (often called Lingua Peano). Cited in 9.7.

———. 1924. Prosta teorja logarytmów. Translated by Samuel Dickstein. *Przegląd matematyczno-fizyczny* 2: 150–152. Originally published in Latino sine Flexione as "Theoria simplice de logarithmos" in *Wiadomości matematyczne* 26 (1924): 53–55. Cited in 4.3.

———. 2002. *L'opera omnia di Giuseppe Peano*. Edited by Clara S. Roero. Turin: Dipartimento di Matematica, Università di Torino. CD-ROM. Contains a reprint of Peano 1903.

Pepłoński, Andrzej. 1995. Polski wywiad wojskowy w okresie bitwy warszawskiej w sierpniu 1920r. *Niepodległość i Pamięć* 1995(2), 39–49. The titles means "Polish military intelligence during the battle of Warsaw in August 1920" and *Independence and Memory*. Cited in 1.1.

Peregrin, Jaroslav, editor. 1999. *Truth and Its Nature (If Any)*. Synthese library 284. Dordrecht: Kluwer Academic Publishers. Contains García-Carpintero 1999.

Petrozolin-Skowrońska, Barbara, et al., editors. 1994. Gimnazjum im. Stefana Żeromskiego. *Encyklopedia Warszawy*: 210. Warsaw: Wydawnictwo Naukowe PWN. Cited in 9.1.

Pieri, Mario. [1908] 1915. *Gieometrja elementarna oparta na pojęciach "punktu" i "kuli."* Translated by Stefan Kwietniewski. Warsaw: Bibljoteka Wektora A3, Skład Główny w Księgarni Gebethnera i Wolffa, Warsaw. Translation of the original version of Pieri [1908] 2007. This item was incorrectly called Pieri 1914 in Marchisotto and Smith 2007. Supported by the Mianowski Fund. Cited in 9.3.

———. [1908] 2007. Elementary geometry based on the notions of point and sphere. In Marchisotto and Smith 2007, chapter 3. Translation of "La Geometria Elementare istituita sulle nozioni di 'punto' e 'sfera,'" *Memorie di matematica e di fisica della Società Italiana delle Scienze* (series 3) 15: 345–450, presented to the Society by Guido Castelnuovo and approved by Corrado Segre, reprinted in Pieri 1980, 455–560. Pieri [1908] 1915 is a Polish translation of the original publication. Cited in 9.3 and 9.4.

———. 1980. *Opere sui fondamenti della matematica.* Edited by the Unione Matematica Italiana, with contributions by the Consiglio Nazionale delle Ricerche. Contains a reprint of the original version of Pieri [1908] 2007. Bologna: Edizioni Cremonese.

Piłatowicz, Józef. 2006. Straszewicz Stefan (1889–1983). *Polski słownik biograficzny* 44(1): 233–236. In Polish. Cited in 9.7.

Piotrowski, Tadeusz, editor. 2004. *The Polish Deportees of World War II: Recollections of Removal to the Soviet Union and Dispersal Throughout the World.* Jefferson, North Carolina: McFarland & Company. Cited in 9.9.

Piotrowski, Walerian. 2003. Hornowski Michał (1893–1966), Rusiecki Antoni Marian (1892–1956), Wojtowicz (Wójtowicz) Władysław Bazar (1874–1942). In Domoradzki, Pawlikowska-Brożek, and Węglowska 2003, 76, 206, 261. Cited in 9.3, 9.7, and 12.3.

Pla i Carrera, Josep. 1984. Alfred Tarski i la lògica contemporània. Primera part. *Butlletí de la Societat Catalana de Matemàtiques* 17: 26–46. MR: Anonymous. In Catalan. Cited in 18.2.

———. 1989. Alfred Tarski i la teoria de conjunts. *Theoria* (Spain) (series 2) 4: 343–417. MR: Anonymous. In Catalan. "Conjunts" means "sets." Cited in 18.2.

Poland. Główny Urząd Statystyczny Rzeczypospolitej Polskiej. 1936. *Mały rocznik statystyczny.* Warsaw: Nakładem Głównego Urzędu Statystycznego. The author name and title mean Main Statistical Office of the Polish Republic, *Concise Statistical Yearbook.* Cited in 13.0.

Poland. Ministerstwo Wyznań Religijnych i Oświecenia Publicznego. 1931. *Program gimnazjum państwowego: wydział matematyczno-przyrodniczy.* Fifth edition. Lwów: Państwowe Wydawnictwo Książek Szkolnych. The author is the Mininstry of Religious Denominations and Public Education and the title means *Program for Government Gimnazjums: Mathematical-Scientific Section.* Cited in 9.2.

———. 1934. *Program nauki w gimnazjach państwowych z polskim językiem nauczania (Tymczasowy).* Biblioteka oświaty i wychowania, 2. Lwów: Państwowe Wydawnictwo Książek Szkolnych. The title means *Program of Instruction in Government Gimnazjums with Polish as the Language of Instruction (Provisional).* The Program was instituted in 1932. This book carries no date, but Polish library records date it 1934. Cited in 9.9.

Poland. President. 1927. Rozporządzenia Prezydenta Rzeczypospolitej z dnia 24 listopada 1927r. o ubezpieczeniu pracowników umysłowych. *Dziennik ustaw Rzeczypospolitej Polskiej* 1927 (numer 106, pozycja 911): 1463–1488. On the Internet at http://isap.sejm.gov.pl/DetailsServlet?id= WDU19271060911 in file D19270911.pdf. The titles mean "Decree of the President of the Republic as of 24 November 1927 concerning insurance of nonmanual workers" and *Journal of Laws of the Republic of Poland.* Analyzed in Tarski [1929] 2014b. Cited in 15.5.

Poland. Sejm. 1932. Ustawa z dnia 11 marca 1932 r. o ustroju szkolnictwa. *Dziennik ustaw Rzczypospolitej Polskiej* 1932 (numer 38, pozycja 389): 639–645. On the Internet at http://isap.sejm.gov.pl/DetailsServlet?id=WDU19320380389, in file D19320389.pdf. The author is the parliament of Poland; the titles mean "Law of 11 March 1932 concerning the educational system" and *Journal of Laws of the Republic of Poland.* Cited in 9.1.

Polonsky, Antony. 2012. *The Jews in Poland and Russia.* Volume 3: *1914 to 2008.* Portland, Oregon: Littman Library of Jewish Civilization. Cited in 1.1.

Polska Akademia Umiejętności. 1948. Alfred Tarski. In "Skład Polskiej Akademii Umiejętności (Stan w Marcu 1948)," *Rocznik Polskiej Akademii Umiejętności* 1946–1947: xli–xlii. The author is the Polish Academy of Art and Sciences in Cracow. Cited in chapter 17.

Polskie Towarzystwo Filozoficzne we Lwowie. 1930–1931. December 1930 minutes. *Ruch filozoficzny* 12: 209–210. The author is the Polish Philosophical Society in Lwów. Cited in 14.2.

———. 1931. *Księga pamiątkowa Polskiego Towarzystwa Filozoficznego we Lwowie 12.II.1904 12.II.1929.* Lwów: Książnica–Atlas. Contains the original version of Łukasiewicz [1931] 1970. The title means *Commemorative Book of the Polish Philosophical Society in Lwów.*

Polskie Towarzystwo Matematyczne. 1927. Roster. *Annales de la Société Polonaise de Mathématique* 6: 135–140. Cited in 8.7.

Polski Zjazd Filozoficzny. 1927. Księga pamiątkowa Pierwszego Polskiego Zjazdu Filozoficznego (Lwów, 1923). *Przegląd filozoficzny* 30: 256–366. The title means, "Commemorative book for the First Polish Philosophical Congress." Cited in chapter 3.

———. 1928. Księga pamiątkowa Drugiego Polskiego Zjazdu Filozoficznego, Warszawa, 1927. *Przegląd filozoficzny* 31(1–2): 1–225. The title means "Commemorative Book of the Second Polish Philosophical Congress." Contains Czeżowski [1927] 1928, Greniewski [1927] 1928, and the original versions of the comments Tarski [1927] 2014a and [1927] 2014b. Described in Hosiassonówna 1928.

———. 1936. Księga pamiątkowa III Polskiego Zjazdu Filozoficznego, Kraków, 1936. *Przegląd filozoficzny* 39(4): 317–548. The title means "Commemorative Book of the Third Polish Philosophical Congress." Contains Ajdukiewicz 1936, Kokoszyńska 1936a, Łukasiewicz 1936, Wilkosz 1936, Zawirski 1936, and the original versions of the comments Tarski [1936] 2014a, ..., [1936] 2014d. Cited in 4.11.

Polskie Zjazd Matematycznego. 1929. *Księga pamiątkowa Pierwszego Polskiego Zjazdu Matematycznego: Lwów, 7–10.IX.1927.* Supplement to *Annales de la Société Polonaise de Mathématique* 6. The title means *Commemorative Book of the First Polish Mathematical Congress.* Contains Tarski [1927] 1929.

Popper, Karl. 1981. Joseph Henry Woodger. *British Journal for the Philosophy of Science* 32: 38–330. Cited in 15.9.

Powszechna Wystawa Krajowa. 1929. *Powszechna Wystawa Krajowa: Poznań, maj–wrzesień 1929.* Poznań: Polskie Towarzystwo Księgarni Kolejowych "Ruch." The title means *The National Exhibition: Poznań, May–September 1929.* Cited in 9.5.

Presburger, Mojżesz. [1929] 1991. On the completeness of a certain system of arithmetic of whole numbers in which addition occurs as the only operation. Translated, with commentary, by Dale Jacquette. *History and Philosophy of Logic* 12: 225–233. Translation of "Über die Vollständigkeit eines gewissen Systems der Arithmetik ganzer Zahlen, in welchem die Addition als einzige Operation hervortritt," in Leja 1930, 92–101, 395, JFM: Fraenkel. Cited in 9.4.

*Przegląd pedagogiczny.* 1929 (volume 48, February). Walne Zebranie Koła Warszawskiego. 187–188. The title means "General assembly of Warsaw circles." Cited in 9.9.

———. 1931 (volume 50). Odczyty o programach matematyki: 23. The title means "Lectures in programs for mathematicians." Cited in 9.9.

Purkert, Walter. 2002. Grundzüge der Mengenlehre: historische Einführung. In Hausdorff 2002, 1–89. Cited in 2.1 and 4.3.

———. 2008. The double life of Felix Hausdorff/Paul Mongré. Translated by Hilde Rowe and David E. Rowe. *Mathematical Intelligencer* 30(4): 36–50. Cited in 4.4.

Raabe, Henryk. 1926. III Gimnazjum Męskie Związku Zawodowego Nauczycielstwa Polskich Szkół Średnich w Warszawie. *Ogniwo* 2(6): 30–33. The title is the name of the school where Tarski taught. Cited in chapter 3 and 9.1.

Rakoczy-Pindor, Krystyna. 2003. Przygoda matematyka z radiem czyli profesor Witold Wilkosz pionierem radiotechniki i radiofonii w Polsce. *Wiadomości Matematyczne* 39: 151–156. The beginning of the title means "Adventure in radio and mathematics." Cited in 15.9.

Rav, Yehuda. 2001. Review of Sinaceur 2001. *Mathematical Reviews* 2001: MR1813395 (2001k: 03004). Cited in 18.2.

Redzik, Adam. 2004. Polish Universities During the Second World War. Logroño, Spain: Encuentros de Historia Comparada Hispano-Polaca. Conference presentation. Cited in 9.9.

Reichenbach, Hans. [1934] 1936. Die Bedeutung des Wahrscheinlichkeitsbegriffs für die Erkenntnis. In International Congress of Philosophy 1936, 163–169. Discussed in Tarski [1934] 1936a and [1934] 2003a. Cited in 16.1.

Reid, Constance. 1982. *Neyman—from Life*. New York: Springer-Verlag. MR: Grimm. Cited in chapter 3.

Réthy, Móricz. 1891. Endlich-gleiche Flächen. *Mathematische Annalen* 38: 405–428. JFM: Kötter. Kötter found a flaw; see Amaldi [1900] 1914, 172. Cited in 4.3.

*Revue de métaphysique et de morale*. 1904 (volume 12, supplement for May, 11–12). Reviews of the first edition of Enriques and Amaldi [1903] 1913 and of Enriques 1900. Cited in 9.2.

Rickey, V. Frederick. 2011. Polish logic from Warsaw to Dublin: the life and work of Jan Łukasiewicz. Manuscript. Cited in 9.4.

Robinson, Raphael M. 1947. On the decomposition of spheres. *Fundamenta Mathematicae* 34: 246–260. MR: Blumberg. Cited in 8.6 and 15.12.

Rodríguez-Consuegra, Francisco. 2005. Tarski's intuitive notion of set. In Sica 2005, 227–266. Cited in 18.2.

Roero, Clara Silvia, editor. 2003. *Le riviste di Giuseppe Peano*. Turin: Dipartimento di Matematica, Università di Torino. CD-ROM. Contains a reprint of Peano 1903.

———. 2010. *Peano e la sua scuola fra matematica, logica e interlingua: Atti del congreso internazionale di studi (Torino, 6–7 Ottobre 2008)*. Turin: Deputazione Subalpina di Storia Patria. Contains Luciano and Roero 2010.

Rojszczak, Artur. 2002. Philosophical background and philosophical content of the semantic definition of truth. *Erkenntnis* 56: 29–62. Cited in 14.2 and 15.11.

Roliński, Adam. 1996. Bolesław Zahorski–"Lubicz" (1887–1922). *A gdy na wojenkę szli Ojczyźnie służyć. Pieśni i piosenki żołnierskie z lat 1914–1918. Antologia*, 482. Cracow: University of Cracow Press. The title means *And When They Went to Battle to Serve the Motherland: Soldier Songs and Lyrics....* Cited in 9.1.

Rosen, Saul. 1985. Alfred Tarski in 1940. *Annals of the History of Computing* 7: 364–365. Cited in 15.12, 16.3, and chapter 17.

Rowe, David E. 2003. Mathematical schools, communities, and networks. In Nye 2003, chapter 6, 113–132. Cited in the preface.

Rusiecki, Antoni M. 1923–1924. Obrazy liczbowe na piątce oparte. *Przegląd matematycznofizyczny* 1: 21–25 and 2: 117–134. The title means "Number pictures based on 5." Cited in 4.3.

———. 1930a. Exercise 1. *Parametr* 1: 35. Krysicki 1930 is a solution. Reprinted in *Młody matematyk* 1(1931, issue 3 supplement): I. Cited in 3.4 and 12.9.

———. 1930b. Osobliwe zadanie. *Parametr* 1: 75, 271–274. The title means "A peculiar problem." Cited in 14.2.

———. 1930c. Zadanie konkursowe: Rekord nie do pobicia. *Parametr* 1, 193–194. The title means, "Contest problem: an unbreakable record." Cited in 12.1.

———. 1930d. III Zadanie konkursowe: Łamigłówka geometryczna. *Parametr* 1: 316. The title means, "Third contest problem: geometric puzzle." Cited in 12.1.

———. 1931. Lingua Peano. *Młody matematyk* 1 (4–5): 49–53. Cited in 9.7.

Rusiecki, Antoni M., editor. 1931a. Dodatek. *Młody matematyk* 1 (3, March 1931): I–V (following page 48). The title means "appendix." This assigned point values to exercises Tarski [1930] 2014b, [1930] 2014c, and [1930] 2014f. Cited in 12.0.

———. 1931b. Solution to Tarski [1930] 2014b. *Młody matematyk* 1: 90. This issue of the journal was distributed with *Parametr* 2 (1931–1932, issue 6–7). The solution was unsigned. Cited in 12.2.

———. 1939. Solution to Tarski [1930] 2014f. *Parametr* 3: 93–94. This was incorrectly labeled as a solution to a different exercise. The solution was unsigned. Cited in 12.6.

Rusiecki, Antoni M., and Stefan Straszewicz, editors. 1930. Dział zadań; Apel; Od redakcji. *Parametr* 1: 116, 229, 317, 323. The titles mean "Problem section," "Appeal," and "From the editors." The articles were unsigned. Cited in 9.7 and 12.0.

———. 1930–1931. Kursy wakacyjne. *Parametr* 1 (1930)(3, March), 111–112; (6, July–August), 224; (1931)(4–5, July), 138. The title means "Vacation courses." The articles were unsigned. Cited in 9.3.

———. 1931. Zadania. *Młody matematyk* 1: 13. The title means "Problems." The article was unsigned. Cited in 12.0.

———. 1931–1932. Od redakcji. *Młody matematyk* 1: 1–2, 97. The title means "From the editors." The article was unsigned. Cited in 9.7 and 12.0.

———. 1932a. Konkurs zadaniowy. *Młody matematyk* 1: 109–111, 154–155. The title means "Problem contest." The article was unsigned. Cited in 7.1 and 12.0.

———. 1932b. Od redakcji. *Parametr* 2: 33, 145. The title means "From the editors." The article was unsigned. Cited in 9.7.

Russell, Bertrand. 1903. *The Principles of Mathematics*. Cambridge, U. K.: Cambridge University Press. JFM: Engel. There are several newer editions and reprints. Cited in 1.2.

Sadowska, Joanna. 2001. *Ku szkole na miarę Drugiej Rzeczypospolitej: geneza, założenia i realizacja reformy Jędrzejewiczowskiej*. Białystok: Wydawnictwo Uniwersytetu w Białymstoku. The title means *Toward Schools, for the Measure of the Second Republic: The Origin, Establishment, and Implementation of the Jędrzejewiczowski Reforms*. Cited in 1.1.

Sally, Judith D., and Paul J. Sally, Jr. 2007. *Roots to Research: A Vertical Development of Mathematical Problems*. Providence: American Mathematical Society. MR: Muldoon. Also reviewed in Holdener 2009. Cited in 8.6.

*San Francisco Chronicle*. 1983 (28 October). Alfred Tarski: Expert Logician, UC Professor. Cited in chapter 17.

*San Francisco Examiner*. 1983 (28 October). Alfred Tarski. Cited in chapter 17.

Scanlan, Michael. 2003. American postulate theorists and Alfred Tarski. *History and Philosophy of Logic* 24: 307–325. Cited in 1.2 and 18.2.

Scazzieri, Roberto, and Raffaella Simili, editors. 2008. *The Migration of Ideas*. Sagamore Beach, Massachusetts: Science History Publications. Contains Galavotti 2008.

Schayer, Wacław. 1924. Życiorys. Handwritten document in the Archive of the University of Warsaw, Sygnatura Schayer Wacław: RP 17204. The title means "Curriculum Vitae." Cited in 9.9.

———. 1932. *Wykaz wykładów i ćwiczeń, pod L. 17204*. In the Archive of the University of Warsaw, Sygnatura Schayer Wacław: RP 17204. The title phrase means *List of Lectures and Recitations*. Schayer's handwritten official record of studies, 1924–1932. Cited in 9.9.

Schinzel, Andrzej. 1974. *Wacław Sierpiński*. Warsaw: Iskry. In Polish. Cited in chapter 3.

———. 2000. Pięćdziesiąt lat Olimpiady Matematycznej. *Wiadomości matematyczne* 36: 155–161. The title means "Fifty years of the Mathematical Olympiad." Cited in 9.9.

———. 2009. The Lvov years of Wacław Sierpiński. In Bojarski, Ławrynowicz, and Prytula 2009, 87–89. Cited in 4.1.

Schmidt am Busch, Hans-Christoph, and Kai F. Wehmeier. 2007. On the relations between Heinrich Scholz and Jan Łukasiewicz. *History and Philosophy of Logic* 28: 67–81. Cited in 9.4.

Scholz, Heinrich. 1957. In memoriam Jan Łukasiewicz. *Archiv für mathematische Logik und Grundlagenforschung* 3: 3–18. Cited in 9.4.

Schönflies, Arthur. 1913. *Entwicklung der Mengenlehre und ihrer Anwendung. Umarbeitung des im VIII. Bande der Jahresbericht der Deutschen Mathematiker-Vereinigung erstatteten Berichts. Erste Hälfte: Allgemeine Theorie der unendlichen Mengen und Theorie der Punktmengen.* Edited by Hans Hahn and Artur Schoenflies. Leipzig: B. G. Teubner. JFM: Vivanti. Cited in 4.3.

Schreiber, Peter. 1996. Felix Hausdorffs paradoxe Kugelzerlegung im Kontext der Entwicklung von Mengenlehre, Masstheorie und Grundlagen der Mathematik. In Brieskorn 1996, volume 1, 135–148. Cited in 8.6.

———. 1999. Über Beziehungen zwischen Heinrich Scholz und polnischen Logikern. *History and Philosophy of Logic* 20: 97–109. MR: Glas. Cited in 9.4.

Schuster, Cynthia A. 1975. Hans Reichenbach. In *Dictionary of Scientific Biography* 1970–, volume 11: 355–359. Cited in 16.1.

Schwabhäuser, Wolfram, Wanda Szmielew, and Alfred Tarski. 1983. *Metamathematische Methoden in der Geometrie.* Givant 1986 item 83$^m$. Berlin: Springer-Verlag. MR: Rautenberg. Cited in 9.4 and 12.11.

Segerberg, Krister. 2009–. The logic of action. In Zalta 2011– at http://plato.stanford.edu/entries/logic-action/. Cited in 15.2.

Sękowska, Magdalena, and Węglowska, Danuta. 2003. Lindenbaum Adolf (1904–1941). In Domoradzki, Pawlikowska-Brożek, and Sękowska 2003, 136. In Polish. Cited in 14.3.

Shilleto, James R. 2012. Personal communication to James T. Smith, June 2012. Cited in 7.1 and 8.5.

Shneiderman, S. L. 1995. Notes for an autobiography. Translated by Fannie Peczenik. College Park, Maryland: University of Maryland Libraries, S. L. and Eileen Shneiderman Collection of Yiddish Books. Cited in chapter 3.

Sica, Giandomenico, editor. 2005. *Essays on the Foundations of Mathematics and Logic.* Monza, Italy: Polimeric International Scientific Publisher. Contains Rodríguez-Consuegra 2005. Cited in 18.2.

Siegmund-Schultze, Reinhard. 2010. On a missed opportunity for collaboration between historians and mathematicians: a biographical avalanche triggered by Professor Ioan James, FRS. *Historia Mathematica* 37: 693–707. Review of James 2009a and some other works. Cited in chapter 17.

Sierpiński, Wacław. 1918. Pewnik p. Zermelo oraz jego rola w teorji mnogości i w analizie.—L'axiome de M. Zermelo et son rôle dans la théorie des ensembles et l'analyse. *Bulletin international de l'Académie des Sciences de Cracovie, classe des sciences mathématiques et naturelles, série A: Sciences mathématiques*: 97–152. JFM: Rosenblatt. In French. Reprinted in Sierpiński 1974–1976, volume 2, 208–255. Cited in 4.1 and 6.3.

———. 1922. Sur l'égalité $2m = 2n$ pour les nombres cardinaux. *Fundamenta Mathematicae* 3: 1–6. JFM: Rosenthal. Reprinted in Sierpiński 1974–1976, volume 2, 417–421. Cited in 4.4.

———. 1923. *Zarys teorji mnogości.* Volume 1: *Liczby pozaskończone.* Second edition. Warsaw: Mianowski Fund. JFM: Lichtenstein. Originally published in 1912. The titles mean *Outline of Set Theory* and *Transfinite Numbers.* Sierpiński [1928] 1950 is a French version. Cited in 4.3 and chapter 5.

———. [1928] 1950. *Leçons sur les nombres transfinis.* Paris: Gauthier-Villars. New printing of the original 1928 edition. French version of Sierpiński 1923. JFM: Feigl.

———. [1929] 1930. Przemówienie. In Leja 1930, 7–10. The title means "Address." Cited in 9.5.

———. 1945. Sur le paradoxe de MM. Banach et Tarski. *Fundamenta Mathematicae* 33: 229–234. MR: Blumberg. Reprinted in Sierpiński 1974–1976, volume 3, 439–444. Cited in 15.12.

———. 1947. Sur un théorème de A. Lindenbaum. *Annals of Mathematics* (series 2) 48: 641–642. MR: Erdős. Reprinted in Sierpiński 1974–1976, volume 3, 469–471. Cited in 15.12.

———. 1954. *On the Congruence of Sets and Their Equivalence by Finite Decomposition.* Lucknow University studies 20: Faculty of Science, session 1948–1949. Lucknow, India: Lucknow University. MR: Halmos. Cited in 4.2 and 4.3.

———. 1974–1976. *Oeuvres choisies.* Three volumes. Edited by Stanisław Hartman and Andrzej Schinzel. Warsaw: PWN—Éditions Scientifiques de Pologne. Contains Kuratowski 1974 and reprints of Sierpiński 1918 , 1922, 1945, and 1947, and Mazurkiewicz and Sierpiński 1914.

Simmons, Keith. 2009. Tarski's logic. In Gabbay and Woods 2009, 511–616. Cited in 18.2.

Sinaceur, Hourya. 1996. Mathématiques et métamathématique du congrès de Paris (1900) au congrès de Nice (1970): Nombres réels et théorie des modèles dans les travaux de Tarski. *Rendiconti del Circolo Matematico di Palermo* (serie 2 supplemento) 44: 113–132. MR: Murawski. Cited in 18.2.

———. 2001. Alfred Tarski: semantic shift, heuristic shift in metamathematics. *Synthese* 126: 49–65. MR: Rav. Cited in 18.2.

———. 2007. Review of Feferman and Feferman 2004. *Notices of the American Mathematical Society* 54: 986–989. Cited in chapter 17.

———. 2009. Tarski's practice and philosophy: between formalism and pragmatism. In Lindström et al. 2009, 357–396. MR: Corcoran. Cited in chapter 17.

Singleterry, Ann M. 1966. Review of the third edition of Tarski [1936] 1995. *Journal of Symbolic Logic* 31: 674–675.

Sinisi, Vito, and Jan Woleński, editors. 1995. *The Heritage of Kazimierz Ajdukiewicz.* Poznań Studies in the philosophy of the sciences and the humanities 40. Amsterdam: Rodopi. Contains Nowak 1995 and a reprint of Giedymin 1978.

Skolimowski, Henryk. 1962. Colourful philosopher. *Polish Perspectives* 5 (8–9): 122–126. Cited in 14.2. About Leon Chwistek.

Śmigielski, Henryk, editor. 1964. *Matematyka i jej zastosowania.* Zeszyty naukowe Politechniki Warszawskiej 99. Warsaw: Wydawnictwa Politechniki Warszawskiej. The title means *Mathematics and Its Applications.* Contains Jakubowski 1964.

Snyder, Timothy. 2010. *Bloodlands: Europe between Hitler and Stalin.* New York: Basic Books. Cited in 9.9.

Sobociński, Bolesław. 1956. In memoriam Jan Łukasiewicz (1878–1956). *Philosophical Studies: Annual Journal of The Philosophical Society, St. Patrick's College, Ireland* 6: 3–49. Contains an erroneous version of Łukasiewicz [1953] 1994. Cited in 9.4.

Sobociński, Leon. 1935. W gimnazjum im. Mikołaja Kopernika w Toruniu. *Gazeta Gdańska* 14: 7–8. The title means "In the Nicolaus Copernicus Gimnazjum in Toruń." Cited in 7.1.

Sosnowski, Paweł. 1923. Państwowy Instytut Pedagogiczny: sprawozdanie. *Przegląd pedagogiczny* 42(1): 74–83. The article title means "National Pedagogical Institute: report." The author was the director of the institute. Cited in chapter 3.

Średniawa, Bronisław. 1961. W dwudziestą rocznicę śmierci prof. Witolda Wilkosza. *Postępy fizyki* 13: 389–391. The title phrase and journal title mean "On the twentieth anniversary of the death..." and *Advances in physics.* Cited in 15.9.

Śrem. Państwowe Gimnazjum im. Generała Józefa Wybickiego. 1928, 1930, 1931. *Sprawozdanie dyrekcji Państwowego Gimnazjum im. Generała Józefa Wybickiego w Śremie.* Volumes for academic years 1919/1920–1927/1928, 1929/1930, and 1930/1931. Śrem: Drukarnia „Gazety Powszechnej" w Poznaniu, Drukarnia Centralna w Śremie. The title means *Report of the directorate of the National Gimnazjum in Śrem named for General Józef Wybicki.* The material for 1928/1929 is contained in Śrem 1929. Cited in 7.1.

Śrem. Państwowe Gimnazjum im. Generała Józefa Wybickiego. 1929. *Sprawozdanie dyrekcji Państwowego Gimnazjum im. Generała Józefa Wybickiego w Śremie za 10-lecie 1918/19–1928/29.* Śrem: Drukarnia „Gazety Powszechnej" w Poznaniu. The title means *Report of the directorate of the National Gimnazjum in Śrem named for General Józef Wybicki, for the ten years 1918/1919–1928/1929.* Overlaps Śrem 1928. Cited in 7.1.

Srzednicki, Jan T. J., and Zbigniew Stachniak, editors. 1998. *Leśniewski's Systems: Protothetic.* Nijhoff international philosophy series 54. Dordrecht: Kluwer Academic Publishers. MR: Murawski. Contains Tarski [1923] 1998.

Steinhaus, Hugo. 1924. O mierzeniu pól płaskich. *Przegląd matematyczno-fizyczny* 2: 24–29. The title means "On measuring area in the plane." Cited in 4.3.

Stone, John C., and Virgil S. Mallory. 1937. *New Plane Geometry.* Chicago: Benj. H. Sanborn & Co. Cited in 9.9.

Stone, Marshall H. 1938. The representation of Boolean algebras. *Bulletin of the American Mathematical Society* 44(1): 807–816. JFM: Rinow. Cited in 8.7.

Straszewicz, Stefan. 1930. Uwagi o t. zw. zadaniach dyskusyjnych w geometrji szkolnej. *Parametr* 1: 53–62. The title means "Remarks on the so-called geometrical discussion problems in school." Cited in 12.0.

Straszewicz, Stefan, and Stefan Kulczycki. [1935] 1948. *Geometria dla III klasy gimnazjum.* Warsaw: Książnica Atlas. The title means *Geometry for the Third Gimnazjum Class.* Cited in 9.7 and 9.9.

Stromberg, Karl. 1979. The Banach–Tarski paradox. *American Mathematical Monthly* 86: 151–161. MR: Bressler. Cited in 4.4 and 8.6.

Suppes, Patrick. 1960. *Axiomatic Set Theory.* Princeton, New Jersey: D. Van Nostrand Company. MR: Rieger. Cited in 1.2.

———. 1988. Philosophical implications of Tarski's work. *Journal of Symbolic Logic* 53: 80–91. MR: Kielkopf. Cited in 18.1.

Suppes, Patrick, Jon Barwise, and Solomon Feferman. 1989. Commemorative meeting for Alfred Tarski: Stanford University, November 7, 1983. In Duren et al. 1989, 393–403. Cited in chapter 17.

Surma, Stanisław J. 1977. Mordchaj Wajsberg: life and work. In Wajsberg 1977, 7–11. Cited in chapter 3.

Szczepański, Janusz. 1995. *Wojna 1920 roku na Mazowszu i Podlasiu.* Warsaw: Pułtusk. The title means *The 1920 War in Mazovia and Podlachia.* Cited in 1.1.

Szczerba, Lesław W. 1986. Tarski and geometry. *Journal of Symbolic Logic* 51: 907–912. Cited in 8.4, 9.3, 12.0, and 18.1.

Szkoła Ziemi Mazowieckiej. [1918] 1927. Document "8-mio Klasowa Szkoła Mazowiecka" dated 18 February 1927. In the Archive of the University of Warsaw, Sygnatura Teitelbaum-Tarski Alfred: RP 2909. Certified handwritten copy of Alfred Tarski's 8 June 1918 certificate of graduation from the Mazowieckie gimnazjum, with a record of his spring 1918 grades. Cited in 1.1.

Szmielew, Wanda. [1938] 1947. On choices from finite sets. *Fundamenta Mathematicae* 34: 75–80. MR: Martin. According to Kordos, Moszyńska, and Szczerba 1977, these results were obtained in 1938. Cited in 9.4 and 14.2.

———. 1955. Elementary properties of Abelian groups. *Fundamenta Mathematicae* 41: 203–271. MR: Lyndon. Cited in 9.4.

Szudy, Józef, editor. 1998. *Born 100 Years Ago: Aleksander Jabłoński (1898–1980).* Toruń: Uniwersytet Mikołaja Kopernika. Contains Frąckowiak 1998.

Szumilewicz-Lachman, Irena. 1994. Zygmunt Zawirski: a short bio-bibliography (28.IX.1882–2.IV.1948). Translated by Feliks Lachman. In Szumilewicz-Lachman, Cohen, and Bergo 1994, Part 1, chapter 5, 33–80. Cited in 15.10.

Szumilewicz-Lachman, Irena, Robert S. Cohen, and Bettina Bergo, editors. 1994. *Zygmunt Zawirski: His Life and Work, with Selected Writings on Time, Logic, and the Methodology of Science.*

Boston studies in the philosophy of science 157. Dordrecht: Kluwer Academic Publishers. Contains Szumilewicz-Lachman 1994.

Szybalski, Wacław. 1999. Maintenance of human-fed live lice in the laboratory and production of Weigl's exanthematous typhus vaccine. In Maramorosch and Mahmood 1999, chapter 8, 161–180. Cited in 2.1.

Tarski, Alfred. See also joint entries after the following single-author entries, and under Banach, Chwiałkowski, Doner, Henkin, Horn, Jónsson, Lindenbaum, Łukasiewicz, Mates, Montague, and Mostowski.

———. [1918] 2014. Curriculum Vitae. In section 16.1 of the present book, Tarski 2014. Translation of the handwritten document "Życiorys" in the Archive of the University of Warsaw, Sygnatura Teitelbaum-Tarski Alfred: RP 2909. Cited in 1.1, chapter 3, and 16.3.

———. 1919–1920. Two documents titled "Do jego Magnificencji Rektora Uniwersytetu Warszawskiego," dated trymestrze I 1919/20 and zimowym 1920/21. Printed forms filled out by hand, in the Archive of the University of Warsaw, Sygnatura Teitelbaum-Tarski Alfred: RP 2909. The titles mean "To his Magnificence, the Rector of the University of Warsaw," autumn trimester 1919/1920 and autumn 1920/1921." Tarski crossed out unused portions, signed them Alfred Tajtelbaum, and submitted them. They are about returning to the university after summer military leave. Cited in 1.1 and chapter 3.

———. [1921] 2014. A contribution to the axiomatics of well-ordered sets. Chapter 2 of the present book, Tarski 2014. Translation of "Przyczynek do aksjomatyki zbioru dobrze uporządkowanego," *Przegląd filozoficzny* 24, 85–94, Givant 1986 item 21, reprinted in Tarski 1986a, volume 1, 1–12. JFM: Chwistek. The author's surname on the original is Tajtelbaum; its title acknowledges the seminar of Prof. Stanisław Leśniewski at the University of Warsaw. Cited in 1.2, 8.2, 16.0, and 16.1.

———. 1921–1926. Do kwestury, letni 1921, zimowy 1921/2, zimowy 1923/4, 1 czerwiec 1924, 27 styczeń 1925, 23 styczeń 1926. Printed forms filled out by hand, in the Archive of the University of Warsaw, Sygnatura Teitelbaum-Tarski Alfred: RP 2909. The title means "To the bursar" with the dates. Receipts for Alfred Tarski's tuition (sometimes waived). The document dated zimowy 1923/4 (autumn 1923/1924) is reproduced in chapter 3 of the present book, Tarski 2014. Cited in chapter 3.

———. 1923a. O wyrazie pierwotnym logistyki. *Przegląd filozoficzny* 26: 68–89. Givant 1986 item 23. Tarski's doctoral dissertation. The author's surname is listed as Tajtelbaum-Tarski. JFM: Rosenthal. Translated into English as Tarski [1923] 1998. Adapted in French as Tarski 1923b and 1924d. Adapted and translated into English in Tarski [1956] 1983, 1–23. Cited in chapter 3 and 16.1.

———. 1923b. Sur le terme primitif de la logistique. *Fundamenta Mathematicae* 4: 196–200. Givant 1986 item 23a. Adaptation of part of Tarski 1923a (the other part is adapted in Tarski 1924d). JFM: Rosenthal. Reprinted in Tarski 1986a, volume 1, 13–20. Cited in chapter 3 and 16.1.

———. [1923] 1998. On the primitive term of logistic. Translation of the Polish original, Tarski 1923a, by Zbigniew Stachniak. In Srzednicki and Stachniak 1998, 43–68. The author's surname is listed as Tajtelbaum-Tarski. Cited in 16.1.

———. 1924a. Une remarque concernant les principes d'arithmétique théorique. *Annales de la Société Polonaise de Mathématique* 3: 150. Givant 1986 item 24$^a$a. Reprinted in Tarski 1986a, volume 4, 535. Cited in 4.3.

———. 1924b. Sur les ensembles finis. *Fundamenta Mathematicae* 6: 45–95. Givant 1986 item 24c. JFM: Rosenthal. Reprinted in Tarski 1986a, volume 1, 65–118. Cited in chapter 3, 4.3, 4.4, 8.3, 9.3, and 14.2.

———. 1924c. Sur les principes de l'arithmétique des nombres ordinaux (transfinis). *Annales de la Société Polonaise des Mathématique* 3: 148–149. Givant 1986 item 24ª. Reprinted in Tarski 1986a, volume 4, 533–534. Cited in 1.1, 4.3, and 8.2.

———. 1924d. Sur les truth-functions au sens de MM. Russell et Whitehead. *Fundamenta Mathematicae* 5: 59–74, Givant 1986 item 24. Adaptation of part of Tarski 1923a (the other part is adapted in Tarski 1923b). JFM: Rosenthal. Reprinted in Tarski 1986a, volume 1, 21–38. The author's surname is listed as Tajtelbaum-Tarski. Cited in chapter 3 and 4.3.

———. 1924e. Sur quelques théorèmes qui équivalent à l'axiome du choix. *Fundamenta Mathematicae* 5: 147–154. Givant 1986 item 24a (which has incorrect publication data). JFM: Rosenthal. Reprinted in Tarski 1986a, volume 1, 39–48. The author's surname is listed as Tajtelbaum-Tarski. Cited in chapter 3, 4.3, 4.4, 8.3, 9.4, 14.2, and 15.12.

———. 1924f. *Wykaz wykładów i ćwiczeń, no. 2909*. In the Archive of the University of Warsaw, Sygnatura Teitelbaum-Tarski Alfred: RP 2909. The title phrase means *List of Lectures and Recitations*. Tarski's handwritten official record of studies, 1918–1924. Cited in 1.1, chapter 3, and 4.1.

———. [1924] 2004. Letter to a Warsaw city government office, 18 March. In Feferman and Feferman 2004, 40. The original handwritten document "Komisariat Rządu na m. st. Warszawa" is in the Archive of the University of Warsaw, Sygnatura Teitelbaum-Tarski Alfred: RP 2909. Cited in chapter 3 and 16.3.

———. [1924] 2014a. Letter to the dean of the Philosophical Faculty, 2 June 1924. In section 16.3 of the present book, Tarski 2014. Translation of handwritten document "Do Pana Dziekana Wydziału Filozoficznego Uniwersytetu Warszawskiego" in the Archive of the University of Warsaw, Signatura Teitelbaum-Tarski Alfred: RP 2909. Cited in chapter 3.

———. [1924] 2014b. On the equivalence of polygons. Chapter 5 of the present book, Tarski 2014. Translation of "O równoważności wielokątów," *Przegląd matematyczno-fizyczny* 2: 47–60, Givant 1986 item 24b, reprinted in Tarski 1986a, volume 1, 49–64. Cited in 4.0, ..., 4.4, 6.0, 8.4, 8.6, 9.2, 9.3, 9.7, 9.8, and 15.7.

———. 1925a. Problème 38. *Fundamenta Mathematicae* 7: 381. Givant 1986 item 25p. The wrong text was printed for this problem in Tarski 1986a; the original text is reproduced in section 4.3 of the present book, Tarski 2014. Cited in 4.3 and 12.0.

———. 1925b. Quelques théorèmes sur les alephs. *Fundamenta Mathematicae* 7: 1–14. Givant 1986 item 25, reprinted in Tarski 1986a, volume 1, 155–170. JFM: Pannwitz. Cited in 8.3 and 9.4.

———. [1925] 2014. Discussion of Łukasiewicz 1925. In section 15.1 of the present book, Tarski 2014. Translation of a discussion in Łukasiewicz 1925, 136: Givant 1986 item 25ᶜ, reprinted in Tarski 1986a, volume 4, 686, and in Jadacki 2003b, 19. Cited in 8.1.

———. 1927a. Sur quelques propriétés caractéristiques des images d'ensembles. *Annales de la Société Polonaise de Mathématique* 6: 127–128. Givant 1986 item 27ª. JFM: Hurewicz. Reprinted in Tarski 1986a, volume 4, 541–543. Abstract of an 18 February talk presented to the Warsaw section of the Polish Mathematical Society. Cited in 8.8.

———. 1927b. Quelques théorèmes généraux sur les images d'ensembles. *Annales de la Société Polonaise de Mathématique* 6: 132–133. Givant 1986 item 27ªa. JFM: Hurewicz. Reprinted in Tarski 1986a, volume 4, 544–545. Abstract of a 9 December talk presented to the Warsaw section of the Polish Mathematical Society. Cited in 8.8.

———. [1927] 1929. Les fondements de la géométrie des corps. In Polski Zjazd Matematyczny 1929, 54–61. Givant 1986 item 29. (Givant reported the pages incorrectly.) Reprinted in Tarski 1986a, volume 1, 225–232. Tarski [1927] 1983 is an English translation and adaptation.

———. [1927] 1983. Foundations of the geometry of solids. In Tarski [1956] 1983, 24–29. Givant 1986 item 29(1). Translated and adapted from Tarski [1927] 1929. Cited in 2.1, 9.4, and 9.7.

———. [1927] 2014a. Discussion of Greniewski [1927] 1928. In section 15.2 of the present book, Tarski 2014. Translation of a discussion in Greniewski [1927] 1928, 160: Givant 1986 item 28°, reprinted in Tarski 1986a, volume 4, 687, and in Jadacki 2003b, 19.

———. [1927] 2014b. Discussion of Czeżowski [1927] 1928. In section 15.3 of the present book, Tarski 2014. Translation of a discussion in Czeżowski [1927] 1928, 167: Givant 1986 item 28°a, reprinted in Jadacki 2003b, 19, and in Tarski 1986a, 549.

———. 1928. Remarques sur les notions fondamentales de la méthodologie des mathématiques. *Annales de la Société Polonaise de Mathématique* 7: 270–273. Givant 1986 item 28ᵃb. Report of a 14 September 1928 presentation in Warsaw. The journal gave that date incorrectly as 1929. Reprinted in Tarski 1986a, volume 4, 552. Tarski [1928] 2010 is an English translation. Cited in 9.5 and 16.1.

———. [1928] 1930. Über Äquivalenz der Mengen in Bezug auf eine beliebige Klasse von Abbildungen. In International Congress of Mathematicians 1930, volume 2, 243–252. JFM: Fraenkel. Givant 1986 item 30b, reprinted in Tarski 1986a, volume 1, 299–310. Cited in 8.8.

———. [1928] 2010. Remarks on fundamental concepts of the methodology of mathematics. Translated by Robert Purdy and Jan Zygmunt. In Béziau 2010, 67–68. Translation of Tarski 1928. Cited in 16.1.

———. [1929] 2014a. An assembly of mathematicians. Chapter 10 of the present book, Tarski 2014. Translation of "Zjazd matematyków," *Ogniwo: organ Związku Zawodowego Nauczycielstwa Polskich Szkół Średnich i biuletyn zarządu głównego Z. Z. N. P. S. Śr.* 9: 401–402, Givant 1986 item 29e, reprinted in Tarski 1986a, volume 1, 251–252. The journal title means *Organ of the Trade Union of Polish Secondary School Teachers and Bulletin of the General Administration* (of the same organization). Cited in 9.4 and 9.5.

———. [1929] 2014b. Marginal note on "Decree of the President of the Republic Concerning Insurance of Nonmanual Workers as of 24 November 1927" In section 15.5 of the present book, Tarski 2014. Translation of "Na marginesie 'Rozporządzenia Prezydenta Rzeczypospolitej o ubezpieczeniu pracownikow umysłowych z dnia 24 listopada 1927 r,'" *Ekonomista, kwartalnik poświęcony nauce i potrzebom życia* 29: 115–119, Givant 1986 item 29d, reprinted in Tarski 1986a, volume 1, 269–273. The decree is Poland President 1927. The journal title means *The Economist: A Quarterly Devoted to Science and the Needs of Life*. Cited in 9.2, 15.0, and 15.5.

———. 1929–1930. Sur les fonctions additives dans les classes abstraites et leur application au problème de la mesure. *Comptes rendus des séances de la Société des Sciences et des Lettres de Varsovie, classe III* 22: 114–117. JFM: Ruziewicz. Givant 1986 item 29b, reprinted in Tarski 1986a, volume 1, 243–248. Cited in 8.6.

———. 1930a. Une contribution à la théorie de la mesure. *Fundamenta Mathematicae* 15: 42–50. Givant 1986 item 30, reprinted in Tarski 1986a, volume 1, 275–286. JFM: Feigl. Cited in 2.1.

———. 1930b. Über definierbare Zahlenmengen. *Annales de la Société Polonaise de Mathématique* 9: 206–207. Givant 1986 item 30ᵃ. Reprinted in Tarski 1986a, volume 4, 560–561. Preliminary version of Tarski [1931] 1983. Cited in 8.6 and 14.2.

———. [1930] 1983a. Fundamental concepts of the methodology of the deductive sciences, I. In Tarski [1956] 1983, 60–109. The original paper, "Fundamentale Begriffe der Methodologie der deduktiven Wissenschaften," *Monatshefte für Mathematik und Physik* 37: 361–404, Givant 1986 item 30e, JFM: Skolem, is reprinted in Tarski 1986a, volume 1, 345–390, and translated into Polish in Tarski 1995–2001, volume 2, 31–92. There is no part II. Cited in 14.2, 15.6, and 16.1.

———. [1930] 1983b. On some fundamental concepts of metamathematics. In Tarski [1956] 1983, 30–37. The original paper, "Über einige fundamentale Begriffe der Metamathematik," *Comptes rendus de la Société des Sciences et des Lettres de Varsovie, Classe III* 23: 22–29,

Givant 1986 item 30c, is reprinted in Tarski 1986a, volume 1, 311–320. Cited in chapter 3, 8.1, 14.2, and 15.6.

———. [1930] 2014a. Exercise: On diluting wine. In section 12.1 of the present book, Tarski 2014. Translation of "Zadanie o rozcieńczaniu wina," *Parametr* 1: 229, Givant 1986 item 30$^p$, reprinted in Tarski 1986a, volume 4, 688.

———. [1930] 2014b. Exercise 83: Iteration of the absolute value symbol. In section 12.2 of the present book, Tarski 2014. Translation of "Nr. 83: Iteracja znaku bezwzględnej wartości," *Parametr* 1: 231, Givant 1986 item 30$^P$a, reprinted in Tarski 1986a, volume 4, 688. Rusiecki 1931a assigned it a point value and Rusiecki 1931b is a solution.

———. [1930] 2014c. Exercise 88: Decomposition into factors. In section 12.3 of the present book, Tarski 2014. Translation of "Nr. 88: Rozkład na czynniki," *Parametr* 1: 277, Givant 1986 item 30$^P$b, reprinted in Iwaszkiewicz 1949b, with a solution, and in Tarski 1986a, volume 4, 689. Rusiecki 1931a assigned it a point value. Cited in 9.7.

———. [1930] 2014d. Exercise 102: Interesting identity. In section 12.4 of the present book, Tarski 2014. Translation of "No. 102: Ciekawa tożsamość," *Parametr* 1: 278, Givant 1986 item 30$^P$c, reprinted in Tarski 1986a, volume 4, 689.

———. [1930] 2014e. Exercise 118. In section 12.5 of the present book, Tarski 2014. Translation of "Nr. 118," *Parametr* 1: 318, Givant 1986 item 30$^P$d, reprinted in Tarski 1986a, volume 4, 689–690.

———. [1930] 2014f. Exercise 136: Equidistant. In section 12.6 of the present book, Tarski 2014. Translation of "Nr. 136: Ekwidystanta," *Parametr* 1: 398, Givant 1986 item 30$^P$e, reprinted in Tarski 1986a, 690. Rusiecki 1931a assigned it a point value. Rusiecki 1939 is a solution.

———. [1930–1931] 2014. On the concept of truth in reference to formalized deductive sciences. In section 15.6 of the present book, Tarski 2014. Translation of "O pojęciu prawdy w odniesieniu do sformalizowanych nauk dedukcyjnych," *Ruch filozoficzny* 12: 210–211, Givant 1986 item 30/31$^a$, reprinted in Tarski 1986a, volume 4, 555–559, and Tarski 1995–2001, volume 1, 3–8. Abstract. Cited in 8.1, 15.0, and 16.2.

———. [1930–1936] 1992. Drei Briefe an Otto Neurath. Edited, with a foreword, by Rudolf Haller. Translated by Jan Tarski. *Grazer philosophische Studien* 43:1–32. Partially translated into Polish in Tarski 1995–2001, volume 1, 206–213. Cited in 15.1 and 16.3.

———. 1931. Neue Resultate und unentschiedene Probleme der Kardinalzahlarithmetik. *Monatshefte für Mathematik und Physik* 38: 23. Givant 1986 item 31$^a$. Reprinted in *Ergebnisse eines mathematischen Kolloquiums* 2 (1932), 12–13; in Menger et al. 1998, 130–131; and in Tarski 1986a, volume 4, 562–563. Report of a 19 February 1930 presentation in Vienna. Cited in 9.5.

———. [1931] 1951. Al ma'alat shivyon-ha-peruk shel metsula (מעל שדידך-חפדדק של מצדלעים על). *Riveon lematematika* 5(1951): 32–38. Hebrew translation by Dov Jarden of the original version of Tarski [1931] 2014a. Givant 1986 item 31b(2). MR: Jerison. Cited in 7.1 and 8.5.

———. [1931] 1952. The degree of equivalence of polygons. In Tarski and Moese 1952, 1–8. Givant 1986 item 31b(1). Draft translation, by Isaak Wirszup, of the original version of Tarski [1931] 2014a. Reprinted in Tarski 1986a, volume 1, 571–580.

———. [1931] 1983. On definable sets of real numbers. In Tarski [1956] 1983, chapter VI, 110–142. Translation of "Sur les ensembles définissables de nombres réels, I," *Fundamenta Mathematicae* 17: 210–239. Givant 1986 item 31, reprinted in Tarski 1986a, vol. 1, 517–548. JFM: Fraenkel. Cited in 8.6 and 15.6.

———. [1931] 2014a. On the degree of equivalence of polygons. Section 7.2 of the present book, Tarski 2014. Translation of "O stopniu równoważności wielokątów," *Młody matematyk* 1 (1931–1932): 37–44, Givant 1986 item 31b, reprinted in 1975 in *Delta* 5 (4): 3–6. Tarski [1931] 1952 is a draft translation of the original paper. The original and that draft were reprinted in Tarski 1986a, volume 1, 561–580. Tarski [1931] 1951 is a Hebrew translation. Cited in 4.3, 7.0, ..., 7.4, 8.5, 9.2, 9.7, 9.8, 12.7, and 12.10.

———. [1931] 2014b. Exercise 167: System of inequalities. In section 12.7 of the present book, Tarski 2014. Translation of "Nr. 167: Układ nierówności," *Młody matematyk* 1 (1931–1932): 46, Givant 1986 item 31$^P$, reprinted in Tarski 1986a, volume 4, 692.

———. [1931] 2014c. Exercise 168: [Another] system of inequalities. In section 12.8 of the present book, Tarski 2014. Translation of "Nr. 168: Układ nierówności," *Młody matematyk* 1 (1931–1932): 46, Givant 1986 item 31$^P$a, reprinted in Tarski 1986a, volume 4, 692.

———. [1931] 2014d. Exercise 169: Cutting a rectangle out of a square. In section 12.9 of the present book, Tarski 2014. Translation of "Nr. 169: Wycięcie prostokąta z kwadratu," *Młody matematyk* 1 (1931–1932): 46, Givant 1986 item 31$^P$b, reprinted in Dubiel 1975 and Tarski 1986a, volume 4, 692.

———. [1931] 2014e. Exercise 170: A theorem on the degree of equivalence of polygons. In section 12.10 of the present book, Tarski 2014. Translation of "Nr. 170: Twierdzenie o stopniu równoważności wielokątów," *Młody matematyk* 1 (1931–1932): 46, Givant 1986 item 31$^P$c, reprinted in Tarski 1986a, volume 4, 693. Cited in 4.3.

———. [1931] 2014f. Exercise 183: Postulate about parallels. In section 12.11 of the present book, Tarski 2014. Translation of "No. 183: Postulat o równoległych," *Parametr* 2 (1931–1932): 78, Givant 1986 item 31/32$^P$, reprinted in Tarski 1986a, volume 4, 691. Cited in 9.3 and 9.4.

———. [1931] 2014g. Exercise 186: Analytic geometry of space. In section 12.12 of the present book, Tarski 2014. Translation of "No. 186: Zadanie z geometrji analitycznej w przestrzeni," *Parametr* 2 (1931–1932): 78, Givant 1986 item 31/32$^P$a, reprinted in Tarski 1986a, volume 4, 691.

———. [1931] 2014h. On a certain system of mathematical logic and the resulting methodological and semantic issues. In section 16.1 of the present book, Tarski 2014. Translation of "O pewnym systemie logiki matematycznej i wynikających z niego zagadnieniach metodologicznych i semantycznych," *Ruch filozoficzny* 12 (1930–1931): 232. Abstract of a 15 April 1931 presentation to the Logic Section of the Warsaw Philosophical Institute. Cited in 14.2.

———. 1932. Der Wahrheitsbegriff in den Sprachen der deduktiven Disziplinen. *Akademie der Wissenschaften in Wien, mathematisch-naturwissenschaftliche Klasse, akademischer Anzeiger* 69: 23–25. Givant 1986 item 32. JFM: Ackermann. Reprinted in Tarski 1986a, volume 1, 613–618. Presented to the Akademie by Hans Hahn. Translated into Polish in Tarski 1995–2001, volume 1, 9–12. Cited in 14.2 and 15.6.

———. [1932] 1952. Further remarks about the degree of equivalence of polygons. In Tarski and Moese 1952, 14–20. Givant 1986 item 31/32a(1). Tarski [1932] 1952 is a draft translation of the original paper. Reprinted in Tarski 1986a, volume 1, 603–612.

———. [1932] 2014a. On geometric properties of Banach's measure. In section 15.7 of the present book, Tarski 2014. Translation of "O własnościach geometrycznych miary Banacha," *Comptes rendus des séances de la Société des Sciences et des Lettres de Varsovie, classe III* 25 (1932–1933): 12, Givant 1986 item 32$^a$. Presented to the Society by Wacław Sierpiński on 23 February 1932. Reprinted in Tarski 1986a, volume 4, 567. Cited in 8.7 and 15.0.

———. [1932] 2014b. Exercise 213: Stereometry. In section 12.13 of the present book, Tarski 2014. Translation of "No. 213: Zadanie ze stereometrji," *Parametr* 2 (1931–1932): 207, Givant 1986 item 31/32$^P$b, reprinted in Tarski 1986a, volume 4, 691.

———. [1932] 2014c. Exercise 214: Arrangement of segments in a plane. In section 12.14 of the present book, Tarski 2014. Translation of "No. 214: Układ dwóch odcinków na płaszczyźnie," *Parametr* 2 (1931–1932): 207, Givant 1986 item 31/32$^P$c, reprinted in Tarski 1986a, volume 4, 691.

———. [1932] 2014d. Remarks on the degree of equivalence of polygons. Section 7.4 of the present book, Tarski 2014. Translation of "Uwagi o stopniu równoważności wielokątów," *Parametr*

2 (1931–1932): 310–314, Givant 1986 item 31/32a. A draft translation by Izaak Wirszup appeared in Tarski and Moese 1952, 15–20, Givant 1986 item 31/32a(1). The original and that draft were reprinted in Tarski 1986a, volume 1, 595–612. Cited in 4.3, 7.0, ..., 7.4, 8.5, 9.7, and 9.8.

———. [1932] 2014e. The theory of the circumference of a circle in the secondary school. Chapter 11 of the present book, Tarski 2014. Translation of "Teorja długości okręgu w szkole średniej," *Parametr* 2 (1931–1932): 257–267, Givant 1986 item 31/32, reprinted in Tarski 1986a, volume 1, 581–594. Cited in 9.7, ..., 9.9.

———. 1933. *Pojęcie prawdy w językach nauk dedukcyjnych*. Prace Towarzystwa Naukowego Warszawskiego, Wydział III—Nauk Matematyczno-fizycznych 34. Givant 1986 item 33$^m$. The title means *The Concept of Truth in the Languages of Deductive Sciences*. Reprinted in Tarski 1995–2001, volume 1, 13–172. Tarski [1933] 1983 is an English translation of the German translation [1933] 1935 which had an added appendix. Tarski [1930–1931] 2014 is a translation of a preliminary version. Cited in 9.3, 15.6, 15.12, 16.2, and 16.3.

———. [1933] 1935. Der Wahrheitsbegriff in den formalisierten Sprachen. Translated by Leopold Blaustein. With a postscript by Alfred Tarski. *Studia Philosophica* 1: 261–405. Translation of Tarski 1933. Givant 1986 item 35b. JFM: Ackermann. Reprinted in Tarski 1986a, volume 2, 51–198. Cited in 15.6 and 15.12.

———. [1933] 1983. The concept of truth in formalized languages. Translated by J. H. Woodger. In Tarski [1956] 1983, chapter 8, 152–278. Givant 1986 item 35b(1). Translation of Tarski [1933] 1935, which was a translation of Tarski 1933. Cited in 9.3, 15.6, 15.10, and 15.11.

———. [1933] 2003. Letter to Kazimierz Twardowski, 12 July. In Jadacki 2003b, 19. Cited in 16.3.

———. 1934. Z badań metodologicznych nad definjowalnością terminów. *Przegląd filozoficzny* 37: 438–460. Givant 1986 item 34. Reprinted in Tarski 1995–2001, volume 2, 114–146. Tarski [1935–1936] 1983 is an English translation of a German translation. Cited in 16.2.

———. [1934] 1936a. Discussion of Reichenbach [1934] 1936 and Zawirski [1934] 1936. In International Congress of Philosophy 1936, 197–198. Givant 1986 item 36$^c$a. (Givant incorrectly described that item as a contribution to a discussion of a paper by Janina Hosiasson-Lindenbaumowa.) Reprinted in Tarski 1986a, volume 4, 695. Reprinted and translated into Polish in Tarski [1934] 2003a. The original text of this discussion is nearly identical to that of Tarski 1935–1936. Cited in 16.1.

———. [1934] 1936b. Discussion of Jørgensen [1934] 1936. In International Congress of Philosophy [1936] 1968, 119. Givant 1986 item 36$^c$. Reprinted in Tarski 1986a, volume 4, 694. Reprinted and translated into Polish in Tarski [1934] 2003b. Cited in 16.1.

———. [1934] 1983. Some observations on the concepts of $\omega$-consistency and $\omega$-completeness. In Tarski [1956] 1983, 279–295. The original paper, "Einige Betrachtungen über die Begriffe der $\omega$-Widerspruchsfreiheit und $\omega$-Vollständigkeit," *Monatshefte für Mathematik und Physik* 40: 97–112, Givant 1986 item 33, JFM: Fraenkel, is reprinted in Tarski 1986a, volume 1, 619–636, and translated into Polish in Tarski 1995–2001, volume 2, pages 93–113. Cited in 16.2.

———. [1934] 2003a. Discussion of Reichenbach [1934] 1936 and Zawirski [1934] 1936. With Polish translation by Mariusz Grygianiec. In Jadacki 2003b, 12–14. Reprint and translation of Tarski [1934] 1936a. Cited in 16.1.

———. [1934] 2003b. Discussion of Jørgensen [1934] 1936. With Polish translation by Mariusz Grygianiec. In Jadacki 2003b, 11. Reprint and translation of Tarski [1934] 1936b. Cited in 16.1.

———. [1934] 2014. Letter to the Minister of Religious Denominations and Public Education, 29 November. Translation of a handwritten document N921/R15345, page 8, preserved in the archival fund No. 14, Ministerstwo Wyznań Religijnych i Oświecenia Publicznego w Warszawie

1917–1939, sygnatura 6216, of the Archiwum Akt Nowych (Central Archives of Modern Records) in Warsaw. In section 16.3 of the present book, Tarski 2014. The original document was transcribed in Jadacki 2003b, 25–26. Cited in 9.1 and 16.3.

———. [1935] 1936a. Grundlegung der wissenschaftlichen Semantik. In International Congress for the Unity of Science 1936, volume 3, 1–8. Givant 1986 item 36f. German version of Tarski 1936a. Reprinted in Tarski 1986a, volume 2, 259–268. Tarski [1935] 1983 is an English translation. Cited in 14.2.

———. [1935] 1936b. Über den Begriff der logischen Folgerung. In International Congress for the Unity of Science 1936, volume 7, 1–11. Givant 1986 item 36g. German version of Tarski 1936b. Reprinted in Tarski 1986a, volume 2, 269–282. Tarski [1935] 2002 and [1956] 1983, 409–430, are English translations. Tarski [1935] 1984 is a Spanish translation. Cited in 14.2 and 16.1.

———. [1935] 1983. The establishment of scientific semantics. In Tarski [1956] 1983, chapter 15, 401–408. Givant 1986 item 36f(1). Translation of Tarski [1935] 1936a.

———. [1935] 1984. Sobre el concepto de consecuencia logica. Translation of Tarski [1935] 1936b by Luiz Vega Reñon. In Castrillo and Vega 1984, section A.2, 178–192. Cited in 16.1.

———. [1935] 2002. On the concept of following logically. Translated from the Polish and German originals Tarski [1935] 1936b and 1936b by Magda Stroińska and David Hitchcock, with their commentaries. *History and Philosophy of Logic* 23: 155–196. MR: Murawski. Cited in the preface and 16.1.

———. [1935] 2014. Two letters to the Ministry of Religious Denominations and Public Education, 25 November and 7 December. In section 16.3 of the present book, Tarski 2014. Translations of handwritten documents, N840/R12478, pages 21–22, and N873/R13421, page 24, preserved in the archival fund No. 14, Ministerstwo Wyznań Religijnych i Oświecenia Publicznego w Warszawie 1917–1939, sygnatura 6216, Archiwum Akt Nowych (Central Archives of Modern Records) in Warsaw. Transcriptions of the original letters were published in Jadacki 2003b, 26–27. Jadacki misstated the date of the first letter. Cited in 9.1 and 16.3.

———. 1935–1936. Wahrscheinlichkeitslehre und mehrwertige Logik. *Erkenntnis* 5: 174–175. Givant 1986 item 35d. The text of this report is nearly identical to that of Tarski [1934] 1936a. Reprinted in Tarski 1986a, volume 2, 199–202. Cited in 16.1.

———. [1935–1936] 1983. Some methodological investigations on the definability of concepts. In Tarski [1956] 1983, 296–319. JFM: Hempel. Givant 1986 item 35c(1). The original paper, "Einige methodologische Untersuchungen über die Definierbarkeit der Begriffe," *Erkenntnis* 5: 80–100, Givant 1986 item 35c, is reprinted in Tarski 1986a, volume 1, 637–659. That, in turn, was a translation of Tarski 1934.

———. 1936a. O ugruntowaniu naukowej semantyki. *Przegląd filozoficzny* 39: 50–57. Givant item 36. Reprinted in Tarski 1995–2001, volume 1, 173–185. Tarski [1935] 1983 is an English translation of the German version Tarski [1935] 1936a. Cited in 14.3 and 16.2.

———. 1936b. O pojęciu wynikania logicznego. *Przegląd filozoficzny* 39: 58–68. Givant 1986 item 36a. Reprinted in Tarski 1995–2001, volume 1, 186–202. Tarski [1935] 2002 is an English translation. In Tarski [1956] 1983, 409–420, is an English translation of the German version Tarski [1935] 1936b. Cited in 14.3, 16.1, and 16.2.

———. [1936] 1995. *Introduction to Logic and to the Methodology of the Deductive Sciences*. Second, revised edition. Translated by Olaf Helmer. New York: Dover Publications. Originally published in 1941 by Oxford University Press, with second and third editions in 1946 and 1965. Tarski 1994 is a fourth edition. Item 41$^m$ in Givant 1986. MR: Frink. Also reviewed in Church 1941–1947 and Singleterry 1966. Expansion and translation of *O logice matematycznej i metodzie dedukcyjnej*, Bibljoteczka matematyczna 3-5, Lwów and Warsaw: Książnica–Atlas, item 36$^m$ in Givant 1986, JFM: Ackermann, also reviewed in Mac Lane 1938. The original title means *On Mathematical Logic and the Deductive Method*. Cited in 9.3, 9.9, 14.2, and 14.3.

———. [1936] 2014a. Discussion of Ajdukiewicz 1936. In section 15.8 of the present book, Tarski 2014. Translation of a discussion in Ajdukiewicz 1936, 337, Givant 1986 item 36°b, reprinted in Tarski 1986a, volume 4, 698, and in Jadacki 2003b, 15.

———. [1936] 2014b. Discussion of Wilkosz 1936. In section 15.9 of the present book, Tarski 2014. Translation of a discussion in Wilkosz 1936, 346–347, Givant 1986 item 36°, reprinted in Tarski 1986a, volume 4, 699, and in Jadacki 2003b, 15.

———. [1936] 2014c. Discussion of Zawirski 1936. In section 15.10 of the present book, Tarski 2014. Translation of a discussion in Zawirski 1936, 350–351, Givant 1986 item 36°d, reprinted in Tarski 1986a, volume 4, 700, and in Jadacki 2003b, 15–16.

———. [1936] 2014d. Discussion of Kokoszyńska 1936a. In section 15.11 of the present book, Tarski 2014. Translation of a discussion in Kokoszyńska 1936a, 425, Givant 1986 item 36°e, reprinted in Tarski 1986a, volume 4, 701, and Tarski 1995–2001, volume 1, 203–205. Cited in 16.2.

———. 1937a. Appendix E. In Woodger 1937, 161–172. Givant 1986 item 37b. Reprinted in Tarski 1986a, volume 2, 335–350. Tarski's first publication in English. Cited in 14.3.

———. 1937b. Sur la méthode déductive. In International Congress of Philosophy 1937, fascicle 6: *Logique et mathématique*, 95–103. Givant 1986 item 37a, reprinted in Tarski 1986a, volume 2, 325–333. Also reviewed in Langford 1938. Tarski's last publication in French. Cited in 14.2 and 14.3.

———. 1937c. Über additive und multiplikative Mengenkörper und Mengenfunktionen. *Comptes rendus des séances de la Société des Sciences et des Lettres de Varsovie, classe III* 30: 151–181. Givant 1986 item 37, reprinted in Tarski 1986a, volume 2, 289–322. JFM: Aumann. Cited in 8.7.

———. [1937] 2003. Two letters to Kazimierz Twardowski, 1 and 24 May. In Jadacki 2003b, 19–20. Cited in 16.3.

———. 1938a. Algebraische Fassung des Massproblems. *Fundamenta Mathematicae* 31: 47–66. Givant 1986 item 38g, reprinted in Tarski 1986a, volume 2, 451–472. JFM: Feigl. Cited in 8.7.

———. 1938b. Über das absolute Mass linearer Punktmengen. *Fundamenta Mathematicae* 30: 218–234. Givant 1986 item 38e, reprinted in Tarski 1986a, volume 2, 425–444. JFM: Radon. Cited in 8.7, 15.7, and items Givant 1986 and Tarski 1986a.

———. 1938c. Review of Mostowski 1938. *Journal of Symbolic Logic* 3: 115–116. Givant 1986 item 38ʳ. Reprinted in Tarski 1986a, volume 4, 704–706. Cited in 14.2.

———. [1938] 1939. Contribution to the discussion. In Barzin [1938] 1939, 55. Givant 1986 item 39ᶜ. Reprinted in Tarski 1986a, volume 4, 707. Translated into Polish in Tarski 1995–2001, volume 2, 472. Cited in 16.1.

———. [1938] 1983. Sentential calculus and topology. In Tarski [1956] 1983, 421–454. The original paper, "Der Aussagenkalkül und die Topologie," *Fundamenta Mathematicae* 31: 103–134, Givant 1986 item 38h, JFM: Rinow, is reprinted in Tarski 1986a, volume 2, 473–506, and translated into Polish in Tarski 1995–2001, volume 2, 158–200. Cited in 16.2.

———. 1939a. New investigations on the completeness of deductive theories. *Journal of Symbolic Logic* 4: 176. Givant 1986 item 39ᵃ. Reprinted in Tarski 1986a, volume 4, 571. Cited in 14.3.

———. 1939b. On undecidable statements in enlarged systems of logic and the concept of truth. *Journal of Symbolic Logic* 4: 105–112. Givant 1986 item 39c. JFM: Skolem. MR: Frink. Translated into Polish in Tarski 1995–2001, volume 1, 214–227. Reprinted in Tarski 1986a, volume 2, 559–568. Cited in 14.3 and 16.2.

———. 1939c. On well-ordered subsets of any set. *Fundamenta Mathematicae* 32: 176–183. Givant 1986 item 39b. JFM: Skolem. Reprinted in Tarski 1986a, volume 2, 549–558. Cited in 14.3.

---. [1939] 2003. Letter to the dean of the Division of Mathematics and Natural Sciences of the University of Warsaw, 16 January. In Jadacki 2003b, 27. Cited in 16.3.

---. [1939–1940] 1995. Some current problems in metamathematics. Edited by Jan Tarski and Jan Woleński. *History and Philosophy of Logic* 16: 159–168. MR: Angelelli. Translated into Polish in Tarski 1995–2001, volume 2, pages 380–395. Cited in 16.2.

---. 1939–1945. Ideale in vollständigen Mengenkörpern. *Fundamenta Mathematicae* 32: 69–86 and 33: 51–65. Givant 1986 items 39 and 45. JFM: Skolem. MR: Birkhoff. Reprinted in Tarski 1986a, volume 2, 507–528, and volume 3, 1–18. The note *Fundamenta Mathematicae* 1945 explains the publication dates. Cited in 8.7 and 14.3.

---. [1940] 1967. *The Completeness of Elementary Algebra and Geometry*. Paris: Centre National de la Recherche Scientifique, Institute Blaise Pascal. Givant 1986 item $67^m$a. Reprint from page proofs of a booklet scheduled for 1940 publication in Paris by Hermann & Cie. but destroyed by war. Reprinted again in Tarski 1986a, volume 4, 289–346, and translated into Polish in Tarski 1995–2001, volume 2, pages 231–252. Cited in 9.4 and 16.2.

---. [1940] 1996–1999. Letter to J. M. Bocheński, 26 April. *Filozofia nauki* 4: 123–125, 7: 197–199. Polish transcription, with an English translation by Jan Tarski. Cited in 16.3.

---. [1942–1947] 1999. Letters to Kurt Gödel, 1942–47. Edited by Jan Tarski. In Woleński and Köhler 1999, 261–273. Translated into Polish in Tarski 1995–2001, volume 2, pages 467–471. Cited in 16.2 and 16.3.

---. 1944. The semantic conception of truth and the foundations of semantics. *Philosophy and Phenomenological Research* 4: 341–375. Givant 1986 item 44a. MR: Martin. Translated into Polish in Tarski 1995–2001, volume 1, 228–282. Reprinted in Tarski 1986a, volume 2, 661–699. Tarski [1944] 1985 is an excerpt. Cited in 16.1 and 16.2.

---. [1944] 1985. Definition. *American Mathematical Monthly* 92: 358. Quotation from Tarski 1944, 359. Cited in 16.1.

---. [1944] 1987. A philosophical letter of Alfred Tarski. Edited by Morton White. *Journal of Philosophy* 84: 28–32. Givant 1986 item $87^t$, with incomplete publication data. MR: Hatcher. Translated into Polish in Tarski 1995–2001, volume 1, 283–291. Cited in 16.2 and 16.3.

---. [1946] 2000. Address at the Princeton University Bicentennial Conference on Problems of Mathematics (December 17–19, 1946). Edited, with an introduction, by Hourya Sinaceur. *Bulletin of Symbolic Logic* 6: 1–44. MR: Dawson. Excerpts were translated into Polish in Tarski 1995–2001, volume 2, 396–413. Cited in 15.12, 16.1, 16.2, and chapter 17.

---. [1946–1947] 2014. Two letters to Wacław Sierpiński, 30 October 1946 and 18 March 1947. In section 15.12 of the present book, Tarski 2014. Translations of handwritten documents, APAN sygn. III-194, in the Archiwum Polskiej Akademii Nauk (Archive of the Polish Academy of Science) in Warsaw. This is their first publication. Cited in 8.6, 15.12, and 16.3.

---. [1947–1955] 2003. Four letters to Tadeusz Kotarbiński, 6 December 1947 and 16 October 1953 from Berkeley, and 7 September 1954 and 10 December 1955 from Amsterdam. In Jadacki 2003b, 20–24. Cited in 16.3.

---. 1948. Axiomatic and algebraic aspects of two theorems on sums of cardinals. *Fundamenta Mathematicae* 35: 79–104. Givant 1986 item 48a. MR: Jónsson. Reprinted in Tarski 1986a, volume 3, 171–198. Cited in 15.12.

---. [1948] 1957. *A Decision Method for Elementary Algebra and Geometry*. Report R-109. Prepared for publication by J. C. C. McKinsey. Second edition. Santa Monica, California: RAND Corporation. Givant item $48^m$. MR: Heyting. Reprinted in Tarski 1986a, volume 3, 297–368, and translated into Polish in Tarski 1995–2001, volume 2, 253–337. Cited in 9.4, 8.5, and 16.2.

---. 1949a. *Cardinal Algebras*. With an appendix, "Cardinal products of isomorphism types," by Bjarni Jónsson and Alfred Tarski. New York: Oxford University Press. MR: Mac Lane. Givant 1986 item $49^m$. Cited in 8.7, 8.8, and 15.12.

———. 1949b. A fixpoint theorem for lattices and its applications: preliminary report. *Bulletin of the American Mathematical Society* 55: 1051–1052, 1192. Givant 1986 item 49ᵃj, reprinted in Tarski 1986a, volume 4, 592. Abstract of a talk presented at the summer meeting of the American Mathematical Society at Boulder, Colorado, 30 August–2 September 1949. Cited in 8.8.

———. 1953. A general method in proofs of undecidability. In Tarski, Mostowski, and Robinson 1953, chapter 1, 1–35. Translated into Polish in Tarski 1995–2001, volume 2, 338–379. Cited in 16.2.

———. 1955. A lattice-theoretical fixpoint theorem and its applications. *Pacific Journal of Mathematics* 5: 285–309. Givant 1986 item 55, reprinted in Tarski 1986a, volume 3, 549–576. MR: Jónsson. Cited in 8.8.

———. [1955] 1993. Sur la théorie des modèles. Translated and edited by Anne Preller. In Bouveresse 1991–1993, volume 2, 137–158. Translated into Polish in Tarski 1995–2001, volume 2, 414–445. Cited in 16.2.

---

Tarski, Alfred. [1956] 1983. *Logic, Semantics, Metamathematics: Papers from 1923 to 1938*. Translated by J. H. Woodger. Second edition, edited with introduction and analytical index by John Corcoran. Indianapolis: Hackett Publishing Company. Givant 1986 item 56ᵐ. MR: Mendelson. Also reviewed in Grattan-Guinness 1989. This contains

- Lindenbaum and Tarski [1935] 1983; Łukasiewicz and Tarski [1930] 1983;
- Tarski [1927] 1983, [1930] 1983a, [1930] 1983b, [1931] 1983, [1933] 1983, [1934] 1983, [1935] 1983, [1935–1936] 1983, [1938] 1983;
- translations of [1935] 1936b and an adaptation of Tarski 1923a.

Cited in 8.1 and 16.1.

---

Tarski, Alfred. [1957] 1959. What is elementary geometry? In Henkin, Suppes, and Tarski 1959, 16–29. Givant 1986 item 59. Reprinted in Tarski 1986a, volume 4, 17–32. MR: Mendelson. Cited in 9.4 and 12.11.

———. [1963] 2003. Letter to Tadeusz Kotarbiński, 7 June. In Jadacki 2003b, 24. Cited in 16.3.

———. 1965a. *Bibliography of Alfred Tarski*. Berkeley: Department of Mathematics, University of California, Berkeley. Superseded by Givant 1986. Cited in chapters 10 and 11, 12.0, and 16.2.

———. 1965b. A simplified formalization of predicate logic with identity. *Archiv für mathematische Logik und Grundlagenforschung* 7: 61–79. Givant 1986 item 65, reprinted in Tarski 1986a, volume 4, 251–272, and translated into Polish in Tarski 1995–2001, volume 2, 201–230. MR: Monk. Cited in 16.2.

———. [1965] 2007. Two unpublished contributions by Alfred Tarski. Edited by Francisco Rodríguez-Consuegra. *History and Philosophy of Logic* 28: 257–264. Cited in 15.10 and 16.1.

———. 1969. Truth and proof. *Scientific American* 220: 63–77. Givant 1986 item 69. Reprinted in Tarski 1986a, volume 4, 399–424. Translated into Polish in Tarski 1995–2001, volume 1, 292–332. Cited in 15.6, 16.1, and 16.2.

———. 1972. O szkicu Feliksa Grossa w *Wiadomościach*. *Wiadomości* 27(1347, 23 January): 6. In English, with Polish commentary by the journal editor. Tarski had written this letter to Feliks Gross urging republication of Gross's article "Trip to the moon: Why?"

———. [1972–1973] 1995. Two letters to Svätoslav Mathé, 21 July 1972 and 14 February 1973. *Organon F* 2: 56–58. The first is in Slovak, translated from Polish by Pavel Cmorej. The second is handwritten in Polish. Cited in 16.3.

> Tarski, Alfred. 1986a. *Collected Papers*. Edited by Steven R. Givant and Ralph McKenzie. Four volumes. Basel: Birkhäuser. Item 81$^m$a in Givant 1986 is a preliminary version. MR: Corcoran. Also reviewed, with corrections, in Maddux 1993. This contains reprints of
> 
> - Givant 1986; Knaster 1927;
> - Horn and Tarski 1948a and 1948b; Kotarbiński [1935] 1955–1956;
> - Lindenbaum and Tarski 1926; Mostowski and Tarski 1939;
> - Tarski 1923b, 1924a...e, 1925b, 1927a...b, [1927] 1929, 1928, [1928] 1930, 1929–1930, 1930a...b, 1931, [1931] 1952, 1932, [1932] 1952, [1933] 1935, [1935] 1936a...b, 1935–1936, 1937a...c, 1938a...c, [1938] 1939, 1939a...c, 1939–1945, [1940] 1967, 1944, 1948, [1948] 1957, 1949b, 1955, [1957] 1959, 1965b, 1969;
> 
> and reprints of the original versions of
> 
> - Banach and Tarski [1924] 2014; Lindenbaum and Tarski [1935] 1983;
> - Łukasiewicz and Tarski [1930] 1983;
> - Tarski [1921] 2014, [1924] 2014b, [1925] 2014, [1927] 1983, [1927] 2014a...b, [1928] 2010, [1929] 2014a...b, [1930] 1983a...b, [1930] 2014a...f, [1930–1931] 2014, [1931] 1983, [1931] 2014a...g, [1932] 2014a...e, [1934] 1936a...b, [1934] 1983, [1935–1936] 1983, [1936] 2014a...d, [1938] 1983.
> 
> Cited in the preface, 1.2, chapter 11, 12.0, 15.3, and 16.0,..., 16.2.

Tarski, Alfred. 1986b. What are logical notions? Edited by John Corcoran. *History and Philosophy of Logic* 7:143–154. Givant 1986 item 87?. MR: Cocchiarella. Translated into Polish in Tarski 1995–2001, volume 2, 446–466. Cited in 16.1.

———. 1994. *Introduction to Logic and to the Methodology of the Deductive Sciences*. Fourth edition. Translated from the Polish by Olaf Helmer. Edited, with a preface and biographical sketch, by Jan Tarski. Oxford logic guides 24. New York: Oxford University Press. MR: Mendelson. Contains J. Tarski 1994a. Tarski 1996 is a Polish translation. For earlier editions, see item Tarski [1936] 1995. Cited in 16.2.

———. 1995–2001. See a box on the next page.

———. 1996. *Wprowadzenie do logiki i do metodologii nauk dedukcyjnych*. Second edition. Translated by Monika Sujczyńska from the fourth English edition, edited and with a preface by Jan Tarski, with a supplement on elements of predicate logic by Witold Marciszewski. Warsaw: Filia Uniwersytetu Warszawskiego, Philomath, Aleph. Translation of Tarski 1994 and J. Tarski 1994a. Reviewed extensively in A. Wójtowicz 1996. The first edition had been issued in 1995 by the same publisher with a different supplement, "Logika a dzielność umysłu"—"Logic and the boldness of mind"—by Marciszewski. See J. Tarski 1996a for a complaint about that. Cited in 16.2.

Tarski, Alfred, and Steven R. Givant. 1987. *A Formalization of Set Theory without Variables*. American Mathematical Society colloquium publication 41. Providence: American Mathematical Society. Givant 1986 item 8–$^m$, with incomplete publication data. MR: Monk. Also reviewed in Monk 1989 and Németi 1990. Cited in 16.2.

———. 1999. Tarski's system of geometry. *Bulletin of Symbolic Logic* 5: 175–214. MR: Pambuccian. Cited in 9.4, 12.11, and 16.1.

> Tarski, Alfred. 1995–2001. *Pisma logiczno-filozoficzne*. Volume 1: *Prawda*. Volume 2: *Metalogika*. Translated and annotated, with introduction, by Jan Zygmunt. With bibliography by Steven Givant, supplemented by Jan Zygmunt. Warsaw: Wydawnictwo Naukowe PWN. The title means *Logical-Philosophical Writings*. Reviewed in Cesarski 1995. This contains Givant 1995, Zygmunt 1995, and
> - reprints of Tarski 1933, 1934, 1936a...b;
> - and of the original versions of Tarski [1930–1931] 2014, [1936] 2014d.
>
> In addition, it contains Polish translations of
> - Tarski 1932, [1938] 1939, 1938–1939, 1939b, [1939–1940] 1995, [1940] 1967, [1942–1947] 1999, 1944, [1944] 1987, [1948] 1957, 1953, [1955] 1993, 1965b, 1969, [1946] 2000, 1986b, parts of Tarski [1930–1936] 1992, [1946] 2000;
>
> and translations of the original versions of
> - Lindenbaum and Tarski [1935] 1983; Łukasiewicz and Tarski [1930] 1983;
> - Tarski [1930] 1983a, [1934] 1983, [1938] 1983, [1938] 1983.
>
> Cited in 16.0,..., 16.2.

> Tarski, Alfred. 2014. *Early Work in Poland—Geometry and Teaching*. Translated and edited by Andrew McFarland, Joanna McFarland, and James T. Smith. New York: Birkhäuser. The present book. This contains
> - Banach and Tarski [1924] 2014; Łukasiewicz [1928–1929] 2014;
> - Moese [1932] 2014;
> - Tarski [1921] 2014, [1924] 2014a...b, [1925] 2014, [1927] 2014a...b, [1929] 2014a...b, [1930] 2014a...f, [1930–1931] 2014, [1931] 2014a...g, [1932] 2014a...e, [1934] 2014, [1935] 2014, [1936] 2014a...d, [1946–1947] 2014,
>
> and a reprint of Tarski 1925a. See the overviews in the preface and in the introductions to Parts One to Four.

Tarski, Alfred, and Henryk Moese. 1952. *Concerning the Degree of Equivalence of Polygons*. Translated by Izaak Wirszup. Chicago: The College, University of Chicago. A reproduced typescript, this contains draft translations Tarski [1931] 1952 and [1932] 1952 of the original versions of Tarski [1931] 2014a and [1932] 2014d and a draft translation of the original version of Moese [1932] 2014. Cited in 7.0,..., 7.4, 8.5, and 9.8.

Tarski, Alfred, Andrzej Mostowski, and Raphael M. Robinson. 1953. *Undecidable Theories*. Amsterdam: North–Holland Publishing Company. Givant 1986 item 53$^{\mathrm{m}}$. MR: Kreisel. Contains Tarski 1953. Cited in 16.2.

Tarski, Alfred, et al. 1971. Letter to Tadeusz Kotarbiński, 24 April. In Jadacki 2003b, 25. There were ten cosigners. Cited in 16.3.

Tarski, Jan. 1994a. A short biographical sketch of Alfred Tarski. In Tarski 1994, xviii–xxi. Tarski 1996 contains a Polish translation. Cited in chapter 17.

———. 1994b. Uwagi o artykule Woleńskiego "Alfred Tarski jako filozof". *Wiadomości matematyczne* 30: 265–269. The title means "Remarks on Wolenski 1987." Cited in 18.2.

———. 1996a. W sprawach wydawniczych. *Filozofia nauki* 4(4): 153–154. The title means "On matters of publishing." Dated 23 March, this letter to the editor is about the first edition of Tarski 1996.

———. 1996b. Philosophy in the creativity of Alfred Tarski. In Hempoliński 1996, 157–159. Cited in chapter 17.

———. 2012. Personal communication to James T. Smith, 20 February. Cited in 14.3.

Tatarkiewicz, Krzysztof. 2003. W sprawie daty śmierci Z. Waraszkiewicza. *Wiadomości matematyczne* 39: 221–222. The title means "On the date of death of Z. Waraszkiewicz." This corrects an error in Derkowska 2001. Cited in 7.1.

Terracini, Alessandro. 1958. Ugo Amaldi. *Atti della Accademia delle Scienze di Torino, classe di scienze fisiche, matematiche e naturali* 92 (1957–1958): 687–695. In Italian. Cited in 9.2.

Thomasson, Amie. 2012. Roman Ingarden. In Zalta 2011– at http://plato.stanford.edu/archives/fall2012/entries/ingarden. Cited in 15.6.

Tomczak, Roman. 1927. *Kalendarz nauczycielski na rok 1927*. Warsaw: Nakładem Zarządu Głównego Związku Polskiego Nauczycielstwa Szkół Powszechnych. The title means *Teacher's Calendar for the Year 1927*. Cited in 9.6.

Toruń, University of. 1995. *Pracownicy nauki i dydaktyki Uniwersytetu Mikołaja Kopernika, 1945–1994: Materiały do biografii*. Edited by Sławomir Kalembki. Toruń: Uniwersytet Mikołaja Kopernika w Toruniu. The initial title phrase means *Scientific and Teaching Staff of the Nicolaus Copernicus University*. Cited in 7.1.

Tricomi, Francesco G. 1962. Matematici Italiani del primo secolo dello stato unitario. *Memorie dell'Accademia delle Scienze di Torino, classe di scienze fisiche, matematiche e naturali* (series 4) 1: 1–120. Augmented by PRISTEM 2005–. Cited in 4.1.

Trzęsicki, Kazimierz. 2010. Greniewski Henryk (1903–1972). Warsaw: Polskie Towarzystwo Logiki i Filozofii Nauki. On the Internet at http://logic.org.pl/doc/historia/hg.pdf, the site of the Polish Society for Logic and Philosophy of Science. In Polish. An abridgement is in Maryniarczyk et al. 2011, volume 1, 451. Cited in 15.2.

Turkowski, Romuald. 1994. Schayer Wacław. *Polski słownik biograficzny* 35: 413–415. In Polish. Cited in 9.9.

Ulam, Stanisław Marcin. 1976. *Adventures of a Mathematician*. New York: Charles Scribner's Sons. MR: Oxtoby. Cited in 14.3.

Van den Dries, Lou. 1988. Alfred Tarski's elimination theory for real closed fields. *Journal of Symbolic Logic* 53: 7–19. MR: Coplakova. Cited in 18.1.

Van Heijenoort, Jean. [1967] 1970. *From Frege to Gödel: A Source Book in Mathematical Logic, 1879–1931*. Cambridge, Massachusetts: Harvard University Press. MR: Dawson. Contains Gödel [1931] 1967 and Padoa [1900] 1901. Cited in 4.1.

Vaught, Robert L. [1971] 1974. Model theory before 1945. In Henkin et al. 1974, 153–172. MR: Anonymous. Cited in 15.6, 18.1, and 18.2.

———. 1986–1987. Alfred Tarski's work in model theory. *Journal of Symbolic Logic* 51: 869–882; 52(4): vii. MR: Baldwin. Cited in 18.1 and 18.2.

Veblen, Oswald. 1911. The foundations of geometry. In Young 1911, 3–54. Cited in 9.4.

Viola, Tullio. 1957. Ugo Amaldi. *Bollettino della Unione Matematica Italiana* (3) 12: 727–730. In Italian. Cited in 9.2.

Vitali, Giuseppe. 1905. *Sul problema della misura dei gruppi di punti di una retta: Nota*. Bologna: Gamberini e Parmeggiani. Five-page pamphlet. Cited in 4.2, 4.4, and 6.0.

*Völkischer Beobachter*. 1938 (North German edition, 2 October, 1–3). Die Reichshauptstadt huldigt dem Befreier des Sudetenlandes. Berlins stürmischer Dank. Wilhelmplatz: ein einziger Orkan jubelnder Begeisterung. Cited in 14.3.

Wagon, Stan. 1993. *The Banach–Tarski Paradox*. Paperback edition. Cambridge: Cambridge University Press. MR: Bzyl. Originally published in 1985; the later edition has some additional information. Cited in 4.2, ..., 4.4, 8.6, and 8.7.

Wajsberg, Mordchaj. 1977. *Logical Works*, edited by Stanisław Surma. Wrocław: Polish Academy of Sciences, Institute of Philosophy and Sociology. Contains Surma 1977.

Wapner, Leonard M. 2005. *The Pea and the Sun: A Mathematical Paradox*. Wellesley, Massachusetts: A K Peters. MR: Plotkin. Also reviewed in Komjáth 2006. Cited in 8.6.

Warsaw, University of. 1918–1919. Document "Uniwersytet Warszawski Sekretariat" dated 8 November 1918 and 5 February 1919. In the Archive of the University of Warsaw, Signatura Teitelbaum-Tarski Alfred: RP 2909. Cited in 1.1.

———. 1921–1939. *Skład uniwersytetu i spis wykładów*. One volume for each year. Warsaw: University of Warsaw. The title means *University Bulletin and Schedule of Lectures*. Cited in 9.3 and 9.7.

Warszawski Instytut Filozoficzny. 1921–1922. Wykłady, odczyty, referaty. *Ruch filozoficzny* 6: 72, 146. The title means "Lectures, readings, papers." Cited in chapter 3.

———. 1931. October 1931 minutes. *Ruch filozoficzny* 12 (1930–1931), 228. Cited in 14.2.

———. 1937. Wykłady, odczyty, referaty. *Ruch filozoficzny* 13: 149. The title means "Lectures, readings, papers." Cited in 14.2.

Waszyński, Edmund. 1996. Professor Dr. Rudolf Weigl (1883–1957) i działalność jego Instytutu Tyfusowego we Lwowie w latach 1939–1944. *Archiwum historii i filozofii medycyny* 59: 77–84. These pages also include the German translation "Professor Dr. Rudolf Weigl (1883–1957) und die Tätigkeit seines Instituts für Fleckfieberforschung in Lwów in den Jahren 1939–1944" and an English summary. Cited in 4.2.

Watt, Richard M. 1979. *Bitter Glory: Poland and Its Fate, 1918 to 1939*. New York: Simon and Schuster. Cited in 1.1 and chapter 3.

Wells, Benjamin. 2007. Alfred Tarski, friend and daemon. *Notices of the American Mathematical Society* 54: 982–984. Cited in chapter 17.

White, Morton Gabriel. 1999. *A philosopher's story*. University Park, Pennsylvania: Pennsylvania State University Press. Cited in 16.3.

Whitehead, Alfred N., and Bertrand Russell. 1910–1913. *Principia Mathematica*. Three volumes. Cambridge, England: The University Press. JFM: Jourdain. Cited in 4.4 and 15.6.

Witt-Hansen, Jobs. 1969. Obituary on Jørgen Jørgensen. *Logique et analyse*, new series, 46: 121–122. Cited in 16.1.

Wilkosz, Witold. 1936. Znaczenie logiki matematycznej dla matematyki i innych nauk ścisłych. In Polski Zjazd Filozoficzny 1936, 343–347. The title means "The significance of mathematical logic in mathematics and other exact sciences." Contains the original version of the comment Tarski [1936] 2014b. Cited in 15.9.

Wilson, Trevor M. 2005. A continuous movement version of the Banach–Tarski paradox: a solution to de Groot's problem. *Journal of Symbolic Logic* 70: 946–952. Cited in 8.6.

Wójcicki, Ryszard, ed. 1981. Maria Kokoszyńska-Lutman (1905–1981). *Studia Logica* 40: 313–314. The article is by the editors; Wójcicki was editor-in-chief. Cited in 14.2.

Wojciechowska, Agnieszka. 2001. Alfred Tarski (1901–1983). *Matematyka: czasopismo dla nauczycieli* 54(3): 132–135. In Polish. Cited in chapter 17.

———. 2011. Alfred Tarski (1901–1983). *Matematyka: czasopismo dla nauczycieli* 64(9): 3–9. In Polish. Cited in chapter 17.

Wójcik, Andrzej. 1985. Alfred Tarski: Życie i dzieło. *Pismo literacko-artystyczne* 4 (1): 174–178. The titles mean "Life and work" and *Literary-Artistic Writings*. Cited in chapter 17.

Wójtowicz, Anna. 1996. Logika dla filozofów nauki. *Filozofia nauki* 4: 117–122. The title means "Logic in the philosophy of science." This is a review of Tarski 1996. Cited in 16.2.

Wojtowicz, Władysław. [1919] 1926. *Zarys geometrji elementarnej do użytku szkół średnich*. Sixth edition. Lwów: Wydawnictwo Zakladu Narodowego im. Ossolinskich. The 1919 edition was published in Warsaw by Gebethner i Wolff. Reviewed in Kwietniewski 1923. The title means *Outline of Elementary Geometry for Use in Secondary Schools*. The third edition has been revised and modernized by editor Jerzy Norwa and published in the series Graϑient—Kufer matematyków, Warsaw: Fundacja Rozwoju Matematyki Polskiej, 2000–2001. Cited in 7.1, 7.2, 7.4, 9.2, 9.3, and 12.11.

Wojtowicz, Władysław, and Stefan Straszewicz. 1923. Od redakcji. *Przegląd matematyczno-fizyczny* 1: 1–2. The title means "From the editors." Cited in 4.3.

Woleński, Jan. 1984. Alfred Tarski (1901–1983). *Studia filozoficzne* (2): 3–8. In Polish. Cited in chapter 17.

———. 1987. Alfred Tarski jako filozof. *Wiadomości matematyczne* 27 (1986–1987): 247–259. MR: Murawski. The title means "Alfred Tarski as a philosopher." J. Tarski 1994b contains remarks about this paper. Cited in 18.2.

———. 1989. *Logic and Philosophy in the Lvov–Warsaw School*. Synthese library 198. Dordrecht: Kluwer Academic Publishers. Cited in the preface.

———. 1990a. *Kotarbiński*. Warsaw: Wiedza Powszechna. Contains Woleński 1990b.

———. 1990b. Życie i twórczość. In Woleński 1990a, 7–15. The title means "Life and work." This is a biographical sketch of Tadeusz Kotarbiński. Cited in 1.1.

———. 1993. Tarski as a philosopher. In Coniglione, Poli, and Woleński 1993, 319–338. This is not a translation of Woleński 1987. Cited in 15.2, 15.10, 15.11, and 18.2.

———. 1994. *Philosophical Logic in Poland*. Synthese library 228. Dordrecht: Kluwer Academic Publishers. Contains Bocheński 1994.

———. 1995a. Mathematical logic in Poland 1900–1939: People, circles, institutions, ideas. *Modern Logic* 5: 363–405. MR: Sinaceur. Reprinted in Woleński 1999, 59–84. This paper has two sections numbered 4. Cited in chapter 3, 9.4, 14.2, and 14.3.

———. 1995b. On Tarski's background. In Hintikka 1995, 331–341. MR: Anonymous. Reprinted in Woleński 1999, 126–133. Cited in chapter 17.

———. 1999. *Essays in the History of Logic and Logical Philosophy*. Dialogikon volume 8. With foreword by Christian Thiel. Cracow: Jagiellonian University Press. Contains reprints of Woleński 1995a and 1995b.

———. 2000–2001. Review of Leśniewski 1992. *Modern Logic* 8: 195–201.

———. 2001a. Alfred Tarski-książę logików: wzór na prawdę. *Polityka* 2281 (20 January): 76–77. The title phrases mean "Prince of logicians: pattern for truth." The journal is a weekly newsmagazine. Cited in chapter 17.

———. 2001b. In defense of the semantic definition of truth. *Synthese* 126: 67–90. Cited in 15.11.

———. 2008. Mathematical logic in Warsaw: 1918–1939. In Ehrenfeucht, Marek, and Srebrny 2008, 30–46. Cited in chapter 3.

———. 2009. Alfred Tarski: Logik z fantazją. *Gazeta wyborcza* (4–5 April): 24–25. The title phrase means "Logician with flair." The journal is a daily newspaper with national circulation. Cited in chapter 17.

Woleński, Jan, and Eckehart Köhler. 1999. *Alfred Tarski and the Vienna Circle: Austro–Polish Connections in Logical Empiricism*. Vienna Circle Institute Yearbook 6. Dordrecht: Kluwer Academic Publishers. Mostly the proceedings of a 1997 congress in Vienna. Contains A. Feferman [1997] 1999, S. Feferman [1997] 1999, and Tarski [1942–1947] 1999. Cited in 16.3 and chapter 17.

Wolna Wszechnica Polska. 1925. *Skład osoby i spis wykładów*. Warsaw: Wolna Wszechnica Polska. The author name and title mean "Free University of Poland" and *List of Personnel and Schedule of Courses*. Cited in chapter 3.

Woodger, Joseph H. 1937. *The Axiomatic Method in Biology*. With appendices by Alfred Tarski and W. F. Floyd. Cambridge, England: Cambridge University Press. JFM: Ackermann. Contains Tarski 1937a. Cited in 14.3.

Woytak, Lidia. 1987. Alfred Tarski—matematyk o języku. *Przegląd Polonijny* 1987 (4): 67–75. The title phrase means "a mathematician on language." The journal title means *Polonia Review*. Cited in chapter 17.

Wynot, Edward D., Jr., 1983. *Warsaw between the World Wars: Profile of the Capital City in a Developing Land, 1918–1939*. East European monograph 129. New York: Columbia University Press. Cited in 1.1, chapter 3, 9.1, and 14.2.

Yanigahara, Kitizi. 1927. On the axiomatic simplicity of the proof of Pythagorean theorem by dissection and addition. *Tôhoku Mathematical Journal* 18: 59–64. JFM: Süss. Cited in 7.1.

Yarden, Tama. 2006. Dov Jarden. On the Internet at http://www.shvoong.com/humanities/religion-studies/243858-dov-jarden/. Cited in 7.1.

Young, J. W. A., editor. 1911. *Monographs on Topics of Modern Mathematics Relevant to the Elementary Field*. London: Longmans, Green and Co. JFM: Jourdain. Reprinted in 1955 by Dover Publications with an introduction by Morris Kline. Contains Veblen 1911.

Zaanen, Adriaan C. 1958. *An Introduction to the Theory of Integration*. Amsterdam: North-Holland Publishing Company. MR: Hildebrandt. Cited in 4.2 and 4.4.

Zagórowski, Zygmunt, editor. 1924–1926. *Spis nauczycieli: Szkół wyższych, średnich, zawodowych, seminarjów nauczycielskich oraz wykaz zakładów naukowych i władz szkolnych*. Two volumes. Lwów: Książnica Polska, Książnica–Atlas. The title means *List of Teachers in Higher, Middle, Vocational, and Teacher Training Colleges and List of Seminars, Scientific Establishments, and Government Schools*. Cited in chapter 3, 9.1, 9.2, and 9.9.

Zahorski, Bolesław. 1920. *Wiersze Piłsudczyka*. Minsk and Warsaw: Wende. The title means *Piłsudski Poems*. Cited in 9.1.

Zakład Ubezpieczeń Społecznych. 1935. Letter to Alfred Tarski, 22 November. In the Archiwum Akt Nowych (Central Archives of Modern Records) in Warsaw. The author name means Social Insurance Institution. Cited in 9.1.

Zalasiewicz, Jan. 2009. The meaning of palaeontology. *Palaeontology Newsletter* 72: 8–13. Cited in 14.1.

Zalta, Edward N., editor. 2011–. *Stanford Encyclopedia of Philosophy*. Stanford: http://plato.stanford.edu/. Internet website. Contains Betti 2011, Cat 2010, Gómez-Torrente 2006, Segerberg 2009–, and Thomasson 2012.

Zawirski, Zygmunt. [1934] 1936. Bedeutung der mehrwehrtigen Logik für die Erkenntnis und ihr Zusammenhang mit der Wahrscheinlichkeitsrechnung. In International Congress of Philosophy 1936, 175–180. Discussed in Tarski [1934] 1936a and [1936] 2003a. Cited in 16.1.

——— . 1936. W sprawie syntezy naukowej. In Polski Zjazd Filosoficzny 1936, 347–352. The title means "Concerning scientific synthesis." Contains the original version of the comment Tarski [1936] 2014c. Cited in 15.10.

Ziemiański, Ignacy. 1933. *Praca kobiet w P.O.W.–Wschód*. Preface and remembrance by Melchior Wańkowicz, cover and archival items by Adolf Buraczewski. Warsaw: Wł. Łazarski. The title means, *Efforts of Women in the Polish Military Organization–East*. Cited in 9.1.

Zupančič, Rihard. 1918–1921. *Poglądowa nauka geometryi dla klasy pierwszej; zarys geometryi dla klasy drugiej; zarys geometryi dla klasy trzeciej*. Three separate volumes in series. Adapted by Ludwik Hordyński from German editions. Lwów: Seyfarth, Ossolinskich. The titles mean *An Intuitive Study of Geometry for the First Class*, and *Outline of Geometry for the Second and Third Classes*. The original volumes were published in Vienna by F. Tempsky. The German form of the author's name is Richard Suppantschitsch. Cited in 9.2.

Zydler, Jan. 1925. *Geometrja w zakresie szkoły średniej.* Fourteenth edition. Warsaw: Wydawnictwo M. Arcta. The title means *Geometry for Secondary Schools.* Cited in 9.2, 12.11, and 13.1.

Zygmunt, Jan. 1991. Mojżesz Presburger: Life and work. *History and Philosophy of Logic* 13: 211–223. MR: Dawson. Cited in 9.3 and 9.4.

———. 1995. Szkic biograficzny. In Tarski 1995–2001, volume 1, vii–xx. Biography of Alfred Tarski. Cited in 9.4 and chapter 17.

———. 2000. Alfred Tarski. *Edukacja filozoficzna* 29: 274–307. In Polish. Cited in chapter 17.

———. 2001. Alfred Tarski. In Mackiewicz 2001, volume 1, 342–375. In Polish. Cited in chapter 17.

———. 2009. Alfred Tarski—logik i metamatematyk. In Dobierzewska-Mozrzymas and Jezierski 2009, 305–327. The title phrase means "logician and metamathematician." Cited in 18.2.

———. 2010. Tarski's first published contribution to general metamathematics. In Béziau 2010, 59–66. Cited in 9.4, 15.6, and 16.1.

# Permissions

The editors acknowledge gratefully the cooperation of several archival institutions that provided access to their resources and permission to publish images or translations of various items in the indicated sections of the present book:

| | | |
|---|---|---|
| Archiwum Akt Nowych (Warsaw) | Tarski [1934] 2014 letter: translation | §16.3 |
| | Tarski [1935] 2014 letter: translation | §16.3 |
| Archiwum Fotografii Ośrodka KARTA (Warsaw) | Photo: Chwiałkowski, from sygnatura OK_0997_0001_024 | §9.9 |
| Archiwum Polskiej Akademii Nauk (Warsaw) | Tarski [1946–1947] 2014 | §15.14 |
| Archiwum Państwowe Miasta Stołecznego Warszawy | Photo: Ul. Koszykowa 51 | §9.1 |
| Archiwum Państwowe w Lublinie | Poster: crucifix | §1.1 |
| Archiwum Uniwersytetu Warszawskiego | Tarski 1924f coursebook | §1.1 |
| | Photo from coursebook Pasenkiewicz 1935 | Ch. 3 |
| | Tarski [1918] 2014 CV: signature, translation | Ch. 3, §16.3 |
| | Tarski 1920 letter: signature | Ch. 3 |
| | Tarski 1921–1926: letter | Ch. 3 |
| | Tarski [1924] 2014a letter: signature, translation | Ch. 3, §16.3 |
| | Tarski [1939] 2003 letter: signature | frontispiece |
| | Photo: Wajsberg, from sygnatura RP15153 | Ch. 3 |
| | Photo: Kuratowski, from sygnatura RP379 | §4.4 |
| | Photo: Presburger, from sygnatura RP16485 | §9.4 |
| | Photo: Mostowski, from sygnatura RP36752 | §9.4 |
| | Photo from coursebook Schayer 1932 | §9.9 |
| | Photo: Hosiassonówna, from sygnatura RP5224 | §14.3 |
| | Photo: Lindenbaum, from sygnatura RP11947 | §14.3 |
| Bancroft Library, University of California, Berkeley | Quotations from Pasenkiewicz 1984 | Ch. 3 |
| | Shipboard menu and cover | §14.3 |

| | | |
|---|---|---|
| Centralna Biblioteka Matematyczna, Instytut Matematyczny, Polska Akademia Nauk (Warsaw) | Photo: 1929 Congress members | §9.5 |
| Książnica Kopernikańska in Toruń | Photo: Henryk Moese *Gazeta Gdańska* 1935.09.26 | 7.1 |
| Mathematisches Forschungsinstitut Oberwolfach: Photo Collection | Photo: Alfred Tarski | Ch. 16 Intro. |
| Narodowe Archiwum Cyfrowe (Warsaw) | Photo: Twardowski | §1.3 |
| | Photo: Jabloński | Ch. 3 |
| | Photo: Sierpiński | §4.1 |
| | Photo: Łukasiewicz | §9.4 |
| | Photo: Wilkosz | §15.9 |
| United States Holocaust Memorial Museum (Washington) | Poster: Trotsky as devil | §1.1 |
| Wikimedia Commons | Photos: Amaldi, Enriques | §9.2 |

Springer, the publisher of this book, extended permission for translating or reproducing images or text from their earlier publications:

| | | |
|---|---|---|
| Translations from Tarski 1986a, *Collected Papers* | Tarski [1921] 2014 | Ch. 2 |
| | Tarski [1924] 2014b | Ch. 5 |
| | Banach–Tarski [1924] 2014 | Ch. 6 |
| | Tarski [1931] 2014a | §7.2 |
| | Tarski [1932] 2014d | §7.4 |
| | Tarski [1929] 2014a | Ch. 10 |
| | Tarski [1932] 2014e | Ch. 11 |
| | Tarski [1930] 2014a–f | §§12.1–6 |
| | Tarski [1931] 2014b–g | §§12.7–12 |
| | Tarski [1932] 2014b–c | §§12.13–1 |
| | Tarski [1925] 2014 | 4 |
| | Tarski [1927] 2014a–b | §15.1 |
| | Tarski [1929] 2014b | §§15.2–3 |
| | Tarski [1930–1931] 2014 | §15.5 |
| | Tarski [1932] 2014 | §15.6 |
| | Tarski [1936] 2014a–d | §15.7 §15.8–11 |
| Quotations from Tarski 1986a, *Collected Papers* | Tarski 1969 | §15.6 |
| From Marchisotto/Smith 2007, *Legacy of Mario Pieri in Geometry and Arithmetic* (Birkhäuser), §3.10.2 | Text and figures | §9.3 |
| From Givant 1991, "Portrait of Alfred Tarski," *Mathematical Intelligencer* | Photos: Tarski, Leśniewski | §1.1 |
| From Givant 1999, "Unifying threads in Alfred Tarski's work," *Mathematical Intelligencer* | Photo: Tarski | Frontispiece |
| | Photo: Szmielew | §9.4 |
| From Purkert 2008, "The double life of Felix Hausdorff/Paul Mongré," *Mathematical Intelligencer* | Photo: Hausdorff | §4.4 |

From Szmuilewicz-Lachman et al. 1994, . . . Photo: Zawirski . . . . . . . . . . . . . . §15.10
*Zygmunt Zawirski* (Kluwer)

Several individuals, publishers, and other organizations graciously extended permission for translating or reproducing images or text from their Internet websites or earlier publications:

| | | |
|---|---|---|
| Prof. Alina Filipowicz-Banach . . . . . . . . . . . . | Photo: Banach . . . . . . . . . . . . . . | §4.2 |
| Galeria Grabek (Lublin, Poland) . . . . . . . . . | Poster: Poznań exhibition . . . . . . | §9.5 |
| Prof. Jacek Juliusz Jadacki . . . . . . . . . . . . . . | Photo: Kotarbiński . . . . . . . . . . . | §1.1 |
| | Photo: Ul. Sułkowski 4 . . . . . . . . | §9.1 |
| Polskie Towarzystwo Matematyczne . . . . . . . | Photo: Mazurkiewicz . . . . . . . . . | §1.1 |
| Uniwersytet Mikołaja Kopernika, . . . . . . . . . | Łukasiewicz [1928–1929] 2014: . . | §15.4 |
| Instytut Filozofii (*Ruch filozoficzny*) | translation | |
| Prof. Jan Tarski . . . . . . . . . . . . . . . . . . . . . . . | Photo: Witkowska sisters . . . . . . | §9.1 |
| | Translations from *Geometrja* . . . . | Ch. 13 |
| | Photo: Kokoszyńska, Tarski . . . . | §14.2 |
| Prof. Andrzej Jerzmanowski . . . . . . . . . . . . . | Translations from *Geometrja* . . . . | Ch. 13 |

The present editors have searched without success for heirs of the third author, Zygmunt Chwiałkowski, of Tarski's [1935] 1946 text *Geometrja*. Its original publishers disappeared in the aftermath of World War II, and we have searched without result for successors. Should anyone have questions about rights to that publication, please contact the senior of the editors, James T. Smith.

# Index of Persons

This book has indexes of persons and of subjects. The subject index starts on page 485. The present index lists persons featured in the main text of the book. It does not cover the entries in the bibliography. An = sign indicates that a person is listed under a different name. Personal dates are supplied if possible. Many biographical sketches are included in this book in boxes or footnotes; those names and the page numbers of the sketches are *italicized* in this index. The book contains many portraits; those names and page numbers are **boldfaced**. Alphabetization of the index ignored spaces, punctuation, and diacritical marks. *Caution*: this convention differs from the Polish standard.

Adalbert, Saint = Wojciech, Saint
Adamowicz, Zofia (1950– ), 413
Addison, John West, Jr. (1930– ), 180, 399
***Ajdukiewicz, Kazimierz*** (1890–1963), *36–37*, 180, 325, 327, 342, 344, 361, 365–366, 370
Alfred = Tarski, Alfred
Allen, Layman Edward (1927– ), 341
***Amaldi, Ugo*** (1875–1957), 45–46, 61–62, 90, *181–182*, 183, 189, 213. *See also the subject index*
Anders, Władysław (1892–1970), 34, 224, 320
Andréka, Hajnal Ilona (1947– ), 414
Anellis, Irving Henry (1946–2013), 401
Apostol, Tom Mike (1923– ), 163
Archimedes (287–212 B.C.E.), 213. *See also the subject index*
Aristotle (384–322 B.C.E.), ix, 362
Auerbach, Walter (1900–1942), 374
Axler, Sheldon Jay (1949– ), xiv

Bacon, Francis (1561–1626), 8
Baley, Stefan (1885–1952), 20
Banach, Katarzyna (1864–19??), 57
Banach, Łucja (????–1954), 57
***Banach, Stefan*** (1892–1945), xi, xiii, 41, 45, 51–**56**, *57*–66, 70, 73–75, 79, 90–123, 165–166, 189, 320, 363–364, 369, 378, 381. *See also the subject index*
Barwińska, Regina = Łukasiewiczowa, Regina
Barwise, Jon = Barwise, Kenneth Jon
Barwise, Kenneth Jon (1942–2000), 407, 419
Beneš, Edvard (1884–1948), 325

Bernstein, Felix (1878–1956), 126. *See also the subject index*
Betti, Arianna (1970– ), 18, 40, 362
Betti, Enrico (1823–1892), 181
Béziau, Jean-Yves (1965– ), 386
Białobrzeski, Czesław (1878–1953), 33
*Błaszczyk, Leon Tadeusz* (1923– ), *320*
Blok, Willem Johannes (1947–2003), 411
*Bocheński, Father Innocentius Maria* (1902–1981), 17, *368–369*, 396
Bocheński, Józef Maria = Bocheński, Father Innocentius Maria
Bolyai, Farkas Wolfgang (1775–1856). *See the subject index*
Bonola, Roberto (1874–1911), 265
Boole, George (1815–1864), vi
Borel, Émile (1871–1956), 33, 65
*Bornstein, Benedykt Leon* (1880–1948), *368*
Borowski, Marian (1879–1938), 12
*Borsuk, Karol* (1905–1982), *379*, 381
Bouveresse, Jacques (1940– ), 389
Brandes, Hans (1883–1965), 126
Braun, Jerzy (1901–1975), 367, 370
Braus, Łucja. = Banach, Łucja
Brentano, Franz (1838–1917), 18
Brouwer, Luitzen Egbertus Jan (1881–1966), v
Burda, Józef (1915?–1992?), 251
Burkhardt, Heinrich Friedrich Karl Ludwig (1861–1914), 189

Caen, Herbert Eugene (1916–1997), 399
Cantor, Georg Ferdinand Ludwig Philipp (1845–1918), vi, 15–16. *See also the subject index*

Carnap, Rudolf (1891–1970), v, 325, 334, 394, 419
Carnielli, Walter Alexandre (1952– ), 401
Carver, Walter Buckingham (1879–1961), 162
Castelnuovo, Guido (1865–1952), 181
Castrillo Criado, Pilar, 388
Cauchy, Augustin Louis (1789–1857), vi
Chang, Chen-Chung (1927– ), 414
Chojnacki, Father Piotr Julian (1897–1969), 367, 370
Chuaqui Kettlun, Rolando Basim (1935–1994), 399, 400
*Chwiałkowski, Zygmunt* (1884–1952?) (С. А. Хвялковскій), 210, 214–**218**, *219*–224, 273–318, 475
*Chwistek, Leon* (1884–1944), 16, 32, 328–**329**, 334, 336, 343
Chwistkowa, Olga, 329
Cmorej, Pavel (1937– ), 398
Coniglione, Francesco (1949– ), 419
Corcoran, John (1937– ), xiv, 38, 389, 403, 407, 418
Couturat, Louis (1868–1914), 36
Curwood, James Oliver (1878–1927), 245
Czarliński, Tadeusz, 252
Czelakowski, Janusz Michal (1949– ), 414
Czerwiński, Sławomir (1885–1931), 226
*Czeżowski, Tadeusz* (1889–1981), 343–*345*
Czubińska, Irena (1961– ), 399

Davies, Norman (1939– ), 3
Davis, Anne C. = Morel, Anne C.
Dawson, John William, Jr. (1944– ), 414
Dedekind, Julius Wilhelm Richard (1831–1916), 238. *See also the subject index*
Dedekind, Richard = Dedekind, Julius Wilhelm Richard
Dehn, Max (1878–1952), 47, 61, 62, 90
Desargues, Girard (1591–1661). *See the subject index*
De Willman-Grabowska, Helena (1870–1957), 367
*De Zolt, Antonio* (1847–1926), 46–*47*. *See also the subject index*
Dickstein, Samuel (1851–1939), 5, 201
Dobierzewska-Mozrzymas, Ewa Zofia (1936– ), 420
Domoradzki, Stanisław, xiv, 406
Doner, John Elliot, 161, 411

*Drewnowski, Jan Franciszek* (1886–1978), *33*–*34*
Dryjski, Albert (1889–1956), 20
Dubiel, Władysław (1934– ), 261
Dunin-Borkowski, Władysław (1884–1922), 329
Duren, Peter (1935– ), 407, 419

Edwards, Paul (1923–2004), 416
Ehrenfeucht, Ewa Krystyna (1938– ) 331, 336, 400, 405
Einstein, Albert (1879–1955), 33, 406
***Enriques, Federigo*** (1871–1946), 45–46, *181*–**182**, 183–184, 186, 188–189, 213. *See also the subject index*
Erdős, Paul (1913–1996), 377–378, 406
Etchemendy, John William (1952– ), 412
Euclid (325?–265? B.C.E.), vi, 45–47, 188,194, 265–267, 278. *See also the subject index*
Eudoxus of Cnidus (410?–355? B.C.E.), 45
Eves, Howard Whitley (1911–2004), 162

Feferman, Anita Burdman (1927– ) and Solomon (1928– ), ix, 3, 36, 38, 174–179, 331–333, 336, 376, 396, 400–404, 409
Feferman, Solomon (1928– ), 167, 361, 407, 414, 416, 419
Fraenkel, Abraham Adolf Halevi (1891–1965), 200, 226
Frege, Friedrich Ludwig Gottlob (1848–1925), vi, ix
Frycz, Stefan, 4
Furman, Grazina Ula, xiv

Gabbay, Dov M. (1945– ), 386
García-Carpintero, Manuel (1957– ), 375
Gawecki, Bolesław Józef (1889–1984), 340
Gerwien, P., 47, 62, 74. *See also the subject index*
Gilewicz, Kazimierz, 245
Givant, Steven Roger (1943– ), ix, xiii, xiv, 38, 159, 162, 188, 244, 267, 390–391, 402–403, 415, 421
Gödel, Kurt (1906–1978), v, ix, 322, 365, 386, 396, 402, 418. *See also the subject index*
Goetz, Arek, xiv
Goldschmidt, Charlotte = Hausdorff, Charlotte
Golińska-Pilarek, Joanna, 401–403, 409
Gómez-Torrente, Mario (1967– ), 415

Grattan-Guinness, Ivor (1941–), v
Greczek, Stefan (1867–1967), 57
*Grelling, Kurt* (1886–1942), 330, *334*, 336
Grelling, Richard (1853–1929), 334
*Greniewski, Henryk* (1903–1972), 341–
   *343*, 348
Gross, Feliks (1906–2006), 397
Grygianiec, Mariusz, 387
Grzymała, Aleksandr, 269
Gupta, Haragauri Narayan, 265

Hadamard, Jacques Salomon (1865–
   1963), 226
Hadwiger, Hugo (1908–1981), 166
Haefeli-Huber, Verena Esther =
   Huber-Dyson, Verena Esther
Hahn, Hans (1879–1934), 393. *See also
   the subject index*
Hájek, Petr (1940– ), 398
Halamon, Perez, 253
Haller, Rudolf (1929–2014), 387
*Hampel, Roman* (1907–1963), 176–*177*
Hardy, Godfrey Harold (1877–1947), vi
Hatcher, William S. (1935–2005), 396
Hausdorff, Charlotte (18??–1942)
**Hausdorff, Felix** (1868–1942), 48, 52–53,
   60, 65–**68**, *69*–71, 75, 95, 109, 115, 163–
   164, 381. *See also the subject index*
Heath, Sir Thomas Little (1861–1940), 265
Heinrich, Władysław (1869–1957), 373
Heller, Celia Stopnicka (1922–2011), 3, 38
Hempel, Carl Gustav (1905–1997), 333–334
Hempoliński, Michał (1930–2005), 407
Henkin, Leon Albert (1921–2006), 389–390,
   404, 414
Hertz, Paul (1881–1940), 386
Hessenberg, Gerhard (1874–1925), 40
**Hilbert, David** (1862–1943), v–vi, 37,
   46–47, 62, 79, 184–**185**, 186, 188, 264,
   348, 388, 419. *See also the subject index.*
Hintikka, Jaakko Kaarlo Juhani (1929– ),
   402, 408
Hitler, Adolf (1889–1945), 330
Hitchcock, David Lancelot (1942– ), xix, 388
Hiż, Henryk (1917–2006), 192, 404
Hodges, Wilfrid Augustine (1941– ), 404, 411
Horn, Alfred (1918–2001), 166
*Hornowski, Michał* (1893–1966), *251*

***Hosiasson-Lindenbaumowa, Janina***,
   (1899–1942), 201, 323, 325–326, 334–
   ***335***, 336, 387
Hosiassonówna, Janina =
   Hosiasson-Lindenbaumowa, Janina
Hrycak, Teodor, 255
Huber-Dyson, Verena Esther (1927– ), 403
Huntington, Edward Vermilye (1874–
   1952), 358. *See also the subject index*
Husserl, Edmund Gustav Albrecht (1859–
   1938), 37, 361

Infeld, Leopold (1898–1968), 38, 59
*Ingarden, Roman* (1893–1970), *361*, 367,
   370–372

***Jabłoński, Aleksander*** (1898–1980),
   33–***34***
Jadacka, Wanda (1974– ), 405
Jadacki, Jacek Juliusz (1946– ), xiv, 3, 38,
   174, 190, 324, 387, 393–394, 405, 415,
James, Ioan Mackenzie (1928– ), 405
Janczewska, Antonina =
   Witkowska, Antonina
Janczewska, Maria = Witkowska, Maria
Janczewski, Julian (????–1918), 174
***Janiszewski, Zygmunt*** (1888–1920), 5–**6**,
   7, *14*, 40, 48–49, 64, 72
*Jarden, Dov* (1911–1986), 126, *130*
Jaroszyńska-Kirchmann, Anna Dorota, xiv
Jasinowski, Bogumił (1883–1969), 20, 370
Jaśkowski, Stanisław (1906–1965), 195,
   198, 208
Jędrzejewicz, Janusz (1885–1951), 215
Jerzmanowski, Andrzej (1946– ), xiv, 221
Jerzmanowski, Wiktor = Schayer, Wacław
Jezierski, Adam (1948– ), 420
Jónsson, Bjarni (1920– ), 379, 412
Jordan, Marie Ennemond Camille
   (1838–1922). *See the subject index*
Jørgensen, Jens Jørgen Frederik Theodor
   (1894–1969), 325, *387*

Kac, Mark (1914–1984), 38, 208
Kaila, Eino (1890–1958), 325
Kalecka, Zofja, 38
*Kalicki, Jan* (1922–1953), *389*
Kalicun-Chodowicki, Bazyli (1878–1942), 184
Kant, Immanuel (1724–1804), 37
Karski, Jan = Kozielewski, Jan Romuald

Katz, Marek = Kac, Mark
Kawaguchi, Akitsugu (1902–1984), 226
Kiselev, Andrei Petrovich (1852–1940) (Киселёв, Андрей Петрович), 180
Klein, Felix (1849–1925), 389
Knaster, Bronisław (1893–1990), 14, 36, 156, 167, 348
Kobrzyński, Zygmunt (1893–1944), 198
Köhler, Eckehart, 396, 400
***Kokoszyńska, Maria*** (1905–1981), 326–***327***, 361, 370, 372, 374–376
Kokoszyńska-Lutmanowa, Maria = Kokoszyńska, Maria
Konarski, Stanisław, 172
Kostant, Ann (1937– ), xiii
Kotarbińska, Janina (1901–1997), 9
***Kotarbiński, Tadeusz Marian*** (1886–1981), 6–***7***, 8–*9*, 12, 17, 32, 34, 36–37, 189, 320, 335, 341–343, 397, 417
Kowalewski, Jan (1892–1965), 14
Kozielewski, Jan Romuald (1914–2000), xiii
Kozłowska, Anna, xiv
Kozłowski (*pseudonym*) = Greniewski, Henryk
Kozłowski, Roman (1889–1977), 321
*Kozłowski, Witold* (1919– ), xiv, 320–*321*, 331, 405
Kreisel, Georg (1923– ), 389
*Krysicki, Włodzimierz* (1905–2001), *261*
Kunugui, Kinjiro (1903–1975), 226
***Kuratowski, Kazimierz*** (1896–1980), 4–5, 14, 33, 35, 40, 48, 70–***72***, 103, 198–201, 328, 333, 340, 361, 382. *See also the subject index*
*Kwietniewski, Stefan* (1874–1940), 5, 183–184, 187–*189*
Kwietniewski, Władysław (1840–1902), 189

Laczkovich, Miklós (1948– ), 63, 164
Lampe, Wiktor (1875–1962), 41, 393
Langford, Cooper Harold (1895–1965), 417
Lapointe, Sandra, 416
Lebesgue, Henri Léon (1875–1941), 14, 33, 52, 75. *See also the subject index*
Legendre, Adrien-Marie (1752–1833), 265
Lejewski, Czesław (1913–2001), 198
Leśniewska, Anna Kazimiera = Sierpińska, Anna Kazimiera
Leśniewska, Zofia (1893–1958), 9
***Leśniewski, Stanisław*** (1886–1939), vii, 6–*9*, 12–***13***, 17, 36–37, 39, 94, 160, 189, 195, 197, 324, 328, 333, 340, 346–349, 357, 362, 386, 417. *See also the subect index*
Leszczyński, Jan (1905–1990), 367
Levi-Civita, Tullio (1873–1941), 181
Lévy, Azriel (1934– ), 161, 412
Lewin Riquelme de la Barrera, Renato Alfredo (1951– ), xiv
Lichtenstein, Leon (1878–1933), 200–201, 226
Lie, Sophus (1842–1899), 181
***Lindenbaum, Adolf*** (1904–1941), 33, 40, 108, 134, 137, 152, 156, 158, 161–165, 199, 326, 331, ***335***–336, 377, 380
Lindström, Sten (1945– ), 407
Lipiński, Edward (1888–1986), 349
Łomnicki, Antoni Marian (1881–1941), 51, *53*, 57, 215, 227
Lorenz, Johann Friedrich (1738–1807), 265
Łoś, Jerzy (1920–1998), 416
Löwenheim, Leopold (1878–1957). *See the subject index*
Lowry, John (1769–1850), 47, 62
Lubicz, Bolesław Zygmunt = Zahorski, Bolesław
***Łukasiewicz, Jan Leopold*** (1878–1976), vii, 5, 7, 9, 12, 17, 33–39, 51, 72, 94, 160, 195–***196***, *197*, 201, 322–325, 331, 337, 339–341, 345, 370–372, 394. *See also the subject index*
Łukasiewiczowa, Regina, 197
Lutman, Roman (1897–1973), 327
Lutman-Kokoszyńska, Maria = Kokoszyńska, Maria
Lutmanowa-Kokoszyńska, Maria = Kokoszyńska, Maria
Luzin, Nikolai Nikolaevich (1883–1950) (Лузин, Николай Николаевич), 49, 200

Mackiewicz, Witold (1941– ), 409
Maddux, Roger Duncan (1948– ), 400
Magidor, Menachem (1946– ), 38
Mahlo, Paul (1883–1971), 126
Malinowski, Grzegorz, 414
Mancosu, Paolo (1961– ), xiv, 416
Marchisotto, Elena Anne Corie (1945– ), xiv
Marciszewski, Witold (1930– ), 392
Marcus, Joseph (1923– ), 172
Martel, Karol (1916– ), 173

Index of Persons  481

Mates, Benson (1919–2009), 389
Mathé, Svätoslav (1943– ), 398
Matuszewska, Maria Irena =
  Maria Mostowska
**Mazurkiewicz, Stefan** (1888–1945), 5–**6**,
  7–9, 12, *14*, 33, 35, 40, 48–49, 62, 64, 72,
  94, 132, 200–201, 379, 394. *See also the
  subject index*
McFarland, Andrew (1978– ), iii–iv, viii, xiv
McFarland, Joanna (1977– ), iii–iv, viii, xiv
McFarland, Maria Anna, ii
McKenzie, Ralph Nelson Whitfield (1941– ),
  ix, 390
McNulty, George Frank (1945– ), 412
*Mehlberg, Henryk* (1904–1979), *344*, 367
Melnick, Norman (1928–1997), 405–406
Mendelson, Elliott (1931– ), 391
Mendez, Edith Prentice, xiv
Menger, Karl (1902–1985), 164, 200, 226, 394
Meschkowski, Herbert (1909–1990), 63
Mianowski, Józef (1804–1879), 189
Mihułowicz, Jerzy (1877–1950), 237
Mnatsakanian, Mamikon A. (1942– ), 163
**Moese, Henryk** (1886–19??), xi, 43, 45,
  125–126, 129–*130*, **131**–134, 142–152,
  155–157, 162, 206, 211, 248, 263
Mongré, Paul = Hausdorff, Felix
Monk, James Donald (1930– ), 390, 412
Monoide, S. P., 399
Montague, Richard Merritt (1930–1971), 333
Montlak, Bronisława, 199
Montlak, Dawid, 199
Moore, Gregory H. (1944– ), 416
Morawiec, Adelina, 403
Morel, Anne C., 167
Morse, Anthony Perry (1911–1984), 381
Mostowska, Maria Irena (1910?– ), 199
**Mostowski, Andrzej Stanisław** (1913–
  1975), 161, **198**–*199*, 328, 331, 335–336,
  380, 389, 413, 416
Murawski, Roman (1949– ), 416

Nagel, Ernest (1901–1985), 325
*Nawroczyński, Bogdan* (1882–1974), 4, *12*
Nelson, Leonard (1882–1927), 37, 334
Németi, István (1942– ), 390
Nerode, Anil (1932– ), 400–401
Nešetřilová, Helena (1948– ), 402
Neumann, János Lajos =
  Von Neumann, John

Neumann, Johann von =
  Von Neumann, John
*Neurath, Otto* (1882–1945), 341, *393*
Neyman, Jerzy (1894–1981), 39
Nikliborc, Władysław (1889–1948), 320
Nikodym, Otton Marcin (1887–1974),
  193, 200
Nowak, Leszek (1943– ), 365

Ovchinnikov, Sergei V., xiv

Paderewski, Ignacy Jan (1860–1941), 12, 197
Pasch, Moritz (1843–1930). *See the subject
  index*
**Pasenkiewicz, Kazimierz** (1897–1995), 31–
  **32**, 33, 36
Patterson, Douglas Eden, 416–417
Pawlikowska-Brożek, Zofia (1941– ), 406
Peano, Giuseppe (1858–1932), vi, 204, 369.
  *See also the subject index*
Peirce, Charles Sanders (1839–1914), vi
Petrażycki, Leon (1867–1931), 7
Pieńkowski, Stefan (1883–1953), 7–8, 394
Pieri, Mario (1860–1913), 184, 186–188, 194
Pigozzi, Don Leonard (1935– ), 411
Piłsudski, Józef Klemens (1867–1935), 4–5,
  8, 10, 14, 175
Pincherle, Salvatore (1853–1936), 181
Piotrowski, Tadeusz (1940– ), 224
Pla i Carrera, Josep (1942– ), 417
Playfair, John (1748–1819). *See the
  subject index*
Płowa, Franciszka (1845–1926), 57
Polonsky, Antony (1940– ), xiv
Popoff, Kyrille (1880–1966) (Попов, Кирил),
  200–201, 226
Porębska-Srebrna, Joanna (1950– ), 403
Preller, Anne, 389
**Presburger, Mojżesz** (1904–1943?), 190,
  195–**196**, 200–201
Prewysz-Kwinto, Zofia = Leśniewska, Zofia
Prussak, Róża = Teitelbaum, Róża
Purdy, Robert, 386
Purkert, Walter (1944– ), 59
Putnam, Hilary Whitehall (1926– ), 389
Pythagoras (569? B.C.–475? B.C.) *See the
  subject index*

Quine, Willard Van Orman (1908–2000),
  333, 397

*Raabe, Henryk Wacław* (1882–1953), 173, *177*, 180
Rav, Yehuda (1930– ), 419
*Reichenbach, Hans* (1891–1953), 325, 334, *387*
Reidemeister, Kurt Werner Friedrich (1893–1910), 393
Réthy, Móricz (1846–1925), 62
*Rickert, Heinrich* (1863–1936), *365*
Rickey, V. Frederick (1941– ), xiv
Robinson, Raphael Mitchell (1911–1995), 165, 381
Rodríguez-Consuegra, Francisco A. (1951– ), 389, 418
Rosen, Saul (1922–1991), 406
Rougier, Louis (1889–1982), 325
Rowe, David E. (1950– ), x
Różycki, Ludomir (1884–1953), 201
*Rudniański, Jarosław* (1921–2008), xiv, *320*
**Rusiecki, Antoni Marjan** (1892–1956), 59, 169, 201, 204–*208*, **209**, 221, 243–248
Russell, Bertrand Arthur William (1872–1970), vi, 335, 418
Ruziewicz, Stanisław (1889–1941), 193

Sadowska, Joanna, 172
Scanlan, Michael (1957?– ), 418
Schayer, Alicja (1931–1944), 221
Schayer, Genowefa = Zając-Jerzmanowska, Genowefa
Schayer, Krystyna (1933– ), 221
*Schayer, Stanisław* (1899–1941), *214*
**Schayer, Wacław Maria** (1905–1959), 169, 190, 208–210, 214–**220**, *221*–224, 273–318
Schinzel, Andrzej Bobola Maria (1937– ), xiv
Schlick, Moritz (1882–1936), 325
Scholz, Heinrich (1884–1956), 197
Schönflies, Arthur Moritz (1853–1928), 60
Schröder, Friedrich Wilhelm Karl Ernst (1841–1902), vi
Schwabhäuser, Wolfram (1931–1985), 267, 390
Scott, Dana Stewart (1932– ), 167, 333
Sergescu, Petre (1893–1954), 226
Shilleto, James Robert (1941– ), xiv, 134, 163
Sica, Giandomenico, 418
Sierpińska, Anna Kazimiera (1888–1961), 49
Sierpiński, Mieczysław (1912–1983), 49

**Sierpiński, Wacław Franciszek** (1882–1969), 7–14, 31, 33, 37, 39–40, **48**–*49*, 54, 57, 60, 62–64, 70, 72, 94, 161, 164–165, 193, 200–201, 226, 320, 335, 339, 376–382, 397. See also the subject index
Simmons, Keith Eric George (1952– ), 418
Sinaceur, Hourya Benis (1940– ), 388, 400–401, 406–407, 418–419
Skabowski, 10
Skolem, Thoralf Albert (1887–1963). See the subject index
Słupecki, Jerzy (1904–1987), 198
Smith, Helen Marie Patteson (1940– ), ii, xiv
Smith, James Thomas (1939– ), iii–iv, xiii, 125, 163, 475
Smoleńska, Agnieszka (1904–1995), 174–**175**
Sobociński, Bolesław (1906–1980), 198
Sosnowski, Paweł (1859–1947), 38
Srebrny, Marian (1947– ), 401–403, 409
Srzednicki, Jan Tadeusz Jerzy (1923–2008), 386
Stachniak, Zbigniew (1953– ), 386
Steckel, Samuel, 245
Steinberg, Witold, 367
Steinhaus, Władysław Hugo Dyonizy (1887–1972), 57, 59, 329
Steinhausówna, Olga = Chwistkowa, Olga
Stępniewski, Jarosław, 367, 370
Stojda, Władysław, 251–253
Stone, Marshall Harvey (1903–1989), 165
Stożek, Włodzimierz (1883–1941), 57, 320
**Straszewicz, Stefan** (1889–1983), 58, 169, 189, 192–193, 204–**209**, *210*, 243–248
Stroińska, Maria Magdalena, xii, 388
Sujczyńska, Monika, 392
Suppantschitsch, Richard = Zupančič, Rihard
Suppes, Patrick (1922– ), 407, 412, 419
Suslin, Mikhail Yakovlevich (1894–1919) (Суслин, Михаил Яковлевич), 64. See also the subject index
Süss, Wilhelm (1895–1958), 126
Świtalski, Kazimierz Stanisław (1886–1962), 201
Szczerba, Lesław Włodzimierz (1938–2011), 162, 267, 412
Szmielew, Borys, 199
**Szmielew, Wanda Montlak** (1918–1976), **198**–*199*, 267, 328, 336, 390, 412

Sztejnbarg-Kamińska, Dina = Kotarbińska, Janina
Szulczyński, Zygmunt (1897–1967), 193

Tajtelbaum, Alfred = Tarski, Alfred
Tajtelbaum-Tarski, Alfred = Tarski, Alfred
Tarska, Ewa Krystyna = Ehrenfeucht, Ewa Krystyna
Tarska, Maria (1902–1990). *See the subject index*
Tarski, Alfred (1901–1983), **i, 13, 327, 384**. *See also the subject index*
Tarski, Jan (1934– ), xiv, 31, 33, 177, 331, 336, 374, 388–389, 391–393, 396, 400, 405, 407–408
Tarski, Janusz Andrzej = Tarski, Jan
Tarski, Wacław (1903–1944), 3, 336
Teitelbaum, Alfred = Tarski, Alfred
Teitelbaum, Ignacy (Isaac) Mayer (1869–1942), 3, 174, 336, 392
Teitelbaum, Róża (1879–1942), 3, 174, 336, 392
Teitelbaum, Wacław = Tarski, Wacław
Thaler, Michael (1935– ), xiv
Thales (c. 624 B.C.E.–c. 546 B.C.E.) *See the subject index*
Tonelli, Leonida (1885–1956), 226
Trotsky, Leon (1879–1940) (Троцкий, Лев), 8, 10
**Twardowski, Kazimierz** (1866–1938), 9, 12, 17–*18*, 36–37, 197, 201, 327–328, 344–345, 373–374, 393, 396, 416

*Ulam, Adam Bruno* (1922–2000), *333–334*
*Ulam, Stanisław Marcin* (1909–1984), *333–334*

Van den Dries, Lou P., 412
Van Heijenoort, Jean Louis Maxime (1912–1986), 400
Vaught, Robert Lawson (1926–2002), 361, 412, 419
Veblen, Oswald (1880–1960). *See the subject index*
Vega Reñón, Luis, 388
Vitali, Giuseppe (1875–1932), 52, 75, 95, 115
Vivanti, Giulio (1859–1949)
Von Neumann, John (1903–1957), 164, 333–334, 378, 381

Voronoy, Georgy Fedoseevich (1868–1908) (Вороной, Георгий Феодосьевич), 49

Wagon, Stanley (1951– ), 54, 63, 66, 164
**Wajsberg, Mordchaj** (1902–1942), 33–*34*, 198, 346–347
Wallace, William (1768–1843), 47, 62
*Waraszkiewicz, Zenon Jan* (1909–1945), *132*, 134, 152, 156, 162
*Warmus, Mieczysław* (1918–2007), *252*
Watt, Richard M. (1930–), 3
Węglowska, Danuta, 406
Weierstrass, Karl Theodor Wilhelm (1815–1897), vi
Weigl, Rudolf Stefan (1883–1957), 57
Wells, Benjamin Franklin, III, 408
Weryho, Wladyslaw (1868–1916), 20
*White, Morton Gabriel* (1917– ), *396*
Whitehead, Alfred North (1861–1947), vi
Wiegner, Adam (1889–1967), 367
**Wilkosz, Witold** (1891–1941), 326, 367, **369**
Wilson, Trevor Miles, 164
*Wirszup, Izaak* (1915–2008), 125–126, *130*, 135, 144, 151, 212
Witkacy = Witkiewicz, Stanisław Ignacy
Witkiewicz, Stanisław Ignacy (1885–1939), 329, 400
Witkowska, Agnieszka = Smoleńska, Agnieszka
Witkowska, Antonina (????–1957), 175
Witkowska, Helena = Woroniecka, Helena
Witkowska, Jadwiga (1909–1988), 174
Witkowska, Józefa Maria = Zahorska, Józefa
Witkowska, Maria (1873–1911), 174
Witkowska, Maria Józefa = Tarska, Maria
Witkowski, Antoni (18??–????), 174
Witkowski, Antoni (1905–1984), 174
Witkowski, Wincenty (1841–1919), 174–175
*Witwicki, Tadeusz* (1902–1970), *361*, 374
Wojciech, Saint (c. 956–997). *See subject index*
Wojciechowska, Agnieszka, 408
Wójcik, Andrzej (1958–2009), 408
Wojeński, Teofil (1890–1963), 173
*Wojtowicz, Władysław Baltazar* (1874–1942), 58, 127–128, 180, 183–184, 188–*189*, 204, 245
Woleński, Jan (1940– ), ix, 341, 374, 388, 396, 400–402, 408–409, 416, 419

*Woodger, Joseph Henry* (1894–1981), 331, 368-*369*
Woroniecka, Helena (1911– ), 174
Woytak, Lidia Teresa (1946– ), 409
Wycech, Czesław (1899–1977), 221

Yanigahara, Kitizi, 126
Young, William Henry (1863–1942), 226

Zaanen, Adriaan Cornelis (1913–2003), 54–55
Zahorska, Józefa (1900–1957), 174, **175**–177, 331, 336
**Zahorski, Bolesław** (1887–1922), **175**-*177*
Zając-Jerzmanowska, Genowefa, 221
Zaremba, Stanisław (1863–1942), 49, 329, 369
Zarzecki, Adam, 208
**Zawirski, Zygmunt Michal** (1882–1948), 325–326, 370-***373***, 387
Zermelo, Ernst Friedrich Ferdinand (1871–1953), 33, 66, 95, 210, 334. *See also the subject index*
Żeromski, Stefan (1864–1925), 173
Zimmermann, Robert von (1824–1898), 18
Znamierowski, Czeslaw (1888–1967), 20
Żółtowski, Adam (1881–1958), 367
Zupančič, Rihard (1878–1949), 180
Zydler, Jan (1867–1934), 180, 219
Zygmund, Antoni (1900–1992), 130
Zygmunt, Jan, xiv, 194–195, 386, 391, 398, 409, 420

# Index of Subjects

This index lists subjects considered in the main text of the book. It does not cover the entries in the bibliography. Some index headings are personal names that identify subject-matter content. For other individuals, consult the index of persons that begins on page 477. Because mathematicians often give commonplace words special meanings, this index includes headings for some adjective as well as substantive phrases. Subheadings under Poland, Tarska, and Tarski are ordered topically. All headings and all other subheadings are alphabetized, ignoring spaces, punctuation, diacritical marks, and initial articles and prepositions.

absolute
  value, exercise on, 250
  *See also under* measure
absolutely measurable. *See under* set
Academy of Sciences
  *See under* Polish; Soviet
action, theory of, 341–342
Adalbert, St. *See* Wojciech
additivity. *See under* measure
adequacy, material, 362–363
algebra
  cardinal, 379–380, 412
  of classes, 358
    symbols for expressions in, 359
  cylindric, 390, 411–412, 417
  ordinal, 411
  universal, 377–382, 411–412
algebraic surface. *See* surfaces, algebraic
almost equivalent. *See under* denumerable decomposition; symbol
alphabetization, xiii
altitude. *See under* parallelogram; rectangle; trapezoid; triangle
Amaldi, Ugo. *See under* axiom system *and in the index of persons*
*American Mathematical Monthly*, 203
Amersfoort. *See under* congresses
Amsterdam, xxii–xxiii
Anders Army, 34, 224, 320
angle
  arm, 264
  bisector, 275
  central, 275
  concave, 289
  convex, 264–265
  full, 278
  interior of, 264–265
  nonconvex, 265
  straight, 278
  vertex of, 264
antinomy
  of heterological concepts, 334, 357
  of the liar, 357, 362
antireflexivity. *See under* axiom; theorem
antisemitism, ix, 8, 10–11, 36, 69, 181, 328, 330–331, 334–336, 387, 401–402
  *See also* Holocaust
antisymmetry. *See under* axiom
apagogic reasoning, 138
apothem, 310
arc, 275
Archimedes, 213. *See also under* axiom
archives, xiv, 377, 473
area
  theory, xi, 45–48, 67, 79–83, 125, 219–220, 222–223, 301–318
  unit of, 304
  *See also under* axiom system; circle; cone; cylinder; parallelogram; polygon; polygonal regions; rectangle; rhombus; trapezoid; triangle
arithmetic
  cardinal, 161, 194, 200, 377–382
  integer, 247, 251, 255
  of natural numbers, 196, 386
  rounding in, 236
  of segments, 216–217, 296
  *See also under* axiom system; completeness; decidability; incompleteness; number: irrational, natural

485

arm. *See under* angle
art theory, 329
assimilation, 38
Association for Symbolic Logic, 334
Austria, xxiii
   Empire, xxii
axiom, 15–16
   antireflexivity, 29
   antisymmetry, 16, 21
   of Archimedes, 45–47, 127, 216
   of choice, vi, 40, 48, 52, 54, 65–67, 71–72, 75, 79, 95, 161, 199, 377, 380–381, 417
   circle, 277–278
   congruence, 186
   continuity, 46, 213, 230, 238–239, 278
   of De Zolt, 46, 48, 51–52, 61, 79, 83, 90, 95
   incidence, 186
   order, 16, 21, 25, 30
   parallel
      Euclid's version, 265
      exercise on, 264–267
      Playfair's version, 265, 267
      Tarski's version, 248, 264–267
   of Pasch, 267
   transitivity, 16, 21
   trichotomy, 16
   well-ordering, 21, 25, 30
axiomatic method. *See* postulate theory
axiom system
   for algebra of classes, 358
   for area, 220, 305
   for arithmetic, 58
   for Boolean logic, Hilbert's, 348
   for geometry
      of Enriques and Amaldi, 183
      Euclid's, 264–267
      Hilbert's, 46–47, 184–186, 194, 264
      Pieri's, 186, 188, 195
      Tarski's, 194–195, 208, 267, 390
      Veblen's, 194
   for inclusion, Huntington's, 358
   independent, 16, 22, 25
   interpretation of, 27
   for magnitude, 51
   for proper inclusion, Tarski's, 358
   for set theory, 199
   relatively weakest, 29–30
   for well-ordering, Tarski's, 15–17, 19–30, 161
   *See also* postulate theory

Baden School. *See* philosophy: neo-Kantian
Banach, Stefan
   space, 57
   *See also under* measure; paradox; theorem *and in the index of persons*
base. *See under* parallelogram; rectangle; trapezoid; triangle
beam of a ship, 315
Berkeley, 125, 339, 376, 397
   Cragmont Avenue, 376
   *See also under* congresses; universities
Berlin, xii–xiii, 330
   *See also under* philosophy; universities
Bernstein, Felix. *See under* theorem *and in the index of persons*
betweenness, 183, 186, 194, 266–267
bibliography
   citations, xii, 421
   conventions, xiii, 421
biographical sketches, xiii
*Biographical Dictionary of Polish Mathematicians*, 428
biology, foundations of, 331, 368–369
Birkhäuser. *See* Springer
bisector. *See under* angle
Bologna, xxii–xxiii. *See also under* congresses
Bolyai, Farkas Wolfgang. *See under* theorem
Bonn. *See under* universities
boundary. *See under* set
bounded. *See under* set
boxed text, xiii, 1, 170, 243, 274, 337, 421
Breslau, xxii–xxiii
broadcasting, 369
Budapest, xxii–xxiii

Cambridge
   England, xxii–xxiii (*see also under* congresses)
   Massachusetts (*see* universities: Harvard)
Cantor, Georg. *See under* theorem *and in the index of persons*
capitalization, xiii
cardinal. *See under* algebra; arithmetic; equivalence
case-ridden arguments, 247, 256–257, 260, 268–270, 272
categorical. *See under* theory
category
   first (*see under* set)
   semantic, 9

Catholicism, 34, 179, 321, 326, 369
   *See also* Tarski: religion
causal relation, 343–344
causality, 324, 343–344
ceiling, 127. *See also under* symbol
center. *See under* circle
central. *See under* angle
Chicago. *See under* congresses; universities
choice. *See under* axiom
chord. *See under* circle
Cipher Bureau. *See* code breaking
circle, 277
   area, 237
   center, 277
   chord, 275
   circumference, 206, 212–213, 222, 229–241
   diameter, 275
   exterior of, 277
   interior of, 277
   line, relationship to a, 278–284
      secant, 282
      tangent, 282
   radius, 277
circles
   intersecting, 216, 277, 280
   relationship of two, 280
   tangent, 280
circle-squaring. *See under* Tarski
circular polygon, 157
circumcenter. *See under* triangle
citations. *See under* bibliography
closed system of theorems, 216–217, 223, 283–284
closure operator, 72
code breaking, 8–10, 14, 49
commensurable. *See under* segments
commutation, law of, 348
compass and straightedge. *See* construction
completeness, 377, 388, 418
   of real arithmetic, 194, 196, 331, 418
computer-assisted design, 261
computer science, 167, 196, 343, 346, 363, 391, 402, 406, 408, 414
concave. *See under* angle
conditional. *See under* probability; sentence
conditions, necessary and sufficient, 291
cone, area and volume of, 237
congresses
   1921–1939 annual Polish secondary-school teachers', 203
   1923 1st Polish Philosophy, Lwów, 41, 53, 324
   1927 1st Polish Mathematics, Lwów, 195, 226, 324
   1927 2nd Polish Philosophy, Warsaw, 195, 323, 341–345
   1928 8th International Mathematicians', Bologna, 167, 226, 324
   1929 Slavic Mathematicians, Warsaw/Poznań, 171, 196, 200–202, 225–227, 322
   1931 2nd Polish Mathematics, Vilnius, 207, 324
   1934 8th International Philosophy, Prague, 177, 324–325, 387
   1934 Unity of Science *Vorkonferenz*, Prague, 177, 324–325
   1935 1st Unity of Science, Paris, 177, 324–325
   1936 2nd Unity of Science, Copenhagen, 326
   1936 3rd Polish Philosophy, Cracow, 323–324, 326, 365–376
   1937 9th International Philosophy, Paris, 324, 326
   1937 3rd Unity of Science, Paris, 324, 326
   1938 4th Unity of Science, Cambridge, 326
   1938 Modern Reasoning, Amersfoort, 324, 330
   1939 5th Unity of Science, Harvard, ix, 324, 326, 333–336
   1946 Princeton Bicentennial, 379, 388, 406
   1955 Reasoning in Mathematics, Paris, 389
   1957 Summer Institute, Cornell, 402–403
   1965 Gödel Theorems, Chicago, 389
   1965 Set Theory, London, 389
   1971 Tarski Symposium, Berkeley, 404–405, 414
   1983 Tarski Commemoration, Stanford, 407, 419
   2001 Tarski Centenary, Warsaw, 405, 413, 415
   2007 State of Jefferson Mathematics, Whiskeytown, xiii
congruent, 65–66, 80, 96, 183, 186, 194
   *See also under* symbol
conjunction
   Tarski's definition of, 39
   *See also under* symbol
connected. *See under* polygonal regions

consequence, 366–367
  direct and indirect, 366–367
  infinitary concept of, 367
  logical, 195, 326, 344, 365, 388–389, 412
  material, 344
consistency proofs, 340–341, 361, 367, 388
construction problems, 188, 216–217, 279–281, 285–291, 301
  compass and straightedge, 285
  set square, 285
contiguity. *See* phenomenon
continuity. *See under* axiom
continuum hypothesis, 200
contrapositive, 291
converse, 291
convex. *See under* angle; polygonal regions
coordinates, 186
Copenhagen
  *See under* congresses; universities
Cornell. *See under* congresses
coryphaeus, 12
Cracow, xxii–xxiii, 38, 57
  Circle (*see under* philosophy)
  *See also under* congresses; Mathematical Society; universities
curriculum
  of Polish schools, 207, 210, 214–215, 222–223, 230–231, 278, 288
  at Third Boys' Gimnazjum, 179–180
curve. *See under* length
cuts
  Dedekind, 230, 238
  method of, 213, 230, 238, 241
cylinder, area and volume of, 237
cylindric. *See under* algebra
Czechoslovakia, 330, 398

daemon, 408
Danzig, 193
decidability, 199, 377, 388, 411
  of real arithmetic, 196, 412, 414, 418, 420
decision procedure. *See* decidability
Dedekind, Richard. *See* cuts
decomposition
  equivalence by, xi
  *See also* denumerable decomposition; finite decomposition
deduction
  theorem, 160
  theory of, 36, 160, 339–341, 344, 346–349

*See also under* definitions
defect. *See under* triangle
definitions, 15–16, 388
  recursive, 165, 167, 340, 359, 362
  in deduction theory, 339, 346–349
degree
  of equivalence (*see under* finite decomposition: into polygons; symbol; tau)
  of irregularity, 157 (*see also under* symbol)
*Delta*, 126
deltoid. *See* kite
denumerable decomposition
  almost equivalent by, 75, 117, 123
  equivalent by, 65–66, 74–76, 95, 115, 163, 377
  into closed sets, 123
  into Lebesgue measurable sets, 123
Desargues, Girard. *See under* theorem
De Zolt, Antonio. *See under* axiom
diagonal of a square, 274, 292–295
diameter. *See under* circle; polygon; symbol
difference. *See under* polygonal regions; sets; symbol
direct. *See under* isometry
disjoint. *See under* sets; symbol
disjunction, 347. *See also under* symbol
disk, 277
  equidecomposibility with a square, 62, 90
displaced persons. *See under* World War II
distance
  equal, between pairs of points, 80
  point to line, 217, 273
  point to point, 217, 273–274
docent, 41
double-subdivision method, 60, 81, 138
draft of a ship, 315
duel, Chwistek's, 329

Economists and Statisticians Association, 349
editorial comments, xii
*Ekonomista*, 349
elementary. *See under* language
empiricism, 370–373, 396, 400
empty. *See under* set; symbol
*Encyclopedia of Unified Science*, 393
Enriques, Federigo. *See under* axiom system *and in the index of persons*
entirety. *See* floor
equality. *See under* symbol
equicomplementability, 47

Index of Subjects

equidecomposibility. *See* finite decomposition: into polygons: equivalence by *and under* finite decomposition: into disjoint sets: equivalence by
equidistance
 ternary, 186
 from three lines, exercises on, 268–271
equidistant curve, exercise on, 256
equivalence
 of axiom systems (*see* postulate theory)
 cardinal, 70
 connective, 346–347
 material, 39, 160
 *See also under* decomposition; denumerable decomposition; finite decomposition; symbol
Erlanger Programm, 389
Euclid. *See under* axiom; axiom system *and in the index of persons*
Europe, Central, maps of
 1914, xxii
 1924, xxiii
event, 341
 generalized, 341–342
excluded middle, law of, 357, 360, 365
exercise, use of the word, 243
exhaustion, method of 45
Exposition, General Polish. *See* congresses: 1929 Slavic Mathematicians'
exterior. *See under* circle

factorization, exercise on, 251–253
figures
 geometric, 80
 numbering of, xx
filament. *See under* Tarski
filter, prime, 165
finite decomposition, 43, 45
 into disjoint sets
  equivalence by, xi, 62, 66, 95–105, 163, 165–166
  in dimension 1, 73
  in dimension ≤ 2, 95, 106–109, 363
  in dimension 2, 53, 58–63, 65, 77–91
  in dimension 3, 90–91, 381
  in dimension ≥ 3, 65, 73–74, 93, 95, 109–112
  transitivity of, 60
 on a sphere, 74, 95, 112–114
 using Lebesgue measurable sets, 123

 into polygons, 341
  equivalence by, xi, 45–47, 59
   degree of, 45, 125–129, 132–158, 162–163, 206, 211–212, 263
   (*see also* tau *and under* symbol)
  methods for, 157
  *See also* area: theory
 into polyhedra, equivalence by, 90
 into spherical polygons, 112
finiteness, definition of, 40–41, 161
first element, 15, 21–22, 25
flea-feeder, 57
floor, 247, 253. *See also under* symbol
Formism. *See* art theory
formula
 definition, 340
 well-formed, 341
 *See also under* quadratic
foundations. *See under* geometry; Leśniewski; mathematics; physics; psychology; theology
fractal, 69
function
 measurable, 54
 of names, 359
 nonextensional, 357
 propositional, 195
 sentential, 358–359
 Tarski's usage, 357
functional 54–56
*Fundamenta Mathematicae*, vii, 14, 62, 64, 93–94

Gdańsk. *See* Danzig
Gdynia, xxiii, 193, 333
*Geometrja* (coauthored by Tarski), xiii, 214–224, 273–318, 475
 contents, 216–219, 223, 274–275, 291, 301
 later editions, 224
 prerequisites, 215–217
 style, 291
geometry
 analytic, 268–271
 based on notion of solid body, 40, 195
 foundations of, 162, 181, 184–188, 194, 199, 267, 390, 412
 non-Euclidean, 188, 265
 texts, 180, 183–184
 transformational, 186, 210
 *See also under* axiom system

Germany, xxiii
   Empire, xxii
Gerwien, P.. *See under* theorem *and in the index of persons*
gimnazjum, xiii, 3, 172
   Kalecka, 38
   Limanowski, 221
   Mazowieckie, 4–5, 12, 392
   State, Number 4, 3, 392
   Third Boys', 37, 173–174, 176–180, 203
   Zamoyski, 218–219, 343
   Żeromski, 177, 320, 328, 331
goals of this book, x, 319
*Głos nauczycielski*, 203
Gödel, Kurt. *See* incompleteness of integer arithmetic *and in the index of persons*
Göttingen, xxii–xxiii
   *See also under* universities
grosz, 318
group
   amenable, 164
   of direct isometries, 55
   measurable, 164
   of rotations, 67–68
   symmetry, 272
   transformation, 164–165, 181
   of translations, 55
   theory, 199
   Vierergruppe, 272
gymnasium. *See* gimnazjum

habilitation. *See under* Tarski: as researcher
Hahn, Hans. *See under* theorem *and in the index of persons*
*Handbook for Self-Education*, 183, 189
Hanover, 224
Harvard. *See under* congresses; universities
Hauber's law, 217, 283
Hausdorff, Felix. *See under* paradox *and in the index of persons*
height. *See under* parallelogram
hemisphere, 129, 133, 146–147
heterological. *See under* antinomy
Hilbert, David.
   problems, 47
   *See also under* axiom system *and in the index of persons*
Holocaust, 34, 57, 69, 334–336, 345, 374
Huntington, Edward V. *See under* axiom system

hyperbolic paraboloid, 268, 271
hyperinflation, 35

idealism, transcendental, 365–367
identity law, 348
implication connective, 30, 347
   *See also under* symbol
incidence, 183, 186
incircle. *See under* triangle
inclusion relation. *See under* axiom system
incommensurable. *See under* segment
incompleteness of integer arithmetic 322, 365, 367, 386, 388, 402
independence. *See under* axiom system; primitive notions
index conventions, xiii, 477, 485
individual, 341
   generalized, 341–342
inductive reasoning. *See under* logic
inequalities, 247, 250, 253, 256, 262
   exercises on, 257–260
inequality. *See under* symbol
inequivalence. *See under* symbol
inference rules, 367
influenza, 14
integer. *See under* arithmetic;
insurance. *See under* mathematics
integral
   Lebesgue, 54–55
   *See also under* symbol
interior, 85
   *See also under* angle; circle; symbol
interpretation. *See under* axiom system
intersection. *See under* circles; symbol
interval, multidimensional, 74, 115
intuitionism, v
invariant. *See under* measure
irrational. *See under* number
irregularity, degree of, 157
isometry, direct, 80

*Jahrbuch über die Fortschritte der Mathematik*. *See* JFM.
Jerusalem, 224
Jews, 3, 5, 8, 34, 36, 38, 57, 69, 130, 173, 181, 330, 333–336, 345, 387, 405, 409
JFM, v, 16, 63, 421
Jordan, Camille. *See under* Peano
*Journal of Symbolic Logic*, 404, 411

Kępno, 130–131, 144, 263
Kiev, xxii–xxiii. *See also under* universities
kite, 157
  area, 313
Königsberg, xxii–xxiii
Koszykowa Street. *See under* Warsaw
Kraków. *See* Cracow
Kristallnacht, 331
Kuratowski, Kazimierz. *See under* theorem *and in the index of persons*

language
  closed, 40
  elementary, 195
  equational, 390
  formal, 357, 362–363
    order of, 363
  natural, 40, 357
  object, 357, 362–363
  philosophy of, 9
  *See also* metalanguage *and* semantics
Latino sine Flexione. *See* Lingua Peano
lattice, complete, 167
Lebesgue, Henri. *See under* measure *and in the index of persons*
length
  of curve, 213
  unit of, 219, 294, 296
  *See also under* segment
Leśniewski, Stanisław
  death, 333
  seminar, 21, 33, 36, 161
  system of foundations, 9, 36, 39, 386
  Tarski, intellectual debt to, 39
  *See also under* logic; Tarski *and in the index of persons*
letterspacing, e x p a n d e d, xii, 380
liceum, 172, 215, 222
limits
  method of, 212, 230–231, 241
  theory of, vi, 212–213, 231, 238–239
line, as primitive notion, 183, 186
Lingua Peano, 204
Lithuania, xxiii
locus, 277
Łódź, xxii–xxiii, 3
  *See also under* university
logarithms, 247
  exercise on, 253

logic, 5, 12, 33
  algebraic, 34, 412, 414
  Aristotelian, 197
  Boolean (*see* deduction: theory of)
  California school of, ix
  deductive, 372
  fundamental concepts of, including those of metamathematics, 36–37, 160, 196, 200, 322–326, 335, 356–363, 367–368, 386, 402, 411–412, 414, 417–419
  game, 340–341
  inductive, 335
  intuitionistic, 34
  Leśniewski's, 9, 39, 160, 324, 349, 386
  modal, 34
  multivalued, 34, 197, 325, 348, 372, 386–387
  nonclassical, v
  philosophical, 37, 160, 327, 339, 345, 404, 412–413, 415–417
  Polish school of, vi–vii, 12, 14, 18, 49, 192, 195, 197, 322, 340, 345, 374, 393, 419
  propositional. *See* deduction: theory of
  sentential. *See* deduction: theory of
  in teaching geometry, 216, 223, 247, 291
  *See also under* Tarski: as researcher
logical
  notion, 372, 389
  positivism, 323, 325, 373, 387, 393
  *See also* truth
logicism, vi
logistic. *See* logic: Leśniewski's; Peano: school of
London, xxii–xxiii
  *See also under* congresses
Löwenheim, Leopold. *See under* theorem
Lublin, xxii–xxiii
Łukasiewicz, Jan
  *See also under* Tarski: as researcher; theorem *and in the index of persons*
Lviv. *See* Lwów
Lwów, xxii–xxiii, 38, 53
  *See also under* congresses; Mathematical Society; mathematics; Philosophical: Society; universities

magnitude. *See under* axiom system
mark, Polish, 35
Marszałkowska Street. *See under* Warsaw
Marxist commentary, 208

*Matematikai és physikai lapok*, 203
*Matematyka: Czasopismo dla nauczycieli*, 208, 248, 408
*Matematyka i szkoła*, 207, 210
material. *See* adequacy *and under* equivalence
*Mathematica*, 274
*Mathematical Gazette*, 203
*Mathematical Reviews*. *See* MR.
Mathematical Society
   Cracow, 57
   Polish, in Lwów, 322–324
   Polish, in Warsaw, 161, 167, 208, 210, 321, 403
mathematics
   characteristically Polish, vi, 4–5, 14, 49, 64, 72, 204, 227, 379
   education
     in Italy, 47, 59, 181–183, 186–188
     in Poland, v, 53, 57–59, 126–127, 130, 137, 169, 171–172, 183–184, 188–189, 191–192, 202–210, 219, 221, 230–231, 241, 243–246, 248, 273–274, 369
       summer courses, 192–193, 207
     in the United States, 130, 350
   foundations of, 5, 14, 49, 60, 64, 184, 368, 390 (*see also under* Leśniewski)
   insurance, 14, 18, 69, 196, 343, 349–356
   Lwów school of, 57
   social history of, x, xii, 223
   *See also under* philosophy
maximal principle, 54, 72
Mazurkiewicz, Stefan. *See under* paradox *and in the index of persons*
mean, arithmetic, 232
meaning, 366, 374–376, 417
measurable. *See under* function; set
measure, 52, 60, 67, 75, 83, 165
   absolute, 166, 364
   Banach's, 165, 363–364
   finitely additive, 52–55, 60, 67
   fully additive, 52
   invariant, 53, 164–165, 363
   Lebesgue, 52, 106, 117, 123, 363
   Lebesgue inner, 76, 122–123
   problem, 51–57, 60, 65, 67, 69, 75, 95, 164–166, 331, 364
   theory, vi, xi, 8, 12, 45, 48, 51–54, 75–76, 164–166, 420
   zero (*See under* set)
   *See also* Peano: Peano–Jordan content
measurement theory, 320
median. *See under* triangle
mereology, 195
   *See also* logic: Leśniewski's
metalanguage, v, 9, 40, 357, 362–363
metalogic, v, 368
metamathematics, v–vi, 341, 388, 419
   *See also* logic: fundamental concepts
metaphysics, 370–372
method. *See* action, theory of
metric, 65
Mianowski Fund, 7, 12, 36, 187–189
Ministry of Religion and Education, 12, 20, 94, 179, 189, 192–193, 197, 204, 207–210, 214–215, 221, 224, 226, 243?, 246, 345, 369, 394–395
Minsk, xxii–xxiii, 10, 174–176
*Młody matematyk*. See *Parametr*
model theory, 34, 160, 389, 412, 414, 418–419
moon, trip to the, 397
Moscow, xxii–xxiii
MR, 421
Munich agreement, 330
Münster, 197. *See also under* universities
mutual convergence. *See under* sequence

names
   personal, xii
   place, xii
nationalism in academia, 402
National Pedagogical Institute, 36–38, 184, 210
National Schoolbook Publisher, 214–215, 273, 276
*Nauka polska*, 189
negation, 347
   Tarski's definition of, 39
   *See also under* symbol
neo-Kantianism. *See under* philosophy
New York City, 332–334, 396
   *See also under* universities
nonconvex. *See under* angle
nonmeasurable. *See under* set
nowhere dense. *See under* set
Nowi Targ, 192
Nowolipki Street. *See under* Warsaw
numbers
   irrational, vi, 49, 217, 222–223, 274, 292–300

Index of Subjects

approximation of, 294–297
arithmetic with, 297–300
equality of, 297
natural, 16
undefinability of, in real arithmetic, 163
ordinal, 16–17, 54, 161, 165
rational, 273, 294
real, 46, 183, 222–223, 238, 274, 418 (*see also under* completeness; decidability)
number theory, 8, 49

*Ogniwo*, 177, 203
Olympiad, 208, 210, 219
ontology, 361
*See also* logic: Leśniewski's
open sentence, 195
*See also* function: sentential
order. *See* betweenness *and under* axiom; language: formal
ordered pair, 72
Order of Polish Rebirth, 221
ordinal. *See under* algebra; number
orientation, 80

π. *See* pi
pansomatism, 407
paradox, 164–166
Banach–Tarski, 53, 57, 62, 90, 93, 164, 378
Hausdorff, 65, 67–69, 73, 109, 164–165
Mazurkiewicz–Sierpiński, 40, 48, 51
parallel
lines, 265, 275
projection, 219
*See also under* axiom; strip
parallelogram, 127–128, 156
altitude, 127–128, 308, 314
area, 45–47, 308–309
base, 308
height, 314
*Parametr* and *Młody matematyk*, 126, 130, 135, 169, 204–213, 408
problem sections, 205, 243–246, 250
difficulty (*see under* Tarski: as educator: exercises)
prizes, 245, 250
parenthesis-free notation, 340–341
Paris, xxii–xxiii,
*See also under* congresses; universities
Pasch, Moritz. *See under* axiom
Peano, Giuseppe, 59

Peano–Jordan content, 166, 364
school of, vi, 51, 323
*See also* Lingua Peano *and in the subject index*
pensions, Polish, 339, 349–356
perimeter. *See under* polygon
*Periodico di matematica*, 203
permissions, 473–475
perpendicular
lines, 216
to two lines, 271
*See also under* segment
phenomenon, 343–345
Philosophical
Institute (*see under* Warsaw)
Society, Polish, in Lwów, 51, 322, 324, 326, 356, 396
philosophy
analytic, 9, 18, 34
Berlin school of, 334
Cracow Circle, 34
deductive, 323
empiricist, 334
history of, 369, 396
neo-Kantian, 365
of mathematics, 181, 188
Polish school of (*see under* logic)
of science, 8, 181, 329, 343–344, 370–373, 387, 412
*See also under* language
physics, foundations of, 368
pi ($\pi$)
approximation, 212, 236
definition, 212, 236, 240
Pieri, Mario. *See under* axiom system *and in the index of persons*
Piłsudski, M/S, 332–333
plan. *See* action, theory of
plane, 186, 216
Playfair, John. *See under* axiom
Płock, iv, xxii–xxiii
point, as primitive notion, 183, 186, 188, 194
point-set theory. *See* topology
Poland, xxiii
unification, vii, 4–5, 8, 10, 49
Military Organization—East, 175 (*see also* Polish: Legion)
economy, 35, 172, 179, 189, 273, 349–356, 395
history and culture, xii, 3, 174, 223

school system, 171–172
  (*see also under* curriculum)
  during World War II
    Army in the East, 224
    clandestine education, xiii, 9, 37, 49, 72,
      132, 197, 199, 208, 210, 221, 345, 373
    government-in-exile, xiii, 224
    underground government, xiii, 32, 221
    People's Party, 221
  after World War II, 72, 177, 379
  *See also under* World War II
Polish
  Academy of Arts and Sciences, 406
  Academy of Sciences, 407–409
  Association of Forced Emigration, 224
  emigré community (Polonia), 397, 404, 409
  Exposition, General, in Poznań, 201–202
  Legion, 14, 329, 369
  spelling, xii
  *See also* parenthesis-free notation;
    Philosophical: Society; Poland *and under*
    congresses; Mathematical Society
Polish–Soviet War, x, 8–11, 14, 32, 34, 49,
  174–177, 210, 221, 401, 405
  Vistula, Miracle of the, 8–10, 49
politics, 7–9, 32, 173, 177, 200, 221, 226,
  328–331, 334–335, 343, 350, 393, 397
Polonia. *See* Polish: emigré community
polygon
  circular, 157
  diameter, 126, 138
  perimeter, 163, 213, 236
  regular, 212–213, 217, 236–237
    area, 310
  similar, 275
  simple, 240
  spherical, 112
  *See also* polygonal regions
polygonal regions, 46, 125
  area, 45–47, 53, 59–60, 77–91, 106–109,
    162–163, 219–220, 304–318
  in the broader sense, 137
  connected, 127, 133, 139
  convex, 127, 133, 138–139, 240
  difference and sum, 302
  *See also* finite decomposition:
    into polygons; polygon
polyhedral region, volume of, 45, 47,
  61, 90–91
positivism. *See under* logical

postulate. *See* axiom
postulate theory, 15–17, 29–30, 51, 162,
  186, 188, 331, 365–368, 418
Poznań, xxii–xxiii, 200–202, 227, 328
  *See also under* universities
*Prace matematyczno-fizyczny*, 189
Prague, xxii–xxiii, 177.
  *See also under* congresses
precedence. *See under* relations
prime filter, 165
primitive notions, 15–16, 80, 183, 186,
  188, 194–195, 346–349
  independence of, 324
Princeton. *See under* congresses; universities
*Principia Mathematica*, 346, 359
prizes for solutions. See under *Parametr*
probability
  conditional, 69, 387
  theory, 14, 66, 197, 325, 334–335, 386–387
problem, use of the word, 243
product, Cartesian. *See under* symbol
projection. *See under* parallel
propaganda in 1920, 8, 10–11
proper inclusion. *See also* set: proper part
  *and under* axiom system; symbol
propositional logic. *See* deduction: theory of
propositional variable, 39, 160
protothetic. *See* logic: Leśniewski's
*Przegląd filozoficzny*, 12, 19–20, 189, 321
*Przegląd matematyczno-fiziczny*, 58–59,
  189, 204, 210
*Przegląd pedagogiczny*, 189, 203, 207, 215
psychology, foundations of 334, 372
Puck, 192–193
Pythagoras. *See under* theorem

quadratic
  formula, 195
  functions, 247, 249, 259–260, 313
quantifier, 195
  elimination, 195–196, 247, 250, 257,
    259–260, 261–262, 412, 414, 418
  existential, 195
  logic, 340
  universal, 39, 160, 195
  *See also under* symbol
quotation marks, 40, 356

radio. *See* broadcasting
radius. *See under* circle

rank. *See under* set
ratio. *See under* segments
rational. *See under* numbers
real. *See under* numbers
rectangle
　altitude, 304
　area, 304–305
　base, 304
　in a square, exercise on, 261–263
recursive. *See under* definitions
reduction to absurdity, 283
refugees. *See* World War II: displaced
Regina, 125
region. *See* polygonal regions; polyhedral region; triangular region
regularity. *See under* theorem
relations
　$\alpha, \beta$-, 66, 70, 75, 103
　precedence, 21
　theory of, 377, 379, 391, 406, 414, 420
relativity. *See under* truth
Reno. *See* universities: of Nevada
rhombus, area of 313
Righteous among the Nations, 57, 345
*Riveon lematematika*, 126, 130
Rockefeller grant. *See under* Tarski: as researcher
rounding. *See under* arithmetic
*Ruch filozoficzny*, 18, 321
*Ruch pedagogiczny*, 203
Russia
　Empire, xxii, 3–4, 210
　Revolution of 1905, 49
　Revolution of 1918, 8, 10
　*See also* Soviet Union
Rysia Street. *See under* Warsaw

San Francisco. *See under* universities
satisfaction, 359–360, 362
scalar, 186
secant line. *See under* circle
segment
　length of, 217
　perpendicular bisector of, 275
　*See also under* arithmetic
segments, 216, 271–272
　commensurable, 217, 291
　exercise on two, 271–272
　incommensurable, 217, 274, 292–297
　ratio of, 296

semantics, 160, 323, 326, 356–363, 374–376, 393, 396, 407, 417, 419
　in computer science, 167, 363, 414
　*See also under* category
sentence, 195, 357–358
　conditional, 368
　open, 195 (*see also* function: sentential)
sentential logic. *See* deduction: theory of
sequence, 231
　exercise on a, 255
　mutually convergent, 231
set
　absolutely measurable, 166, 364
　boundary, 105, 112, 119
　bounded, 84
　empty, 81
　first-category, 70, 105
　measurable, 52–53, 76, 106
　of measure zero, 70, 75
　nonmeasurable, 52, 74–75, 123
　nowhere dense, 70, 105
　ordered, 16, 21
　of Peano–Jordan content zero, 70, 105
　proper part of, 80, 358
　rank of, 381
　square (*see under* construction problems)
　well-ordered, 12, 15–21, 58, 161, 165, 331
　*See also* sets; set theory
sets
　difference of, 80
　disjoint, 80
　identical, 80
　sum of, 80
　union of, 80
　*See also under* symbol
set theory, vi–vii, 5–8, 12, 33, 40, 49, 60, 64, 69, 72, 79, 328, 331, 333, 335, 379, 412, 417, 420
　geometric application of, 33, 39–41, 43, 45, 53, 58–59, 162, 186
　without variables, 390
　Zermelo's, 17
　*See also* algebra: of classes *and under* axiom system
side of a line, 264
Sierpiński, Wacław. *See under* paradox *and in the index of persons*
similarity, 219–220, 301, 316–318
singleton. *See under* symbol
Skolem, Thoralf. *See under* theorem

Slovakia, 398
snow job, 399
Social Insurance Institution, 179, 343, 351, 395
Society of Sciences and Letters. *See under* Warsaw
Society of Scientific Courses, 210
solid body. *See under* geometry
Sonoma. *See under* universities
Soviet Union, xxiii, 10, 226
   Academy of Sciences, 200
sphere, 186–188, 195, 208
   area and volume, 237–238
   *See also under* strip
Springer, xiv, 474
square, set (*see under* construction problems)
Stanford. *See under* congresses; universities
St. Petersburg, xxii–xxiii
straight. *See under* angle
strip
   parallel, 129, 145, 152
   spherical, 146
      area of, 147
subset. *See under* symbol
Sułkowski Street. *See under* Warsaw
sum. *See under* polygonal regions; sets
summer courses. *See under* mathematics: education
superset. *See under* symbol
surfaces, algebraic 181, 189
Suslin's problem, 382
   *See also in the index of persons*
symbol
   almost-equivalence by denumerable decomposition, 75, 117 .......... $A \underset{p}{=} B$
   betweenness, 267 ............... $\beta OPQ$
   ceiling, 127 ........................ $\lceil x \rceil$
   congruence of sets, 66 ............ $A \cong B$
   conjunction, 39 .................. $\Phi \& \Psi$
   degree of equivalence, 126, 137 .... $\sigma(A,B)$
      Tarski's functions, 127 .......... $\tau(x)$
           163 .......... $\tau_m(x)$
           156 .......... $T(\varphi)$
   degree of irregularity, 157 ......... $\rho(W)$
   diameter of a set, 126, 138 ......... $\delta(A)$
   difference of sets, 80 ............. $A - B$
   disjointness, 80 ................... $A \,][\, B$
   disjunction, 347 ................. $A \Phi \Psi$
           358 ................. $\Phi \vee \Psi$
   empty set, 93 ...................... $\phi$

equality, 21, 80 ................. $x = y$
equivalence by denumerable decomposition, 66, 96 .......... $A \underset{d}{=} B$
equivalence by finite decomposition, 60 .............. $A \equiv B$
           66 .............. $A \underset{f}{=} B$
equivalence connective, 39 ....... $\Phi \Leftrightarrow \Psi$
existential quantifier, 195 ....... $\exists x \, \Phi(x)$
floor, 253 ..................... $\mathrm{E}x, \lfloor x \rfloor$
image of a set by a function, 99 ...... $\varphi(A)$
implication, connective, 30 ........ $\Phi \supset \Psi$
           347 ....... $C\Phi\Psi$
inequality, 21 ................... $x \neq y$
inequivalence of sets, 81 ......... $A \not\equiv B$
integral, 54 ...................... $\int$
interior of a set, 93 ................ $A^\circ$
intersection of sets, 93 ............. $A \cap B$
limit, 229 ....................... lim
minimum number of strips of width $x$,
        129, 145 .......... $\sigma_w(x)$
        129 .............. $\sigma_W(x)$
        147 ......... $\sigma_k(x), \sigma_{kw}(x)$
negation
   of a formula, 39 .................. $\neg f$
           347 ............... $Nf$
           358 ............... $\sim f$
   of a relation, 21 .................. $R'$
pi, 212, 236 ....................... $\pi$
product, Cartesian, 40 ............ $A \times B$
proper part of, 80 ................ $A \subsetneq B$
           358 ................ $A \subset B$
singleton, 93 ..................... $\{x\}$
subset, 93 ....................... $A \subseteq B$
superset, 93 ..................... $A \supseteq B$
union (sum) of sets, 80, 93 . $A \cup B$, also $\cup$
unique existence, 16 ........ $(\exists! x) \Phi(x)$
universal quantifier, 195 ....... $(\forall x) \Phi(x)$
        358 ........ $\Pi x \, \Phi(x)$
width of a set, 133, 152 ............ $\omega(A)$
*See also under* translation
symmetry, 247, 250, 253, 256–257, 262, 271–272

T, condition, 362
$\tau$, T, functions. *See* tau
tangent. *See under* circle
Tarska, Maria
   childhood and family, 174
   in the Polish–Soviet War, 175, 405
   marriage, 174

Index of Subjects

  as educator, 174–176, 326
  pregnancies, 177, 331
  life and character, 174–176, 179, 320–
    321, 325, 328, 330, 396
  portraits, 175, 329, 400
  reunited with Alfred, 336, 376
Tarski, Alfred
  childhood and youth, 3–5, 392, 400
  languages, 3–4, 321, 331, 369, 396, 406
  biology, interest in, xi, 4–5, 331, 392
  university student
    first years, x, 6–8, 12, 19, 40
    enrollment record, 6
    military service, 5, 8
    first published paper, 12, 15, 19–30
    first scientific presentation, 36, 160, 322
    doctoral study, ix, 31, 33–36, 40,
      48, 58, 77, 160–161
    dissertation, 38, 160, 386
  physics, interest in, 393
  nationalism, 4, 8, 38
  religion, 38, 174, 179, 321, 405
  name changes, 8, 12, 15, 19–20, 38–
    39, 174, 328, 392
  signatures, i, 38, 41, 392–393
  as educator, v, vii, x–xi, 43, 45, 59, 125,
    225–227, 229–241, 350, 401, 405
    first teaching jobs, 36, 38, 41
    1929 congress report, 225–227
    at the Third Boys' (= Żeromski) Gim-
      nazjum 127, 173, 179–180, 320, 405
    training teachers, 190–191
    exercises, published
      xi, 243–272
      difficulty of, 246–248
    geometry text (see *Geometrja*)
  as researcher
    threads and filaments, 159
    habilitation, 41, 161, 190
    logic and set theory, 160–162, 339
      Leśniewski, intellectual debt to, 40
    applying set theory to geometry, 43,
      45, 58, 64–66
    circle-squaring problem, 62–63, 90, 164
    with Łukasiewicz
      as assistant in seminar, 160, 190–192,
        194–198, 247, 267, 322–323, 326,
        328, 331, 335, 356, 419
      as collaborator, 160, 346
      as adjunkt, 191–192, 198, 328, 395

  Rockefeller grant, 177, 325, 394
  logic course, introductory, 191–192,
    326, 331, 404
  logic text, 192, 217, 223, 283,
    391–392, 407
  publication delays, 196, 267, 377, 380, 393
  philosophical position, 389, 402,
    407–408, 413, 416–419
    pragmatism, 407
  salary, 179
  search for professorship, 326, 328, 333
  workload, 176, 180, 190–194, 319, 326, 331
  political activity (*see under* politics)
  recreation, 176, 192–193, 201, 227, 320, 398
  emigration, 165, 167, 330–334,
    389, 396, 408
  family
    parents', 3, 36, 336, 392, 404
    wedding, 174
    Tarski's, 174, 176–179, 336,
      376, 395–396, 403
    apartments, 176–179, 203, 395
    reunited, 336, 376
  in Berkeley, 336, 376–377, 389,
    396, 406–407
  personality, 320, 399, 401,
    403, 407, 409, 411
  portraits, i, 13, 329, 384, 400, 407–409
  writing style, 17, 126, 222–223, 246, 380
  obituaries
    *San Francisco Chronicle*, 405–406
    *San Francisco Examiner*, 407.
  biography, ix–x, 3, 399–409
  bibliography, ix–x, 385–398, 390–391, 421,
    other collections, 369, 385, 390–391, 415,
      436, 464–466
    translations into Polish, 385, 391
  archives, ix, 201, 333, 392–397,
    409, 473–474
  *See also under* axiom; axiom system;
    conjunction; function; Leśniewski;
    *Geometrja*; negation; paradox;
    symbols; theorem
tau ($\tau, \tau_n$, T), 127–129, 132–134, 142–145,
  148–152, 154–158, 163, 211, 263
  *See also under* symbol: degree of
    equivalence
teachers' journals, Polish, 203–205
  See also *Delta*; *Matematyka i szkoła*;
    *Matematyka*; Mianowski Fund

teachers' organizations, Polish, 203, 207, 219, 221
   Trade Union of Polish Secondary School Teachers, 173, 177, 225
Thales. *See under* theorem
theology, foundations of, 368
theorems, 15–16, 357, 361
   antireflexivity, 22–23
   Banach's fixed-point, 57
   Banach's, on measure, 53–55, 60, 71, 83–84, 91, 106, 363–364
   Banach–Tarski, 64, 90, 93
   Bolyai–Gerwien, 47, 125
   Cantor–Bernstein, 66, 70, 74–75, 167, 381
   deduction (*see* deduction, theory of)
   of Desargues, 186
   Gerwien's, about spherical polygons, 74
   Gödel's (*see* incompleteness)
   Hahn–Banach, 54–55
   Kuratowski's, 70–71, 103, 167
   of Łukasiewicz, 30
   of Pythagoras, 46, 128, 217, 222
   regularity, 76, 119
   Skolem–Löwenheim, 419
      upward 196
   Tarski's fixpoint, 167
   of Thales, 219, 301
   triangle similarity, 301
theory, 195
   categorical, 324, 388
   decidable (*see* decidability)
   undecidable, 412, 420
thread. *See under* Tarski, Alfred.
time
   continuity of, 344
   theory of, 344, 373
topology, 5, 14, 49, 60, 69, 72, 80, 195, 335, 379
torus, 54
transitivity. *See under* axiom
translation, xii, xiv
   conventions, xii, 273, 356, 388
   of symbols, xii
trapezoid
   altitude, 310
   area, 310
   base, 310
   right, 289
triangle
   altitude, 309
   area, 309
   base, 309
   circumcenter, 275
   defect, 265
   incircle, 275
   inequality, 250
   median, 279
triangular region, 46
triangulation, 46
trichotomy. *See under* axiom
trigonometry, 213, 230, 232–241
   tables, 189, 212, 236
truth
   Aristotelian concept, 362
   bearer, 362
   classical concept, 362
   function, 39
   logical, 344, 371–372, 387, 396
   relativity of, 374–376
   semantic concept of, 362
   theory of, v, 40, 160, 322–323, 333, 356–363, 374–376, 387, 399, 404, 407, 412, 419
   value, 39
tuberculosis, 5
tuition. *See under* universities: Warsaw
type theory, 329, 342, 358, 363
typhus, 5
   vaccine production, 57
*Twardowski, Pan*, 201

ultrafilter. *See* filter, prime
undecidability. *See under* theory
undefined notion. *See* primitive notions
underground. *See under* Poland: during World War II
union. *See under* sets; symbol
unique existence, 291
   *See also under* symbol
unit. *See under* area; length
Unity of Science. *See under* congresses
universal. *See under* algebra; symbol
universities and similar institutions
   Berkeley, ix, xiii–xiv, 198–199, 336, 376, 400, 404, 407
   Berlin 387, 393
   Bonn, 69
   Chicago 125–126, 130, 212, 334, 344, 426
   Copenhagen, 325–326, 387
   Cracow 343, 368–369, 373
   Göttingen, vi, 14, 36, 53, 189, 329, 334

Harvard, 324, 333, 388, 396
Kiev, 208
Leipzig, 69
Lwów, vii, 4, 9, 14, 18, 36, 38, 49, 53, 57, 177, 197, 327–329, 333, 344–345, 361, 368, 373
Lwów Polytechnic 51, 53, 57, 72, 373
Münster, 197, 331
of Nevada at Reno, 382
New York, City College of, 334, 379, 396, 406
Paris, 14, 329, 373, 378
Poznań, 201, 328, 345, 373
Princeton Institute for Advanced Study, 333–334, 379
Rome, 181
of San Francisco, 408
San Francisco State, ix, xiv
Sonoma State, xiii
Stanford, 400, 407
Turin, vi, 59, 369
Vienna, 199, 322, 324
  Mathematisches Kolloquium, 324
Warsaw, vii, 3–9, 12, 14, 21, 31–38, 49, 72, 132, 176, 189, 190–191, 194–199, 208, 210, 214, 221, 267, 321, 326, 328, 335, 345, 350, 379, 392–396, 404
  academic year, 7, 190
  tuition, 7–8, 35
Warsaw Free, 41, 210
Warsaw Polytechnic, 200–201, 210
Yale, 341
Zurich, 14, 189, 199, 210
Urelemente, 199

Veblen, Oswald. *See under* axiom system
venia legendi. *See* habilitation
Versailles, treaty of, 10
vertex. *See under* angle
Vienna, xxii–xxiii, 18, 177, 191, 200, 322–325, 345, 356, 393
  Circle, 325, 327, 331, 341, 369–372, 374, 393, 396, 400, 419
  *See also under* universities
Vierergruppe. *See under* group
Vilnius, xxii–xxiii, 10, 174
  *See also under* congresses
Vistula, Miracle of the. *See* Polish–Soviet War

*Völkischer Beobachter*, 330
volume. *See under* cone; cylinder; polyhedral region

Warsaw, ix, xxii–xxiii, 3, 8–10, 14, 49, 72, 130, 132, 171–179, 189, 196–201, 208–210, 214, 218–221, 273, 321–322, 330–331, 335–336, 343, 392, 404–405, 415
  history and culture, 171–172, 330, 401, 404
  Koszykowa Street, 3, 35, 178, 394, 404
  Marszałkowska Street, 173
  Nowolipki Street, 173
  Philosophical Institute, 9, 12, 36, 38, 321–322, 340, 346, 386
  Rysia Street, 173
  Smolna Street, 219
  Society of Sciences and Letters, 164, 321–322, 356, 363, 415
  Sułkowski Street, 176, 178–179
  Żoliborz district, 176–177, 179
  *See also under* congresses; Mathematical Society; universities
weakest, relatively. *See under* axiom system
*Wektor*, 189
well-ordering. *See under* axiom; axiom system; set
wff. *See under* formula
*Wff 'N Proof*. *See* logic: game
Whiskeytown. *See under* congresses
white space, xii
width, 133, 152. *See also under* symbol
Wilno. *See* Vilnius
wine, exercise on diluting, 249–250
Wojciech's Bookstore, St., 204, 207
World War I, ix, xii, xxii, 4–5, 10, 14, 32, 37, 49, 72, 132, 171–175, 329, 334, 369
World War II, ix, xiii, 14, 32, 34, 37, 49, 53, 57, 69, 72, 132, 161, 334–336, 343–345, 373–376, 404
  displaced-person camps, 197, 224
  *See also under* Poland
Wrocław. *See* Breslau.

Zakopane, xxii–xxiii, 176, 329
Zermelo, Ernst. *See under* set theory *and in the index of persons*
zloty, 35
Żoliborz district. *See under* Warsaw
Zurich. *See under* universities